Lecture Notes in Physics

Springer
Berlin
Heidelberg
New York
Barcelona
Hong Kong
London
Milan
Paris
Singapore
Tokyo

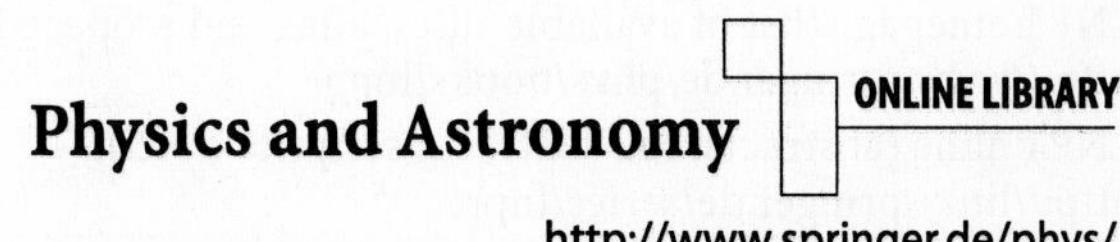

http://www.springer.de/phys/

Editorial Policy

The series *Lecture Notes in Physics* (LNP), founded in 1969, reports new developments in physics research and teaching – quickly, informally but with a high quality. Manuscripts to be considered for publication are topical volumes consisting of a limited number of contributions, carefully edited and closely related to each other. Each contribution should contain at least partly original and previously unpublished material, be written in a clear, pedagogical style and aimed at a broader readership, especially graduate students and nonspecialist researchers wishing to familiarize themselves with the topic concerned. For this reason, traditional proceedings cannot be considered for this series though volumes to appear in this series are often based on material presented at conferences, workshops and schools (in exceptional cases the original papers and/or those not included in the printed book may be added on an accompanying CD ROM, together with the abstracts of posters and other material suitable for publication, e.g. large tables, colour pictures, program codes, etc.).

Acceptance

A project can only be accepted tentatively for publication, by both the editorial board and the publisher, following thorough examination of the material submitted. The book proposal sent to the publisher should consist at least of a preliminary table of contents outlining the structure of the book together with abstracts of all contributions to be included. Final acceptance is issued by the series editor in charge, in consultation with the publisher, only after receiving the complete manuscript. Final acceptance, possibly requiring minor corrections, usually follows the tentative acceptance unless the final manuscript differs significantly from expectations (project outline). In particular, the series editors are entitled to reject individual contributions if they do not meet the high quality standards of this series. The final manuscript must be camera-ready, and should include both an informative introduction and a sufficiently detailed subject index.

Contractual Aspects

Publication in LNP is free of charge. There is no formal contract, no royalties are paid, and no bulk orders are required, although special discounts are offered in this case. The volume editors receive jointly 30 free copies for their personal use and are entitled, as are the contributing authors, to purchase Springer books at a reduced rate. The publisher secures the copyright for each volume. As a rule, no reprints of individual contributions can be supplied.

Manuscript Submission

The manuscript in its final and approved version must be submitted in camera-ready form. The corresponding electronic source files are also required for the production process, in particular the online version. Technical assistance in compiling the final manuscript can be provided by the publisher‘s production editor(s), especially with regard to the publisher's own Latex macro package which has been specially designed for this series.

Online Version/ LNP Homepage

LNP homepage (list of available titles, aims and scope, editorial contacts etc.):
http://www.springer.de/phys/books/lnpp/

LNP online (abstracts, full-texts, subscriptions etc.):
http://link.springer.de/series/lnpp/

Thorsten Pöschel Stefan Luding (Eds.)

Granular Gases

Editors

Thorsten Pöschel
Charité, Institut für Biochemie
Humboldt-Universität zu Berlin
Monbijoustrasse 2
10117 Berlin, Germany

Stefan Luding
Institut für Computeranwendungen
Abt. I Physik
Universität Stuttgart
Pfaffenwaldring 27
70569 Stuttgart, Germany

Cover picture: Snapshot of an event driven simulation of a cooling Granular Gas.
Author: Thomas Schwager, Humboldt University Berlin

Library of Congress Cataloging-in-Publication Data applied for.

Die Deutsche Bibliothek - CIP-Einheitsaufnahme

Granular gases / Thorsten Pöschel ; Stefan Luding (ed.). - Berlin ;
Heidelberg ; New York ; Barcelona ; Hong Kong ; London ; Milan ; Paris
; Singapore ; Tokyo : Springer, 2001
(Lecture notes in physics ; 564) (Physics and astronomy online library)
ISBN 3-540-41458-4

ISSN 0075-8450
ISBN 3-540-41458-4 Springer-Verlag Berlin Heidelberg New York

Springer-Verlag Berlin Heidelberg New York
a member of BertelsmannSpringer Science+Business Media GmbH

http://www.springer.de

Typesetting: Camera-ready by the authors/editors
Cover design: *design & production*, Heidelberg

Printed on acid-free paper
SPIN: 10791831 55/3141/du - 5 4 3 2 1 0

Preface

"Granular Material" as a subject of physical research has a long standing history. Eminent scientists such as Coulomb, Faraday, Hagen, Huygens, Rayleigh, Reynolds, and others contributed to the body of research. Following the early days, however, the physical approach gave way to engineering disciplines. Among other reasons, the enormous advances in computing power rekindled the scientific interest of physicists. In the last decade large-scale simulations of granular systems have given new insights into the nature of granular materials – and continue to do so.

In the strict definition, "Granular Gases" are dilute granular systems, i.e., many-particle systems, in which the mean free path of the particles is much larger than the typical particle size. This condition implies that the duration of particle contacts is much shorter than the mean flight time. In contrast to molecular gases, in Granular Gases the particle collisions occur dissipatively, i.e., in each binary collision the particles lose part of the kinetic energy of their relative motion. The dissipation of kinetic energy causes a series of nontrivial effects, such as the formation of clusters and other spatial structures, non-Maxwellian velocity distributions, anomalous diffusion, correlations in the velocity field, characteristic shock waves, and others. In the dilute limit and with the assumption of weak dissipation the formal similarity of natural gases and Granular Gases allows us to also describe the latter using the powerful tools of classical Statistical Mechanics, Kinetics, and Hydrodynamics. Granular Gases provide, more than any other granular system, the possibility of conducting simultaneously experimental and theoretical studies and numerical simulations as well. Part I of the volume comprises rigorous theoretical results for the dilute limit.

Such theoretical descriptions of a Granular Gas as a many-particle system requires comprehensive understanding of the details of collisions of only two isolated grains. Without this knowledge it would be impossible to deduce the macroscopic properties of Granular Gases. The properties of binary collisions and simple one-dimensional models are the subject of Part II.

Part III contains experimental investigations of Granular Gases: In order to maintain a steady gaseous state under gravity conditions, the material has to be excited externally to balance the loss of kinetic energy due to inelastic particle collisions. Ideally, the energy feed would be homogeneous and isotropic throughout the system, which is certainly difficult to achieve in an experiment. For the experimental investigation of Granular Gases feeding energy by mechanical vibrations of the container has proven to be a good compromise – to date all earthborn experiments make use of vibrated containers.

Large-scale Granular Gases in steady state are found in astrophysical systems as discussed in Part IV. The rings around the outer planets of the Solar System have been intensively studied for centuries and lots of data exist from earthbound and missile observations. Although there are still many open questions, several aspects of the rich structure of planetary rings are well understood. Therefore, planetary rings offer a wide field for practical application of Granular Gas theory and may, thus, serve as test systems to validate theoretical results.

The last part (Part V) of the book is devoted to possible generalizations with respect to increased density. High-density Granular Gases, in a broader sense, can be defined as systems with instantaneous binary collisions, i.e., only by the condition that the contact duration is thought to be negligible and, hence, does not affect the overall properties of the many-particle system. From the theoretical point of view this definition is more problematic since the assumption of Molecular Chaos as a prerequisite of simple kinetic theory may be violated. Nevertheless, many interesting phenomena occur in systems of increased density and there is a need to thus generalize the theory. The motivation for extending the gas picture to higher density is to bridge the still-existing gap between Kinetic Theory and fluid dynamics approaches, which are very successfully applied in engineering.

The articles in this volume are selected contributions of the conference "Granular Gases" held from March 8 to 12, 1999 in Bad Honnef. Many of the authors are renowned experts in the field and have decisively shaped the present state of knowledge. Therefore, the articles in this volume cover a large part of the present knowledge about Granular Gases, ranging from investigations of two-particle interactions to Kinetic Theory of dilute many-particle systems and numerical simulations of rather dense systems.

In order to find a wide readership not only among experts, but also among undergraduate and graduate students, the authors have been asked to present their material in a self-contained and pedagogical way. All contributions have been carefully refereed.

The editors are grateful to all authors for their contributions and to the referees whose comments considerably improved the manuscripts. Claudia Kunze, Marlies Parsons, and Simon Renard are acknowledged for support in typesetting. Finally, the Wilhelm-und-Else-Heraeus-Stiftung is acknowledged for financial support.

Berlin, Stuttgart,
September 2000

Thorsten Pöschel
Stefan Luding

Contents

III. Vibrated Granular Media

IV. Granular Astrophysics

V. Towards Dense Granular Systems

I

Kinetic Theory and Hydrodynamics

Kinetic Theory of Granular Gases

Twan P.C. van Noije and Matthieu H. Ernst

Instituut voor Theoretische Fysica, Universiteit Utrecht, Postbus 80006, 3508 TA Utrecht, The Netherlands. e-mail: M.H.Ernst@phys.uu.nl

Abstract. Granular fluids are many-body systems with very short range and strongly repulsive interactions that dissipate energy upon collision. The basic models are soft and hard inelastic spheres, which may or may not be driven by external deterministic or stochastic forces, as a means of maintaining a steady state. Starting from the Liouville equation and BBGKY hierarchy for such systems, we derive the Boltzmann and ring kinetic equations for the limiting case of instantaneous hard sphere interactions. The ring kinetic equation provides the basis from which fluctuating hydrodynamics and mode coupling theory can be derived. In the second part of this article non-Gaussian properties, such as cumulants and high energy tails, of the single-particle velocity distribution are studied for homogeneous granular fluids of inelastic hard spheres or disks, based on the Enskog-Boltzmann equation for the undriven and randomly driven case. The velocity distribution in the randomly driven steady state exhibits a high energy tail $\sim \exp(-Ac^{3/2})$, where c is the velocity scaled by the thermal velocity and $A \sim 1/\sqrt{\epsilon}$ with ϵ the inelasticity. The results are compared with molecular dynamics simulations, as well as direct Monte Carlo simulations of the Boltzmann equation.

1 Introduction

The basic concepts that rapid granular flows can be considered as a collection of particles with short range interactions, moving ballistically and suffering instantaneous and inelastic binary collisions, are formulated in Haff's seminal article *Grain flow as a fluid mechanical phenomenon* [1]. This paper has laid the foundations for using the methods of kinetic theory in this field, which have been further developed to calculate the granular temperature, pressure, distribution functions and transport coefficients in several classic papers in the eighties [2–5], and a rapid flow of further results has continued in the nineties [6–14].

Granular fluids can be considered as dense fluids with very short range and strongly repulsive interactions. The basic models are inelastic soft and hard spheres, in which the particles follow free particle trajectories until a collision occurs. In the first model of *soft* inelastic spheres the particles interact with a rather stiff elastic repulsion, and loose energy during contact through a frictional force, which is proportional to their relative velocities [15–19]. In the second model of inelastic *hard* spheres (IHS) [20–27] there is an instantaneous loss of kinetic energy on contact.

Kinematics and dynamics of *soft* inelastic spheres have been widely discussed in the literature (for a review see [28]), but the appropriate kinetic equations for the single particle distribution function are barely discussed. For *hard* spheres, on the other hand, kinetic theory is well developed [2–10, 12–14], and based on the Enskog-Boltzmann equation. Inherent to this description is the molecular chaos assumption of uncorrelated binary collisions. However, as shown in the literature [25, 27] there exist long range spatial correlations in density and flow fields, which cannot be understood on the basis of a mean field type kinetic equation, like the Boltzmann or Enskog-Boltzmann equation.

The goal of this article is to construct the Liouville equation, i.e. the time evolution equation for the N-particle density in phase space for fluids with dissipative interactions, where the flows in phase space are not volume conserving, but contracting. This will be done for inelastic soft and hard sphere interactions. Once the equation has been formulated, the methods of kinetic theory can be applied to derive kinetic equations, to study the extent to which collective effects are of importance at a fundamental level of description, to correct for the breakdown of the molecular chaos assumption, and to account for the effects of dynamic correlations.

In the last 35 years many-body theories have been developed to account for these dynamic correlations in systems of microscopic particles obeying the standard conservation laws. The fundamental concept to describe these dynamic correlations are 'ring collisions', i.e. sequences of correlated binary collisions, which lead to the so called *ring kinetic theory* [29–33]. This ring kinetic theory for systems of smooth elastic hard spheres has been at the basis of all major developments in nonequilibrium statistical mechanics over the last three decades: it explains the breakdown of the virial expansion for transport coefficients and their logarithmic density dependence [31], the algebraic long time tails of the velocity autocorrelation function and similar current-current correlation functions [29, 30, 32, 34], the non-analytic dispersion relations for sound propagation and for relaxation of hydrodynamic excitations [33, 34], the breakdown of the Navier-Stokes equations in two-dimensional (2D) fluids at very long times and the non-existence of linear transport coefficients in 2D, as well as the positive deviations of 3D transport coefficients from the Enskog theory for elastic hard spheres, see Ref. [35]. Moreover, it explains the existence of long range spatial correlations in nonequilibrium stationary states [36], driven by reservoirs which impose shear rates or temperature gradients, or in driven diffusive systems. Such systems violate the conditions of detailed balance and the stationary states are non-Gibbsian states [36, 37].

In this article we will extend the mean field type Enskog-Boltzmann equation for rapid granular flows by including ring collisions. In particular we consider the model of inelastic hard disks or spheres. The kinetic theory for inelastic hard spheres will be developed in close analogy with that for elastic ones. In Sec. 2 the rather singular streaming operators $S_t(\boldsymbol{x})$, which generate the phase space trajectories of the N-hard sphere system, and formulate them in terms of binary collision operators, denoted by T and $\overline{T}$, following the original derivation in Ref. [38]. This enables us to introduce a pseudo-Liouville equation for the N-particle phase space distribution function, and obtain the Bogoliubov-Born-Green-Kirkwood-Yvon (BBGKY) hierarchy for the reduced distribution functions. The same results for the binary collision operators have been derived independently by Brey et al. [8]. The Liouville equation and the corresponding BBGKY hierarchy form the standard starting point for deriving kinetic equations, such as the Boltzmann equation and the ring kinetic equation.

After having derived the Boltzmann and ring kinetic equations in Sec. 3, we study in Sec. 4 the solutions of the Boltzmann equation for freely evolving and uniformly heated granular gases. Most theories for rapid granular flows are based on the assumption that the particle velocities are distributed according to a Gaussian or Maxwellian distribution. Since granular particles collide inelastically, this assumption is not obvious. In fact, granular systems are typically in far from equilibrium states in the sense that an external driving force is necessary to maintain a stationary or periodic state. Only in systems of nearly elastic particles, states resembling thermal equilibrium may be expected.

Deviations from Gaussian behavior in rapid granular flows have been studied in several contexts. Jenkins and Richman [5] have analyzed the Enskog-Boltzmann equation for inelastic hard disk fluids under uniform shear by introducing an anisotropic Maxwellian velocity distribution, which captures the most important features of the distribution measured in molecular dynamics (MD) simulations [39]. Using MD simulations, Goldhirsch et al. [40] measured the flatness or kurtosis of the velocity distribution function (VDF) in an undriven fluid of inelastic hard disks, and found a broadening of the VDF when transitions to shearing or clustered states occurred. Brey et al. [41] measured higher cumulants for the same system in the homogeneous cooling state (HCS) through direct Monte Carlo simulation of the Boltzmann equation, assuming that the Boltzmann equation remains meaningful for large inelasticity, $\epsilon = 1 - \alpha^2$, where α is the coefficient of normal restitution. These investigators have compared their simulation results with the theoretical results of our theory [12]. Several investigators performed computer simulations of gas-fluidized or vibrated beds of grains, measured cumulants of the VDF [42], and observed power law behavior of the high energy tails [42, 43].

Sela and Goldhirsch [13] have numerically obtained a perturbative solution of the Boltzmann equation for inelastic hard spheres to orders of $\mathcal{O}(\epsilon)$, $\mathcal{O}(\epsilon k)$, $\mathcal{O}(k^2)$, where the wave number k is proportional to the gradients. The order $\mathcal{O}(\epsilon)$ estimates the deviation from Gaussian behavior of the homogeneous solution and contributes to the rate of homogeneous cooling.

Unfortunately, there also exist publications that are totally incorrect [44]. The authors claim to derive an exact equation for the rate of change of the velocity moments in a freely evolving IHS fluid. Their results come from a term that yields, when properly evaluated, an irrelevant contribution of relative order $\mathcal{O}(1/V)$, which vanishes for large volume V. The relevant contributions of $\mathcal{O}(1)$ are erroneously set equal to zero.

The VDF has been studied for another type of granular flows, namely randomly driven IHS fluids, where each particle is subject to a random white noise. This system possesses a spatially homogeneous steady state. The randomly driven IHS fluid models properties of rapidly vibrated granular layers [45–48] and gas-fluidized beds of grains [42].

Peng and Ohta [48], and Pagonabarraga et al. [49] have carried out molecular dynamics simulations of two-dimensional randomly driven IHS fluids, and measured the high energy tails and fourth cumulants of the VDF. Recently, Montanero and Santos [50] have measured the same properties by direct Monte Carlo simulation of the Boltzmann equation for randomly driven IHS gases, where the Boltzmann collision term is supplemented with a Fokker-Planck diffusion term to account for the random accelerations.

In Sec. 4 we use the moment method of Goldshtein and Shapiro [6] to solve the Enskog-Boltzmann equation, and study the non-Gaussian properties of the VDF [12]. Furthermore, we employ an asymptotic method of Krook and Wu [51] to calculate the high energy tail of the VDF for the undriven [10] and the randomly driven [12] spatially homogeneous IHS fluid. The results are compared with existing simulation results, obtained either from molecular dynamics or direct Monte Carlo simulations of the Boltzmann equation. In the concluding section 5 we highlight the most important results and fundamental problems.

The above description concentrates mostly on the dominant part of the distribution function or of its velocity moments. Conceptually the most simple procedure to derive the macroscopic equations for granular gases or fluids from the Boltzmann or Boltzmann-Enskog equation, from the Revised Enskog Theory [52] or from model-Boltzmann equations [8], is to use the Chapman-Enskog method [6, 14, 53, 54]. This yields in a systematic manner the normal solution of the kinetic equation as well as the macroscopic equations for the granular gases and fluids, first at the Euler level, next at the Navier-Stokes level, etc. with explicit expressions for the transport coefficients.

Apart from these investigations it is also of interest to study the effects of different inelastic collision models, like hard or soft [18, 55, 62], and smooth or rough spheres [56], or models with velocity dependent coefficients of normal and tangential restitution [57]. These subjects have been amply discussed in the literature, cited above, and will not be reviewed here.

2 Liouville Equation

2.1 Hamiltonian Systems

To give a didactic presentation we start from the equation for Hamiltonian systems, and then systematically extend it to include nonconservative interactions such as frictional and stochastic forces.

In nonequilibrium statistical mechanics one can calculate the mean value of a dynamical variable $A(\boldsymbol{x})$ from either of the following expressions,

$$\int \mathrm{d}\boldsymbol{x}\rho(\boldsymbol{x},0)A(\boldsymbol{x}(t)) = \int \mathrm{d}\boldsymbol{x}\rho(\boldsymbol{x},t)A(\boldsymbol{x}) \ , \tag{1}$$

where $\boldsymbol{x} = \{x_1, x_2, \ldots x_N\}$ with $x_i = \{\boldsymbol{r}_i, \boldsymbol{v}_i\}$ is a point in the N-particle phase space. On the left hand side the time dependence is assigned to the dynamical variable $A(\boldsymbol{x}(t)) \equiv S_t(\boldsymbol{x})A(\boldsymbol{x})$ with $S_t(\boldsymbol{x})$ the time evolution or streaming operator, and on the right hand side to the N-particle distribution function $\rho(\boldsymbol{x},t)$. This can be done by considering expression (1) as an inner product. The time evolution of the N-particle distribution function is then given by

$$\rho(\boldsymbol{x},t) = S_t^{\dagger}\rho(\boldsymbol{x},0) \ , \tag{2}$$

where $S_t^{\dagger}(\boldsymbol{x})$ is the adjoint of $S_t(\boldsymbol{x})$.

If the interactions are conservative and additive, the force between the pair (ij) is $\boldsymbol{F}_{ij} = -\partial V(r_{ij})/\partial \boldsymbol{r}_{ij}$, where $V(r)$ is the pair potential, and the streaming operator is given by

$$S_t(\boldsymbol{x}) = \exp[tL(\boldsymbol{x})] = \exp[t\sum_i L_i^0 - t\sum_{i<j}\theta(ij)] \ , \tag{3}$$

where $L(\boldsymbol{x})\cdots = \{H(\boldsymbol{x}),\ldots\}$ is the Poisson bracket with the Hamiltonian. This yields

$$\begin{aligned} L_i^0 &= \boldsymbol{v}_i \cdot \frac{\partial}{\partial \boldsymbol{r}_i} \\ \theta(ij) &= \frac{1}{m}\frac{\partial V(r_{ij})}{\partial \boldsymbol{r}_{ij}} \cdot \left(\frac{\partial}{\partial \boldsymbol{v}_i} - \frac{\partial}{\partial \boldsymbol{v}_j}\right) \ . \end{aligned} \tag{4}$$

In this case $S_t(\boldsymbol{x})$ is a unitary operator, $S_t^{\dagger}(\boldsymbol{x}) = S_{-t}(\boldsymbol{x})$, and $L^{\dagger} = -L$. The free streaming operator $S_t^0(\boldsymbol{x}) = \exp[t\sum_i L_i^0]$ generates the free particle trajectories,

$$S_t^0(\boldsymbol{x})A(\boldsymbol{r}_i, \boldsymbol{v}_i) = A(\boldsymbol{r}_i + \boldsymbol{v}_i t, \boldsymbol{v}_i) \ . \tag{5}$$

The Liouville equation, which is the evolution equation for the phase space density for conservative systems, follows from Eq. (2), and reads

$$(\partial_t + \sum_i L_i^0)\rho(\boldsymbol{x},t) = \sum_{i<j}\theta(ij)\rho(\boldsymbol{x},t) \ , \tag{6}$$

which is an expression of the incompressibility of the flow in phase space.

An equivalent representation of the time evolution of the system can be given in terms of reduced s-particle distribution functions ($s = 1, 2, \ldots$), defined as

$$f_{12\ldots s}(t) \equiv f^{(s)}(x_1, x_2, \ldots x_s, t) = \frac{N!}{(N-s)!} \int \mathrm{d}x_{s+1} \ldots \mathrm{d}x_N \rho(\boldsymbol{x}, t) \ , \quad (7)$$

where $\rho(\boldsymbol{x}, t)$ is normalized to unity. Integrating (6) over $x_{s+1} \ldots x_N$ yields the BBGKY hierarchy for the reduced distribution functions. We only quote the first two equations of the hierarchy,

$$\begin{aligned} (\partial_t + L_1^0) f_1 &= \int \mathrm{d}x_2 \theta(12) f_{12} \\ \left[\partial_t + L_1^0 + L_2^0 - \theta(12)\right] f_{12} &= \int \mathrm{d}x_3 [\theta(13) + \theta(23)] f_{123} \ . \end{aligned} \quad (8)$$

This set of equations is an open hierarchy, which expresses the time evolution of the s-particle distribution function in terms of the $(s+1)$-th function.

It is a minor generalization to include external conservative force fields. In that case the infinitesimal generator L_i^0 of the free particle motion should be replaced by

$$L_i^0 = \boldsymbol{v}_i \cdot \frac{\partial}{\partial \boldsymbol{r}_i} + \boldsymbol{a}_i \cdot \frac{\partial}{\partial \boldsymbol{v}_i} \ , \quad (9)$$

where $\boldsymbol{a}_i$ is an external conservative force per unit mass, acting on the i-th particle.

2.2 Frictional and Stochastic Forces

Our next goal is to generalize this result to include frictional or stochastic forces, exerted by external sources or by interparticle interactions in complex fluids. Suppose the N-particle system is described by the equations of motion

$$\frac{\mathrm{d}\boldsymbol{r}_i}{\mathrm{d}t} = \boldsymbol{v}_i \ , \quad \text{and} \quad \frac{\mathrm{d}\boldsymbol{v}_i}{\mathrm{d}t} = \boldsymbol{a}_i + \hat{\boldsymbol{\xi}}_i \ , \quad (10)$$

where $\boldsymbol{a}_i$ and $\hat{\boldsymbol{\xi}}_i$ are respectively the systematic and stochastic force per unit mass. The total force is vanishing, i.e. $\sum_i \boldsymbol{a}_i = \boldsymbol{0}$ and $\sum_i \hat{\boldsymbol{\xi}}_i = \boldsymbol{0}$. Then, the average momentum density satisfies a local conservation law, implying that the local flow velocity $\boldsymbol{u}(\boldsymbol{r}, t)$ is a slowly varying macroscopic field. Consequently, the model qualifies as a fluid. The systematic forces may be conservative or frictional. The stochastic ones may be caused by external sources, or by interparticle interactions, as used to model complex fluids.

In the present article we mainly focus on *external* noise, which is most relevant in the case of granular fluids. If the correlation time t_{noise} of the random noise is much shorter than the mean free time between collisions, t_0, then $\hat{\boldsymbol{\xi}}_i(t)$ can be considered as independent Gaussian white noise with zero mean and autocorrelation

$$\overline{\hat{\xi}_{i\alpha}(t)\hat{\xi}_{j\beta}(t')} = \xi_0^2 \delta_{ij}\delta_{\alpha\beta}\delta(t-t') \ , \tag{11}$$

where Greek indices denote Cartesian components. The overline indicates an average over the noise source. The derivation of the equation of motion for the N-particle distribution $\rho(\boldsymbol{x},t)$ follows the standard methods [58, 59]. To do so, we write $\rho(\boldsymbol{x},t) = \langle\delta(\boldsymbol{x} - \hat{\boldsymbol{x}}(t))\rangle$, where $\hat{\boldsymbol{x}}(t)$ denotes the trajectory started at the phase point $\hat{\boldsymbol{x}} = \hat{\boldsymbol{x}}(0)$, and $\langle\dots\rangle$ denotes an average over an initial distribution $\rho(\hat{\boldsymbol{x}},0)$ and over the noise source.

To calculate the change $\delta\rho(\boldsymbol{x},t)$ during a short time δt, which is *large* compared to t_{noise}, we write the increment $\delta\hat{x}_i$ of the phase $\hat{x}_i = \{\hat{\boldsymbol{r}}_i, \hat{\boldsymbol{v}}_i\}$ as

$$\delta\hat{x}_i = \hat{\alpha}_i \delta t + \delta\hat{\omega}_i \ , \tag{12}$$

where $\hat{\alpha}_i = \{\hat{\boldsymbol{v}}_i, \hat{\boldsymbol{a}}_i\}$ and $\delta\hat{\omega}_i = \{\boldsymbol{0}, \delta\hat{\boldsymbol{w}}_i(t)\}$ with

$$\delta\hat{\boldsymbol{w}}_i(t) = \int_t^{t+\delta t} \mathrm{d}t' \hat{\boldsymbol{\xi}}_i(t') \ . \tag{13}$$

The autocorrelation of the noise source,

$$\overline{\delta\hat{w}_{i\alpha}(t)\delta\hat{w}_{j\beta}(t)} = \xi_0^2 \delta_{ij}\delta_{\alpha\beta}\delta t \ , \tag{14}$$

follows directly from (11). The last equality implies formally that $\delta\omega_i$ is a quantity of $\mathcal{O}(\sqrt{\delta t})$.

Next, we consider the distribution function $\rho(\boldsymbol{x}, t+\delta t) = \langle\delta(\boldsymbol{x} - \hat{\boldsymbol{x}}(t) - \delta\hat{\boldsymbol{x}}(t))\rangle$, expand it in powers of δt, and find to linear order in δt,

$$\begin{aligned}\rho(\boldsymbol{x}, t+\delta t) \simeq \rho(\boldsymbol{x},t) &- \sum_i \frac{\partial}{\partial x_i} \cdot \overline{\delta\hat{x}_i} \rho(\boldsymbol{x},t) \\ &+ \frac{1}{2}\sum_{ij} \frac{\partial^2}{\partial x_i \partial x_j} : \overline{\delta\hat{x}_i \delta\hat{x}_j} \rho(\boldsymbol{x},t) \ . \end{aligned} \tag{15}$$

The first jump moment $\overline{\delta\hat{x}_i} = \{\boldsymbol{v}_i, \boldsymbol{a}_i\}\delta t$; the second one, $\overline{\delta\hat{x}_i\delta\hat{x}_j}$, has only nonvanishing components in the velocity directions, given by (14). By defining the 'time derivative', $\partial_t\rho$, as the limiting value of $\delta\rho/\delta t$ under the restriction that δt is large compared to t_{noise}, we obtain the multi-variate Fokker-Planck equation, referred to as Liouville equation in the present context,

$$\partial_t \rho = -\sum_i \left[\frac{\partial}{\partial \boldsymbol{r}_i} \cdot \boldsymbol{v}_i + \frac{\partial}{\partial \boldsymbol{v}_i} \cdot \boldsymbol{a}_i - \frac{1}{2}\xi_0^2 \frac{\partial}{\partial \boldsymbol{v}_i} \cdot \frac{\partial}{\partial \boldsymbol{v}_i} \right] \rho \ . \tag{16}$$

Here $\boldsymbol{a}_i$ describes the deterministic forces, and the diffusion term in velocity space accounts for the effects of Gaussian white noise. The Liouville equation

(16) does *not* describe an incompressible flow in phase space, as its Hamiltonian counterpart (6) does, but a contracting flow as the forces are dissipative. If ξ_0 is vanishing and $\boldsymbol{a}_i$ is conservative, we recover the standard Liouville equation (6) for Hamiltonian systems. The deterministic forces may include external and interparticle forces. These forces may be conservative, impulsive or frictional (in which case $\boldsymbol{a}_i$ may depend on the velocity of the particles). If the conservative force contains an external potential, it yields directly the result (9). If the forces are impulsive, the method of 'pseudo-pair forces' with $T(ij)$ and $\overline{T}(ij)$ operators can be used, which will be discussed in the next section.

Eq. (16) with vanishing noise strength, $\xi_0 = 0$, can be applied directly to the model of soft inelastic spheres, which was used first in 1986 by Walton and Braun [15] to model granular fluids, and has been frequently used afterwards in computer simulations [16–18, 60]. In this model the interparticle forces $m\boldsymbol{a}_i = \sum_{j(\neq i)} \boldsymbol{F}_{ij}$ with $\boldsymbol{F}_{ij} = \boldsymbol{F}_{ij}^C + \boldsymbol{F}_{ij}^D$ contain a conservative and a dissipative or frictional part, i.e.

$$\begin{aligned} \boldsymbol{F}_{ij}^C &= Y(\sigma - r_{ij})\hat{\boldsymbol{r}}_{ij}\vartheta(\sigma - r_{ij}) \\ \boldsymbol{F}_{ij}^D &= -\gamma_n \boldsymbol{v}_{ij} \cdot \hat{\boldsymbol{r}}_{ij}\hat{\boldsymbol{r}}_{ij}\vartheta(\sigma - r_{ij}) \ , \end{aligned} \tag{17}$$

where $\vartheta(\sigma - r_{ij}) = 1$ for $r < \sigma$ and $\vartheta(\sigma - r_{ij}) = 0$ for $r > \sigma$. Here σ is the diameter of the soft sphere, $\hat{\boldsymbol{a}}$ is a unit vector along $\boldsymbol{a}$, the relative distance is $\boldsymbol{r}_{ij} = \boldsymbol{r}_i - \boldsymbol{r}_j$ with length r_{ij}, the relative velocity $\boldsymbol{v}_{ij} = \boldsymbol{v}_i - \boldsymbol{v}_j$. Moreover, Y is the Young modulus and γ_n the coefficient of (normal) friction. The frictional force above is directed normal to the surface. Sometimes one also considers a tangential friction,

$$\boldsymbol{F}_{ij}^{D\prime} = -\gamma_t(\boldsymbol{v}_{ij} - \boldsymbol{v}_{ij} \cdot \hat{\boldsymbol{r}}_{ij}\hat{\boldsymbol{r}}_{ij})\vartheta(\sigma - r_{ij}) \ . \tag{18}$$

The forces in (17) and (18) are only nonvanishing when the particles are in contact. A convenient starting point to develop kinetic theory for systems of soft inelastic spheres, would be the BBGKY hierarchy, which can be derived as in subsection 2.1. Here we only quote the first hierarchy equation

$$\begin{aligned} \left(\partial_t + \boldsymbol{v}_1 \cdot \frac{\partial}{\partial \boldsymbol{r}_1}\right) f(x_1, t) = \frac{1}{m}\frac{\partial}{\partial \boldsymbol{v}_1} \cdot \int d\boldsymbol{v}_2 \int d\boldsymbol{r}_{12}\vartheta(\sigma - r_{12})\hat{\boldsymbol{r}}_{ij} \times \\ [\gamma_n \boldsymbol{v}_{12} \cdot \hat{\boldsymbol{r}}_{12} - Y(\sigma - r_{ij})]\, f^{(2)}(x_1, x_2, t) \ . \end{aligned} \tag{19}$$

However, the derivation of a proper Boltzmann equation for soft inelastic spheres, as well as explicit results for viscosities and heat conductivity seem to be scarce [61, 62].

The external noise, described below (10), is referred to as 'homogeneous heating'. It has been used to describe steady states in granular flows, where energy has to be supplied to the system, to compensate for the energy loss due to inelastic collisions. It has been employed extensively in analytic studies and computer simulations of randomly driven granular flows [12, 27, 45, 46, 48, 49, 63]. Physical realizations of such models may be collections of pucks floating on air tables [64], or steel balls and glass beads fluidized by vertically vibrating plates covered with a single or many layers of beads [65–68].

2.3 Impulsive Forces and Binary Collision Expansions

If the interactions have a hard core, as for inelastic hard spheres, the derivatives of the potential in (4) are ill-defined, but the trajectories of hard spheres in phase space are well-defined. They consist of free streaming and instantaneous collisions, with well-defined collision rules.

For an explicit description we need to further specify the *inelastic hard sphere* (IHS) model. The interparticle interactions are modeled by instantaneous collisions as in the case of elastic hard spheres (EHS). During a collision momentum will be transferred instantaneously along the line joining the centers of the two colliding particles, indicated by the vector $\boldsymbol{\sigma}$ pointing from the center of particle 2 to that of particle 1. The reflection law, $\boldsymbol{v}_{12}^{*}\cdot\hat{\boldsymbol{\sigma}} = -\alpha\boldsymbol{v}_{12}\cdot\hat{\boldsymbol{\sigma}}$, is inelastic, as illustrated in Fig. 1 for colliding spheres of diameter σ.

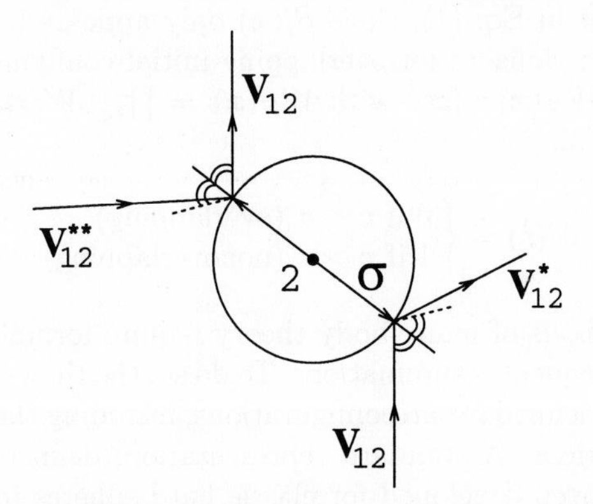

Fig. 1. Direct and restituting inelastic collisions with reflection law $\boldsymbol{v}_{12}^{*}\cdot\hat{\boldsymbol{\sigma}} = -\alpha\boldsymbol{v}_{12}\cdot\hat{\boldsymbol{\sigma}}$ where the restitution coefficient α satisfies $0 < \alpha < 1$. The dashed lines refer to elastic collisions. The action sphere around particle 2 has a radius σ.

Inelastic hard spheres of equal mass change their velocities upon collision as

$$\begin{aligned} \boldsymbol{v}_1^{*} &= \boldsymbol{v}_1 - \tfrac{1}{2}(1+\alpha)(\boldsymbol{v}_{12}\cdot\hat{\boldsymbol{\sigma}})\hat{\boldsymbol{\sigma}} \\ \boldsymbol{v}_2^{*} &= \boldsymbol{v}_2 + \tfrac{1}{2}(1+\alpha)(\boldsymbol{v}_{12}\cdot\hat{\boldsymbol{\sigma}})\hat{\boldsymbol{\sigma}}\ , \end{aligned} \tag{20}$$

where $\hat{\boldsymbol{\sigma}} = \boldsymbol{\sigma}/\sigma$ denotes a unit vector. The total momentum of the two particles is conserved in a collision, but the total energy is not. Per collision an amount $\frac{1}{4}m\epsilon(\boldsymbol{v}_{12}\cdot\hat{\boldsymbol{\sigma}})^2$ is lost, where the coefficient of inelasticity is defined as $\epsilon = 1 - \alpha^2$. The inelastic collisions are described by a coefficient of normal restitution α $(0 < \alpha < 1)$, which is assumed here to be independent of the relative velocity of impact.

The restituting (precollision) velocities $(\boldsymbol{v}_1^{**}, \boldsymbol{v}_2^{**})$ leading to $(\boldsymbol{v}_1, \boldsymbol{v}_2)$, are found by inverting collision rule (20) and given by (see Fig. 1)

$$\begin{aligned} \boldsymbol{v}_1^{**} &= \boldsymbol{v}_1 - \tfrac{1}{2}(1+\alpha^{-1})(\boldsymbol{v}_{12}\cdot\hat{\boldsymbol{\sigma}})\hat{\boldsymbol{\sigma}} \\ \boldsymbol{v}_2^{**} &= \boldsymbol{v}_2 + \tfrac{1}{2}(1+\alpha^{-1})(\boldsymbol{v}_{12}\cdot\hat{\boldsymbol{\sigma}})\hat{\boldsymbol{\sigma}} \ . \end{aligned} \tag{21}$$

Note that this inversion is not possible for $\alpha = 0$.

The IHS model incorporates the most fundamental feature of the dissipative dynamics of rapid granular flows, namely the irreversible loss of kinetic energy in collisions. Moreover, it is a many-body system with well-defined and relatively simple dynamics, to which the many-body methods of nonequilibrium statistical mechanics can be applied.

Next, we construct the streaming operators S_t for IHS fluids. The results described in this section have been derived in [8, 11, 63]. For particles with hard core interactions the dynamics is undefined for physically inaccessible configurations, where the particles are overlapping. Such configurations have a vanishing weight in Eq. (1), since $S_t(\boldsymbol{x})$ only appears in the combination $\rho(\boldsymbol{x}, 0)S_t(\boldsymbol{x})$ which vanishes for overlapping initial configurations. So, it suffices to consider $W_N(\boldsymbol{x})S_t(\boldsymbol{x})$, with $W_N(\boldsymbol{x}) = \prod_{i<j} W(r_{ij})$ where $W(r)$ is the overlap function:

$$W(r) = \begin{cases} 0 \text{ if } r < \sigma \text{ (overlapping)} \\ 1 \text{ if } r > \sigma \text{ (nonoverlapping).} \end{cases} \tag{22}$$

However, the methods of many-body theory require formal perturbation expansions and subsequent resummations. To do so, the time evolution operator $S_t(\boldsymbol{x})$ needs to be defined for *all* configurations, including the unphysical overlapping configurations. A standard representation, defined for all points in phase space, has been developed for elastic hard spheres in Ref. [38], and is based on the binary collision expansion of $S_t(\boldsymbol{x})$ in terms of binary collision operators $T(ij)$.

The binary collision operator is defined in terms of two-body dynamics through the time displacement operator $S_t(12)$ as

$$S_t(12) = S_t^0(12) + \int_0^t \mathrm{d}\tau S_\tau^0(12) T(12) S_{t-\tau}^0(12) \ . \tag{23}$$

Following the argument of Ref. [38] for the case of elastic hard spheres step by step, the operator $T(12)$ for inelastic hard spheres is constructed as

$$T(12) = \sigma^{d-1} \int_{\boldsymbol{v}_{12}\cdot\hat{\boldsymbol{\sigma}}<0} \mathrm{d}\hat{\boldsymbol{\sigma}} |\boldsymbol{v}_{12}\cdot\hat{\boldsymbol{\sigma}}| \delta(\boldsymbol{r}_{12} - \boldsymbol{\sigma})(b_\sigma^* - 1) \ , \tag{24}$$

where b_σ^* is an operator that replaces all (precollision) velocities $\boldsymbol{v}_i$ $(i = 1, 2)$ appearing to its right by postcollision velocities $\boldsymbol{v}_i^*$.

The operator $T(12)$ is defined for overlapping and nonoverlapping configurations of two hard spheres. It extends the definition of $S_t(12)$ to all points in phase space. In the ensemble average considered in Eq. (1), the overlap function $W_N(\boldsymbol{x})$ contains a factor $W(r_{12})$ which vanishes whenever $r_{12} < \sigma$. The generator $S_t(12)$ for two-particle dynamics is only defined for *positive* times. The conservation laws imply $T(12)(a(\boldsymbol{v}_1) + a(\boldsymbol{v}_2)) = 0$, where $a(\boldsymbol{v})$ is a collisional invariant with $a(\boldsymbol{v}) = \{1, \boldsymbol{v}, v^2\}$ for elastic hard spheres, and $a(\boldsymbol{v}) = \{1, \boldsymbol{v}\}$ for inelastic hard spheres.

Combinations of T operators and free streaming operators S_t^0, preceded by appropriate combinations of overlap functions, can be used to construct the time displacement operators S_t for dynamical variables of the many-body problem. To discuss the Liouville equation and describe the time evolution of the reduced distribution functions we need in addition to consider the adjoint time displacement operator, $S_t^\dagger$, which is defined through the innerproduct (1). This yields $S_t^{0\dagger} = S_{-t}^0$, and the adjoint $\overline{T}$ of T is constructed as

$$\overline{T}(12) = \sigma^{d-1} \int_{\boldsymbol{v}_{12}\cdot\hat{\boldsymbol{\sigma}}>0} \mathrm{d}\hat{\boldsymbol{\sigma}} (\boldsymbol{v}_{12} \cdot \hat{\boldsymbol{\sigma}}) \left(\frac{1}{\alpha^2} \delta(\boldsymbol{r}_{12} - \boldsymbol{\sigma}) b_\sigma^{**} - \delta(\boldsymbol{r}_{12} + \boldsymbol{\sigma}) \right) . \quad (25)$$

Here b_σ^{**} acts on the velocities $\boldsymbol{v}_i$ $(i = 1, 2)$ to its right and replaces them by the restituting velocities, $\boldsymbol{v}_i^{**}$, as defined in collision rule (21).

A useful property of $\overline{T}$ is

$$\int \mathrm{d}x_1 \mathrm{d}x_2 (a(\boldsymbol{v}_1) + a(\boldsymbol{v}_2)) \overline{T}(12) B(x_1, x_2) = 0 , \quad (26)$$

where $a(\boldsymbol{v})$ is a collisional invariant, and $B(x_1, x_2)$ an arbitrary function of x_1 and x_2.

The time displacement operators $S_t(12)$ can be put in a more convenient form by using the property, $T(12)S_t^0(12)T(12) = 0$, valid for any $t > 0$. It also holds with T replaced by $\overline{T}$. This relation expresses the fact that two hard spheres cannot collide more than once with only free propagation in between. Using this property, the time displacement operators can be written as

$$\begin{aligned} W(12)S_t(12) &= W(12) \exp[tL^0(12) + tT(12)] \\ S_t^\dagger(12)W(12) &= \exp[-tL^0(1,2) + t\overline{T}(12)]W(12) . \end{aligned} \quad (27)$$

This can readily be generalized to the full N-particle system, and be represented in the compact form of pseudo-streaming operators,

$$\begin{aligned} W_N(\boldsymbol{x})S_t(\boldsymbol{x}) &= W_N(\boldsymbol{x}) \exp[tL^0(\boldsymbol{x}) + t\sum_{i<j} T(ij)] \\ S_t^\dagger(\boldsymbol{x})W_N(\boldsymbol{x}) &= \exp[-tL^0(\boldsymbol{x}) + t\sum_{i<j} \overline{T}(ij)]W_N(\boldsymbol{x}) , \end{aligned} \quad (28)$$

with $L^0(\boldsymbol{x}) = \sum_i L_i^0$ the free particle streaming operator. The time evolution operators are defined everywhere in phase space, and the overlap function gives a vanishing weight to unphysical configurations, provided that $W_N(\boldsymbol{x})$ appears to the left of T operators, or to the right of $\overline{T}$ operators.

Similar results for the binary collision operators T and $\overline{T}$ for inelastic hard spheres have been derived independently by Brey et al. [8].

The time evolution of the N-particle distribution function $\rho(\boldsymbol{x}, t)$ is given by the pseudo-Liouville equation. For conservative Hamiltonian systems this equation is the Liouville equation which is an expression of the incompressibility of the flow in phase space. In the case of dissipative systems which by definition are time irreversible, the phase space volumes are contracted along the flow. According to Eqs. (2) and (28) the time evolution of the distribution function for inelastic hard spheres is given by the pseudo-Liouville equation

$$[\partial_t + L^0(\boldsymbol{x})]\rho(\boldsymbol{x}, t) = \sum_{i<j} \overline{T}(ij)\rho(\boldsymbol{x}, t) \ . \tag{29}$$

The binary collision operators $\{T(ij), \overline{T}(ij)\}$ for elastic or inelastic hard sphere fluids account for the impulsive hard sphere interactions, and may be considered as 'pseudo-pair forces'. Hence, the name pseudo-Liouville equation.

An equivalent representation of the time evolution of the system can be given in terms of reduced s-particle distribution functions, defined in (7). Integration of Eq. (29) and using (26) for $a(\boldsymbol{v}) = 1$ yields the BBGKY hierarchy for the reduced distribution functions. We only quote the first two hierarchy equations for elastic and inelastic hard spheres, in the absence of external driving forces ($L_i^0 = \boldsymbol{v}_i \cdot \partial / \partial \boldsymbol{r}_i$),

$$\begin{aligned} (\partial_t + L_1^0) f_1 &= \int \mathrm{d}x_2 \overline{T}(12) f_{12} \\ \left[\partial_t + L_1^0 + L_2^0 - \overline{T}(12)\right] f_{12} &= \int \mathrm{d}x_3 [\overline{T}(13) + \overline{T}(23)] f_{123} \ . \end{aligned} \tag{30}$$

The pseudo-Liouville equation and related BBGKY hierarchy are the *exact* evolution equations for elastic ($\alpha = 1$) and inelastic hard spheres, which are *not driven* by external forces.

In subsection 2.2 we have extended the Liouville equation (16) to include stochastic forces, in order to describe some cases of driven systems. We want to apply these general results to *randomly* driven IHS fluids, where the forces contain only impulsive interactions between inelastic hard spheres, already expressed in binary collision operators in (24) and (25). So, we simply apply the Liouville equation (16) to impulsive forces. The pseudo-Liouville equation for the randomly driven IHS fluid can then be written in the same form as (29) with the generator L_i^0 replaced by

$$L_i^0 = \boldsymbol{v}_i \cdot \frac{\partial}{\partial \boldsymbol{r}_i} - \frac{1}{2}\xi_0^2 \frac{\partial}{\partial \boldsymbol{v}_i} \cdot \frac{\partial}{\partial \boldsymbol{v}_i} \ . \tag{31}$$

The BBGKY hierarchy for this case has again the same functional form as (30) with L_i^0 from (31).

One should be aware that the derivation of the Fokker-Planck equation (16) requires that $\hat{\boldsymbol{a}}_i\delta t$ in (12) is a small quantity of $\mathcal{O}(\delta t)$, which requires a *bounded* acceleration $\hat{\boldsymbol{a}}_i$. This is not the case for the 'pseudo-forces' $T(ij)$ and $\overline{T}(ij)$ (representing impulsive forces), which contain Dirac delta functions. However, Eqs. (29) and (30) in combination with (31) for IHS fluids, subject to external noise are expected to hold 'inside averages'. In fact, the derivation of the Fokker-Planck equation (16) implies (i) that the correlation time t_{noise} of the random forces is much shorter than the mean free time, and (ii) that the random forces are a small perturbation on the unperturbed hard sphere trajectories, or equivalently, that the changes of velocities caused by the random forces are much smaller than the changes caused by the hard sphere collisions. In zeroth order approximation one expects the latter condition to be satisfied if the energy supplied by the random kicks is small compared to the mean kinetic energy of the particles.

However, these conditions cannot be satisfied uniformly for all colliding pairs, because for any given t_{noise} there always exists a range of *small* relative velocities, where the actual time between two successive collisions of a single particle is of the same order of magnitude as t_{noise}. These nonuniformities become more important with increasing inelasticities, where the relative velocities after collision are substantially reduced. However, as the collision frequency, collisional dissipation rate, temperature, pressure, and transport fluxes all involve powers of velocities, we expect that the nonuniform convergence of the $\boldsymbol{v}$ integrations will not create any new singularities or show qualitatively different behavior, as long as the *inelasticities* are *small.*

3 Boltzmann and Ring Kinetic Equations

In this section we discuss the Boltzmann equation, the Enskog-Boltzmann equation, and the ring kinetic equations for IHS systems. In the literature on kinetic theory of inelastic hard spheres the first equation of the BBGKY hierarchy has been derived intuitively [3–6] and used as a starting point to obtain the Enskog-Boltzmann equation for the single-particle distribution function. Using the explicit expression (25) for $\overline{T}(12)$ the first hierarchy equation can be written in full detail as

$$(\partial_t + L_1^0) f(\boldsymbol{r}_1, \boldsymbol{v}_1, t) = \sigma^{d-1} \int \mathrm{d}\boldsymbol{v}_2 \int_{\boldsymbol{v}_{12}\cdot\hat{\boldsymbol{\sigma}}>0} \mathrm{d}\hat{\boldsymbol{\sigma}} (\boldsymbol{v}_{12}\cdot\hat{\boldsymbol{\sigma}}) \times$$

$$\left\{ \frac{1}{\alpha^2} f^{(2)}(\boldsymbol{r}_1, \boldsymbol{v}_1^{**}, \boldsymbol{r}_1 - \boldsymbol{\sigma}, \boldsymbol{v}_2^{**}, t) - f^{(2)}(\boldsymbol{r}_1, \boldsymbol{v}_1, \boldsymbol{r}_1 + \boldsymbol{\sigma}, \boldsymbol{v}_2, t) \right\} . \quad (32)$$

This equation contains the pair distribution function $f^{(2)}$ of two particles just *before* the hard sphere collision, i.e. $r_{12} = \sigma + 0$.

Consider first the case of the freely evolving IHS fluid, where $L_i^0 = \boldsymbol{v}_i \cdot \partial/\partial \boldsymbol{r}_i$. The corresponding equation for the rate of change of an average is obtained by multiplying the first hierarchy equation in (30) with $\int \mathrm{d}\boldsymbol{v}_1 \psi(\boldsymbol{v}_1)$, and using the adjoint of $\overline{T}(12)$ to find

$$\begin{aligned} &\frac{\partial}{\partial t} \int \mathrm{d}\boldsymbol{v}_1 \psi(\boldsymbol{v}_1) f(\boldsymbol{r}_1, \boldsymbol{v}_1, t) + \frac{\partial}{\partial \boldsymbol{r}_1} \cdot \int \mathrm{d}\boldsymbol{v}_1 \boldsymbol{v}_1 \psi(\boldsymbol{v}_1) f(\boldsymbol{r}_1, \boldsymbol{v}_1, t) \\ &= \int \mathrm{d}\boldsymbol{v}_1 \int \mathrm{d}x_2 f_{12} T(12) \psi(\boldsymbol{v}_1) \\ &= \sigma^{d-1} \int\!\!\int \mathrm{d}\boldsymbol{v}_1 \mathrm{d}\boldsymbol{v}_2 \int_{\boldsymbol{v}_{12}\cdot\hat{\boldsymbol{\sigma}}>0} \mathrm{d}\hat{\boldsymbol{\sigma}} (\boldsymbol{v}_{12} \cdot \hat{\boldsymbol{\sigma}}) \times \\ &\qquad f^{(2)}(\boldsymbol{r}_1, \boldsymbol{v}_1, \boldsymbol{r}_1 + \boldsymbol{\sigma}, \boldsymbol{v}_2, t)[\psi(\boldsymbol{v}_1^*) - \psi(\boldsymbol{v}_1)], \end{aligned} \tag{33}$$

In the case of the randomly driven IHS fluid, there appears an additional source term on the right hand side of Eq. (33), given by

$$S_\psi(\boldsymbol{r}_1, \boldsymbol{v}_1, t) = \frac{1}{2}\xi_0^2 \int \mathrm{d}\boldsymbol{v}_1 f(\boldsymbol{r}_1, \boldsymbol{v}_1, t) \left(\frac{\partial}{\partial \boldsymbol{v}_1}\right)^2 \psi(\boldsymbol{v}_1) \ . \tag{34}$$

In case $\psi(\boldsymbol{v}_1) = \{1, \boldsymbol{v}_1\}$ this source term vanishes, but for $\psi(\boldsymbol{v}_1) = \frac{1}{2}mv_1^2$ it is nonvanishing, $S(\boldsymbol{r}_1, \boldsymbol{v}_1, t) = (d/2)mn(\boldsymbol{r}, t)\xi_0^2$. Equation (33) is the starting point for deriving macroscopic conservation laws, hydrodynamic equations, as well as the starting point of the Chapman-Enskog method to calculate transport coefficients for IHS fluids [13, 14]. The original kinetic theory studies [2–5] are based on Grad's 13-moment method, and start from the 'adjoint' equation.

We return again to the kinetic equations. In order to derive a closed equation for the single-particle distribution function f a closure relation is required to express f_{12} in terms of f. The basic method to do so is Boltzmann's molecular chaos assumption, requiring that the velocities of two particles, just before collisions, are *uncorrelated*, or equivalently, that the pair distribution function just before collision factorizes. This assumption implies that the time evolution of the single-particle distribution function is only determined by sequences of *uncorrelated* binary collisions, whereas sequences of *correlated* binary collisions, e.g. (12) (13) (23), can be neglected. To state this even more emphatically: as far as the time evolution of $f(x, t)$ is concerned, it is assumed that no particle collides more than once with any other particle in its whole time evolution. It is clear that this assumption can only be correct for low densities. Here we assume that Boltzmann's assumption of molecular chaos also holds for a dilute gas of *inelastic* hard spheres, at least for *small* inelasticity, $\epsilon = 1 - \alpha^2$.

When the density increases the molecular chaos assumption in elastic hard sphere systems is *violated* [31], due to the increasing importance of ring collisions, which create dynamic correlations between the velocities of particles. Ring kinetic theory [31, 33] takes these correlated collisions into account. The same scenario applies to inelastic hard spheres.

Here we simply illustrate how the ring kinetic theory for elastic hard spheres can be transferred directly to the case of inelastic hard spheres. This will be done by deriving the Boltzmann equation and the ring kinetic equation for the IHS gases.

The Boltzmann equation is obtained from the first hierarchy equation by keeping only terms to dominant order in the density. This implies the fundamental assumption of molecular chaos for dilute gases, expressing the absence of dynamic correlations between the velocities of two colliding particles just *before* collision, i.e. at $|\boldsymbol{r}_{12}| = \sigma + 0$,

$$f_{12} = f(\boldsymbol{r}_1, \boldsymbol{v}_1, t) f(\boldsymbol{r}_2, \boldsymbol{v}_2, t) \ . \tag{35}$$

Furthermore, in the low density limit the spatial separation between the colliding particles can be neglected, and the binary collision operator $\overline{T}(12)$, entering in the BBGKY hierarchy, reduces to

$$\overline{T}(12) = \delta(\boldsymbol{r}_{12})\overline{T}_0(12) = \delta(\boldsymbol{r}_{12})\,\sigma^{d-1} \int_{\boldsymbol{v}_{12}\cdot\hat{\sigma}>0} \mathrm{d}\hat{\boldsymbol{\sigma}}(\boldsymbol{v}_{12}\cdot\hat{\boldsymbol{\sigma}}) \left(\frac{1}{\alpha^2} b_\sigma^{**} - 1\right) .$$

Then the nonlinear Boltzmann equation for a dilute gas of inelastic hard spheres becomes

$$(\partial_t + L_1^0) f(\boldsymbol{r}_1, \boldsymbol{v}_1, t) = \sigma^{d-1} \int \mathrm{d}\boldsymbol{v}_2 \int_{\boldsymbol{v}_{12}\cdot\hat{\sigma}>0} \mathrm{d}\hat{\boldsymbol{\sigma}}(\boldsymbol{v}_{12}\cdot\hat{\boldsymbol{\sigma}}) \times$$
$$\left\{\frac{1}{\alpha^2} f(\boldsymbol{r}_1, \boldsymbol{v}_1^{**}, t) f(\mathbf{r}_1, \boldsymbol{v}_2^{**}, t) - f(\boldsymbol{r}_1, \boldsymbol{v}_1, t) f(\boldsymbol{r}_1, \boldsymbol{v}_2, t)\right\} \equiv I(f, f) \ . \tag{36}$$

There are two significant differences with the Boltzmann equation for the elastic case: (i) the occurrence of $1/\alpha^2$ in the gain term on the right hand side of (36); one factor $1/\alpha$ comes from the Jacobian $\mathrm{d}\boldsymbol{v}_1^{**}\mathrm{d}\boldsymbol{v}_2^{**} = (1/\alpha)\mathrm{d}\boldsymbol{v}_1\mathrm{d}\boldsymbol{v}_2$ and the other one from the reflection law $\boldsymbol{v}_{12}^{**}\cdot\hat{\boldsymbol{\sigma}} = -(1/\alpha)\boldsymbol{v}_{12}\cdot\hat{\boldsymbol{\sigma}}$ (see Fig. 1). (ii) In the inelastic case, the *restituting* precollision velocities, which yield $(\boldsymbol{v}_1, \boldsymbol{v}_2)$ as postcollision velocities, are *different* from the postcollision velocities $(\boldsymbol{v}_1^*, \boldsymbol{v}_2^*)$, which result from the *direct* precollision velocities $(\boldsymbol{v}_1, \boldsymbol{v}_2)$. In the elastic case $(\alpha = 1)$ the relation $\boldsymbol{v}_i^* = \boldsymbol{v}_i^{**}$ holds.

There exists also a semi-phenomenological extension of the Boltzmann equation to liquid densities, the Enskog-Boltzmann equation. The Enskog-Boltzmann equation for inelastic hard spheres is obtained by replacing f_{12} in the first hierarchy equation of (30) by $\chi(\boldsymbol{r}_1, \boldsymbol{r}_2|n) f_1 f_2$, where $\chi(\boldsymbol{r}_1, \boldsymbol{r}_2|n)$ is the local pair correlation function of elastic hard spheres in a spatially nonuniform equilibrium state. This version of the molecular chaos assumption still neglects the velocity correlations, built up by sequences of correlated binary collisions, but does account for static short range correlations, caused by excluded volume effects.

As the density increases the contributions of correlated collision sequences to the collision term on the right hand side of (32) become more and more important. The most simple sequence of correlated collisions are the so called *ring* collisions; for example (12)(13)(14)...(23′)(24′)...(12), ending with a recollision of the pair (12), which was involved in the first collision. In the intermediate time particle 1 collides, say, s times with s different particles (3,4,...), and particle 2 collides s' times with *another* set of s' different particles (3′,4′,...). When particles 1 and 2 are about to recollide, they are dynamically correlated through their collision history, and the molecular chaos assumption (35) is no longer valid, i.e. $g_{12} \equiv f_{12} - f_1 f_2 \neq 0$.

A simple way to take these correlations into account at moderate densities has been given in Refs. [33, 69]. The method is based on a cluster expansion of the s-particle distribution functions, defined recursively as

$$\begin{aligned} f_{12} &= f_1 f_2 + g_{12} \\ f_{123} &= f_1 f_2 f_3 + f_1 g_{23} + f_2 g_{13} + f_3 g_{12} + g_{123} \, , \end{aligned} \tag{37}$$

etc. Here g_{12} accounts for pair correlations, g_{123} for triplet correlations, etc. The molecular chaos assumption implies $g_{12} = 0$, which is equivalent to (35). The basic assumption to obtain the ring kinetic equation is that the pair correlations are dominant and higher order terms in (37) can be neglected, i.e. $g_{123} = g_{1234} = \cdots = 0$.

Substitution of Eq. (37) into (30) and elimination of $\partial f_i / \partial t$ $(i = 1, 2)$ from the second hierarchy equation using the first one, yields the *ring kinetic equations* for inelastic hard spheres:

$$\begin{aligned} (\partial_t + L_1^0) f_1 &= \int \mathrm{d}x_2 \overline{T}(12)(f_1 f_2 + g_{12}) \\ \big[\partial_t + L_1^0 + L_2^0 - \overline{T}(12) - (1 + \mathcal{P}_{12}) &\int \mathrm{d}x_3 \overline{T}(13)(1 + \mathcal{P}_{13}) f_3 \big] g_{12} \\ &= \overline{T}(12) f_1 f_2 \, . \end{aligned} \tag{38}$$

Here $\mathcal{P}_{ij}$ is a permutation operator that interchanges the particle labels i and j. The second equation is the so called *repeated ring* equation for the pair correlation function. If the operator $\overline{T}(12)$ on the left hand side of the second equation is deleted, one obtains the simple *ring* approximation. Formally solving this equation for g_{12} yields an expression in terms of the single-particle distribution functions f_i $(i = 1, 2, 3)$, and subsequent substitution into the first hierarchy equation above yields the generalized Boltzmann equation in ring approximation. For a more detailed discussion of the collision sequences taken into account by Eqs. (38) we refer to the original literature [33, 69].

The kinetic equations (38) constitute the ring kinetic equations for inelastic hard spheres. All standard results for elastic hard spheres are recovered by setting the restitution coefficient $\alpha = 1$.

What is our motivation for studying the ring kinetic equations for undriven and driven IHS fluids? In Ref. [11] the ring equations have been analyzed to establish a more fundamental justification of the basic assumption upon which the phenomenological theory of fluctuating hydrodynamics of Ref. [25] has been built, namely the assumption that a fluctuation-dissipation theorem holds also for the so called homogeneous cooling state (to be discussed in the next section).

We consider first the freely evolving case, where $L_i^0 = \boldsymbol{v}_i \cdot \partial/\partial \boldsymbol{r}_i$ in (38). For elastic spheres one can show that the long time and long wavelength behavior of the ring collisional integral gives in *general* the same results as mode coupling theories of fluctuating hydrodynamic theories for current correlation functions, in their common region of validity, i.e. at low densities. Such a general proof is not yet available for inelastic hard spheres. However as a first step in the justification of the phenomenological theory [25], we have calculated from ring kinetic theory the structure factor $\langle |u_\perp(\boldsymbol{k},t)|^2 \rangle$ of the tranverse velocity field, which depends only on the shear modes. The resulting expression turns out to be the same as the result obtained from fluctuating hydrodynamic theory [25].

In Ref. [27] a mode coupling theory has been proposed for the uniformly heated IHS fluid, to calculate renormalized collision frequency, temperature, pressure etc. in the nonequilibrium steady state. It would be of interest to justify and derive these intuitive estimates also on the basis of the ring kinetic equations (38), but this program has not yet been carried out.

4 Homogeneous Solution of the Boltzmann Equation

4.1 Homogeneous Cooling State

In the previous sections we have constructed the fundamental equations of nonequilibrium statistical mechanics for complex fluids, and used standard methods of kinetic theory to derive the Boltzmann and ring kinetic equations for IHS fluids with and without external noise.

To solve the Boltzmann equation one uses the Chapman-Enskog expansion around a reference state. For the elastic hard sphere fluid this is the local equilibrium state; for the freely evolving IHS fluid, this is the homogeneous cooling state (HCS); for the uniformly heated IHS fluid, this is a nonequilibrium steady state (NESS). Having determined the single particle distribution function in the reference state, one calculates the next term in the Chapman-Enskog expansion, which is proportional to the gradients of temperature $\boldsymbol{\nabla} T$, of flow field $\boldsymbol{\nabla} \boldsymbol{u}$, and of density $\boldsymbol{\nabla} n$. This term yields the corresponding transport coefficients. For a systematic discussion we refer to [4, 13, 14].

In the present section, we focus on the distribution function for the HCS, which is the zeroth order solution of the Boltzmann equation, and consider first the freely evolving IHS gas. Our starting point is the nonlinear Enskog-Boltzmann equation for the single-particle distribution function $f(\boldsymbol{r}, \boldsymbol{v}, t)$ in a dense system of inelastic hard spheres in d dimensions, as discussed below

(36). This is essentially a mean field result, because we have neglected the correlations between velocities of particles that are aiming to collide. In the absence of external forces, the Enskog-Boltzmann equation admits a *homogeneous* solution $f(\boldsymbol{v},t)$, which obeys the following equation,

$$\partial_t f(\boldsymbol{v}_1,t) = \chi\sigma^{d-1}\int \mathrm{d}\boldsymbol{v}_2 \int^{(+)} \mathrm{d}\hat{\boldsymbol{\sigma}}(\boldsymbol{v}_{12}\cdot\hat{\boldsymbol{\sigma}}) \left\{\frac{1}{\alpha^2} f(\boldsymbol{v}_1^{**},t) f(\boldsymbol{v}_2^{**},t) - f(\boldsymbol{v}_1,t) f(\boldsymbol{v}_2,t)\right\} \equiv \chi I(f,f)\ , \tag{39}$$

where $\chi = g(r=\sigma)$ is the pair distribution function of hard spheres or disks at contact, and the superscript $^{(+)}$ on the $\hat{\boldsymbol{\sigma}}$ integration denotes the condition $\boldsymbol{v}_{12}\cdot\hat{\boldsymbol{\sigma}} > 0$.

Note that for the spatially homogeneous case, the only difference between the Enskog-Boltzmann equation for dense systems and the Boltzmann equation for dilute systems, is the presence of the factor $\chi(n)$. It accounts for the increased collision frequency in dense systems, caused by excluded volume effects.

For undriven IHS fluids, Goldshtein and Shapiro [6] have shown that Eq. (40) admits an isotropic scaling solution, describing the homogeneous cooling state, with a single-particle distribution function depending on time only through the temperature $T(t)$ as

$$f(\boldsymbol{v},t) = \frac{n}{v_0^d(t)}\tilde{f}\left(\frac{v}{v_0(t)}\right)\ , \tag{40}$$

where n is the constant density. Here the thermal velocity $v_0(t)$ is defined in terms of the temperature by $T(t) = \frac{1}{2}mv_0^2(t)$, with

$$\tfrac{1}{2}dnT(t) = \textstyle\int \mathrm{d}\boldsymbol{v}\tfrac{1}{2}mv^2 f(\boldsymbol{v},t)\ , \tag{41}$$

and m the particle mass. To obtain the rate of change of the temperature, we multiply (40) with $\psi = \frac{1}{2}mv_1^2$, and integrate over $\boldsymbol{v}_1$. The result after rescaling to dimensionless variables, $\boldsymbol{c}_i = \boldsymbol{v}_i/v_0$, is

$$\frac{\mathrm{d}T}{\mathrm{d}t} = -\frac{\mu_2}{d}m\chi n\sigma^{d-1}v_0^3 \equiv -2\gamma\omega_0 T\ , \tag{42}$$

where

$$\mu_p \equiv -\int \mathrm{d}\boldsymbol{c}_1 c_1^p \tilde{I}(\tilde{f},\tilde{f})$$

$$\tilde{I}(\tilde{f},\tilde{f}) \equiv \int \mathrm{d}\boldsymbol{c}_2 \int' \mathrm{d}\hat{\boldsymbol{\sigma}}(\boldsymbol{c}_{12}\cdot\hat{\boldsymbol{\sigma}})\left\{\frac{1}{\alpha^2}\tilde{f}(c_1^{**})\tilde{f}(c_2^{**}) - \tilde{f}(c_1)\tilde{f}(c_2)\right\}\ . \tag{43}$$

In (42) ω_0 is the Enskog collision frequency for elastic hard spheres, defined as the average loss term in Eq. (40) in thermal equilibrium, i.e.

$$\omega_0 = \chi n\sigma^{d-1}v_0 \int \mathrm{d}\boldsymbol{c}_1\mathrm{d}\boldsymbol{c}_2 \int' \mathrm{d}\hat{\boldsymbol{\sigma}}(\boldsymbol{c}_{12}\cdot\hat{\boldsymbol{\sigma}})\phi(c_1)\phi(c_2) = \frac{\Omega_d}{\sqrt{2\pi}}\chi n\sigma^{d-1}v_0. \tag{44}$$

In the first equality we have introduced dimensionless variables, where $\phi(c) = \pi^{-d/2}\exp(-c^2)$ is the Maxwellian. For technical details of the derivation we refer to Ref. [12]. Furthermore, the second equality in (42) defines the *time independent* dimensionless cooling rate as $\gamma \equiv (\sqrt{2\pi}/d\Omega_d)\mu_2$, where $\Omega_d = 2\pi^{d/2}/\Gamma(d/2)$ is the surface area of a d-dimensional unit sphere. With the help of Eqs. (40) and (42), it can be shown that the scaling form $\tilde{f}(c)$ satisfies the integral equation

$$\frac{\mu_2}{d}\left(d + c_1\frac{\mathrm{d}}{\mathrm{d}c_1}\right)\tilde{f}(c_1) = \tilde{I}(\tilde{f},\tilde{f}) , \tag{45}$$

where μ_2 still depends on the unknown $\tilde{f}$.

The moment method for solving this equation is described in [6, 12], and yields in first approximation

$$\tilde{f}(c) = \phi(c)\left\{1 + a_2 S_2(c^2) + \dots\right\} , \tag{46}$$

where $S_2(x)$ is a Sonine polynomial, defined as

$$S_2(x) = \frac{1}{2}x^2 - \frac{1}{2}(d+2)x + \frac{1}{8}d(d+2) . \tag{47}$$

The coefficients a_2 is proportional to the fourth cumulant of the scaling distribution $\tilde{f}(c)$. Calculations, given in Ref. [12], yield then

$$\begin{aligned} \mu_2 &= \tfrac{1}{2}(1-\alpha^2)\tfrac{\Omega_d}{\sqrt{2\pi}}\left\{1 + \tfrac{3}{16}a_2\right\} \\ a_2 &= \frac{16(1-\alpha)(1-2\alpha^2)}{9 + 24d + 8\alpha d - 41\alpha + 30(1-\alpha)\alpha^2} . \end{aligned} \tag{48}$$

This result is plotted in Fig. 2 as a function of α.

The plot shows that the homogeneous scaling form is well approximated by a Maxwellian for a large range of coefficients of restitution (say $0.6 \lesssim \alpha < 1$). For these values we have $|a_2| \lesssim 0.04$ in three dimensions and $|a_2| \lesssim 0.024$ in two dimensions. In a qualitative sense, the same conclusions were formulated also in Ref. [6].

To obtain the time dependence of the temperature, it is convenient to introduce the new time variable τ representing the average number of collisions suffered per particle in a time t, and defined as $\mathrm{d}\tau = \omega_0(T(t))\mathrm{d}t$, where ω_0 is calculated in (44). This yields

$$\tau = \frac{1}{\gamma}\ln(1 + \gamma t/t_0) . \tag{49}$$

Here $t_0 = 1/\omega_0(T_0)$ is the mean free time at the initial temperature $T(0) = T_0$. In the literature [1] a similar relation for $\tau(t)$ with γ replaced by $\gamma_0 = (1-\alpha^2)/2d$ has been derived by approximating $\tilde{f}(c)$ by a Maxwellian $\phi(c)$. Next we find from Eq. (42) Haff's homogeneous cooling law,

$$T(t) = T_0\exp(-2\gamma\tau) = \frac{T_0}{(1+\gamma t/t_0)^2} . \tag{50}$$

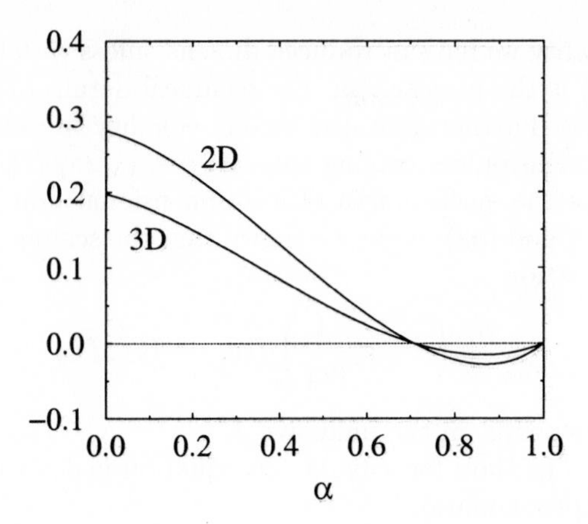

Fig. 2. Fourth cumulant a_2 versus α for the homogeneous cooling solution in a freely evolving fluid.

The moment expansion method allows us to calculate corrections to the dimensionless damping rate γ_0 and to the collision frequency ω_0 with the result

$$\begin{aligned} \gamma &= \gamma_0 \left\{1 + \tfrac{3}{16} a_2\right\} \\ \omega &= \omega_0 \left\{1 - \tfrac{1}{16} a_2\right\} , \end{aligned} \qquad (51)$$

where the Enskog frequency ω_0 is defined in (44). Sela and Goldhirsch [13] have performed a numerical perturbation expansion of the Boltzmann equation to first order in $\epsilon = 1 - \alpha^2$ and found the result $\gamma = \gamma_0(1 - 0.0258\epsilon + \mathcal{O}(\epsilon^2))$, which is close to the result $\gamma = \gamma_0(1 - 3\epsilon/128 + \mathcal{O}(\epsilon^2)) = \gamma_0(1 - 0.0234\epsilon + \mathcal{O}(\epsilon^2))$, obtained here. Since the contribution from a_2 in (51) to γ and ω are small for all α, they are very well approximated by γ_0 and ω_0, respectively.

So far, molecular dynamics simulations have not been able to obtain sufficient statistical accuracy to determine the fourth moment of the velocity distribution in undriven granular fluids. Such measurements are possible, however, by means of the direct simulation Monte Carlo (DSMC) method for the Enskog-Boltzmann equation. Using this method, Brey et al. [41] have solved the nonlinear Boltzmann equation (40) for homogeneously cooling inelastic hard spheres ($d = 3$) and measured the fourth and sixth moment of the distribution $\tilde{f}(c)$. The measured temperature decay shows no significant deviations of the cooling rate from its value γ_0. Fig. 5 of Ref. [41] compares their simulation data for the fourth cumulant a_2 with our prediction (48), and shows quantitative agreement. Moreover, the approximation (46) shows good agreement with the simulation data for the functional form of $\tilde{f}$ (see Figs. 7 and 8 of Ref. [41]). This second Sonine approximation is qualitatively similar to the form presented in Fig. 3 of Ref. [13], calculated numerically to order $\mathcal{O}(\epsilon)$. Of course, the good agreement with DSMC simulations does *not* show that the Boltzmann equation gives a valid description of IHS gases, but only confirms that the calculations presented here are correct. However, the good agreement between DSMC and MD simulations of a very dilute 2D IHS gas, reported by Brey [70], do indeed indicate that the Boltzmann equation for dilute IHS gases is correct up to moderate inelasticities ($\alpha \gtrsim 0.7$).

4.2 NESS for Heated Fluids

Next, we consider the Enskog-Boltzmann equation for randomly driven IHS fluids, which admits an isotropic stationary solution, $f(\boldsymbol{v}) = (n/v_0^d)\tilde{f}(v/v_0)$, and we will determine its properties. The Enskog-Boltzmann equation for the single-particle distribution function $f(\boldsymbol{r}, \boldsymbol{v}, t)$ of a system heated in this way is modified with a Fokker-Planck diffusion term (see e.g. Ref. [58, 59]), representing the change of the distribution function caused by the small random kicks. It reads in the spatially homogeneous case,

$$\partial_t f(\boldsymbol{v}_1, t) = \chi I(f, f) + \frac{\xi_0^2}{2}\left(\frac{\partial}{\partial \boldsymbol{v}_1}\right)^2 f(\boldsymbol{v}_1, t) \;, \tag{52}$$

as follows from Eq. (32) with L_i^0 given in (31). The diffusion coefficient ξ_0^2 in velocity space is proportional to the rate of energy input $\frac{d}{2}\xi_0^2$ per unit mass. The equation for the temperature balance can be derived from Eq. (52) in a similar fashion as in Eq. (42) for the cooling granular fluid, and reads

$$\frac{\mathrm{d}T}{\mathrm{d}t} = m\xi_0^2 - 2\gamma\omega_0 T \;. \tag{53}$$

We look for a stationary solution of (52), where the heating exactly balances the loss of energy due to collisions, and the temperature becomes time independent. Again it is convenient to introduce a rescaled distribution function

$$f(\boldsymbol{v}) = \frac{n}{v_0^d}\tilde{f}\left(\frac{v}{v_0}\right) \;, \tag{54}$$

where now the thermal velocity v_0 and temperature $T = \frac{1}{2}mv_0^2$ are time independent. The distribution function can again be expanded in Sonine polynomials as in (46), and a_2 for the heated case is given by

$$a_2 = \frac{16(1-\alpha)(1-2\alpha^2)}{73 + 56d - 24\alpha d - 105\alpha + 30(1-\alpha)\alpha^2} \;. \tag{55}$$

This function is shown in Fig. 4.2 for the two- and three-dimensional case. Again we find only small corrections to a Maxwellian distribution ($a_2 <$ 0.086 in two dimensions and 0.067 in three). Therefore to a good approximation, μ_2 and γ in (42) are given by their zeroth order approximations,

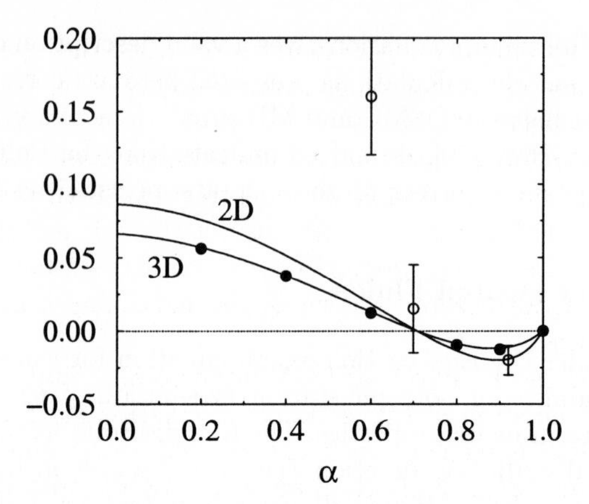

Fig. 3. Fourth cumulant a_2 versus α for the stationary state of a randomly driven system. Good agreement is found with results of direct Monte Carlo psimulation of Eq.52 (filled circles, courtesy of J.M. Montanero and A. Santos [50]). MD simulation results [49] (open circles, obtained at a packing fraction $\phi = 0.63$ for $\alpha \geq 0.7$ and at $\phi = 0.55$ for $\alpha = 0.6$) show deviations at higher inelasticities, indicating that the assumption of molecular chaos, underlying Eq. (52), breaks down.

$\mu_2 \simeq d\Omega_d\gamma_0/\sqrt{2\pi}$ and $\gamma \simeq \gamma_0 = (1-\alpha^2)/2d$, and the stationary temperature is found from the stationary solution of (53) and (44) as

$$T_0 = m\left(\frac{d\xi_0^2\sqrt{\pi}}{(1-\alpha^2)\Omega_d\chi n\sigma^{d-1}}\right)^{2/3} . \tag{56}$$

Next, we compare theoretical results with computer simulations. A detailed comparison between molecular dynamics results [27, 48] and our theoretical predictions for the temperature and collision frequency in randomly driven fluids of inelastic hard disks has already been discussed in Ref. [27]. There is good agreement for the time dependent temperature, obtained from (53), and its stationary value (56), for different types of initial conditions [27] and for all times. The simulation results for the collision frequency show that the Enskog prediction (44) gives reasonable values for $\alpha > 0.6$. At larger inelasticities, important deviations from (44) are observed [49]. They are caused by recollisions between particles, which are not taken into account in the Enskog-Boltzmann equation. This clearly shows that the molecular chaos assumption, used to derive the mean-field type Boltzmann equations, is violated at larger inelasticities ($\alpha < 0.7$) and at the high densities considered here.

It should be mentioned however, that the assumption of molecular chaos is violated as well in elastic hard sphere fluids at higher densities. Molecular dynamics measurements of transport coefficients of elastic hard spheres show sizeable positive deviations from the Enskog predictions [35]. These results were obtained by integrating the long time tails of the Green-Kubo current correlation functions, which result from recollisions or, equivalently, from mode coupling effects.

Moreover, the fourth cumulant a_2 has been measured in MD simulations of inelastic hard disks, and shows good agreement with the theoretical predictions for $\alpha \gtrsim 0.7$. For $\alpha \lesssim 0.7$, the deviations between the results of MD simulations and the theoretical predictions in Fig. 4.2 become rather large. This confirms again the breakdown of the molecular chaos assumption at large inelasticities.

Unpublished results obtained by Montanero and Santos [50] from direct Monte Carlo simulation of the Boltzmann equation for the heated IHS gas show quantitative agreement with our predictions for the fourth cumulant (55) for all values of α. This confirms that the asymptotic analysis of Eq. (52), carried out in Ref. [12], is indeed correct.

4.3 High Energy Tails

To discuss the high energy tail of the VDF we use an asymptotic method for solving the nonlinear Boltzmann equation, employed by Krook and Wu [51] to study the formation of Maxwellian tails in elastic systems with rapidly decreasing differential scattering cross sections at large impact energies. In doing so we obtain the tail distributions for the *undriven* and the *randomly driven* case. In this section we will derive the asymptotic solution of the Enskog-Boltzmann equation for high velocities for freely evolving and randomly driven granular fluids. Esipov and Pöschel [10] have given a similar derivation for an undriven gas and found a high energy tail $\tilde{f}(c) \sim \exp(-Ac)$. The derivation in both cases proceeds along similar lines and is based on the asymptotic methods developed in Ref. [51].

We consider first the undriven IHS fluid. If particle 1 is a fast particle ($c_1 \gg 1$), the dominant contributions to the collision integral come from collisions where particle 2 is typically in the thermal range, so that $\boldsymbol{c}_{12}$ in the collision integral $\widetilde{I}(\tilde{f}, \tilde{f})$ in (43) can be replaced by $\boldsymbol{c}_1$. The gain term $\widetilde{I}_g$ of the collision integral $\widetilde{I}$ can then be neglected with respect to the loss term $\widetilde{I}_l$, as will be verified a posteriori. The collision integral $\widetilde{I}(\tilde{f}, \tilde{f})$ then reduces to $\widetilde{I}_l \approx -\beta_1 c_1 \tilde{f}(c_1)$, where the remaining angular integral yields $\beta_1 = \pi^{(d-1)/2}/\Gamma(\frac{1}{2}(d+1))$ [12] and Eq. (45) simplifies to

$$\frac{\mu_2}{d}\left(d + c\frac{\mathrm{d}}{\mathrm{d}c}\right)\tilde{f}(c) = -\beta_1 c \tilde{f}(c) \,, \tag{57}$$

where μ_2 is given by (48). The first term on the left hand side can be neglected with respect to the right hand side, and the large c solution has the form

$$\tilde{f}(c) \sim \mathcal{A} \exp\left(-\frac{\beta_1 d}{\mu_2} c\right) \,, \tag{58}$$

where $\mathcal{A}$ is an undetermined integration constant. For $c \gg 1$ this solution corresponds to a tail which is overpopulated when compared to $\exp(-c^2)$ if $c \gtrsim 1/\epsilon$.

To determine the high energy tail of $\tilde{f}(c)$ for the randomly driven system, we proceed in a similar fashion and use (53) to write Eq. (52) as

$$\tilde{I}(\tilde{f},\tilde{f}) + \frac{\mu_2}{2d}\left(\frac{\partial}{\partial \boldsymbol{c}_1}\right)^2 \tilde{f}(c_1) = 0 \, . \tag{59}$$

For large velocities c_1, the collision integral can again be replaced by $-\beta_1 c_1 \tilde{f}(c_1)$, and Eq. (59) reduces to

$$-\beta_1 c \tilde{f}(c) + \frac{\mu_2}{2d}\left(\frac{\mathrm{d}^2}{\mathrm{d}c^2} + \frac{d-1}{c}\frac{\mathrm{d}}{\mathrm{d}c}\right)\tilde{f}(c) = 0 \, , \tag{60}$$

where we have used isotropy of the distribution function. Inserting solutions of the form $\tilde{f}(c) \propto \exp(-Ac^B)$, where the coefficients A and B are determined self-consistently, we obtain the large c solution with $B = \frac{3}{2}$ and $A = \frac{2}{3}\sqrt{2d\beta_1/\mu_2}$, which is the only solution that vanishes for $c \to \infty$. Again we find an enhanced population for high energies. The technical details of these asymptotic estimates are given in Ref. [12].

To conclude, we have shown that $\tilde{f}(c) \sim \exp(-Ac^B)$ is a consistent large c solution of the Boltzmann Eqs. (45) and (52) with $B = 1$ and A given in (58) for the undriven fluid, and $B = \frac{3}{2}$ and A given below (60) for the randomly driven fluid. The predicted behavior $\tilde{f}(c) \sim \exp(-Ac)$ in the former case for the high energy tail has also been confirmed in DSMC simulations [71], where quantitative agreement was found for A, as given in (58).

Finally, evidence for the high energy tail $\sim \exp(-Ac^{3/2})$ in the steady state distribution function for the heated case can also been found in the simulations by Peng and Ohta [48], but their statistical accuracy is too low to make any quantitative comparison. Comparison with the high energy tails, measured in Ref. [50] from direct Monte Carlo simulation of the Boltzmann equation, show the correctness of our asymptotic solution of the Boltzmann equation. There exist also laboratory experiments [67, 68] in which overpopulated tails have been observed in vibrated granular layers.

5 Conclusion

In this article we have discussed the basis for the nonequilibrium statistical mechanics of granular fluids with or without external noise, modeled as soft or hard inelastic spheres. In the absence of stochastic forces, the N-particle dynamics is described by a generalized version of the Liouville equation for the phase space density $\rho(\boldsymbol{x}, t)$, which does not describe incompressible, but contracting flows. In the presence of stochastic driving forces, the evolution equation for $\rho(\boldsymbol{x}, t)$ is the multivariate Fokker-Planck equation. Once these generalized 'Liouville' equations have been obtained, one can derive the BBGKY hierarchy for the reduced distribution functions, which form the basic starting point to derive kinetic equations, such as the Boltzmann and ring kinetic equations, as well as the modified Boltzmann equations with a Fokker-Planck diffusion term $\frac{1}{2}\xi_0^2(\partial/\partial \boldsymbol{v})^2$ added to account for external noise. The limiting case of the soft inelastic sphere model with very stiff elastic repulsion approaches the inelastic hard sphere model, which has instantaneous interactions. The Boltzmann and ring kinetic equations for the latter case have been derived in terms of binary collision operators T and $\overline{T}$, analogous to the case of elastic hard spheres.

In the case of external stochastic forces, the standard derivation of the Fokker-Planck diffusion term remains valid, as long as the deterministic forces vary sufficiently slowly, i.e. remain bounded. Then, the assumption of white noise (vanishing correlation time) is meaningful. However, the derivation of the Liouville equation in systems subject to white noise *and* to hard sphere interactions, which are formally defined as a limiting case of vanishing contact duration, involves two limiting operations, to be taken in the proper order. The *first* limit is that the correlation time of the external noise vanishes ($t_{\text{noise}} \to 0$). The second limit is that the contact duration of the soft sphere collisions vanishes (elastic modulus $Y \to \infty$). In the ad hoc derivation, presented here, we have simply combined the instantaneous T and $\overline{T}$ operators with the Fokker-Planck diffusion term, and we expect this to give a faithful description of the evolution of the single particle distribution function, at least at small inelasticities.

As an application of the Boltzmann equation we have analyzed the non-Gaussian behavior of the solution of this equation in systems with and without external noise, and compared the results with those of event driven molecular dynamics simulations in N-particle systems and with those obtained from direct simulations Monte Carlo (DSMC) to solve the nonlinear Boltzmann equation.

The agreement between the results of kinetic theory with those of fluctuating hydrodynamics, and with those of MD or DSMC simulation methods is in general quite good, at least at small inelasticities. This confirms that the kinetic and hydrodynamic theories, discussed here, give sensible descriptions of rapid granular flows for the idealized models, discussed in this article.

References

1. P. K. Haff, J. Fluid Mech. **134**, 401 (1983).
2. J. T. Jenkins and S. B Savage, J. Fluid Mech. **130**, 187 (1983).
3. C. K. K. Lun, S. B. Savage, D. J. Jeffrey and N. Chepurniy, J. Fluid Mech. **140**, 223 (1984).
4. J. T. Jenkins and M. W. Richman, Arch. Rat. Mech. Anal. **87**, 355 (1985).
5. J. T. Jenkins and M. W. Richman, J. Fluid Mech. **192**, 313 (1988).
6. A. Goldshtein and M. Shapiro, J. Fluid Mech. **282**, 75 (1995).

7. N. Sela, I. Goldhirsch and S. H. Noskowicz, Phys. Fluids **8**, 2337 (1996).
8. J. J. Brey, F. Moreno and J. W. Dufty, Phys. Rev. E **54**, 445 (1996); J. J. Brey, J. W. Dufty and A. Santos, J. Stat. Phys. **87**, 1051 (1997).
9. P. Deltour and J.-L. Barrat, J. Phys. I **7**, 137 (1997).
10. S. E. Esipov and T. Pöschel, J. Stat. Phys. **86**, 1385 (1997).
11. T. P. C. van Noije, M. H. Ernst and R. Brito, Physica A **251**, 266 (1998).
12. T. P. C. van Noije and M. H. Ernst, Granular Matter **1**, 57 (1998), cond-mat/9803042.
13. N. Sela and I. Goldhirsch, J. Fluid. Mech. **361**, 41 (1998).
14. J. J. Brey, J. W. Dufty, C. S. Kim and A. Santos, Phys. Rev. E **58**, 4638 (1998).
15. O. R. Walton and R. L. Braun, J. Rheol. **30**, 949 (1986); Acta Mech. **63**, 73 (1986).
16. C. S. Campbell, Annu. Rev. Fluid Mech. **22**, 57 (1990).
17. S. Chen, Y. Deng, X. Nie and Y. Tu, cond-mat/9804235.
18. H. J. Herrmann, Physica A **191**, 263 (1992).
19. S. Luding, E. Clement, A. Blumen, J. Rajchenbach and J. Duran, Phys. Rev. E **50**, 4113 (1994).
20. C. S. Campbell and C. E. Brennen, J. Fluid Mech. **151**, 167 (1985).
21. M. A. Hopkins and M. Y. Louge, Phys. Fluids A **3**, 47 (1991).
22. I. Goldhirsch and G. Zanetti, Phys. Rev. Lett. **70** 1619 (1993).
23. S. McNamara and W. R. Young, Phys. Rev. E **50**, R28 (1994).
24. S. McNamara and W. R. Young, Phys. Rev. E **53**, 5089 (1996).
25. T. P. C. van Noije, M. H. Ernst, R. Brito and J. A. G. Orza, Phys. Rev. Lett. **79**, 411 (1997); T. P. C. van Noije, R. Brito and M. H. Ernst, Phys. Rev. E **57**, R4891 (1998).
26. R. Brito and M. H. Ernst, Europhys. Lett. **43**, 497 (1998); Int. J. Mod. Phys. C **9**, 1339 (1998).
27. T. P. C. van Noije, M. H. Ernst, E. Trizac and I. Pagonabarraga, Phys. Rev. E **59**, 4326 (1999).
28. S. Luding, *Die Physik kohäsionsloser granularer Medien*, Habilitationsschrift, Stuttgart University (Logos Verlag, Berlin, 1998).
29. B. J. Alder and T. E. Wainwright, Phys. Rev. Lett. **18**, 988 (1967); B. J. Alder and W. E. Alley, Phys. Today **37**, 56 (1984).
30. E. G. D. Cohen, Phys. Today **37**, 64 (1984); Am. J. Phys. **61**, 524 (1993).
31. J. R. Dorfman and H. van Beijeren, *The Kinetic Theory of Gases*, in *Statistical Mechanics, Part B: Time-Dependent Processes*, Ed. B. J. Berne (Plenum Press, New York, 1977), Chap. 3.
32. J. R. Dorfman and E. G. D. Cohen, Phys. Rev. Lett. **25**, 1257 (1970).
33. M. H. Ernst and J. R. Dorfman, Physica **61**, 157 (1972).
34. Y. Pomeau and P. Resibois, Phys. Rep. **19C**, 63 (1975).
35. J.-P. Hansen and I. R. McDonald, *Theory of simple liquids* (Academic Press, London, 1986).
36. J. R. Dorfman, T. R. Kirkpatrick and J. Sengers, Ann. Rev. Phys. Chem. **45**, 213 (1994).
37. G. Grinstein, D.-H. Lee and S. Sachdev, Phys. Rev. Lett. **64**, 1927 (1990); G. Grinstein, J. Appl. Phys. **69**, 5441 (1991).
38. M. H. Ernst, J. R. Dorfman, W. R. Hoegy and J. M. J. van Leeuwen, Physica **45**, 127 (1969).

39. I. Goldhirsch and M-L. Tan, Phys. Fluids **8**, 1752 (1996).
40. I. Goldhirsch, M-L. Tan and G. Zanetti, J. Scient. Comp. **8**, 1 (1993).
41. J. J. Brey, M. J. Ruiz-Montero and D. Cubero, Phys. Rev. E **54**, 3664 (1996).
42. K. Ichiki and H. Hayakawa, Phys. Rev. E **52**, 658 (1995); **57**, 1990 (1998).
43. Y-h. Taguchi and H. Takayasu, Europhys. Lett. **30**, 499 (1995).
44. J. M. Salazar and L. Brenig, J. Plasma Phys. **59**, 639 (1998); Phys. Rev. E **59**, 2093 (1999).
45. D. R. Williams and F. C. MacKintosh, Phys. Rev. E **54**, R9 (1996).
46. A. Puglisi, V. Loreto, U. Marini Bettolo Marconi, A. Petri and A. Vulpiani, Phys. Rev. Lett. **81**, 3848 (1998), cond-mat/9810059.
47. M. R. Swift, M. Boamfă, S. J. Cornell and A. Maritan, Phys. Rev. Lett. **80**, 4410 (1998).
48. G. Peng and T. Ohta, Phys. Rev. E **58**, 4737 (1998).
49. I. Pagonabarraga, E. Trizac, T. P. C. van Noije and M. H. Ernst, preprint, June 1999.
50. J. M. Montanero and A. Santos, unpublished.
51. M. Krook and T. T. Wu, Phys. Rev. Lett. **36**, 1107 (1975).
52. H. van Beijeren and M. H. Ernst, Physica A**68**.437 (1973); **70**, 225 (1973). These articles derive of the RET as an exact short time limit from the pseudo-Liouville equation (see here Eq. (29) for $\alpha = 1$). It gives, when compared to the earlier phenomenological Enskog theory, important corrections to static and dynamic structure factors in the theory of neutron scattering, and the Onsager relations for transport coefficients follow from RET, and not from the earlier Enskog theory. The Navier-Stokes transport coefficients in a single component elastic hard sphere fluid on the other hand agree in both theories.
53. S. Chapman and T. G. Cowling, *The Mathematical Theory of Non-uniform Gases* (Cambridge University Press, 1970).
54. V. Garzo and J. W. Dufty, Phys. Rev. E **59**, 5895 (1999).
55. M. H. Ernst, in: *Dynamics: Models and Kinetic Methods for Nonequilibrium Many-Body Systems*, ed. J. Karkheck (Kluwer academic Publ., Dordrecht, NL, 1999); accessible through website ftp://ftp.phys.uu.nl/pub/LEIDEN-ASI/
56. M. Huthmann, T. Aspelmeier and A. Zippelius, Phys. Rev. E (July 1999); T. Aspelmeier, M. Huthmann, and A. Zippelius, *Free cooling of particles with rotational degrees of freedom*, (in this volume, page 31).
57. N. V. Brilliantov and T. Poeschel, *Granular Gases with Impact-velocity Dependent Restitution Coefficient*, (in this volume, page 100).
58. N. G. van Kampen, *Stochastic Processes in Physics and Chemistry*, (North-Holland, Amsterdam, 1992).
59. C. W. Gardiner, *Handbook of stochastic methods for physics, chemistry and the natural sciences* (Springer-Verlag, Berlin, 1994).
60. C. Bizon, M. D. Shattuck, J. B. Swift and H. L Swinney, cond-mat/9904132.
61. C. K. K. Lun and S. B. Savage, Acta Mech. **63**, 15 (1986).
62. H. Hwang and K. Hutter, *Continuum Mech. and Thermodynamics* **7**, 357 (1995).
63. T. P. C. van Noije, Ph.D. thesis, Utrecht University, 1999.
64. L. Oger, C. Annic, D. Bideau, R. Dai and S. B. Savage, J. Stat. Phys. **82**, 1047 (1996).
65. F. Melo, P. B. Umbanhowar and H. L. Swinney, Phys. Rev. Lett. **72**, 172 (1994); **75**, 3838 (1995).

66. J. M. Huntley, in *Dynamics: models and kinetic methods for nonequilibrium many-body systems*, Ed. J. Karkheck (Kluwer Academic Publishers, Dordrecht, 1999).
67. J. S. Olafsen and J. S. Urbach, Phys. Rev. Lett. **81**, 4369 (1998); cond-mat/9905173.
68. W. Losert, D. G. W Cooper, J. Delour, A. Kudrolli and J. P. Gollub, cond-mat/9901203.
69. J. R. Dorfman and E. G. D. Cohen, J. Math. Phys. **8**, 282 (1967).
70. J. J. Brey and D. Cubero, *Hydrodynamic transport coefficients of Granular Gases*, (in this volume, page 59).
71. J. J. Brey, D. Cubero and M. J. Ruiz-Montero, Phys. Rev. E **59**, 1256 (1999).

Free Cooling of Particles with Rotational Degrees of Freedom

Timo Aspelmeier, Martin Huthmann, and Annette Zippelius

Institut für Theoretische Physik, Universität Göttingen,
D-37073 Göttingen, Germany.
e-mail: annette@Theorie.Physik.UNI-Goettingen.DE

Abstract. Free cooling of granular materials is analyzed on the basis of a pseudo-Liouville operator. Exchange of translational and rotational energy requires surface roughness for spherical grains, but occurs for non-spherical grains, like needles, even if they are perfectly smooth. Based on the assumption of a homogeneous cooling state, we derive an approximate analytical theory. It predicts that cooling of both rough spheres and smooth needles proceeds in two stages: An exponentially fast decay to a state with stationary ratio of translational and rotational energy and a subsequent algebraic decay of the total energy. These results are confirmed by simulations for large systems of moderate density. For higher densities, we observe deviations from the homogeneous state as well as large-scale structures in the velocity field. We study non-Gaussian distributions of the momenta perturbatively and observe a breakdown of the expansion for particular values of surface roughness and normal restitution.

1 Introduction

The hard-sphere model has been a very useful reference system for our understanding of classical liquids [1]. As far as static correlations are concerned, an analytical expression for the pair correlation is available [2] and provides a good first approximation for particles interacting via smooth-potential functions. The hard-sphere model is even more important for the dynamics, because it allows for approximate analytical solutions, based on the Boltzmann equation and its generalization by Enskog to account for a finite particle diameter and pair correlations at contact [3]. The model has the additional advantage that it is particularly well suited for numerical simulations [4] and in fact many of the important phenomena of dense liquids have been observed first in simulations of hard spheres. Examples are the discovery of long-time tails [5] and two-dimensional solids [6].

Not surprisingly, the model has become very popular also in the context of granular media, which are characterized by inelastic collisions of their constituents. Focusing on the rapid-flow regime, where kinetic theory should apply, generalized Boltzmann- and Enskog equations have been formulated and solved approximately [7–11]. The success of the Boltzmann-Enskog equation in classical fluids is based on the linearisation of the collision operator around local equilibrium. The resulting linear hermitean operator can then be treated by standard methods of functional analysis [12–14]. For inelastic systems no analog of the local equilibrium distribution is known. In many studies, including the present one, a homogeneity assumption is made, which is known to be unstable for dense and large enough systems and long times [15]. Hence the analysis is restricted to small and intermediate densities. Alternatively, one may restrict oneself to almost elastic collisions and expand around the elastic case.

Kinetic theory of rough, inelastic, circular disks was first discussed by Jenkins and Richman [8]. These authors introduced two temperatures, one for the translational and one for the rotational degrees of freedom, and studied deviations from a two-temperature Maxwellian distribution, using Grad's moment expansion. Subsequently Lun and Savage [9, 10] extended the approach to rough, inelastic spheres. A set of conservation equations and constitutive relations was derived from the Boltzmann equation, assuming small inelasticity and surface roughness. Goldshtein and Shapiro [11] discuss in detail the homogeneous cooling state of rough spheres. They determine the asymptotic ratio of rotational to translational energy as a function of surface roughness and coefficient of normal restitution. Hydrodynamic equations and constitutive relations are derived with help of the Enskog expansion. More recently, event-driven simulations of rough spheres have been performed by McNamara and Luding [16]. They investigate free cooling as a function of arbitrary surface roughness and normal restitution and compare their results to an approximate kinetic theory [17, 18].

Most analytical and numerical studies of kinetic phenomena have concentrated on spherical objects so far[1]. The question then arises, which of the results are specific to spherical objects and which are generic for inelastically colliding particles. A single collision of two arbitrarily shaped, but convex objects is quite difficult to describe analytically [21], set aside the problem of an ensemble of colliding grains. In this paper we have chosen the simplest non spherical objects, needles, which allow for an analytical, albeit approximate solution and large scale simulations [22].

The paper is organized as follows. In Sec. 2 we introduce the time evolution operator. For pedagogical reasons we first discuss smooth potentials and recall the formalism of a pseudo-Liouville operator for elastic, hard-core collisions. Subsequently the formalism is extended to inelastic, rough spheres and needles. The homogeneous cooling state is introduced in Sec. 3. We present results for both spheres and needles, assuming a Maxwellian distribution for linear and angular momenta. We show with simulations that for dense systems of needles the assumption of homogeneity breaks down. Corrections to a Gaussian approximation are discussed in Sec. 4. Finally in Sec. 5 we summarize results and present conclusions. Some details of the calculation are delegated to appendices.

---

[1] Exceptions are computer simulations of polygonal particles [19] and cellular automata models [20]

2 The Liouville Operator

We are interested in macroscopic properties of systems of many particles which are themselves meso- or macroscopic, i.e. behave according to the laws of classical mechanics as opposed to quantum mechanics. In addition, our systems are *granular* so energy is not conserved. This means that they can not be treated with Hamiltonian mechanics. We will present here a formalism based on the Liouville operator that enables us nevertheless to derive properties of the system under consideration.

We consider two different models: The first is a system of spheres of diameter d and the second is one of (infinitely) thin rods or needles of length L. In order to keep the discussion as transparent as possible, the formalism of the (pseudo) Liouville operator will be demonstrated for Hamiltonian systems with smooth potentials first, for hard core potentials next, and finally for granular spheres and needles. It is interesting to note that both cases, spheres and needles, are analytically tractable so that comparisons between different geometrical particle shapes are possible.

2.1 Smooth Potentials

We consider a system of N classical particles of mass m in a volume V, interacting through pairwise potentials. The system is characterized by its total energy

$$H = \sum_{i=1}^{N} \frac{\boldsymbol{p}_i^2}{2m} + \sum_{i<j} U(\boldsymbol{r}_i - \boldsymbol{r}_j) \tag{1}$$

in terms of particle momenta $\boldsymbol{p}_i$ and coordinates $\boldsymbol{r}_i$. The time evolution of an observable $f(\Gamma)$, which is a function of phase space variables $\Gamma := \{\boldsymbol{r}_i, \boldsymbol{p}_i\}$, but does not depend on time explicitly, is given in terms of the Poisson bracket by

$$\frac{df}{dt} = \{H, f\} =: i\mathcal{L} f \ . \tag{2}$$

This defines the Liouville operator $\mathcal{L}$. The time evolution of f can then formally be written in terms of $\mathcal{L}$: $f(t) = e^{i\mathcal{L}t} f(0)$.

We decompose the Liouville operator $\mathcal{L} = \mathcal{L}_0 + \mathcal{L}_{\text{inter}}$ into a free-streaming part $\mathcal{L}_0$ and an operator $\mathcal{L}_{\text{inter}}$, which accounts for interactions. The definition of the Poisson bracket,

$$\{H, f\} = \sum_j \left(\frac{\partial f}{\partial \boldsymbol{r}_j} \cdot \frac{\partial H}{\partial \boldsymbol{p}_j} - \frac{\partial f}{\partial \boldsymbol{p}_j} \cdot \frac{\partial H}{\partial \boldsymbol{r}_j} \right) , \tag{3}$$

thus yields

$$i\mathcal{L}_0 = \sum_j \frac{\boldsymbol{p}_j}{m} \cdot \frac{\partial}{\partial \boldsymbol{r}_j} \quad \text{and} \quad i\mathcal{L}_{\text{inter}} = \sum_{k<j} \frac{\partial U}{\partial \boldsymbol{r}_{kj}} \cdot \left(\frac{\partial}{\partial \boldsymbol{p}_j} - \frac{\partial}{\partial \boldsymbol{p}_k} \right) \tag{4}$$

with $\boldsymbol{r}_{ij} = \boldsymbol{r}_i - \boldsymbol{r}_j$.

2.2 Elastic Hard-Core Interactions

A pseudo-Liouville operator for hard-core collisions has been formulated by Ernst et al. [23] and has been applied by many groups [24] to study the dynamic evolution of a gas of hard spheres. Collisions are instantaneous and characterized by collision rules. In a collision of two particles, numbered 1 and 2, their pre-collisional velocities $\boldsymbol{v}_1 = \boldsymbol{p}_1/m$ and $\boldsymbol{v}_2 = \boldsymbol{p}_2/m$ are changed instantaneously to their post-collisional values $\boldsymbol{v}_1'$ and $\boldsymbol{v}_2'$ according to

$$\begin{aligned} \boldsymbol{v}_1' &= \boldsymbol{v}_1 - (\boldsymbol{v}_{12} \cdot \hat{\boldsymbol{r}}_{12})\hat{\boldsymbol{r}}_{12} \\ \boldsymbol{v}_2' &= \boldsymbol{v}_2 + (\boldsymbol{v}_{12} \cdot \hat{\boldsymbol{r}}_{12})\hat{\boldsymbol{r}}_{12} \; . \end{aligned} \tag{5}$$

We have denoted the relative velocity by $\boldsymbol{v}_{12} = \boldsymbol{v}_1 - \boldsymbol{v}_2$, and $\hat{\boldsymbol{r}} = \boldsymbol{r}/|\boldsymbol{r}|$. The free-streaming part of the Liouville operator remains unchanged, whereas the part which accounts for interactions has to be modified because the potential is no longer differentiable in the limit of hard-core interactions. As a consequence, $\mathcal{L}$ is no longer self adjoint as it is for systems with smooth potentials. This is why it is called a pseudo-Liouville operator for hard-core systems. For the same reason we will need *two* Liouville operators below, one for forward and one for backward time evolution.

In order to construct the pseudo Liouville operator, we consider the change of a dynamical variable due to a collision of just two particles. What we need is an operator $\mathcal{T}_+^{(12)}$ that

- gives the change of an observable through a collision when integrated over a short time interval containing the collision time (since the hard core interaction is non-differentiable, we have to resort to integrating over the collision instead of looking at the derivatives directly),
- only acts at the time of contact,
- only acts when the particles are approaching but not when they are receding.

The second requirement can be satisfied by $\mathcal{T}_+^{(12)} \propto \delta(|\boldsymbol{r}_{12}| - d)$, the third one demands $\mathcal{T}_+^{(12)} \propto \Theta(-\frac{d}{dt}|\boldsymbol{r}_{12}|)$, where $\Theta(\cdot)$ is the usual Heaviside step function. In order to satisfy the first point, we use an operator $b_+^{(12)}$ which is defined by its action on an observable f according to

$$b_+^{(12)} f(\boldsymbol{v}_1, \boldsymbol{v}_2) = f(\boldsymbol{v}_1', \boldsymbol{v}_2') \; , \tag{6}$$

i.e. it simply replaces all velocities according to Eqs. (5). The operator $\mathcal{T}_+^{(12)}$ should give the *change* induced by a collision, so that $\mathcal{T}_+^{(12)} \propto b_+^{(12)} - 1$. We collect the three terms and make sure to include a prefactor which is chosen such that the integration of an observable over a short time interval around the collision time yields the change of the observable, as induced by the collision rules (5). The complete expression for $\mathcal{T}_+^{(12)}$ is thus

$$i\mathcal{T}_+^{(12)} = \left|\frac{d}{dt}|\boldsymbol{r}_{12}|\right| \delta(|\boldsymbol{r}_{12}| - d)\Theta\left(-\frac{d}{dt}|\boldsymbol{r}_{12}|\right)(b_+^{(12)} - 1) \,. \tag{7}$$

Since the probability that three or more particles touch at precisely the same instant is zero, we only need to consider two-particle collisions and find for the time-evolution operator for the system of elastically colliding hard spheres:

$$f(t) = e^{i(\mathcal{L}_0 + \mathcal{L}_\pm)t} f(0) \quad \text{for } t \gtrless 0 \,, \quad \text{with} \tag{8}$$

$$i\mathcal{L}_\pm = \sum_{i<j} i\mathcal{T}_\pm^{(ij)} = \sum_{i<j} \left|\frac{d}{dt}|\boldsymbol{r}_{ji}|\right| \delta(|\boldsymbol{r}_{ji}| - d)\Theta\left(\mp\frac{d}{dt}|\boldsymbol{r}_{ji}|\right)(b_\pm^{(ij)} - 1). \tag{9}$$

The negative time evolution is given by $\mathcal{L}_-$, and $b_-^{(ij)}$ is the operator that replaces post-collisional velocities by pre-collisional ones.

Extension to rough spheres. Hard-core models of elastically colliding spheres have been extended to include *rotational* degrees of freedom and *surface roughness* [8,14]. Rotational degrees of freedom offer the possibility to describe molecules with internal degrees of freedom and surface roughness is needed to transfer energy from the translational degrees of freedom to the rotational ones.

We only discuss the simplest case of identical spheres of mass m, moment of inertia I and diameter d. Translational motion is characterized by the center-of-mass velocities $\boldsymbol{v}_i$ and rotational motion by the angular velocities $\boldsymbol{\omega}_i$. Let the surface normal $\hat{\boldsymbol{r}}_{12}$ at the point of contact point from sphere 2 to sphere 1. The important quantity to model the collision is the relative velocity of the point of contact:

$$\boldsymbol{V} = (\boldsymbol{v}_1 - \frac{d}{2}\boldsymbol{\omega}_1 \times \hat{\boldsymbol{r}}_{12}) - (\boldsymbol{v}_2 + \frac{d}{2}\boldsymbol{\omega}_2 \times \hat{\boldsymbol{r}}_{12}) \,. \tag{10}$$

There are two contributions, firstly the center of mass velocity of each sphere, and secondly the contributions from the rotations of each sphere. The minus sign in the first parenthesis stems from the fact that the surface normal, as it was defined, points outwards for sphere 2 and inwards for sphere 1.

Now we can specify the collision rules. Primed variables always denote quantities immediately after the collision; unprimed variables denote pre-collisional quantities:

$$\begin{aligned} \hat{\boldsymbol{r}}_{12} \cdot \boldsymbol{V}' &= -\hat{\boldsymbol{r}}_{12} \cdot \boldsymbol{V} \\ \hat{\boldsymbol{r}}_{12} \times \boldsymbol{V}' &= -e_t\, \hat{\boldsymbol{r}}_{12} \times \boldsymbol{V} \,. \end{aligned} \tag{11}$$

As we are still dealing with elastic spheres, energy conservation requires $e_t = +1$, corresponding to perfectly rough spheres, where the tangential velocity component is completely reversed. Perfectly smooth spheres $e_t = -1$ are also compatible with energy conservation, but reduce to the above simple case of spheres without rotational degrees of freedom, because during collision the angular velocities remain unchanged. Later, we will also admit other values for e_t.

Eqs. (11) form three linearly independent equations. In addition, total momentum is conserved,

$$\boldsymbol{v}_1' + \boldsymbol{v}_2' = \boldsymbol{v}_1 + \boldsymbol{v}_2 \ , \tag{12}$$

and forces during a collision can only act at the point of contact. Therefore there is no torque with respect to this point and consequently we have conserved angular momentum (also with respect to the point of contact) for *both* particles involved:

$$\begin{aligned} \frac{md}{2}\hat{\boldsymbol{r}}_{12} \times (\boldsymbol{v}_1' - \boldsymbol{v}_1) + I(\boldsymbol{\omega}_1' - \boldsymbol{\omega}_1) &= \boldsymbol{0} \\ \frac{md}{2}\hat{\boldsymbol{r}}_{12} \times (\boldsymbol{v}_2' - \boldsymbol{v}_2) - I(\boldsymbol{\omega}_2' - \boldsymbol{\omega}_2) &= \boldsymbol{0} \ . \end{aligned} \tag{13}$$

Altogether we have 12 independent equations for 12 unknowns, namely the four vectors $\boldsymbol{v}_i'$ and $\boldsymbol{\omega}_i'$ with three components each. Solving for these, we obtain:

$$\begin{aligned} \boldsymbol{v}_1' &= \boldsymbol{v}_1 - \eta_t \boldsymbol{v}_{12} - (\eta_n - \eta_t)(\hat{\boldsymbol{r}}_{12} \cdot \boldsymbol{v}_{12})\hat{\boldsymbol{r}}_{12} - \eta_t \frac{d}{2}\hat{\boldsymbol{r}}_{12} \times (\boldsymbol{\omega}_1 + \boldsymbol{\omega}_2) \\ \boldsymbol{v}_2' &= \boldsymbol{v}_2 + \eta_t \boldsymbol{v}_{12} + (\eta_n - \eta_t)(\hat{\boldsymbol{r}}_{12} \cdot \boldsymbol{v}_{12})\hat{\boldsymbol{r}}_{12} + \eta_t \frac{d}{2}\hat{\boldsymbol{r}}_{12} \times (\boldsymbol{\omega}_1 + \boldsymbol{\omega}_2) \\ \boldsymbol{\omega}_1' &= \boldsymbol{\omega}_1 + \frac{2}{dq}\eta_t \hat{\boldsymbol{r}}_{12} \times \boldsymbol{v}_{12} + \frac{\eta_t}{q}\hat{\boldsymbol{r}}_{12} \times (\hat{\boldsymbol{r}}_{12} \times (\boldsymbol{\omega}_1 + \boldsymbol{\omega}_2)) \\ \boldsymbol{\omega}_2' &= \boldsymbol{\omega}_2 + \frac{2}{dq}\eta_t \hat{\boldsymbol{r}}_{12} \times \boldsymbol{v}_{12} + \frac{\eta_t}{q}\hat{\boldsymbol{r}}_{12} \times (\hat{\boldsymbol{r}}_{12} \times (\boldsymbol{\omega}_1 + \boldsymbol{\omega}_2)) \ . \end{aligned} \tag{14}$$

The dimensionless constant $q = 4I/(md^2)$ abbreviates a frequently appearing combination of factors. We have also introduced two parameters η_n and η_t, because we anticipate the more general collision rules for the inelastic case. For elastically colliding spheres, we simply have $\eta_n = 1$ and $\eta_t = q/(1+q)$ for perfectly rough and $\eta_t = 0$ for perfectly smooth spheres.

The pseudo-Liouville operator for elastically colliding rough spheres is still given by Eq. (9) but the operator $b_+^{(ij)}$ now replaces linear *and* angular velocities according to Eqs. (14).

Extension to rough needles. Elastic collisions of hard needles have been discussed by Frenkel et al. [25]. It is straightforward to rephrase their results in terms of a pseudo-Liouville operator [22]. The free streaming part of the Liouville operator is derived from the kinetic energy of the Hamiltonian according to the general rules of classical dynamics. Note, however, that for needles, one of the moments of inertia is zero; this implies that the angular-momentum component along the corresponding axis, which points along the orientation of the needle, is also always zero. Therefore, rotations about this axis can be ignored, and $\boldsymbol{\omega}$ has only two components, both perpendicular to the orientation of the needle. The center of mass coordinate of needle i will be denoted by $\boldsymbol{r}_i$ and its orientation by the unit vector $\boldsymbol{u}_i$. The moments of inertia perpendicular to $\boldsymbol{u}_i$ are equal due to symmetry and will be denoted by I.

The formulation of the collision rules proceeds in close analogy to rough spheres. First we determine the conditions of contact. The unit vectors $\boldsymbol{u}_1$ and $\boldsymbol{u}_2$ span a plane E_{12} with normal

$$\boldsymbol{u}_\perp = \frac{\boldsymbol{u}_1 \times \boldsymbol{u}_2}{|\boldsymbol{u}_1 \times \boldsymbol{u}_2|} \; . \tag{15}$$

We decompose $\boldsymbol{r}_{12} = \boldsymbol{r}_1 - \boldsymbol{r}_2$ into a component perpendicular $\boldsymbol{r}_{12}^\perp = (\boldsymbol{r}_{12}\boldsymbol{u}_\perp)\boldsymbol{u}_\perp$ and parallel $\boldsymbol{r}_{12}^\parallel = (s_{12}\boldsymbol{u}_1 - s_{21}\boldsymbol{u}_2)$ to E_{12} (see Fig. 1). The rods are in contact if $\boldsymbol{r}_{12}\boldsymbol{u}^\perp = 0$ and simultaneously $|s_{12}| < L/2$ and $|s_{21}| < L/2$.

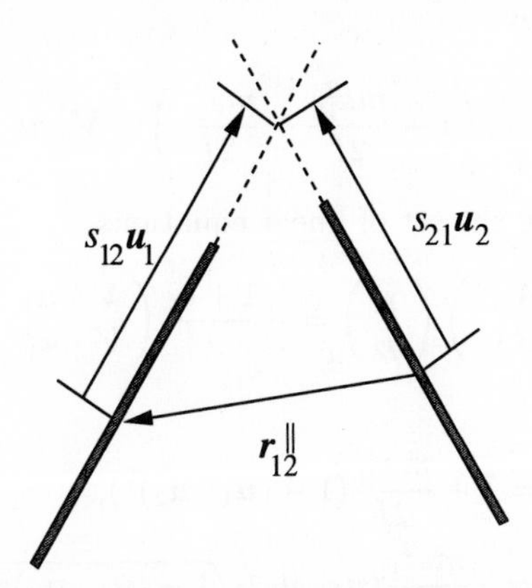

Fig. 1. Configuration of two needles projected in the plane spanned by the unit vectors $\boldsymbol{u}_1$ and $\boldsymbol{u}_2$

The relative velocity of the point of contact is given by

$$\boldsymbol{V} = \boldsymbol{v}_{12} + s_{12}\dot{\boldsymbol{u}}_1 - s_{21}\dot{\boldsymbol{u}}_2 \; . \tag{16}$$

It is useful to introduce a set of normalized basis vectors

$$\boldsymbol{u}_1, \quad \boldsymbol{u}_1^\perp = (\boldsymbol{u}_2 - (\boldsymbol{u}_1 \cdot \boldsymbol{u}_2)\boldsymbol{u}_1)/\sqrt{1 - (\boldsymbol{u}_1 \cdot \boldsymbol{u}_2)^2}, \quad \text{and} \quad \boldsymbol{u}_\perp \tag{17}$$

with $\boldsymbol{u}_\perp$ as defined in Eq. (15). Total momentum conservation is given by (12) and conservation of angular momentum with respect to the contact point reads

$$\boldsymbol{\omega}_1' = \boldsymbol{\omega}_1 + \frac{ms_{12}}{I}\boldsymbol{u}_1 \times (\boldsymbol{v}_1' - \boldsymbol{v}_1) \quad \text{and} \quad \boldsymbol{\omega}_2' = \boldsymbol{\omega}_2 + \frac{ms_{21}}{I}\boldsymbol{u}_2 \times (\boldsymbol{v}_2' - \boldsymbol{v}_2) \ . \tag{18}$$

Three additional equations follow from the change in the relative velocity of the contact point, which is modeled in close analogy to the case of rough spheres:

$$\boldsymbol{V}' \cdot \boldsymbol{u}_\perp = -\boldsymbol{V} \cdot \boldsymbol{u}_\perp, \quad \boldsymbol{V}' \cdot \boldsymbol{u}_1 = -e_t \boldsymbol{V} \cdot \boldsymbol{u}_1, \quad \text{and} \quad \boldsymbol{V}' \cdot \boldsymbol{u}_2 = -e_t \boldsymbol{V} \cdot \boldsymbol{u}_2 \ . \tag{19}$$

Again, energy conservation implies $e_t = \pm 1$, corresponding to either perfectly rough or perfectly smooth needles (see also Eq. (32)). Solving for $\boldsymbol{v}_i'$ and $\boldsymbol{\omega}_i'$, we obtain after a lengthy calculation:

$$\boldsymbol{v}_1' = \boldsymbol{v}_1 + \Delta \boldsymbol{v} \quad \text{and} \quad \boldsymbol{v}_2' = \boldsymbol{v}_2 - \Delta \boldsymbol{v} \tag{20}$$

and $\boldsymbol{\omega}_1'$, $\boldsymbol{\omega}_2'$ given by Eq. (18). The change in velocity $\Delta \boldsymbol{v}$ can be decomposed with respect to the basis defined above, $\Delta \boldsymbol{v} = \gamma_1 \boldsymbol{u}_1 + \gamma_2 \boldsymbol{u}_1^\perp + \alpha \boldsymbol{u}_\perp$. The coefficient α is given by

$$\alpha = -\left(1 + \frac{ms_{12}^2}{2I} + \frac{ms_{21}^2}{2I}\right)^{-1} \boldsymbol{V} \cdot \boldsymbol{u}_\perp \ , \tag{21}$$

while γ_1 and γ_2 satisfy the set of linear equations

$$\begin{pmatrix} A & B \\ B & C \end{pmatrix} \begin{pmatrix} \gamma_1 \\ \gamma_2 \end{pmatrix} = -\frac{1+e_t}{2} \begin{pmatrix} \boldsymbol{V} \cdot \boldsymbol{u}_1 \\ \boldsymbol{V} \cdot \boldsymbol{u}_1^\perp \end{pmatrix} \tag{22}$$

with

$$\begin{aligned} A &= 1 + \frac{ms_{21}^2}{2I}(1 - (\boldsymbol{u}_1 \cdot \boldsymbol{u}_2)^2), \\ B &= -\frac{ms_{21}^2}{2I}(\boldsymbol{u}_1 \cdot \boldsymbol{u}_2)\sqrt{1 - (\boldsymbol{u}_1 \cdot \boldsymbol{u}_2)^2}, \\ C &= 1 + \frac{ms_{12}^2}{2I} + \frac{ms_{21}^2}{2I}(\boldsymbol{u}_1 \cdot \boldsymbol{u}_2)^2 \ . \end{aligned} \tag{23}$$

The Liouville operator for two needles must obey the same basic requirements as for spheres. The only changes are in the condition for a collision to take place,

$$iT_+^{(12)} \propto \Theta(L/2 - |s_{12}|)\Theta(L/2 - |s_{21}|)\delta(|\boldsymbol{r}_{12}^\perp| - 0^+) \ , \tag{24}$$

and in the condition that the two particles are approaching,

$$iT_+^{(12)} \propto \Theta\left(-\frac{d}{dt}|\boldsymbol{r}_{12}^\perp|\right) \ . \tag{25}$$

Collecting terms and choosing the correct prefactor gives the result

$$iT_+^{(12)} = \left|\frac{d}{dt}|\boldsymbol{r}_{12}^{\perp}|\right| \Theta\left(-\frac{d}{dt}|\boldsymbol{r}_{12}^{\perp}|\right)$$
$$\times\, \Theta(L/2 - |s_{12}|)\Theta(L/2 - |s_{21}|)\delta(|\boldsymbol{r}_{12}^{\perp}| - 0^+)(b_+^{(12)} - 1). \quad (26)$$

The operator $b_+^{(12)}$ replaces all velocities according to Eqs. (18) and (20).

2.3 Inelastic Collision

The collision rules for rough spheres and needles are easily generalized to inelastic collisions. This will allow us to set up a formulation of the dynamics of inelastically colliding grains in terms of a pseudo-Liouville operator.

Rough spheres. Energy dissipation is modeled by normal and tangential restitution. The collision rules imply for the change in the relative velocity of the points of contact:

$$\begin{aligned} \hat{\boldsymbol{r}}_{12} \cdot \boldsymbol{V}' &= -e_n\, \hat{\boldsymbol{r}}_{12} \cdot \boldsymbol{V} \\ \hat{\boldsymbol{r}}_{12} \times \boldsymbol{V}' &= -e_t\, \hat{\boldsymbol{r}}_{12} \times \boldsymbol{V}\,. \end{aligned} \quad (27)$$

The first of these equations describes the reduction of the normal-velocity component by a non-negative factor e_n. This is the well-known normal restitution. The second equation tries to describe surface roughness and friction in that it imposes a reduction or even a reversal of the tangential velocity component. It is motivated by the picture of small "bumps" on the surface which become hooked when the surfaces are very close. For all $-1 < e_t < +1$ dissipation is present. The change in energy is given by

$$\Delta E = -m\Big[\frac{1-e_n^2}{4}(\hat{\boldsymbol{r}}_{12} \cdot \boldsymbol{v}_{12})^2 +$$
$$\frac{1-e_t^2}{4}\frac{q}{1+q}\Big(\boldsymbol{v}_{12} - (\hat{\boldsymbol{r}}_{12} \cdot \boldsymbol{v}_{12})\hat{\boldsymbol{r}}_{12} - \frac{d}{2}\hat{\boldsymbol{r}}_{12} \times (\boldsymbol{\omega}_1 + \boldsymbol{\omega}_2)\Big)^2\Big]. \quad (28)$$

Within the parameter range $0 \le e_n \le 1$ and $-1 \le e_t \le 1$, energy is only lost and never gained in a single collision.

The conservation laws for linear and angular momenta are unchanged, so we obtain the same set of equations for the post-collisional velocities as Eqs. (14), with however different parameter values

$$\eta_n = \frac{1+e_n}{2} \quad \text{and} \quad \eta_t = \frac{q}{1+q}\frac{e_t+1}{2}\,. \quad (29)$$

Later we will need the inversion of Eqs. (14), i.e. for given post-collisional velocities we want to determine the pre-collisional ones. This is simply done by replacing e_t by $1/e_t$ and e_n by $1/e_n$ in Eqs. (14). The pre-collisional velocities obtained from post-collisional ones will in the following be denoted by $\boldsymbol{v}_1'', \boldsymbol{v}_2'', \boldsymbol{\omega}_1''$ and $\boldsymbol{\omega}_2''$.

Rough needles. For hard needles we introduce normal and tangential restitution according to:

$$\boldsymbol{V}' \cdot \boldsymbol{u}_\perp = -e_n \boldsymbol{V} \cdot \boldsymbol{u}_\perp, \quad \boldsymbol{V}' \cdot \boldsymbol{u}_1 = -e_t \boldsymbol{V} \cdot \boldsymbol{u}_1, \quad \text{and} \quad \boldsymbol{V}' \cdot \boldsymbol{u}_2 = -e_t \boldsymbol{V} \cdot \boldsymbol{u}_2 \,. \tag{30}$$

The conservation laws for linear and angular momenta are the same as for the elastic case, so that one arrives at the same set of Eqs. (20), the only change affecting the parameter

$$\alpha = -\frac{1+e_n}{2}\left(1 + \frac{m s_{12}^2}{2I} + \frac{m s_{21}^2}{2I}\right)^{-1} \boldsymbol{V} \cdot \boldsymbol{u}_\perp \,. \tag{31}$$

The energy loss for needles is given by

$$\begin{aligned} \Delta E = -m\frac{1-e_t^2}{4}\left(\frac{C(\boldsymbol{V}\cdot\boldsymbol{u}_1)^2 - 2B(\boldsymbol{V}\cdot\boldsymbol{u}_1)(\boldsymbol{V}\cdot\boldsymbol{u}_1^\perp) + A(\boldsymbol{V}\cdot\boldsymbol{u}_1^\perp)^2}{AC-B^2}\right) \\ - m\frac{1-e_n^2}{4}\left(1 + \frac{m s_{12}^2}{2I} + \frac{m s_{21}^2}{2I}\right)^{-1} (\boldsymbol{V}\cdot\boldsymbol{u}_\perp)^2. \end{aligned} \tag{32}$$

It can be checked with Eqs. (23) that the first term is less than 0 if and only if $-1 \le e_t \le 1$. Obviously, the second term is also less than 0 if $0 \le e_n \le 1$. Our method of modeling granular collisions of needles is therefore consistent with the constraint that energy may not be gained in a single collision.

2.4 Time Evolution of the Distribution Function

We will be interested in ensemble averages of observables $f(\Gamma)$ at a time t defined by:

$$\langle f\rangle(t) = \int d\Gamma\, \rho(\Gamma;0) f(\Gamma;t) = \int d\Gamma\, \rho(\Gamma;t) f(\Gamma) \,. \tag{33}$$

Here $\rho(\Gamma;t)$ is the N-particle distribution function at time t. The average can either be taken over the *initial* distribution $\rho(\Gamma;0)$ at time 0, the observable being propagated to time t, or equivalently over the distribution $\rho(\Gamma;t)$ at time t with the unchanged observable $f(\Gamma)$. We write Eq. (33) as

$$\langle f\rangle(t) = \int d\Gamma\, \rho(\Gamma;0) e^{i\mathcal{L}t} f(\Gamma) =: \int d\Gamma\, \left(e^{i\overline{\mathcal{L}}t}\rho(\Gamma;0)\right) f(\Gamma) \,, \tag{34}$$

to define the time-evolution operator $\overline{\mathcal{L}}$ which describes the time evolution of ρ [26]. To determine $\overline{\mathcal{L}}$ explicitly, we take the derivative of Eq. (34) at time $t=0$ for simplicity,

$$\begin{aligned} \partial_t \langle f\rangle(t)\big|_{t=0} &= \int d\Gamma\, \rho(\Gamma;0) i\mathcal{L} f(\Gamma) \\ &= \int d\Gamma\, \left(\partial_t \rho(\Gamma;t)\big|_{t=0}\right) f(\Gamma) = \int d\Gamma\, \left(i\overline{\mathcal{L}}\rho(\Gamma;0)\right) f(\Gamma) \,. \end{aligned} \tag{35}$$

The time-evolution operator of the density due to free streaming, $\overline{\mathcal{L}}_0$, is easily calculated by integration by parts and we get $\overline{\mathcal{L}}_0 = -\mathcal{L}_0$. To find an expression for the time-evolution operator of the density due to collisions $\overline{\mathcal{T}}_+^{(12)}$ for spheres, we use Eq. (35). Phase-space coordinates before collision are denoted by Γ, after collision by $\Gamma' = b_+^{(12)}\Gamma$ so that

$$\int d\Gamma \rho(\Gamma;0) i\mathcal{T}_+^{(12)} f(\Gamma) = \int d\Gamma \rho(\Gamma;0)\delta(|\boldsymbol{r}_{12}| - d)\Theta\left(-\frac{d}{dt}|\boldsymbol{r}_{12}|\right)\left|\frac{d}{dt}|\boldsymbol{r}_{12}|\right|(f(\Gamma') - f(\Gamma)). \tag{36}$$

In the first term on the right hand side we make a coordinate transformation to the variables after collision with Jacobian $\mathcal{J} := \left|\frac{\partial\Gamma}{\partial\Gamma'}\right|$. We use the inverse operator of $b_+^{(12)}$, namely $b_-^{(12)}\Gamma' = \Gamma''$. Here the coordinates before collision in terms of the coordinates after collision are denoted by $\Gamma'' = \Gamma(\Gamma')$. We note that $\frac{d}{dt}|\boldsymbol{r}_{12}| = \boldsymbol{v}_{12}\hat{\boldsymbol{r}}_{12}$ and rewrite the first term

$$\int d\Gamma \rho(\Gamma;0)\delta(|\boldsymbol{r}_{12}| - d)\Theta\left(-\frac{d}{dt}|\boldsymbol{r}_{12}|\right)\left|\frac{d}{dt}|\boldsymbol{r}_{12}|\right| f(\Gamma') = \int d\Gamma' \mathcal{J} \rho(\Gamma'';t)\delta(|\boldsymbol{r}_{12}| - a)\Theta(-\boldsymbol{v}_{12}''\cdot\hat{\boldsymbol{r}}_{12})\,|\boldsymbol{v}_{12}''\cdot\hat{\boldsymbol{r}}_{12}|\, f(\Gamma') \tag{37}$$

Next, we rename Γ' to Γ and make use of $\boldsymbol{v}_{nm}''\hat{\boldsymbol{r}}_{nm} = -\frac{1}{e_n}(\boldsymbol{v}_{nm}\hat{\boldsymbol{r}}_{nm})$. This allows us to identify the time-evolution operator of the distribution function, $\overline{\mathcal{T}}_+^{(12)}$, by:

$$i\overline{\mathcal{T}}_+^{(12)} = \delta(|\boldsymbol{r}_{12}| - d)\left|\frac{d}{dt}|\boldsymbol{r}_{12}|\right|\left(\Theta\left(\frac{d}{dt}|\boldsymbol{r}_{12}|\right)\frac{\mathcal{J}}{e_n}b_-^{(12)} - \Theta\left(-\frac{d}{dt}|\boldsymbol{r}_{12}|\right)\right). \tag{38}$$

It is common to rewrite Eq. (38) by multiplying it with $\int d\boldsymbol{\sigma}\delta(\boldsymbol{\sigma} - \boldsymbol{r}_{12})$ so that we can replace $\boldsymbol{r}_{12}$ by $\boldsymbol{\sigma}$ in Eq. (38). In the second term the integral transformation $\boldsymbol{\sigma} \to -\boldsymbol{\sigma}$ is done and we integrate over $|\boldsymbol{\sigma}|$. We obtain in D dimensions

$$i\overline{\mathcal{T}}_+^{(12)} = d^{D-1}\int_{\boldsymbol{v}_{12}\cdot\hat{\boldsymbol{\sigma}}>0} d\hat{\sigma}\,(\boldsymbol{v}_{12}\cdot\hat{\boldsymbol{\sigma}})\left(\frac{\mathcal{J}}{e_n}\delta(\boldsymbol{r}_{12} - d\hat{\boldsymbol{\sigma}})b_-^{(12)} - \delta(\boldsymbol{r}_{12} + d\hat{\boldsymbol{\sigma}})\right). \tag{39}$$

Finally, we note that $t = 0$ is not special since we have only chosen it for the sake of simplicity. Hence we have derived the time-evolution operator for the N-particle distribution function $\rho(\Gamma;t)$ which is given by the pseudo-Liouville equation

$$\partial_t\rho(\Gamma,t) = i\left(-\mathcal{L}_0(\Gamma) + \sum_{i<j}\overline{\mathcal{T}}_+^{(ij)}\right)\rho(\Gamma,t)\,. \tag{40}$$

A similar procedure yields the time evolution operator for the distribution of needles.

3 Homogeneous Cooling State

We are interested in the time evolution of a gas of freely cooling rough spheres or needles which is dominated by two-particle collisions, as discussed in the previous section. We aim at a description in terms of macroscopic quantities and focus on the decay in time of the average kinetic energy of translation and rotation, defined as

$$\begin{aligned} \langle E_{\mathrm{tr}}\rangle(t) &= \frac{m}{2N}\sum_i \int d\Gamma\,\rho(\Gamma;t)\boldsymbol{v}_i^2 =: \frac{D_{\mathrm{tr}}}{2}T_{\mathrm{tr}}(t)\,, \\ \langle E_{\mathrm{rot}}\rangle(t) &= \frac{I}{2N}\sum_i \int d\Gamma\,\rho(\Gamma;t)\boldsymbol{\omega}_i^2 =: \frac{D_{\mathrm{rot}}}{2}T_{\mathrm{rot}}(t)\,. \end{aligned} \tag{41}$$

Here D_{tr} and D_{rot} denote the total number of translational and rotational degrees of freedom respectively. It is impossible to compute the above expectation values exactly and we have to resort to approximations. We assume that the N-particle probability distribution $\rho(\Gamma,t)$ is homogeneous in space and depends on time only via the average kinetic energy of translation and rotation:

$$\rho_{\mathrm{HCS}}(\Gamma;t)_{\mathrm{HCS}} \sim W(\boldsymbol{r}_1,\ldots,\boldsymbol{r}_N)\,\tilde{\rho}\left(\{\boldsymbol{v}_i,\boldsymbol{\omega}_i\};T_{\mathrm{tr}}(t),T_{\mathrm{rot}}(t)\right)\,. \tag{42}$$

The function $W(\boldsymbol{r}_1,\ldots,\boldsymbol{r}_N)$ gives zero weight to configurations with overlapping particles and 1 otherwise. Needles have vanishing volume in configuration space, so that $W\equiv 1$ for needles. We shall furthermore assume that $\tilde{\rho}$ factors neglecting correlations of the velocities of different particles. In the simplest approximation we take $\tilde{\rho}$ to be Gaussian in all its momentum variables

$$\tilde{\rho}\left(\{\boldsymbol{v}_i,\boldsymbol{\omega}_i\};T_{\mathrm{tr}}(t),T_{\mathrm{rot}}(t)\right) \propto \exp\left[-N\left(\frac{E_{\mathrm{tr}}}{T_{\mathrm{tr}}(t)}+\frac{E_{\mathrm{rot}}}{T_{\mathrm{rot}}(t)}\right)\right]\,. \tag{43}$$

In the next section, we shall discuss non Gaussian distributions and shall compute corrections perturbatively.

To determine the time dependence of $T_{\mathrm{tr}}(t)$ and $T_{\mathrm{rot}}(t)$ we take time derivatives of Eqs. (41) and use the identity $\frac{d}{dt}\langle f\rangle(t)=\int d\Gamma(\frac{d}{dt}\rho(\Gamma,t))f(\Gamma)=\int d\Gamma(i\overline{\mathcal{L}}\rho(\Gamma,t))f(\Gamma)=\int d\Gamma\rho(\Gamma,t)i\mathcal{L}f(\Gamma)$. Then $\rho(\Gamma,t)$ is replaced by $\rho_{\mathrm{HCS}}(\Gamma;t)$, resulting in

$$\begin{aligned} \frac{d}{dt}T_{\mathrm{tr}}(t) &= \frac{2}{D_{\mathrm{tr}}}\int d\Gamma\,\rho_{\mathrm{HCS}}(\Gamma;t)i\mathcal{L}E_{\mathrm{tr}} = \frac{2}{D_{\mathrm{tr}}}\langle i\mathcal{L}E_{\mathrm{tr}}\rangle_{\mathrm{HCS}} \quad \text{and} \\ \frac{d}{dt}T_{\mathrm{rot}}(t) &= \frac{2}{D_{\mathrm{rot}}}\int d\Gamma\,\rho_{\mathrm{HCS}}(\Gamma;t)i\mathcal{L}E_{\mathrm{rot}} = \frac{2}{D_{\mathrm{rot}}}\langle i\mathcal{L}E_{\mathrm{rot}}\rangle_{\mathrm{HCS}}\,. \end{aligned} \tag{44}$$

All that remains to be done are high dimensional phase-space integrals, the details of which are delegated to appendices A and B, for spheres and needles.

3.1 Results for Spheres

After integration over phase space has been performed (see Appendix A for details), we find

$$\frac{D_{\text{tr}}}{2}\frac{d}{dt}T_{\text{tr}}(t) = \langle i\mathcal{L}E_{\text{tr}}\rangle_{\text{HCS}} = -GAT_{\text{tr}}^{3/2} + GBT_{\text{tr}}^{1/2}T_{\text{rot}} \,,$$
$$\frac{D_{\text{rot}}}{2}\frac{d}{dt}T_{\text{rot}}(t) = \langle i\mathcal{L}E_{\text{rot}}\rangle_{\text{HCS}} = GBT_{\text{tr}}^{3/2} - GCT_{\text{tr}}^{1/2}T_{\text{rot}} \,, \tag{45}$$

with the always positive constants A, B, C, and G depending on space dimensionality D. In two dimensions, the constants in Eqs. (45) are given by

$$G = 4d\frac{N}{V}\sqrt{\frac{\pi}{m}}g(d), \qquad A = \frac{1-e_n^2}{4} + \frac{\eta_t}{2}(1-\eta_t),$$
$$B = \frac{\eta_t^2}{2q}, \qquad C = \frac{\eta_t}{2q}\left(1 - \frac{\eta_t}{q}\right) \,, \tag{46}$$

and in three dimensions they read

$$G = 8d^2\frac{N}{V}\sqrt{\frac{\pi}{m}}g(d), \qquad A = \frac{1-e_n^2}{4} + \eta_t(1-\eta_t),$$
$$B = \frac{\eta_t^2}{q}, \qquad C = \frac{\eta_t}{q}\left(1 - \frac{\eta_t}{q}\right) \,. \tag{47}$$

The pair correlation function at contact, $g(d)$, is defined in the usual way [1]. A detailed discussion of these results, and in particular the dependence of free cooling on e_n and e_t, can be found in [18].

The Enskog value [14, 28] of the collision frequency ω_E, i.e. the average number of collisions which a particle suffers per unit time in D dimensions is given by

$$\omega_E := S_D\frac{N}{V}g(d)d^{D-1}\sqrt{\frac{T_{\text{tr}}(t)}{\pi m}} \,. \tag{48}$$

S_D is the surface of a unit sphere in D dimensions. Note that always $\omega_E \propto GT^{1/2}$. We define dimensionless time τ by $d\tau = \omega_E dt$ so that τ counts the collisions that on average each particle has suffered until time t. In a simulation this would simply be done by counting the number of collisions. The functional dependence of the two temperatures on τ is determined by

$$\frac{d}{d\tau}T_{\text{tr}} = -aT_{\text{tr}} + bT_{\text{rot}}, \tag{49}$$
$$\frac{d}{d\tau}T_{\text{rot}} = bT_{\text{tr}} - cT_{\text{rot}} \tag{50}$$

with properly defined a, b, c. Eq. (49) has a simple interpretation: In a given short interval Δt a number of $\Delta\tau$ collisions occur. Due to these collisions, the translational energy decreases by an amount given by the first term, but there is also a gain term, reflecting that rotational energy is transfered to translational energy. The solution of Eqs. (49,50) can be written as

$$T_{\text{tr}} = c_1 K_+ \exp(-\lambda_+ t) + c_2 K_- \exp(-\lambda_- t) \,, \tag{51}$$

$$T_{\text{rot}} = c_1 \exp(-\lambda_+ t) + c_2 \exp(-\lambda_- t) \,, \tag{52}$$

$$K_\pm = \frac{1}{2b}(c - a \pm \sqrt{(c-a)^2 + 4b^2}) \,, \tag{53}$$

$$\lambda_\pm = \frac{1}{2}\left(c + a \mp \sqrt{(c-a)^2 + 4b^2}\right) \,. \tag{54}$$

The constants c_1 and c_2 are determined by the initial conditions and $\lambda_- > \lambda_+ > 0$ holds for all e_t, e_n. Hence for long times the ratio of $T_{\text{tr}}/T_{\text{rot}}$ is determined by K_+.

We now assume that the ratio $T_{\text{tr}}/T_{\text{rot}}$ has reached its asymptotic value K_+ for some $\tau > \tau_0$ or equivalently $t > t_0$ and substitute $T_{\text{rot}} = T_{\text{tr}}/K_+$ into Eq. (45) we obtain

$$\frac{d}{dt} T_{\text{tr}} = -F T_{\text{tr}}^{3/2}. \tag{55}$$

The resulting equation is of the same functional form as for homogeneous cooling of smooth spheres, except for the coefficient F, which contains all the dependence on system parameters. Its solution is given by

$$T_{\text{tr}} = \frac{T_{\text{tr}}(t_0)}{\left[1 + T_{\text{tr}}(t_0)^{1/2}(F/2)(t - t_0)\right]^2} \sim \frac{1}{(Ft/2)^2} \,, \tag{56}$$

Haff's [27] law of homogeneous cooling. We have determined two time scales, first an exponentially fast decay (measuring time in collisions) towards a state where we find a constant ratio of translational and rotational energy. As long as dissipation is small, we can approximate the Enskog-collision frequency for sufficiently short times by its initial value $\omega(t) \sim \omega(0)$ so that we find exponential behavior also in real time. The second stage of relaxation is characterized by a slow, algebraic decay of both energies, such that their ratio remains constant. These two time regimes are clearly seen in the numerical solution of Eq. (46) for initial conditions $T_{\text{tr}}(0) = 1$ and $T_{\text{rot}}(0) = 0$, i.e. a system prepared in an equilibrium state of perfectly smooth spheres. We show in Fig. 2(a) in a double logarithmic plot the time dependence of the total energy $E = \frac{3}{2}(T_{\text{tr}}(t) + T_{\text{rot}}(t))$ and the ratio $T_{\text{tr}}(t)/T_{\text{rot}}(t)$. Time is plotted in units of $\frac{2}{3} G T_{\text{tr}}^{1/2}(0)$. We have chosen $e_n = 0.9$ and $e_t = -0.8$.

3.2 Results for Needles

In the case of needles we restrict ourselves to the case of perfectly smooth needles, i.e. $e_t = -1$. After some lengthy algebra, presented in appendix B,

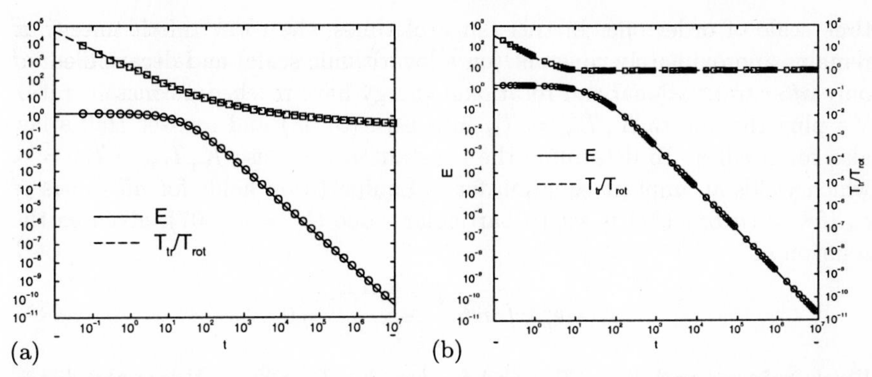

Fig. 2. (a) Theoretical prediction (lines) for spheres for the total energy $E = \frac{3}{2}(T_{\mathrm{tr}}(t) + T_{\mathrm{rot}}(t))$ and the ratio $T_{\mathrm{tr}}(t)/T_{\mathrm{rot}}(t)$ versus time. Time is plotted in units of $\frac{2}{3}GT_{\mathrm{tr}}^{1/2}(0)$. We have chosen $e_n = 0.9$ and $e_t = -0.8$. The symbols represent data of a simulation of 1000 particles in a box of length 16 d.
(b) The same as (a) for needles: The total kinetic energy $E = (3/2)(T_{\mathrm{tr}} + T_{\mathrm{rot}})$ and $T_{\mathrm{tr}}/T_{\mathrm{rot}}$ are plotted vs. time in units of $\gamma_n\sqrt{T_{\mathrm{tr}}(0)}$. The simulation data are from a system of 10^4 needles in a box of length 24 L with $e_n = 0.8$.

Eq. (44) can be cast in the following form

$$\frac{2\dot{T}_{\mathrm{tr}}}{\gamma_n T_{\mathrm{tr}}^{3/2}(1+e_n)} = -\int_{\square} d^2r \frac{(1+\frac{T_{\mathrm{rot}}}{T_{\mathrm{tr}}}kr^2)^{1/2}}{1+kr^2} + \frac{1+e_n}{2}\int_{\square} d^2r \frac{(1+\frac{T_{\mathrm{rot}}}{T_{\mathrm{tr}}}kr^2)^{3/2}}{(1+kr^2)^2}, \quad (57)$$

$$\frac{4\dot{T}_{\mathrm{rot}}}{3\gamma_n T_{\mathrm{tr}}^{3/2}(1+e_n)} = -\int_{\square} d^2r \frac{\frac{T_{\mathrm{rot}}}{T_{\mathrm{tr}}}kr^2(1+\frac{T_{\mathrm{rot}}}{T_{\mathrm{tr}}}kr^2)^{1/2}}{1+kr^2} + \frac{1+e_n}{2}\int_{\square} d^2r \frac{kr^2(1+\frac{T_{\mathrm{rot}}}{T_{\mathrm{tr}}}kr^2)^{3/2}}{(1+kr^2)^2}, \quad (58)$$

with $\gamma_n = (2NL^2\sqrt{\pi})/(3V\sqrt{m})$ and $k = (mL^2)/(2I)$. The two-dimensional integration extends over a square of unit length, centered at the origin.

In Fig. 2(b) we plot the numerical solution of Eqs. (57,58) for $e_n = 0.8$ and $k = 6$ ($k = 6$ corresponds to a homogeneous mass distribution along the rod) as a function of time in units of $\gamma_n\sqrt{T_{\mathrm{tr}}(0)}$. In addition we have performed simulations of a system of 10000 needles, confined to a box of length 24 L. We show the total kinetic energy $E = \frac{3}{2}T_{\mathrm{tr}} + T_{\mathrm{rot}}$ (in units of $T_{\mathrm{tr}}(\tau = 0)$) and the ratio $T_{\mathrm{tr}}/T_{\mathrm{rot}}$. Analytical theory and simulation are found to agree within a few percent over eight orders of magnitude in time. ($T_{\mathrm{rot}}(0) = 0$ has been chosen as initial condition). For needles, we observe an even clearer separation of time scales. The decay of $T_{\mathrm{tr}}/T_{\mathrm{rot}}$ to a constant value K_+ happens on a time scale of order one. In this range of times, the total kinetic energy E remains approximately constant (on a logarithmic scale) and decays like t^{-2} only *after* translational and rotational energy have reached a constant ratio. We plug the ansatz $K_+ T_{\text{rot}} = T_{\text{tr}}$ into Eqs. (57,58) and recover Haff's law also for needles. To determine the constant K_+ we use $K_+ \dot{T}_{\text{rot}} - \dot{T}_{\text{tr}} = 0$, which yields an implicit equation for c. Equipartition holds for *all* values of e_n if $k = (mL^2)/(2I)$ is set to particular value ($k^* = 4.3607$), given as the solution of

$$(1 - e_n^2) \int_{\square} d^2 r \frac{1 - \frac{3}{2} k^* r^2}{\sqrt{1 + k^* r^2}} = 0 \ .$$

For $k < k^*$ we find $T_{\text{tr}} < T_{\text{rot}}$ and for $k > k^*$, $T_{\text{tr}} > T_{\text{rot}}$. Hence the distribution of mass along the rods determines the asymptotic ratio of rotational and translational energy, including equipartition as a special case.

3.3 Breakdown of Homogeneity in Dense Systems of Needles

It is well known that dense and large systems of inelastically colliding spheres exhibit clustering so that the assumption of homogeneity breaks down and deviations from Haff's law of homogeneous cooling are observed [15, 30]. To investigate inhomogeneities for dense systems of needles we measure hydrodynamic quantities, i.e we define local variables as the density field, the translational and rotational flow field and the local rotational and translational kinetic energy. In order to take local averages over small volumes, we divide the simulation box into cells whose sizes are small compared to the box size but large enough to give a reasonable statistics. We choose the cell size such that on average about 25 needles are in each cell. For each cell indexed by α, we compute the number density $\rho_\alpha := \frac{1}{V_{\text{cell}}} \sum_{i \in \text{cell}_\alpha} 1 = \langle 1 \rangle_\alpha$, the translational energy per particle $\rho_\alpha E_\alpha^{\text{tr}} = \langle \frac{m}{2} \boldsymbol{v}_i^2 \rangle_\alpha$, and the hydrodynamic temperature $T_\alpha^{\text{tr}} = E_\alpha^{\text{tr}} - m \boldsymbol{U}_\alpha^2 / 2$, defined by fluctuations around the flow field $\rho_\alpha \boldsymbol{U}_\alpha = \langle \boldsymbol{v} \rangle_\alpha$. The corresponding observables of the rotational degrees of freedom are the rotational energy per particle E_α^{rot}, the hydrodynamic rotational temperature T_α^{rot}, and the rotational flow field $\boldsymbol{\Omega}_\alpha$.

To check for spatial clustering, we compare the statistics of fluctuations of the local density, velocity, and translational energy for elastic and inelastic systems. As an example, we show in Fig. 3 the histogram of the deviation of the local density $\delta\rho_\alpha = \rho_\alpha / n - 1$. We performed simulations of a dense and large system of 20000 needles confined to a volume with linear dimension $12L$ and $e_n = 0.9$. The initial distribution is uniform, corresponding to the equilibrium state of an elastic system. As the system develops in time with particles colliding inelastically, we observe that the distribution broadens, a clear indication that regions of large density have developed. Histograms of the local translational and rotational energies look very similar.

Inelastic hard spheres without surface roughness tend to move more and more parallel so that large scale structures in the velocity field develop. In

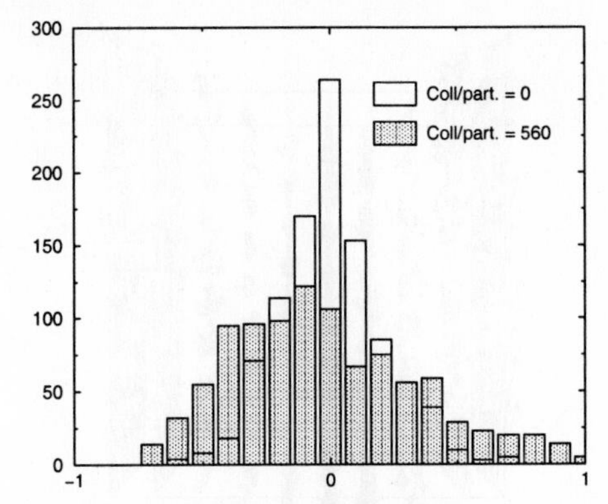

Fig. 3. Histogram of density fluctuations in the initial state and after 560 Collisions per particle. It is obvious that regions with high density have developed.

such a state, most of the kinetic energy is found in the energy of the flow field, whereas the energy of the fluctuations around the flow field is small. A quantitative measure for this effect [29] is the ratio of the total energy of the flow to the total internal energy of fluctuations: $S_{\rm tr} := (\sum_\alpha \frac{m}{2}\rho_\alpha \boldsymbol{U}_\alpha^2)/(\sum_\alpha \rho_\alpha T_\alpha^{\rm tr})$ and the analogous quantity $S_{\rm rot}$ for the rotational degrees of freedom. In Fig. 4 we show $S_{\rm tr}$ and $S_{\rm rot}$ as a function of time, measured in collisions per particle. We observe an increase of $S_{\rm tr}$ by a factor of 50, whereas $S_{\rm rot}$ increases only by about 50 %. Hence the large scale structures in the flow field are much more pronounced for the translational velocity.

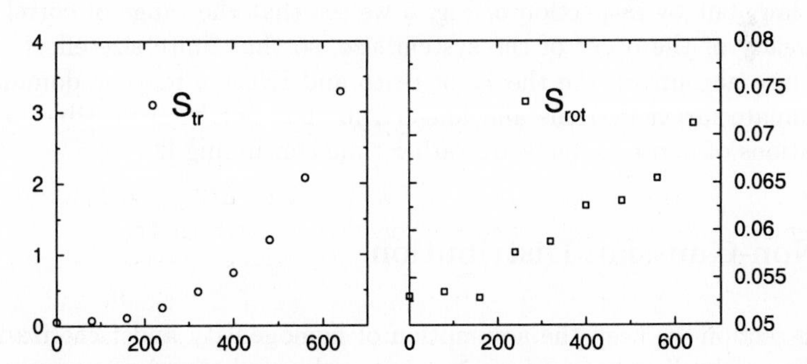

Fig. 4. Ratio $S_{\rm tr}$ ($S_{\rm rot}$) of the local macroscopic energy to the local temperature for the translational (rotational) degrees of freedom as a function of the number of collisions per particle.

In Fig. 5 we show the flow field after 600 collisions per particle. We observe two shear bands (note the periodic boundary conditions) in which the flow field is to a large degree aligned. In periodic boundary conditions, stable shear bands have to be aligned with the walls of the the box.

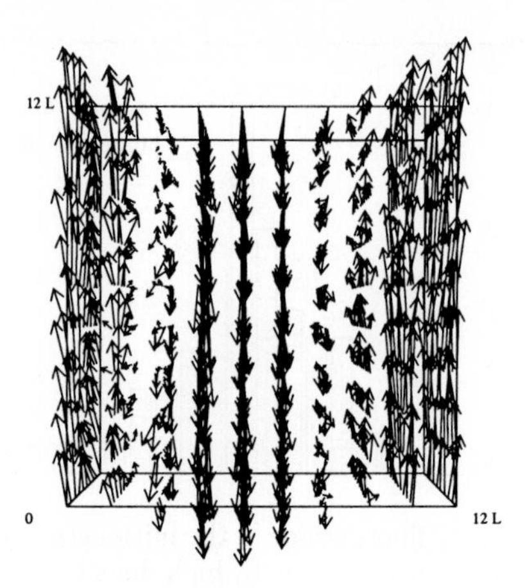

Fig. 5. Flow field after 600 collisions per particle

How does the organization of the flow field influence the decay of the average energy in the system? Brito and Ernst [30] have suggested a generalized Haff's law to describe the time dependence of the kinetic energy of smooth inelastically colliding spheres even in the non-homogeneous state. They found that in the late state where one finds a well developed flow field, the energy decays like $\tau^{-D/2}$ in D dimensions. As in section 3.1, τ is the average number of collisions suffered by a particle within time t. In Fig. 6 we compare the data of the simulation with the solution of Eqs. (57,58), and in the inset we plot $T_{\rm tr}$ as a function of τ and compare it to $\tau^{-3/2}$. We can not confirm a $\tau^{-3/2}$ law, but by inspection of Fig. 5 we see that the range of correlations are already of the order of the system size, so that finite size effects – not taken into account in the theory of Brito and Ernst – may be dominating. To simulate larger systems and longer runs has not been possible because simulations of dense systems are rather time consuming [22].

4 Non-Gaussian Distribution

In this section we keep the assumption of homogeneity and factorization of the N-particle distribution function, but go beyond the approximation of a purely Gaussian state. Initially the system is prepared in a Gaussian state, so that deviations from the Gaussian should be small for short times and perturbation theory can be used to check the range of validity of the Gaussian approximation. We expand the one particle distribution function in generalized Laguerre polynomials (for a definition see [31]) around the Gaussian with time dependent variances. We define an average velocity $v_0(t) = \sqrt{2T_{\rm tr}(t)/m}$ and $\omega_0(t) = \sqrt{2T_{\rm rot}(t)/I}$ and scale linear velocities by $v_0(t)$ and angular velocities by $\omega_0(t)$. The general ansatz for the N-particle distribution function

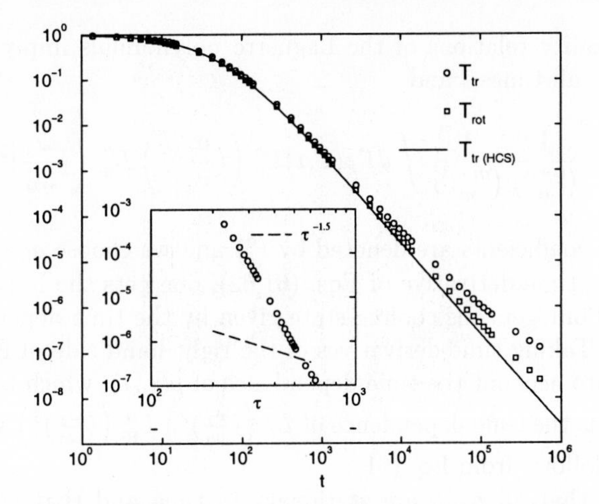

Fig. 6. Data of simulations for translational and rotational temperature as a function of time (units of $\gamma_n\sqrt{T_{\rm tr}(0)}$) are compared to the numerical solution of Eq. (57,58) and to $\tau^{-3/2}$. The inset shows $T_{\rm tr}$ as a function of τ and $\tau^{-3/2}$.

of the homogeneous cooling state then reads

$$\rho(\Gamma,t) \sim W(\boldsymbol{r}_1,...\boldsymbol{r}_N)\prod_{i=1}^{N}\rho_i(\boldsymbol{v}_i,\boldsymbol{\omega}_i,t)\ , \quad \text{and}$$

$$\rho_i(\boldsymbol{v}_i,\boldsymbol{\omega}_i,t) = \frac{1}{Z(t)}\exp\left(-\left(\frac{\boldsymbol{v}_i}{v_0(t)}\right)^2-\left(\frac{\boldsymbol{\omega}_i}{\omega_0(t)}\right)^2\right)$$
$$\sum_{n,m=0}^{\infty} a_{n,m}(t)L_n^{\alpha}\left(\left(\frac{\boldsymbol{v}_i}{v_0(t)}\right)^2\right)L_m^{\beta}\left(\left(\frac{\boldsymbol{\omega}_i}{\omega_0(t)}\right)^2\right)\ . \quad (59)$$

We have introduced the abbreviations $\alpha = D_{\rm tr}/2-1$ and $\beta = D_{\rm rot}/2-1$. The average linear and angular velocities, $v_0(t)$ and $\omega_0(t)$, are time dependent and so are the coefficients $a_{n,m}(t)$ of the double expansion. At time $t=0$ the system is equilibrated with temperature T so that $\frac{m}{2}v_0^2 = \frac{D_{\rm tr}}{2}T$ and $\frac{I}{2}\omega_0^2 = \frac{D_{\rm rot}}{2}T$ and hence $a_{n,m}(t)=0$.

The factor $Z(t)$ follows from the proper normalization, $\int d\boldsymbol{v}_i d\boldsymbol{\omega}_i \rho_i = 1$,

$$Z(t) = v_0^{D_{\rm tr}}\omega_0^{D_{\rm rot}}\sqrt{\pi}^{D_{\rm tr}}\sqrt{\pi}^{D_{\rm rot}}a_{0,0}\ , \quad (60)$$

and we require that $v_0(t)$ and $\omega_0(t)$ be determined by the conditions

$$\int d\Gamma \boldsymbol{v}_1^2\rho(\Gamma,t) = \frac{D_{\rm tr}}{2}v_0^2(t) \quad \text{and} \quad \int d\Gamma\boldsymbol{\omega}_1^2\rho(\Gamma,t) = \frac{D_{\rm rot}}{2}\omega_0^2(t)\ . \quad (61)$$

The orthogonality relations of the Laguerre polynomials imply $a_{1,0}(t) = a_{0,1}(t) = 0$ for all times t and

$$a_{n,m}(t) = \frac{1}{\binom{n+\alpha}{n}} \frac{1}{\binom{m+\beta}{m}} \int d\Gamma \rho(\Gamma, t) L_n^\alpha \left((\frac{v_1}{v_0})^2 \right) L_m^\beta \left((\frac{\omega_1}{\omega_0})^2 \right) . \quad (62)$$

The binomial coefficients are denoted by $\binom{a}{b}$ and we choose $a_{0,0} = 1$.

Taking the time derivative of Eqs. (61,62), one gets the full time dependence of the homogeneous cooling state given by the time dependence of all its momenta. Taking time derivatives of the right hand side of Eq. (62), one has to take into account the time dependence of $\rho(\Gamma, t)$, which is determined by $\overline{\mathcal{L}}$, as well as the time dependence of $L_n^\alpha \left((\frac{v_1}{v_0})^2 \right) L_m^\beta \left((\frac{\omega_1}{\omega_0})^2 \right)$ via $v_0(t)$ and $\omega_0(t)$, which follows from Eq. (61).

Assuming that all $a_{n,m}$ are stationary in time and that $v_0/\omega_0 = \mu$ is constant we get an infinitely large, nonlinear system of equations. To make further progress we truncate the expansion in Eq. (59) and take into account only $a_{n,m}$ for $n+m \leq 2$. We also neglect in the system of equations products of different $a_{n,m}$, which we assume to be of higher order. We show results for $a_{0,2}$ in Fig. 7 for fixed $e_n = 0.9$ as a function of e_t. Deviations from the Gaussian vanish for perfectly smooth spheres and are found to increase dramatically for $e_t \to -0.9$. Deviations from the Gaussian distribution are also small for perfectly rough spheres which is unexpected, because rotational degrees of freedom are coupled to translational ones and $e_n = 0.9$. In fact deviations stay small for a broad range of values of $e_t \gtrsim -0.75$. For $e_t \lesssim -0.75$, we don't consider it meaningful to plot the theoretical result, once a divergence of $a_{0,2}$ has occurred. We measured $a_{0,2}$ in simulations of small systems. Thereby we avoid clustering but have to bear with poor statistics. The simulations confirm the increase of $a_{0,2}$ around $e_t = -0.7$ in agreement with the perturbation expansion.

Goldshtein and Shapiro [11] propose a similar set of momentum equations but they solve it only to lowest order, resulting therefore in the same asymptotic ratio μ as given in Eq. (53).

5 Conclusion

Two simple models of granular particles with rotational degrees of freedom are discussed: Rough spheres or discs and, as an example for non-spherical particles, needles. We focus on the simplest collision rules, which allow for a transfer of translational energy to rotational degrees of freedom. For spheres this is achieved by tangential restitution (in addition to normal restitution), for needles normal restitution is sufficient. We show that the time evolution can be formulated in terms of a pseudo Liouville operator, thereby generalizing previous work on elastic collisions to inelastic ones. The presented

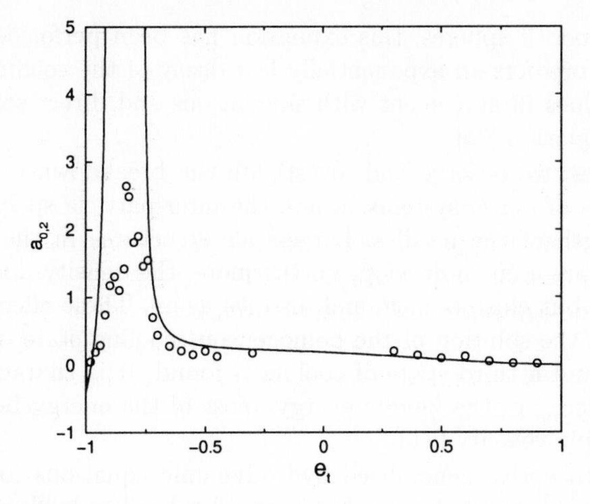

Fig. 7. Coefficient $a_{0,2}$ for $e_n = 0.9$ as a function of e_t. Theory (straight line) and simulations (circles) are compared.

formalism is general enough to include more realistic collision rules, for example Coulomb friction for small angles of impact and tangential restitution for large angles. Work along those lines can be found in [32].

The computation of non-equilibrium expectation values, like e.g. the relaxation of kinetic energy, require approximations. These are formulated for the N-particle distribution function, which we assume to be homogeneous in space and to depend on time only via the average translational and rotational energy, T_{tr} and T_{rot}. The distribution of linear and angular momenta is expanded in Laguerre polynomials around a Gaussian state with time dependent widths, T_{tr} and T_{rot}. The zeroth-order approximation, i.e. a pure Gaussian, leads to two coupled differential equations for the two temperatures. In both systems, spheres and needles, the relaxation of translational and rotational kinetic energy is characterized by two time scales: (1) An exponentially fast decay towards a state with constant ratio of translational to rotational energy and (2) an algebraically slow decay of the energy, such that the above ratio keeps constant in time. The theoretically predicted cooling dynamics is supported by computer simulations of systems of small or moderate density, where no shearing or cluster instability is observed and the system remains homogeneous [18, 22]

To study deviations from the Gaussian state, we restrict ourselves to rough spheres and truncate the expansion in Laguerre polynomials, keeping the first three terms (the first order term for smooth spheres has already been computed in [33]). This perturbative approach is shown to break down for certain values of e_t and e_n, where deviations from the Gaussian are shown to diverge. These results are confirmed by simulations. We indicate how a more general expansion with time dependent coefficients can be achieved.

For totally smooth spheres, this expansion has been performed up to fifth order [34]. It predicts an exponentially fast decay of the coefficients to their stationary values in agreement with simulations and direct solutions of the Boltzmann equation [35].

For needles, we observe and investigate the breakdown of homogeneity in simulations of dense systems, where the inter-particle spacing is smaller than the length of the needles. Large-scale structures in the translational velocity field are seen to develop. Furthermore, the density does not remain homogeneous but clusters form and dissolve again. These effects lead to deviations from the solution of the homogeneous cooling state on the longest times scales and a third stage of cooling is found. It is characterized by an even slower decay of the kinetic energy, most of the energy being stored in the macroscopic velocity field.

We plan to derive generalized hydrodynamic equations for grains with rotational degrees of freedom and in particular hard rods. Such a set of hydrodynamic equations could serve as a starting point for a stability analysis, similar to the work of Brito and Ernst [30] for smooth spheres.

Acknowledgements

This work has been supported by the DFG through SFB 345 and Grants Zi209/5-1 and Zi209/6-1.

Appendix

A Calculations for Spheres

In this appendix we explain, as an example, the main steps to calculate $\langle i\mathcal{L}E_{\mathrm{tr}}\rangle_{\mathrm{HCS}}$ of Eq. (44) in 2D. We define the configuration integral

$$Q^N := \int \left(\prod_{i=1}^{N} d\boldsymbol{r}_i\right) W(\boldsymbol{r}_1,\dots,\boldsymbol{r}_N)\ . \tag{63}$$

The properly normalized N-particle distribution function for the HCS-state reads

$$\rho_{\mathrm{HCS}}(\Gamma;t) = \frac{1}{Q^N} W(\boldsymbol{r}_1,\dots,\boldsymbol{r}_N) \left(\frac{m}{2\pi T_{\mathrm{tr}}(t)}\right)^N \left(\frac{I}{2\pi T_{\mathrm{rot}}(t)}\right)^{N/2} \times \exp\left[-\sum_{i=1}^{N}\left(\frac{m}{2T_{\mathrm{tr}}(t)}\boldsymbol{v}_i^2 + \frac{I}{2T_{\mathrm{rot}}(t)}\omega_i^2\right)\right]\ . \tag{64}$$

The angular velocity is a scalar in two dimensions, but a vector in more than two dimensions. Free streaming does not change the energy, so we have to take into account only the collision operator $\mathcal{L}_+$ and compute

$$\langle i\mathcal{L}_+ E_{\rm tr}\rangle_{\rm HCS} = \frac{1}{2}\sum_{i\neq j}\int d\Gamma\, \rho_{\rm HCS}(\Gamma;t) i\mathcal{T}_+^{(ij)} \frac{1}{N}\sum_{k=1}^{N}\frac{m}{2}\boldsymbol{v}_k^2 =$$
$$\frac{1}{2N}\sum_{i\neq j}\int d\Gamma\, \rho_{\rm HCS}(\Gamma;t) i\mathcal{T}_+^{(ij)} \frac{m}{2}\left(\boldsymbol{v}_i^2+\boldsymbol{v}_j^2\right) . \quad (65)$$

The binary collision operator $\mathcal{T}_+^{(ij)}$ gives a contribution only if either $k=i$ or if $k=j$. Next, we introduce two δ-functions,

$$\langle i\mathcal{L}_+ E_{\rm tr}\rangle_{\rm HCS} = \frac{1}{2N}\sum_{i\neq j}\int d\Gamma \int d\boldsymbol{R}_1 d\boldsymbol{R}_2 \delta(\boldsymbol{R}_1-\boldsymbol{r}_i)\delta(\boldsymbol{R}_2-\boldsymbol{r}_j)$$
$$\rho_{\rm HCS}(\Gamma;t) i\mathcal{T}_+^{(ij)} \frac{m}{2}\left(\boldsymbol{v}_i^2+\boldsymbol{v}_j^2\right) , \quad (66)$$

which allows us to replace $\boldsymbol{r}_i$ by $\boldsymbol{R}_1$ and $\boldsymbol{r}_j$ by $\boldsymbol{R}_2$ in $\mathcal{T}_+^{(ij)}$. We define the pair correlation function $g(|\boldsymbol{R}_1-\boldsymbol{R}_2|)$ by

$$\frac{N}{V^2} g(|\boldsymbol{R}_1-\boldsymbol{R}_2|) :=$$
$$\frac{1}{N}\sum_{i\neq j}\frac{1}{Q^N}\int\prod_{k=1}^{N} d\boldsymbol{r}_k W(\boldsymbol{r}_1,\ldots,\boldsymbol{r}_N)\delta(\boldsymbol{R}_1-\boldsymbol{r}_i)\delta(\boldsymbol{R}_2-\boldsymbol{r}_j) . \quad (67)$$

Eq. (67) is used to rewrite Eq. (66) in terms of the pair correlation function. Integration over all velocities and angular velocities with index k and $i\neq k\neq j$ gives 1 due to normalization. We get

$$\langle i\mathcal{L}_+ E_{\rm tr}\rangle_{\rm HCS} = \frac{N}{2V^2}\left(\frac{m}{2\pi T_{\rm tr}(t)}\right)^2 \frac{I}{2\pi T_{\rm rot}(t)}\int d\omega_1 d\omega_2 d\boldsymbol{R}_1 d\boldsymbol{R}_2 d\boldsymbol{v}_1 d\boldsymbol{v}_2$$
$$\exp\left(-\frac{m}{2T_{\rm tr}(t)}(\boldsymbol{v}_1^2+\boldsymbol{v}_2^2)-\frac{I}{2T_{\rm rot}(t)}(\omega_1^2+\omega_2^2)\right)$$
$$g(\boldsymbol{r})\,|\boldsymbol{v}_{12}\cdot\hat{\boldsymbol{r}}|\,\Theta\left(-\boldsymbol{v}_{12}\cdot\hat{\boldsymbol{r}}\right)\delta\left(|\boldsymbol{r}|-d\right)\Delta E_{\rm tr} . \quad (68)$$

The loss of translational energy of two colliding particles is denoted by $\Delta E_{\rm tr}$. We use the abbreviation $\boldsymbol{R}_1-\boldsymbol{R}_2=\boldsymbol{r}=r\hat{\boldsymbol{r}}$ and neglect non-contributing terms linear in Ω so that $\Delta E_{\rm tr}$ is given by

$$\Delta E_{\rm tr} = \frac{m}{2}\left[2\eta_t(\eta_t-1)(\boldsymbol{v}_{12}^2-(\boldsymbol{v}_{12}\cdot\hat{\boldsymbol{r}})^2)-\right.$$
$$\left.(1/2)(1-e_n^2)(\boldsymbol{v}_{12}\cdot\hat{\boldsymbol{r}})^2+(1/2)\eta_t^2 d^2(\omega_1+\omega_2)^2\right] . \quad (69)$$

To perform the remaining integrations we substitute

$$\Omega = \frac{1}{\sqrt{2}}(\omega_1 + \omega_2), \quad \omega = \frac{1}{\sqrt{2}}(\omega_1 - \omega_2), \tag{70}$$

$$\boldsymbol{V} = \frac{1}{\sqrt{2}}(\boldsymbol{v}_1 + \boldsymbol{v}_2), \quad \boldsymbol{v} = \frac{1}{\sqrt{2}}(\boldsymbol{v}_1 - \boldsymbol{v}_2), \tag{71}$$

$$\boldsymbol{r} = \boldsymbol{R}_1 - \boldsymbol{R}_2, \qquad \boldsymbol{R} = \boldsymbol{R}_1. \tag{72}$$

The Jacobian determinant for the above transformation is 1. Integrations over ω, $\boldsymbol{V}$ and $\boldsymbol{R}$ all give 1 due to normalization. We are left with

$$\begin{aligned}\langle i\mathcal{L}_+ E_{\mathrm{tr}}\rangle_{\mathrm{HCS}} = \frac{N}{V}\frac{m}{2\pi T_{\mathrm{tr}}(t)}\left(\frac{2I}{2\pi T_{\mathrm{rot}}(t)}\right)^{1/2} \int d\Omega d\boldsymbol{r} d\boldsymbol{v} \\ \exp\left(-\frac{m\boldsymbol{v}^2}{2T_{\mathrm{tr}}(t)} - \frac{I\Omega^2}{2T_{\mathrm{rot}}(t)}\right) g(\boldsymbol{r})\, |\boldsymbol{v}\cdot\hat{\boldsymbol{r}}|\, \Theta\left(-\boldsymbol{v}\cdot\hat{\boldsymbol{r}}\right) \delta\left(|\boldsymbol{r}| - d\right) \\ \frac{m}{2}\left[2\eta_t(\eta_t - 1)(\boldsymbol{v}^2 - (\boldsymbol{v}\cdot\hat{\boldsymbol{r}})^2) - (1/2)(1-e_n^2)(\boldsymbol{v}\cdot\hat{\boldsymbol{r}})^2 + (1/2)\eta_t^2 d^2\Omega^2\right].\end{aligned}$$

The integration over $|\boldsymbol{r}|$ yields $dg(d)$. Choosing e.g. $\boldsymbol{r}$ to point along the x-axis, the integrals over linear and angular velocities can easily be done as moments of a Gaussian distribution. The result is independent of $\hat{\boldsymbol{r}}$, so that the integration over $\hat{\boldsymbol{r}}$ gives 2π. Finally we obtain the result of Eq. (46).

B Calculations for Needles

In this appendix, we present some of the detailed calculations for needles. As a first step, we express the orientation of the rods in spherical coordinates $\boldsymbol{u}_i = (\sin(\theta_i)\cos(\phi_i), \sin(\theta_i)\sin(\phi_i), \cos(\theta_i))$. The canonical momenta (translational and rotational) are then given by

$$\boldsymbol{p}_i = m\boldsymbol{v}_i\,, \qquad p_{\theta_i} = I\dot{\theta}_i\,, \qquad p_{\phi_i} = I\dot{\phi}_i \sin^2\theta\,. \tag{73}$$

In the following calculation it will be necessary to express $\dot{\boldsymbol{u}}_i$ in terms of canonical momenta

$$\dot{\boldsymbol{u}}_i = \frac{p_{\theta_i}}{I}\boldsymbol{e}_{\theta_i} + \frac{p_{\phi_i}}{\sin\theta_i I}\boldsymbol{e}_{\phi_i}\,. \tag{74}$$

$\boldsymbol{e}_{\theta_i}$ and $\boldsymbol{e}_{\phi_i}$ are orthogonal unit vectors in θ_i and ϕ_i direction. The kinetic energies per particle are then given by

$$E_{\mathrm{tr}} = \frac{1}{N}\sum_{i=1}^{N}\frac{1}{2m}\boldsymbol{p}_i^2, \quad E_{\mathrm{rot}} = \frac{1}{N}\sum_{i=1}^{N}\frac{1}{2I}p_{\theta_i}^2 + \frac{1}{2I\sin^2\theta_i}p_{\phi_i}^2\,. \tag{75}$$

We want to calculate non-equilibrium expectation values with the normalized probability distribution given in Eq. (43). We consider again as an example the translation energy per particle E_{tr}.

$$\langle i\mathcal{L}_+ E_{\text{tr}} \rangle = \frac{1}{V^N} \frac{1}{(4\pi)^N} \frac{1}{(2\pi M T_{\text{trans}})^{3N/2}} \frac{1}{(2\pi I T_{\text{rot}})^N}$$
$$\frac{1}{2} \sum_{m \neq n} \int \prod_{j=1}^{N} d\boldsymbol{r}_j \, d\phi_j \, d\theta_j \, d\boldsymbol{p}_j \, dp_{\theta_j} \, dp_{\phi_j}$$
$$\exp[-N E_{\text{tr}}/T_{\text{tr}}(t) - N E_{\text{rot}}/T_{\text{rot}}(t)] i\mathcal{T}_+^{(nm)} E_{\text{tr}} \,. \quad (76)$$

Similar to the calculation for the spheres we see that the binary collision operator $\mathcal{T}_+^{(nm)}$ gives a contribution only if either $i = n$ or if $i = m$. We can sum over $N(N-1)$ identical integrals and get

$$\frac{N-1}{2V^2} \frac{1}{(4\pi)^2} \frac{1}{(2\pi m T_{\text{tr}})^3} \frac{1}{(2\pi I T \text{rot})^2} \int \prod_{j=1}^{2} d\boldsymbol{r}_j \, d\phi_j \, d\theta_j \, d\boldsymbol{p}_j \, dp_{\theta_j} \, dp_{\phi_j}$$
$$\exp[-E_{\text{tr}}^{12}/T_{\text{tr}}(t) - E_{\text{rot}}^{12}/T_{\text{rot}}(t)] \left| \frac{d}{dt} |\boldsymbol{r}_{12}^{\perp}| \right| \Theta\left(-\frac{d}{dt}|\boldsymbol{r}_{12}^{\perp}|\right)$$
$$\Theta(L/2 - |s_{12}|) \Theta(L/2 - |s_{21}|) \delta(|\boldsymbol{r}_{12}^{\perp}| - 0^+) \Delta E_{\text{tr}}^{12} \,. \quad (77)$$

E_{tr}^{12} (E_{rot}^{12}) is the sum of the translational (rotational) kinetic energy of particle 1 and 2 and with $\Delta E_{\text{tr}}^{12}$ we denote the change of the translational kinetic energy of particle 1 and 2 in a collision:

$$\Delta E_{\text{tr}}^{12} = \frac{(\boldsymbol{p}_1 - \boldsymbol{p}_2) \cdot \Delta \boldsymbol{p}}{m} + \frac{\Delta \boldsymbol{p}^2}{m} \,, \quad (78)$$

$$\Delta \boldsymbol{p} = -\frac{1 + e_n}{2} \frac{1}{\frac{1}{m} + \frac{s_{12}^2}{2I} + \frac{s_{21}^2}{2I}} (\boldsymbol{V} \cdot \boldsymbol{u}_\perp) \boldsymbol{u}_\perp \,. \quad (79)$$

$\boldsymbol{V}$ is the relative velocity of the contact points defined in Eq. (16).

We introduce relative coordinates $\boldsymbol{r}_{12} = \boldsymbol{r}_1 - \boldsymbol{r}_2$ and $\boldsymbol{r} = \boldsymbol{r}_1$ and the variables

$$z := \boldsymbol{r}_{12} \cdot \boldsymbol{u}_\perp \,,$$
$$a := \boldsymbol{r}_{12} \cdot \boldsymbol{u}_1 - \frac{\boldsymbol{u}_1 \cdot \boldsymbol{u}_2}{\sqrt{1 - (\boldsymbol{u}_1 \cdot \boldsymbol{u}_2)^2}} \boldsymbol{r}_{12} \cdot \boldsymbol{u}_1^{\perp} = -s_{12} \,,$$
$$b := \frac{1}{\sqrt{1 - (\boldsymbol{u}_1 \cdot \boldsymbol{u}_2)^2}} \boldsymbol{r}_{12} \cdot \boldsymbol{u}_1^{\perp} = s_{21} \,.$$

The Jacobian of the transformation is given by $\sqrt{1 - (\boldsymbol{u}_1 \cdot \boldsymbol{u}_2)^2}$. We remark that $\frac{d}{dt} |\boldsymbol{r}_{12}^{\perp}| = \boldsymbol{V} \cdot \boldsymbol{u}_\perp \text{sign}(\boldsymbol{r}_{12} \cdot \boldsymbol{u}^{\perp})$ and we find again the relative velocity of the contact points $\boldsymbol{V} = \frac{\boldsymbol{p}_{12}}{m} - a\dot{\boldsymbol{u}}_1 - b\dot{\boldsymbol{u}}_2$ given in the new coordinates.

Integration over $\boldsymbol{r}$ gives V and integration over z gives the sum of two Θ–functions $\Theta(\pm \boldsymbol{V} \cdot \boldsymbol{u}_\perp)$. This reflects the fact that if one particle touches the other from 'above', the sign of the relative velocity of the contact point has to be negative, if the particle touches from 'below' the velocity has to be positive. Next one introduces relative and center of mass momenta as well as dimensionless variables:

$$\chi := \frac{1}{\sqrt{2mT_{\text{trans}}}}(\boldsymbol{p}_1 - \boldsymbol{p}_2)\ , \qquad \gamma := \frac{1}{\sqrt{2mT_{\text{trans}}}}(\boldsymbol{p}_1 + \boldsymbol{p}_2)\ ,$$
$$\tilde{p}_{\theta_i} := \frac{p_{\theta_i}}{\sqrt{IT_{\text{rot}}}}\ , \qquad \tilde{p}_{\phi_i} := \frac{p_{\phi_i}}{\sqrt{IT_{\text{rot}}}\sin\theta_i}\ .$$

The integration over γ can be done and the result is proportional to

$$\sum_{p=\pm 1} \int da\, db\, d\phi_1 \sin\theta_1 d\theta_1\, d\phi_2 \sin\theta_2 d\theta_2\, d\chi\, d\tilde{p}_{\theta_1}\, d\tilde{p}_{\phi_1} d\tilde{p}_{\theta_2}\, d\tilde{p}_{\phi_2}$$
$$\sqrt{1-(\boldsymbol{u}_1 \cdot \boldsymbol{u}_2)^2} \exp[-\frac{1}{2}(\chi^2 + \tilde{p}_{\phi_1}^2 + \tilde{p}_{\phi_2}^2 + \tilde{p}_{\theta_1}^2 + \tilde{p}_{\theta_2}^2)]$$
$$\left|\tilde{\boldsymbol{V}} \cdot \boldsymbol{u}_\perp\right| \Theta\left(p\left|\tilde{\boldsymbol{V}} \cdot \boldsymbol{u}_\perp\right|\right) \Theta(|a| - L/2)\Theta(|b| - L/2)\Delta E_{12}\ , \qquad (80)$$

all expressed in new variables and $\dot{\boldsymbol{u}}$ by Eq. (74), e.g.

$$\tilde{\boldsymbol{V}} = \sqrt{2T_{\text{trans}}/m}\chi - a\sqrt{T_{\text{rot}}/I}(\tilde{p}_{\theta_1}\boldsymbol{e}_{\theta_1} + \tilde{p}_{\phi_1}\boldsymbol{e}_{\phi_1}) -$$
$$b\sqrt{T_{\text{rot}}/I}(\tilde{p}_{\theta_2}\boldsymbol{e}_{\theta_2} + \tilde{p}_{\phi_2}\boldsymbol{e}_{\phi_2})\ . \qquad (81)$$

We want to perform the remaining Gaussian integrals, but we have expressed different terms either in $(\boldsymbol{u}_i^\perp, \boldsymbol{u}^\perp)$ defined according to Eq. (17) with $i = 1, 2$ or in $(\boldsymbol{e}_{\theta_i}, \boldsymbol{e}_{\phi_i})$. It is useful to note that $(\boldsymbol{u}_i^\perp, \boldsymbol{u}^\perp)$ and $(\boldsymbol{e}_{\theta_i}, \boldsymbol{e}_{\phi_i})$ are *two different* orthonormal basis sets in the plane perpendicular to $\boldsymbol{u}_i$, so that we can make a orthogonal coordinate transformation from one system to the other. The variables $\tilde{p}_{\theta_i}$ and $\tilde{p}_{\phi_1}$ are now standard normally distributed and after a orthogonal coordinate transformation the new coordinates will again be standard normally distributed. This means we can equivalently write $(\tilde{p}_{\theta_i}\boldsymbol{e}_{\theta_i} + \tilde{p}_{\phi_i}\boldsymbol{e}_{\phi_i})$ as $(v_i \boldsymbol{u}_i^\perp + w_i \boldsymbol{u}^\perp)$ with standard normally distributed variables v_i and w_i. With this definition of v_i and w_i, we are able to evaluate for example terms of the form $(\tilde{p}_{\theta_1}\boldsymbol{e}_{\theta_1} + \tilde{p}_{\phi_1}\boldsymbol{e}_{\phi_1}) \cdot \boldsymbol{u}_\perp \equiv (v_1 \boldsymbol{u}_1^\perp + w_1 \boldsymbol{u}_\perp) \cdot \boldsymbol{u}_\perp = w_1$, where we used $\boldsymbol{u}_1^\perp \cdot \boldsymbol{u}^\perp = \boldsymbol{u}_2^\perp \cdot \boldsymbol{u}^\perp = 0$. We can integrate freely over v_1 and v_2 and the two components of χ perpendicular to $\boldsymbol{u}_\perp$. We introduce

$d\Omega_i := d\phi_i \sin(\theta_i) d\theta_i$ and the intermediate result reads

$$\sum_{p=\pm 1} \frac{N-1}{2V} \frac{1}{(4\pi)^2} \frac{1}{(2\pi)^{(3/2)}} \int da\, db\, d\boldsymbol{s}\, d\Omega_1 d\Omega_2 \exp(-\frac{1}{2}\boldsymbol{s}^2) \sqrt{1-(\boldsymbol{u}_1 \cdot \boldsymbol{u}_2)^2}$$

$$|\boldsymbol{G}\cdot\boldsymbol{s}|\Theta(p\boldsymbol{G}\cdot\boldsymbol{s}) \Big[-s_1 \sqrt{\frac{2T_{\mathrm{tr}}}{m}} \frac{1+e_n}{2} \frac{1}{\frac{1}{m}+\frac{a^2}{2I}+\frac{b^2}{2I}} \boldsymbol{G}\cdot\boldsymbol{s} +$$

$$\frac{1}{m}\left(\frac{1+e_n}{2}\right)^2 \left(\frac{1}{\frac{1}{m}+\frac{a^2}{2I}+\frac{b^2}{2I}}\right)^2 (\boldsymbol{G}\cdot\boldsymbol{s})^2 \Big] . \quad (82)$$

We introduced the vectors $\boldsymbol{s} := (s_1, s_2, s_3) := (\boldsymbol{\chi}\cdot\boldsymbol{u}_\perp, w_1, w_2)$ and $\boldsymbol{G} = \left(\sqrt{\frac{2T_{\mathrm{tr}}}{m}}, -a\sqrt{\frac{T_{\mathrm{rot}}}{I}}, -b\sqrt{\frac{T_{\mathrm{rot}}}{I}}\right)$.

We can now perform the integral over $\boldsymbol{s}$. We sketch here only how this is done. We want to integrate

$$\int d\boldsymbol{s}\, \exp(-\frac{1}{2}\boldsymbol{s}^2)\Theta(\pm\boldsymbol{G}\cdot\boldsymbol{s})|\boldsymbol{G}\cdot\boldsymbol{s}|(\boldsymbol{G}\cdot\boldsymbol{s})s_1 . \quad (83)$$

Let $(\boldsymbol{e}_1, \boldsymbol{e}_2, \boldsymbol{e}_3)$ be the original coordinate system and we define a coordinate system $(\boldsymbol{e}_x, \boldsymbol{e}_y, \boldsymbol{e}_z)$ in which the z-axis is parallel to $\boldsymbol{G}$ and we decompose $\boldsymbol{s}$ in this coordinate system $\boldsymbol{s} = (s_x, s_y, s_z)$. Then Eq. (83) reads

$$\int ds_x ds_y ds_z \exp(-\frac{1}{2}(s_x^2+s_y^2+s_z^2))\Theta(\pm s_z)$$
$$|\boldsymbol{G}||s_z||\boldsymbol{G}|s_z \left[(s_x\boldsymbol{e}_x + s_y\boldsymbol{e}_y + s_z\boldsymbol{e}_z)\cdot\boldsymbol{e}_1\right] . \quad (84)$$

Only the term proportional to $s_z\boldsymbol{e}_z$ contributes and the Gaussian integral can easily be performed. Using that $|\boldsymbol{G}|\boldsymbol{e}_z = \boldsymbol{G}$ we write $|\boldsymbol{G}|\boldsymbol{e}_z\cdot\boldsymbol{e}_1 = \boldsymbol{G}\cdot\boldsymbol{e}_1 = G_1$ and we end up with the result $4\pi|\boldsymbol{G}|G_1$. Only the integrals over Ω_1 and Ω_2 have to be done with standard techniques. All other integrals are performed similarly and the results are quoted in the main text.

References

1. For a review on classical liquids see e.g. J.-P. Hansen and I. R. McDonald *Theory of Simple Liquids*, Academic Press (1986).
2. E. Thiele, J. Chem. Phys. **39**, 474 (1963); M. S. Wertheim, Phys. Rev. Lett. **10**, 321 (1963); J. Math. Phys. **5**, 634 (1964); L. Verlet and D. Levesque, Mol. Phys. **46**, 969 (1982).
3. J. L. Lebowitz, J. K. Percus and J. Sykes, Phys. Rev. **188**, 487 (1996); H. H. U. Konijnendijk and J. M. van Leeuwen, Physica **64**, 342 (1973); P. M. Furtado, G. F. Mazenko and S. Yip, Phys. Rev. **A12**, 1653 (1975); H. van Beijeren and M. H. Ernst, J. Stat. Phys. **21**, 125 (1979).
4. B. J. Alder, D. M. Gass and T. E. Wainwright, J. Chem. Phys. **53**, 3813 (1970).

5. J. J. Erpenbeck and W. W. Wood, J. Stat. Phys. **24**, 455 (1981).
6. B. J. Alder and T. E. Wainwright, Phys. Rev. **127**, 359 (1962).
7. J. T. Jenkins and S. B. Savage, J. Fluid Mech.**130**, 187 (1983); C. K. K. Lun, S. B. Savage, D. J. Jeffrey and N. Chepurniy, J. Fluid Mech.**140**, 223 (1984). A. Goldshtein and M. Shapiro, J. Fluid Mech.**282**, 75 (1995).
8. J. T. Jenkins and M. W. Richman, Phys. of Fluids **28**, 3485 (1985).
9. C. K. K. Lun and S. B. Savage, J. Appl. Mech. **54**, 47 (1987).
10. C. K. K. Lun, J. Fluid Mech. **233**, 539 (1991).
11. A. Goldshtein and M. Shapiro, J. Fluid Mech. **282**, 75 (1995).
12. H. Grad *Principles of the kinetic theory of gases* in *Handbuch der Physik*, ed. S. F. Fluegge, Springer (1958).
13. L. Waldmann *Transporterscheinungen in Gasen von mittlerem Druck* in *Handbuch der Physik*, ed. S. F. Fluegge, Springer (1958).
14. S. Chapman and T. G. Cowling, *The Mathematical Theory of Nonuniform Gases*, Cambridge University Press, London (1960).
15. I. Goldhirsch and G. Zanetti, Phys. Rev. Lett. **70**, 1619 (1993).
16. S. McNamara and S. Luding, Phys. Rev. E **58**, 2247 (1998).
17. M. Huthmann and A. Zippelius, Phys. Rev. E **56**, R6275 (1997).
18. S. Luding, M. Huthmann, S. McNamara, A. Zippelius, Phys. Rev. E **58**, 3416 (1998).
19. O. Walton, in *Energy and Technology Review*, edited by A. J. Poggio, (Lawrence Livermore National Laboratory, Livermore, CA) (1988); M. A. Hopkins and H. Shen in *Micromechanics of Granular Materials*, eds. M. Satake and J. T. Jenkins, Amsterdam, Elsevier, (1987).
20. G. W. Baxter and R. P. Behringer, Phys. Rev. **A42**, 1017 (1990).
21. B. Brogliato, *Nonsmooth Impact mechanics*, Springer 1996.
22. M. Huthmann, T. Aspelmeier and A. Zippelius, Phys. Rev. E **60** , 654 (1999).
23. M. H. Ernst, J. R. Dorfmann, W. R. Hoegy, and J. M. J. van Leeuwen, Physica **45**, 127 (1969).
24. P. Resibois and J. L. Lebowitz, J. Stat. Phys. **12**, 483 (1975); P. Resibois, J. Stat. Phys. **13**, 393 (1975); E. Leutheusser, J. Phys. **C15**, 2801 (1982).
25. D. Frenkel and J. F. Maguire, Mol. Phys. **49**, 503 (1983).
26. It was shown in [23] and [28] that the time evolution operator can be represented in the form $\exp(i\overline{\mathcal{L}}t)$ without generating overlap configurations.
27. P. K. Haff, Journ. of Fluid. Mech. **134**, 401 (1983).
28. T. P. C. van Noije, M. H. Ernst, R. Brito, Physica A **251** 266 (1998).
29. S. McNamara and W. Young, Phys. Rev. E **53**, 5089 (1996).
30. R. Brito and M. H. Ernst, Europhysics Letters **43**, 497 (1998).
31. W. Magnus, F. Oberhettinger, and R. P. Soni, *Formulas and Theorems for the Special Functions of Mathematical Physics*, Springer (1966).
32. O. Herbst, M. Huthmann, and A. Zippelius (preprint) (1999).
33. T. P. C. van Noije, M. H. Ernst, Granular Matter, **1**, 57 (1998).
34. M. Huthmann, J. A. G. Orza, and R. Brito (unpublished).
35. J. Javier Brey, M. J. Ruiz Montero and D. Cubero, Phys. Rev. E, **54** 3664 (1996).

Hydrodynamic Transport Coefficients of Granular Gases

J. Javier Brey and David Cubero

Area de Física Teórica. Facultad de Física. Universidad de Sevilla. Apartado de Correos 1065, 41080 Sevilla, Spain. e-mail: brey@cica.es

Abstract. Some transport properties of granular gases are investigated. Starting from a kinetic theory level of description, the hydrodynamic transport equations to Navier-Stokes order are presented. The equations are derived by means of the Chapman-Enskog procedure. To test the existence of a normal solution and the possibility of a hydrodynamic description, the theoretical predictions are compared with numerical simulations of the underlying kinetic equation for small deviations around the reference homogeneous state. An excellent agreement is found for all the range of dissipation in collisions considered. Similar analysis is presented for self-diffusion and Brownian motion. In the former case, also Molecular Dynamics results are shown to agree with the theoretical predictions. Quantitative and also qualitative differences with the elastic limit are discussed.

1 Introduction

Granular media in the so-called rapid flow regime are often described by means of continuum hydrodynamic equations [1]. The possibility of such a macroscopic description for systems with inelastic collisions is suggested by analogy with normal fluids. Dissipation in collisions is accounted for by introducing a source term in the evolution equation for the temperature. As a consequence, there is no homogeneous steady equilibrium state, but the simplest solution is given by a uniform system cooling constantly in time. Nevertheless, the justification for a hydrodynamic description, the form of the corresponding transport equations, the explicit expressions of the transport coefficients appearing in them, and the range of validity of the theory, require a detailed derivation from a more fundamental microscopic basis. As it is the case for molecular systems, the kinetic theory provides the right starting description from which the above questions can be addressed.

The simplest possibility of modeling granular flows at the particle level is as a system composed by identical smooth hard spheres or disks which collide inelastically. Moreover, the coefficient of restitution is supposed to be independent of the velocities of the colliding particles. The general formalism based on the (pseudo-)Liouville equation and also the Boltzmann and Enskog equations are easily generalized to the inelastic case [2]. In fact, several derivations of the transport equations to Navier-Stokes order for inelastic systems by applying the Chapman-Enskog method to the kinetic equations have been presented in the last decade or so [3–6]. Nevertheless, the technical difficulties following from the inelasticity in collisions has led to the introduction of approximations not required in the elastic case. These approximations restricted the validity of the resulting equations to the low dissipation or quasi-elastic limit.

Recently [7, 8], the above analysis is extended to arbitrary inelasticity, and the hydrodynamic fluxes and transport coefficients have been determined as functions of the coefficient of restitution. Also the cooling rate in the equation for the temperature has been analyzed to second order in the gradients, and its linear part computed explicitly. It has been found that the linear second order contributions give very small corrections to the linearized equations, so that they can be accurately neglected in linear analysis. Consequently, it is likely that the same happens with the nonlinear in the gradients part and the relevant contribution of the cooling rate to the transport equations be simply given by its zeroth order in the gradients limit.

Of course, the analogy between rapid granular flows and molecular fluids can be extended to many other transport situations. Two particularly simple cases, allowing detailed analysis are self-diffusion and Brownian motion [9, 10]. Both processes can be considered in the low density limit, in which they are described by the Boltzmann-Lorentz equation. Their study has attracted much attention in molecular gases due to their simplicity, their close relationship with experimental situations, and also because of the possibility of a direct comparison of the theoretical predictions with computer experiments. A great deal of relevant information about dynamical processes in gases has been obtained from the analysis of diffusion data. The same is expected to happen for granular flows, especially taking into account the existence of phenomena such as density clustering, compaction, and segregation [11] that seem to be closely related to diffusion. In fact, self-diffusion in granular systems has already been the subject previous works. Macroscopic flows [12] and vertically vibrated systems [13, 14] have been considered experimentally. Also, Molecular Dynamics simulations have been used to compute the self-diffusion coefficient in a sheared cell [15]. In this context, let us point out that knowing the explicit expression of the diffusion coefficient can be a necessary ingredient in order to determine the granular temperature in three–dimensional flows when using techniques which do not have enough time resolution to measure the temperature distribution directly.

The motion of a Brownian particle in a molecular fluid is described by the Fokker-Planck equation, that for a dilute gas can be derived from the Boltzmann-Lorentz equation in the limit of asymptotically large relative mass for the tagged particle [9, 16]. The generalization to the inelastic case has been also considered [17]. Some interesting quantitative and also qualitative differences occur, but the point we want to stress here is that a hydrodynamic description, characterized by a diffusion equation, still holds.

The direct simulation Monte Carlo (DSMC) method [18] provides a way for testing numerically the (analytical) theoretical predictions derived from the Boltzmann equation. Particularly interesting is the possibility of verifying the validity of a hydrodynamic description for granular systems, beyond the quasielastic limit, a point that has been a topic of interest and controversy [19–21]. Here we will report results obtained by applying the method to both the nonlinear inelastic Boltzmann equation and to the Boltzmann-Lorentz equation.

In the context of the simulation of the Boltzmann-Lorentz equation, it is worth to insist on the nature of the DSMC method as initially formulated. This method was not proposed to describe the dynamics of the particles in a low density gas, but to provide a numerical solution of an integro-differential equation, namely the nonlinear Boltzmann equation. The particles in the simulation do not correspond to real particles in the system. The number of the former can be as large as wanted and, nevertheless, one still remains in the low density limit, since it is the Boltzmann equation what is being simulated. Of course, it is possible to modify the "rules" of the simulation algorithm trying to incorporate physical effects that are not accounted for in the Boltzmann equation, but we believe this is outside the spirit of the original DSMC method. Here, we will modify the algorithm as to adjust it to the Boltzmann-Lorentz equation. This means that the distribution function of the bath, which determines the fluid "seen" by the tagged particle, is an input for the numerical simulation (as it is for the own kinetic equation). In this way, we are just numerically solving the considered kinetic equation.

The aim of this paper is to offer a short review of some recent results obtained in relation with the three above mentioned problems: Navier-Stokes transport coefficients, self-diffusion, and Brownian motion, in a granular gas. More concretely, we will focus on the possibility of a hydrodynamic description and the reasons why it is expected to be valid. Simulations of the kinetic equations will be compared with the theoretical predictions derived by assuming the existence of a Chapman-Enskog normal solution, in which the existence of the hydrodynamic level of description is inherent.

2 Navier-Stokes Transport Coefficients

We consider a system of smooth hard spheres (d=3) or disks (d=2) of mass m and diameter σ. The particles collide inelastically and the dissipation in collisions is characterized by a constant coefficient of normal restitution α. In the low density limit, the time evolution of the one-particle distribution function of the gas, $f(\mathbf{r},\mathbf{v},t)$, is assumed to be described by the Boltzmann equation [2, 6]

$$\left(\frac{\partial}{\partial t}+\boldsymbol{v}_1\cdot\boldsymbol{\nabla}\right)f(\boldsymbol{r},\boldsymbol{v}_1,t)=J[\boldsymbol{r},\boldsymbol{v}_1|f(t)]\ , \tag{1}$$

where J is the (inelastic) Boltzmann collision operator,

$$J[\boldsymbol{r},\boldsymbol{v}_1|f(t)] = \sigma^{d-1}\int d\boldsymbol{v}_2 \int d\widehat{\boldsymbol{\sigma}}\, \Theta(\widehat{\boldsymbol{\sigma}}\cdot\boldsymbol{g})(\widehat{\boldsymbol{\sigma}}\cdot\boldsymbol{g})[\alpha^{-2}f(\boldsymbol{r},\boldsymbol{v}_1',t)f(\boldsymbol{r},\boldsymbol{v}_2',t) - f(\boldsymbol{r},\boldsymbol{v}_1,t)f(\boldsymbol{r},\boldsymbol{v}_2,t)] \,. \tag{2}$$

Here $\widehat{\boldsymbol{\sigma}}$ is a unit vector along the line joining the centers of particles 2 and 1 at contact, away from the former, $\boldsymbol{g} = \boldsymbol{v}_1 - \boldsymbol{v}_2$ is the relative velocity, and Θ is the Heaviside step function. The velocities $\boldsymbol{v}_1', \boldsymbol{v}_2'$ are the precollisional velocities leading after collision to velocities $\boldsymbol{v}_1, \boldsymbol{v}_2$. They are given by

$$\boldsymbol{v}_1' = \boldsymbol{v}_1 - \frac{1+\alpha}{2\alpha}(\widehat{\boldsymbol{\sigma}}\cdot\boldsymbol{g})\widehat{\boldsymbol{\sigma}} \,, \quad \text{and} \quad \boldsymbol{v}_2' = \boldsymbol{v}_2 + \frac{1+\alpha}{2\alpha}(\widehat{\boldsymbol{\sigma}}\cdot\boldsymbol{g})\widehat{\boldsymbol{\sigma}} \,. \tag{3}$$

The macroscopic balance equations are obtained from Eq. (1) by multiplying with 1, $m\boldsymbol{v}_1$, and mv_1^2 and integrating over $\boldsymbol{v}_1$,

$$\partial_t n + \boldsymbol{\nabla}\cdot(n\boldsymbol{u}) = 0 \,, \tag{4a}$$

$$\partial_t \boldsymbol{u} + \boldsymbol{u}\cdot\boldsymbol{\nabla}\boldsymbol{u} + (nm)^{-1}\boldsymbol{\nabla}\cdot\mathsf{P} = 0 \,, \tag{4b}$$

$$\partial_t T + \mathbf{u}\cdot\boldsymbol{\nabla}T + 2(dnk_B)^{-1}(\mathsf{P} : \boldsymbol{\nabla}\boldsymbol{u} + \boldsymbol{\nabla}\cdot\boldsymbol{q}) + T\zeta = 0 \,. \tag{4c}$$

The local particle number density n, flow velocity $\boldsymbol{u}$, and temperature T are defined in the usual way,

$$n(\boldsymbol{r},t) = \int d\boldsymbol{v}\, f(\boldsymbol{r},\boldsymbol{v},t) \,, \tag{5a}$$

$$n(\boldsymbol{r},t)\boldsymbol{u}(\boldsymbol{r},t) = \int d\boldsymbol{v}\, \boldsymbol{v} f(\boldsymbol{r},\boldsymbol{v},t) \,, \tag{5b}$$

$$\frac{d}{2}n(\boldsymbol{r},t)k_B T(\boldsymbol{r},t) = \int d\boldsymbol{v}\, \frac{mV^2}{2} f(\boldsymbol{r},\boldsymbol{v},t) \,, \tag{5c}$$

where k_B is the Boltzmann constant and $\boldsymbol{V}(\boldsymbol{r},t) = \boldsymbol{v} - \boldsymbol{u}(\boldsymbol{r},t)$. In Eqs. (4a)–(4c) the pressure tensor P, and the heat flux $\boldsymbol{q}$ are given by

$$\mathsf{P}(\boldsymbol{r},t) = \int d\boldsymbol{v}\, m\boldsymbol{V}\boldsymbol{V} f(\boldsymbol{r},\boldsymbol{v},t) \,, \tag{6}$$

$$\boldsymbol{q}(\boldsymbol{r},t) = \int d\boldsymbol{v}\, \frac{mV^2}{2}\boldsymbol{V} f(\boldsymbol{r},\boldsymbol{v},t) \,. \tag{7}$$

Finally, the cooling rate ζ in the equation for the temperature (4c) takes into account the energy dissipation in collisions, and it is a nonlinear functional of the distribution function,

$$\zeta(\boldsymbol{r},t) = \frac{(1-\alpha^2)m\pi^{\frac{d-1}{2}}\sigma^{d-1}}{4d\Gamma\left(\frac{d+3}{2}\right)nk_B T}\int d\boldsymbol{v}_1 \int d\boldsymbol{v}_2\, g^3 f(\boldsymbol{r},\boldsymbol{v}_1,t)f(\boldsymbol{r},\boldsymbol{v}_2,t) \,. \tag{8}$$

Macroscopic balance equations similar to Eqs. (4a)–(4c) have been derived many times in the literature [3–5]. Of course, they only become closed hydrodynamic equations once the pressure tensor, the heat flux, and the cooling rate are expressed as functionals of the macroscopic fields. In principle this can be achieved by means of a Chapman-Enskog expansion, in the same spirit as for elastic molecular gases. Nevertheless, the complexity introduced by the energy dissipation in collisions has led to the introduction of some additional approximations, restricting the validity of the results to the small inelasticity limit. Only very recently explicit expressions for the fluxes to first order in the gradients as explicit functions of the coefficient of restitution have been obtained [7]. The expressions read

$$P_{ij} = nk_BT\delta_{ij} - \eta(\nabla_i u_j + \nabla_j u_i - \frac{2}{d}\delta_{ij}\boldsymbol{\nabla}\cdot\boldsymbol{u}) , \tag{9}$$

$$\boldsymbol{q} = -\kappa\boldsymbol{\nabla}T - \mu\boldsymbol{\nabla}n , \tag{10}$$

where η is the shear viscosity, κ the thermal conductivity, and μ a new transport coefficient, which has no analogue in elastic gases, coupling density gradient and heat flux. These transport coefficients are given by

$$\eta^*(\alpha) \equiv \frac{\eta(\alpha)}{\eta_0} = \left[\nu_1^*(\alpha) - \frac{\zeta^*(\alpha)}{2}\right]^{-1} , \tag{11}$$

$$\kappa^*(\alpha) \equiv \frac{\kappa(\alpha)}{\kappa_0} = [\nu_2^*(\alpha) - \frac{2d}{d-1}\zeta^*(\alpha)]^{-1}[1+c^*(\alpha)] , \tag{12}$$

$$\begin{aligned}\mu^*(\alpha) &\equiv \frac{n}{T\kappa_0}\mu(\alpha) \\ &= 2\zeta^*(\alpha)\left[\kappa^*(\alpha) + \frac{(d-1)c^*(\alpha)}{2d\zeta^*(\alpha)}\right]\left[\frac{2(d-1)}{d}\nu_2^*(\alpha) - 3\zeta^*(\alpha)\right]^{-1} .\end{aligned} \tag{13}$$

In the above expressions

$$\eta_0 = \frac{2+d}{8}\Gamma(d/2)\pi^{-\frac{d-1}{2}}(mk_BT)^{1/2}\sigma^{-(d-1)} , \quad \text{and} \tag{14}$$

$$\kappa_0 = \frac{d(d+2)^2}{16(d-1)}\Gamma(d/2)\pi^{-\frac{d-1}{2}}k_B\left(\frac{k_BT}{m}\right)^{1/2}\sigma^{-(d-1)} \tag{15}$$

are the values in a molecular gas of the shear viscosity and thermal conductivity, respectively. The dimensionless functions of the coefficient of restitution introduced in Eqs. (11)–(13) have the expressions

$$\zeta^*(\alpha) = \frac{2+d}{4d}(1-\alpha^2)\left[1+\frac{3}{32}c^*(\alpha)\right] , \tag{16}$$

$$\nu_1^*(\alpha) = \frac{(3-3\alpha+2d)(1+\alpha)}{4d}\left[1-\frac{1}{64}c^*(\alpha)\right] , \tag{17}$$

$$\nu_2^* = \frac{1+\alpha}{d-1}\left[\frac{d-1}{2} + \frac{3(d+8)(1-\alpha)}{16} + \frac{4+5d-3(4-d)\alpha}{1024}c^*(\alpha)\right], \quad (18)$$

$$c^*(\alpha) = \frac{32(1-\alpha)(1-2\alpha^2)}{9+24d+(8d-41)\alpha+30\alpha^2(1-\alpha)} . \quad (19)$$

The energy sink term has the form $\zeta = \zeta^{(0)} + \zeta^{(2)}$, where $\zeta^{(0)}$ denotes the zeroth order in the gradients contribution,

$$\zeta^{(0)} = \zeta^* \frac{nk_BT}{\eta_0} , \quad (20)$$

while $\zeta^{(2)}$ is of second order in the gradients. Only its linear in the gradients part, $\zeta_l^{(2)}$, must be considered for the linear analysis we will present in the following. It has the form

$$\zeta_l^{(2)} = \zeta_1 \nabla^2 T + \zeta_2 \nabla^2 n . \quad (21)$$

The expressions for the transport coefficient ζ_1 and ζ_2 are quite involved and not particularly relevant for the purposes here, since they give contributions to the transport equations that can be accurately neglected [7].

A point to be noted is that the above expressions for the transport coefficients have been obtained in the so-called first Sonine approximation, in which the distribution function of the gas is expanded in Sonine polynomials and only the first corrections to the Gaussian giving contributions to the several fluxes are retained. This approximation is also usual in molecular gases where it has been proved to be quite accurate. It is expected to hold also for inelastic systems since the reference state is Gaussian with very good approximation [6, 7, 22]. Let us also point out that the above analysis has been very recently extended to the revised Enskog kinetic theory for hard spheres [2, 23], providing then a macroscopic description at higher densities [8].

The granular hydrodynamic equations admit a solution describing the homogeneous cooling state (HCS), characterized by uniform fields and a time dependent temperature $T_H(t)$ obeying the equation

$$\frac{\partial}{\partial t} T_H(t) = -\zeta^{(0)}(t) T_H(t) . \quad (22)$$

We want to investigate the validity of the hydrodynamic description for states close to the HCS. Then we define deviations by

$$n(\boldsymbol{r},t) = n + \delta n(\boldsymbol{r},t), \; \boldsymbol{u}(\boldsymbol{r},t) = \delta \boldsymbol{u}(\boldsymbol{r},t), \; T(\boldsymbol{r},t) = T_H(t) + \delta T(\boldsymbol{r},t), \quad (23)$$

where n is the average density of the system. Linearization of Eqs. (4a)–(4c) about the HCS leads to partial differential equations with time dependent coefficients which are not suitable for a direct linear stability analysis. This is a consequence of the time dependence of the reference state, the HCS, and can be eliminated through a change of time and space variables, and a scaling of the hydrodynamic fields. We define

$$\boldsymbol{l} = \frac{\nu_H(t)}{2} v_H^{-1}(t)\boldsymbol{r}, \quad \tau = \frac{1}{2}\int_0^t dt'\, \nu_H(t') \,, \tag{24}$$

where $\nu_H(t) = nk_B T_H/\eta_0(T_H)$ is a characteristic frequency, and $v_H(t) = (k_B T_H/m)^{1/2}$ is the thermal velocity. Note that the length scale transformation is time independent. The scaled fields are

$$\rho(\boldsymbol{l},\tau) = \frac{\delta n(\boldsymbol{l},\tau)}{n}, \quad \boldsymbol{\omega}(\boldsymbol{l},\tau) = \frac{\delta \boldsymbol{u}(\boldsymbol{l},\tau)}{v_H(\tau)}, \quad \theta(\boldsymbol{l},\tau) = \frac{\delta T(\boldsymbol{l},\tau)}{T_H(\tau)} \,. \tag{25}$$

In the remainder of this Section, we will restrict ourselves to the particular case of hard spheres for the sake of simplicity, i. e. we take $d = 3$. When the new variables and fields are used, the linearized hydrodynamic equations become

$$\partial_\tau \rho_{\boldsymbol{k}} + ikw_{\boldsymbol{k}\parallel} = 0 \,, \tag{26}$$

$$\left(\partial_\tau - \zeta^* + \frac{2}{3}\eta^* k^2\right) w_{\boldsymbol{k}\parallel} + ik\theta_{\boldsymbol{k}} + ik\rho_{\boldsymbol{k}} = 0 \,, \tag{27}$$

$$\left(\partial_\tau - \zeta^* + \frac{1}{2}\eta^* k^2\right) \boldsymbol{w}_{\boldsymbol{k}\perp} = 0 \,, \tag{28}$$

$$\left(\partial_\tau + \zeta^* + \frac{5}{4}\kappa^* k^2\right) \theta_{\boldsymbol{k}} + \left(2\zeta^* + \frac{5}{4}\mu^* k^2\right) \rho_{\boldsymbol{k}} + \frac{2}{3} ikw_{\boldsymbol{k}\parallel} = 0 \,. \tag{29}$$

We have introduced the Fourier transformed of the hydrodynamic fields defined by

$$\rho_{\boldsymbol{k}}(\tau) = \int d\boldsymbol{l}\, e^{-i\boldsymbol{k}\cdot\boldsymbol{l}} \rho(\boldsymbol{l},\tau) \,, \tag{30}$$

and so on. Besides, $w_{\boldsymbol{k}\parallel}$ and $\boldsymbol{w}_{\boldsymbol{k}\perp}$ are the longitudinal and transversal components of the velocity field relative to the wave vector $\boldsymbol{k}$, respectively. From Eq. (28), the time evolution of the transversal components of the velocity field is directly obtained

$$\boldsymbol{w}_{\boldsymbol{k}\perp}(\tau) = \boldsymbol{w}_{\boldsymbol{k}\perp}(0) e^{s_\perp \tau} \,, \tag{31}$$

where the eigenvalue $s_\perp$ associated to these "shear modes" is

$$s_\perp = \zeta^* - \frac{1}{2}\eta^* k^2 \,. \tag{32}$$

Taking into account the definitions in Eq. (25) and that from Eq. (22) it follows that $T_H(\tau) = T_H(0)\exp(-2\zeta^*\tau)$, Eq. (31) leads to

$$\boldsymbol{u}_{\boldsymbol{k}\perp}(\tau) = \boldsymbol{u}_{\boldsymbol{k}\perp}(0) e^{-\frac{\eta^* k^2 \tau}{2}} \,, \tag{33}$$

i.e. perturbations of the transversal component of the velocity always decay in time. Let us remark that an exponential behavior in the reduced variable τ translates into an algebraic decay in the actual time t. It is easily seen from Eqs. (22) and (24) that

$$e^{\tau s} = \left(1 + \frac{t}{t_0}\right)^{s/\zeta^*} , \tag{34}$$

with $t_0^{-1} = \zeta^* \nu_H(0)$.

The above hydrodynamic description has been derived from the (inelastic) Boltzmann equation by assuming that the generalization of the Chapman–Enskog method can be used in order to obtain a normal solution to the kinetic equation. This requires a clear separation between the time scale governing the kinetic excitations of the system and the much larger time scale on which the macroscopic fields change in time. This separation is well established in the case of molecular gases, but the situation is more complicated when the collisions are not elastic. The time evolution of the hydrodynamic fields is not determined only by their spatial gradients, but there is another time scale for the temperature, set up by the inelasticity of the system through the homogeneous cooling rate $\zeta^{(0)}$. In this sense, it could be said that there are two hydrodynamic time scales in rapid granular flows: one associated to the macroscopic gradients and another one following directly from dissipation in collisions. Nevertheless, the point is not whether these scales are or are not separated one from the other, but whether they are both much larger than the one associated to the (microscopic) kinetic excitations.

A direct check of the accuracy of the hydrodynamic equations is provided by the comparison of solutions of them with solutions to the Boltzmann equation in which no hydrodynamic concepts have been introduced externally. Numerical solutions of the Boltzmann equation can be constructed by means of the direct simulation Monte Carlo (DSMC) method [18]. The general idea of the method is to generate a Markov process which mimics the dynamical processes described by the kinetic equation. Inelasticity in collisions is incorporated just by changing the expressions of the postcollisional velocities as compared with the elastic case. Since the details of the method have been discussed many times in the literature and can be found in Ref. [18], they will not be given here.

The kind of initial conditions we have considered corresponds to small amplitude perturbations about the HCS reached by a freely evolving granular gas, so that the linearized hydrodynamic equations (26)–(29) are expected to hold. The easiest macroscopic perturbation one can think of consists in an initial harmonic perturbation of the transversal component of the velocity field given by

$$u_y(x, 0) = u_0 \sin(q_0 x) , \tag{35}$$

where $u_0 = 0.1\sqrt{2} v_H(0)$ and $q_0 = 2\pi/L$, L being the size of the system in the x-direction. Along this direction, periodic boundary conditions are applied in the simulation of the Boltzmann equation. According to Eq. (31), the time evolution of the perturbation follows the law:

$$u_y(x,\tau) = u_\tau \sin(q_0 x), \quad u_\tau = u_0 e^{-\eta^* k_0^2 \tau/2} \,. \tag{36}$$

Here k_0 is the dimensionless reduced wavenumber corresponding to q_0, i.e. $k_0 = 2\nu_H^{-1} v_H q_0$. The simulation results show the qualitative behavior described above, i.e. the transversal component of the velocity flow u_y has a profile along the x-direction that can be accurately fitted by a sine function with an amplitude decreasing exponentially with the scaled time τ [24]. This provides a numerical value for the reduced shear viscosity η^* in Eq. (36). The results for several values of the restitution coefficient α are compared with the theoretical prediction given by Eq. (11) in Fig. 1. It is seen that there is a fairly good agreement over the wide range of α values considered, along which the variation of the shear viscosity coefficient is of the order of 20%. Close inspection of Fig. 1 indicates a small but systematic discrepancy

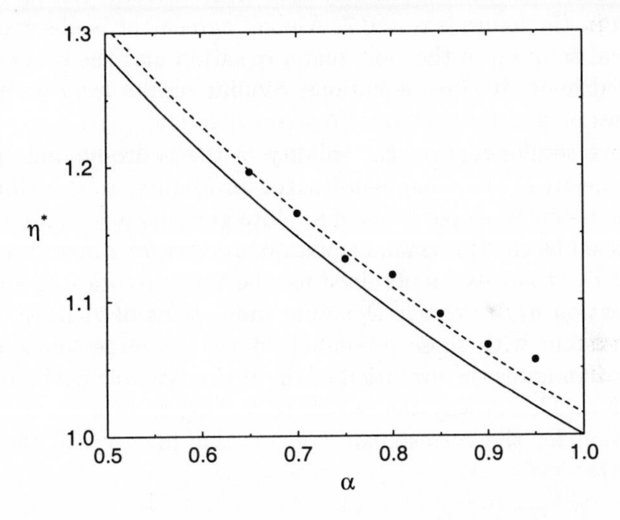

Fig. 1. Reduced shear viscosity η^* as a function of the coefficient of restitution α. The solid line is the theoretical prediction obtained by the Chapman–Enskog method in the first Sonine approximation and the points are from the DSMC method. The dashed line has been obtained by using a correction factor derived in the elastic limit $\alpha = 1$

between theory and simulation. Results from the latter always lay above the theoretical curve. This seems to be a consequence of the use of the first Sonine approximation upon deriving the hydrodynamic transport coefficients. In the elastic limit, a correction factor of the order of 1.016 to the first Sonine expression for the shear viscosity η_0 has been obtained [25]. The dashed curve in Fig. 1 has been constructed by using this factor for all α.

Since the time evolution of the density also depends on the values of other hydrodynamic fields, a more complicated initial perturbation is

$$\rho(x,0) = \rho_0 \sin(q_0 x) . \tag{37}$$

Now the solution of the linearized hydrodynamic equations can not be written down in a simple way. The time evolution of the amplitudes of the reduced density, temperature, and longitudinal component of the flow field is given by the linear combination of three exponentials, in the reduced time scale. The values of the relaxation times in the exponents are the roots of a cubic equation defining the dispersion relations [7].

In Fig. 2 we present the time evolution of the Fourier amplitudes of the hydrodynamic fields after introducing at $t = 0$ a perturbation given by Eq. (37) with $\rho_0 = 0.1$ (Note a factor of 2 of difference between the amplitude of the sine function and the corresponding Fourier component). The longitudinal component of the velocity field has not been plotted since it remains very small, below the noise level. The coefficient of restitution in the simulations shown in the figure is $\alpha = 0.7$. A good agreement is observed between the numerical solution of the Boltzmann equation and the predictions from the linearized hydrodynamic equations. Similar results have been found for several values of α in the interval $0.7 \leq \alpha \leq 0.95$ [24].

The above results confirm the validity of the hydrodynamic picture, as derived by means of the Chapman-Enskog procedure, to describe the time evolution of the macroscopic fields of a dilute granular gas whose time evolution is governed by the Boltzmann equation, at least for states close the HCS. Let us point out that results obtained for the time evolution of a linear density perturbation by Molecular Dynamic simulations of inelastic hard disks [26] are consistent with those presented above. For large times, one of the eigenmodes dominates the evolution of the hydrodynamic fields in the linear approximation. This is why the curves in the figure become straight lines. The expression for the corresponding eigenvalue involves all the transport coefficients [7].

3 Self-Diffusion

Let us now consider that some of the particles in the gas are labeled, but are otherwise identical to the others. The tagged particles will be described by the one particle distribution function $f_s(\boldsymbol{r}, \boldsymbol{v}, t)$. It will be assumed that the gas as a whole is in the HCS, and its distribution function will be denoted by $f_H(\boldsymbol{v}, t)$. Then, f_s obeys the Boltzmann-Lorentz equation,

$$(\partial_t + \boldsymbol{v}_1 \cdot \boldsymbol{\nabla}) f_s(\boldsymbol{r}, \boldsymbol{v}_1, t) = \sigma^{d-1} \int d\boldsymbol{v}_2 \int d\widehat{\boldsymbol{\sigma}}\, \Theta(\widehat{\boldsymbol{\sigma}} \cdot \boldsymbol{g})(\widehat{\boldsymbol{\sigma}} \cdot \boldsymbol{g}) \times [\alpha^{-2} f_s(\boldsymbol{r}, \boldsymbol{v}_1', t) f_H(\boldsymbol{v}_2', t) - f_s(\boldsymbol{r}, \boldsymbol{v}_1, t) f_H(\boldsymbol{v}_2, t)] . \tag{38}$$

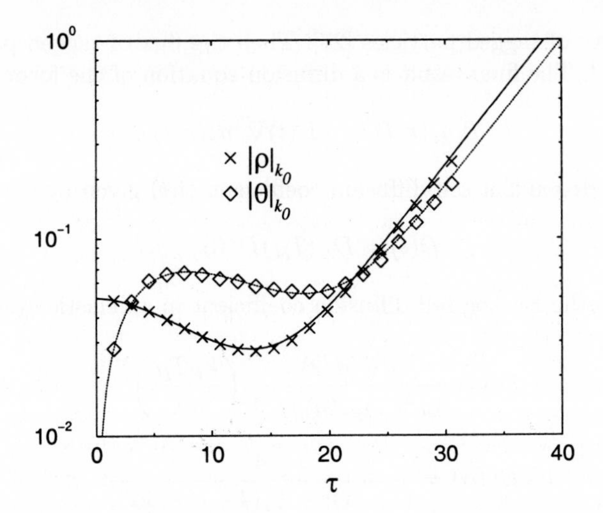

Fig. 2. Time evolution of Fourier components of the scaled density ρ and temperature θ, following an harmonic perturbation of the density. The symbols are from DSMC method and the lines from the linear hydrodynamic equations. All quantities are measured in the dimensionless units defined in the main text

Here we are using the same notation as in Eqs. (1)–(2).

The Boltzmann-Lorentz equation is based on the same hypothesis as the nonlinear Boltzmann equation and is restricted to the low density limit. Higher densities can be addressed by means of the (inelastic) Enskog equation [2] or, in the particular case of self-diffusion, the Enskog-Lorentz equation. For homogeneous systems, the only difference between the Enskog and Boltzmann description of self-diffusion is in the presence of the equilibrium pair correlation function of the system at distance σ, $g_e(n)$, as a factor in front of the collision integral [10]. Therefore, one can translate results obtained for the Boltzmann-Lorentz equation into results for Enskog-Lorentz equation by simply substituting σ^{d-1} by $g_e(n)\sigma^{d-1}$.

The density of tagged particles

$$n_s(\boldsymbol{r},t) = \int d\boldsymbol{v}\, f_s(\boldsymbol{r},\boldsymbol{v},t) \tag{39}$$

obeys the conservation law

$$\partial_t n_s(\boldsymbol{r},t) = -\boldsymbol{\nabla}\cdot \boldsymbol{J}_s(\boldsymbol{r},t)\ , \tag{40}$$

where $\boldsymbol{J}_s$ is the flux of tagged particles

$$\boldsymbol{J}_s(\boldsymbol{r},t) = \int d\boldsymbol{v}\, \boldsymbol{v} f_s(\boldsymbol{r},\boldsymbol{v},t)\ . \tag{41}$$

By using the Chapman-Enskog method it is possible to obtain a normal solution to the Enskog-Lorentz equation valid to first order in the gradient of the density of tagged particles [27]. Then the flux of tagged particles can be computed. The final result is a diffusion equation of the form

$$\partial_t n_s(\boldsymbol{r},t) = -D(t)\nabla^2 n_s(\boldsymbol{r},t) \ , \tag{42}$$

with a time dependent self-diffusion coefficient $D(t)$ given by

$$D(t) = D_E(T_H)D^*(\alpha) \ , \tag{43}$$

where D_E is the Enskog self-diffusion coefficient in an elastic system,

$$D_E = \frac{d\Gamma(d/2)}{4\pi^{\frac{d-1}{2}} n g_e(n)\sigma^{d-1}} \left(\frac{k_B T_H}{m}\right)^{1/2} \tag{44}$$

and

$$D^*(\alpha) = \frac{4}{(1+\alpha)^2 - \frac{c^*}{32}(4+\alpha-3\alpha^2)} \ . \tag{45}$$

Of course, the Boltzmann limit is obtained by taking $g_e(n) = 1$. The self-diffusion coefficient in Eq. (42) depend on time through the granular temperature $T_H(t)$. This time dependence can be eliminated by using again the time and space scales defined by Eq. (24). To get a simpler result, instead of the frequency ν_H given below Eq. (24) we use here

$$\nu_0 = \frac{2k_B T}{mD_E(T_H)} \ . \tag{46}$$

In the reduced variables Eq. (42) reads

$$\partial_\tau \rho_s(\boldsymbol{l},\tau) = D^*(\alpha)\nabla_l^2 \rho_s(\boldsymbol{l},\tau) \ , \tag{47}$$

where $\rho_s = n_s/n$. Now we have a diffusion equation with a constant diffusion coefficient. It follows that the mean square displacement of the scaled position $\boldsymbol{l}$ of the tagged particles is given by

$$\langle(\Delta\boldsymbol{l})^2;\tau\rangle = 2dD^*(\alpha)\tau \ . \tag{48}$$

To test the applicability of the Chapman-Enskog procedure to the inelastic self-diffusion problem, the DSMC method has been applied to the Boltzmann–Lorentz equation for hard spheres. As indicated in Eq. (38), the distribution function of the complete gas is required for input. Consistently with the theory we have developed, the distribution is taken to be the homogeneous cooling solution to the Boltzmann equation. More concretely, we have used the expression obtained in the first Sonine approximation [6, 28]. As a consequence, the problem of the clustering instability of the HCS for large wavelengths and strong dissipation [22, 29] can not be addressed in these simulations.

Two different procedures have been used to measure numerically the self-diffusion coefficient. In the first one, the scaled mean square displacement

$\langle(\Delta l)^2\rangle$ was measured as a function of the reduced time τ. After a short transient time, a linear behavior was found, in agreement with Eq. (48). The slope of the straight line fitting the numerical data provides the value for D^*. In the second method, an initial density of tagged particles perturbation was introduced,

$$n_s(x,0) = n_0[1+\sin(q_0 x)] \ , \tag{49}$$

where, again, $q_0 = 2\pi/L$ and periodic boundary conditions along the x-direction are employed. Notice that, contrary to the linear analysis of the Navier-Stokes equations presented in the previous Section, now we are not restricted to the small perturbation limit, and the precise value of n_0 is not relevant here. The diffusion equation predicts that

$$n_s(x,\tau) = n_0\left[1+e^{-s_D\tau}\sin(q_0 x)\right] \ , \tag{50}$$

with $s_D = D^* k_0^2$. Again k_0 denotes the dimensionless wave number corresponding to q_0. Then, by following the time decay of the amplitude of the sine perturbation, a numerical value for D^* follows. An example of the time evolution of the density profile is given in Fig. 3, where it is plotted at three different times. The coefficient of normal restitution is $\alpha = 0.95$. It is seen that the perturbation it is accurately described by a sine function with a time dependent amplitude.

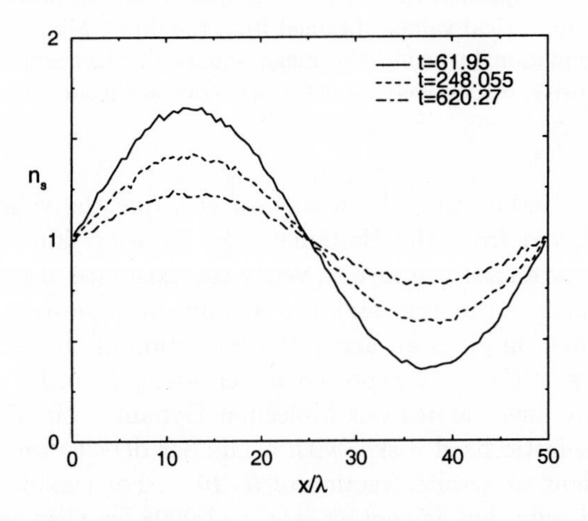

Fig. 3. Density profile along the x-direction at three different times following a sine initial perturbation with a wavelength determined by the size of the system. The density is normalized with the average density, and length is measured in units of the mean free path $\lambda = (\sqrt{2}n_0\pi\sigma^2)^{-1}$. The indicated times have been scaled as indicated in Eq. (24), but with the frequency ν_0

Fig. 4 shows the comparison of the two simulation experiments we have described with the theoretical prediction given by Eq. (45) for several values of the coefficient of restitution. A fairly good agreement is observed for all the range of values of α considered, namely $0.6 \leq \alpha \leq 1$.

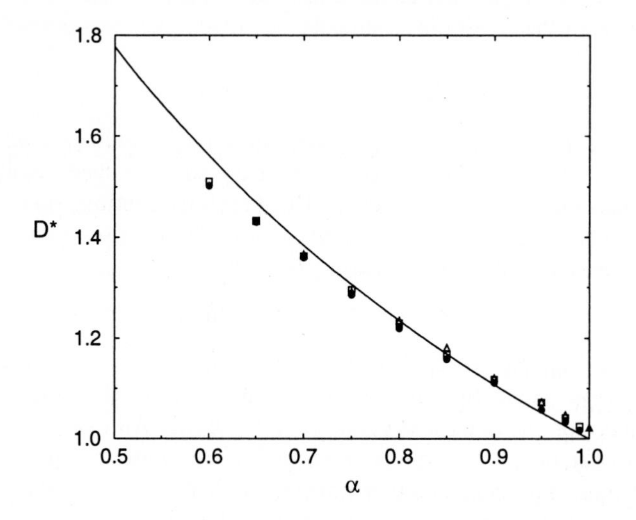

Fig. 4. Reduced self-diffusion coefficient D^* as a function of the coefficient of normal restitution α. The solid line is the theoretical prediction derived from the Boltzmann-Lorentz equation by using the Chapman-Enskog procedure, the circles and squares are numerical values obtained from the direct Monte Carlo simulation of the kinetic equation by using the mean square displacement and a sine perturbation in density, respectively, and the triangles are from Molecular Dynamics simulations

The DSMC method provides a way for checking the validity of theoretical results derived from the Boltzmann (or Enskog) kinetic equation. In particular, we have used it above to verify the existence of normal or hydrodynamic solutions. A different and also fundamental question is whether the own kinetic equation gives an accurate description of the time evolution of the system of particles it is expected to represent. In order to address this second issue, we have carried out Molecular Dynamics simulations of a system of 6400 inelastic hard disks, with a number density $n\sigma^2 = 6.25 \cdot 10^{-4}$, that is equivalent to a solid fraction of $5 \cdot 10^{-4}$. For this density, the equilibrium pair distribution at contact is $g_e \simeq 1.0008$, so that we are clearly in the low density region. The quantity we computed was the mean square displacement or, more precisely, $m_D = (4D_E)^{-1}\partial_t \langle (\Delta \boldsymbol{r})^2 \rangle$. After a few collisions per particle, this quantity reached a time independent plateau, that according to Eq. (42) should correspond to the diffusion regime and give the value of the reduced self-diffusion coefficient. The values for D^* obtained in this way have been also included in Fig. 4. The agreement with the theoretical prediction is again very good. Quite interestingly, the Molecular Dynamics data coincide on the scale of the figure with the numerical solution of the

Boltzmann-Lorentz equation obtained by the DSMC method. This strongly suggests that the small discrepancy between the Chapman-Enskog solution and the Molecular Dynamics results is due to the introduction of approximations when carrying out the former and, in particular, to the use of the first Sonine approximation. The situation seems to be similar to the one found in the study of the shear viscosity coefficient (see Fig. 1 and the discussion about it).

Two points deserve some additional comments before closing this Section. Firstly, in the Molecular Dynamics simulations there is no control on the state of the gas system, and it can develop cluster instabilities under the appropriate conditions. Therefore, the fact that self-diffusion behavior was observed, means that the results obtained by applying the Chapman-Enskog procedure to the Boltzmann-Lorentz equation are relevant, in the sense that a system of inelastic hard particles presents time and space windows inside which hydrodynamics provides an accurate description of the internal fluxes of particles. On the other hand, it is true than when the system becomes very dissipative, instabilities develops very soon in large systems and the "hydrodynamic window" may become very narrow. For this reason, the smallest value of the restitution coefficient for which we report a value of the diffusion coefficient from Molecular Dynamics is $\alpha = 0.7$. Note that for this dissipation the self-diffusion coefficient has already increased more than 30% above the elastic limit value.

The second comment refers to the consideration of hard spheres in the DSMC method, while hard disks were used in Molecular Dynamics for efficiency reasons. It may appear as surprising that the results agree for different dimension of the system. However, it is seen in Eq. (45) that the only dependence of D^* on d occurs through c^*, and the term containing it is negligible as compared with $(1+\alpha)^2$.

4 Brownian Motion

Now let us consider a tagged particle of mass m immersed in a low density gas of particles with mass m_g, i.e. the tagged particle is not mechanically equivalent to the gas particles. We will assume that all particles are hard spheres or disks colliding inelastically. The coefficient of normal restitution for the gas particle collisions will be denoted by α_g while for collisions of the tagged particle (with the gas particles) α will be used. As in our study of self-diffusion in the previous section, the gas will be supposed to be in the HCS.

The probability density for the tagged particle, $f_s(\boldsymbol{r}, \boldsymbol{v}, t)$, will obey the Boltzmann-Lorentz equation (38). In this Section the case of a very massive tagged particle will be studied. In the limit $\Delta \equiv m_g/m \to 0$, the equation reduces to leading order to a Fokker-Planck equation [17],

$$(\partial_t + \boldsymbol{v} \cdot \boldsymbol{\nabla}) f_s(\boldsymbol{r}, \boldsymbol{v}, t) = \mathcal{L}[T_H(t)] f_s(\boldsymbol{r}, \boldsymbol{v}, t) \, , \tag{51}$$

$$\mathcal{L}(T_H) = \gamma_e(T_H)a(\alpha)\frac{\partial}{\partial \boldsymbol{v}} \cdot \left[\boldsymbol{v} + \frac{k_B T_H}{m} a(\alpha)\frac{\partial}{\partial \boldsymbol{v}}\right] , \tag{52}$$

where γ_e is the same friction coefficients as for elastic systems, except as a function of the time dependent temperature of the gas $T_H(t)$,

$$\gamma_e(T_H) = \frac{4\pi^{\frac{d-1}{2}}\sigma_0^{d-1} n_g \Delta^{1/2}}{d\Gamma\left(\frac{d}{2}\right)} \left(\frac{2k_B T_H}{m}\right)^{1/2} . \tag{53}$$

Here n_g is the density of the gas, $\sigma_0 = (\sigma + \sigma_g)/2$, and $a = (1+\alpha)/2$. This latter quantity contains all the influence of the inelasticity of collision between the Brownian particle and the fluid particles. Its presence inside the square brackets in Eq. (52) implies that the usual fluctuation-dissipation for elastic particles is modified. An important and nontrivial feature is that the derivation of the above Fokker-Planck equation requires, in addition to the above-mentioned limit $\Delta \to 0$, that also $\alpha_g \to 1$, in such a way that

$$\epsilon_0 \equiv \frac{\zeta^{(0)}(t)}{2a\gamma_e(t)} \to \text{constant} < 1 . \tag{54}$$

Again $\zeta^{(0)}(t)$ is the cooling rate for the HCS, given in Eq. (20). Therefore, the validity of the Fokker-Planck equation, as derived from the inelastic Boltzmann-Lorentz equation is restricted to the limit of weak dissipation in the gas, although there is no limitation on the inelasticity of collisions of the tagged particle.

It is convenient to define a temperature $T(t)$ of the Brownian particle from its velocity fluctuations by

$$\frac{d}{2}k_B T = \int d\boldsymbol{r} \int d\boldsymbol{v}\, \frac{1}{2}m(\boldsymbol{v}-\boldsymbol{u})^2 f_s , \tag{55}$$

where $\boldsymbol{u}$ is the spatial average of the macroscopic velocity field of the particle,

$$\boldsymbol{u}(t) = \int d\boldsymbol{r} \int d\boldsymbol{v}\, \boldsymbol{v} f_s . \tag{56}$$

Equation (51) has the following two properties [17]:

1. In the long time limit the temperature of the tagged particle approaches the same cooling rate as the surrounding gas, but both temperatures remain different and their ratio approaches a constant,

$$\lim_{t\to\infty} \frac{T(t)}{T_H(t)} = \frac{a(\alpha)}{1-\epsilon_0} . \tag{57}$$

   Then, the asymptotic temperature of the tagged particles can be larger or smaller than that of the gas depending on whether it is $(1-\alpha)/2 < \epsilon_0 < 1$ or $0 < \epsilon_0 < (1-\alpha)/2$.

2. The long time limit of the probability distribution of the Brownian particle is Gaussian, even though the distribution of the surrounding gas is not,

$$f_s(\boldsymbol{r}, \boldsymbol{v}, t) \to f_{s,H}(\boldsymbol{v}, t) = \Omega^{-1} \frac{\tilde{v}_0^{-d}(t)}{\pi^{d/2}} e^{-\frac{v^2}{\tilde{v}_0^2(t)}} , \tag{58}$$

where Ω is the volume of the system and $\tilde{v}_0(t) = [2k_B T(t)/m]^{1/2}$.

Let us write the Fokker-Planck equation in Fourier space,

$$(\partial_t + i\boldsymbol{q} \cdot \boldsymbol{v}) f_s(\boldsymbol{q}, \boldsymbol{v}, t) = \mathcal{L}[T_H(t)] f_s(\boldsymbol{q}, \boldsymbol{v}, t) , \tag{59}$$

where

$$f_s(\boldsymbol{q}, \boldsymbol{v}, t) = \int d\boldsymbol{r}\, e^{-i\boldsymbol{q}\cdot\boldsymbol{r}} f_s(\boldsymbol{r}, \boldsymbol{v}, t) . \tag{60}$$

The eigenvalues of the operator $\mathcal{L} - i\boldsymbol{q} \cdot \boldsymbol{v}$ have the form [9, 10, 30]

$$\lambda_n = -a(\alpha)\gamma_e[T_H(t)] \sum_{i=1}^{d} n_i - D_e[T_H(t)] q^2 , \tag{61}$$

$n_i = 0, 1, 2, 3, \cdots$. The coefficient D_e is the same as the elastic diffusion coefficient but with the time dependent temperature $T_H(t)$,

$$D_e(T_H) = \frac{k_B T_H}{m\gamma_e(T_H)} . \tag{62}$$

In the elastic limit, the λ_n do not depend on time and define the modes of the system. There is a diffusive hydrodynamic mode given by $-D_e q^2$, which corresponds to $n_i = 0, i = 1, \cdots, d$, and an infinite set of kinetic modes, decaying much faster. The problem now is that the frequencies λ_n depend on time and they do not characterize the time evolution of the system. This is a direct consequence of the dissipation in collisions. Nevertheless, it is possible to transform the Fokker-Planck equation into one with time independent coefficients by using dimensionless variables. This is similar to the transformations used in the two previous sections for the Navier-Stokes transport coefficients and for the self-diffusion equation. We introduce

$$\tau = a(1 - \epsilon_0) \int_0^t dt' \gamma_e(t'), \quad \boldsymbol{k} = \boldsymbol{q} \frac{\widehat{v}_0(t)}{a(\alpha)(1 - \epsilon_0)\gamma_e(T_H)} , \quad \widehat{\boldsymbol{v}} = \frac{\boldsymbol{v}}{\widehat{v}_0(t)} , \tag{63}$$

where

$$\widehat{v}_0^2 = \frac{2k_B T_H(t) a(\alpha)}{m(1 - \epsilon_0)} . \tag{64}$$

The Fokker-Planck equation becomes

$$(\partial_\tau + i\boldsymbol{k} \cdot \widehat{\boldsymbol{v}}) \widehat{f}_s = \widehat{\mathcal{L}} \widehat{f}_s . \tag{65}$$

Now the reduced density distribution is

$$\widehat{f}_s(\boldsymbol{k},\widehat{\boldsymbol{v}},\tau) = \Omega \widehat{v}_0^d(t) f_s(\boldsymbol{q},\boldsymbol{v},t) \tag{66}$$

and

$$\widehat{\mathcal{L}} \equiv (1-\epsilon_0)\frac{\partial}{\partial \widehat{\boldsymbol{v}}} \cdot \left(\widehat{\boldsymbol{v}} + \frac{1}{2}\frac{\partial}{\partial \widehat{\boldsymbol{v}}} \right) . \tag{67}$$

The eigenvalues of the operator $\widehat{\mathcal{L}} + i\boldsymbol{k} \cdot \widehat{\boldsymbol{v}}$ are given by

$$\widehat{\lambda}_n = -(1-\epsilon_0)\sum_{i=1}^{d} n_i - \frac{D_e^*}{(1-\epsilon_0)^2} k^2 , \tag{68}$$

where

$$D_e^* = \frac{1-\epsilon_0}{2} . \tag{69}$$

In this dimensionless form we easily identify kinetic modes and a diffusive mode. As compared with the elastic limit, in which $\epsilon_0 = 0$, the kinetic modes have been slowed by a factor $(1-\epsilon_0)$ while the diffusion mode has been enhanced by a factor $(1-\epsilon_0)^{-2}$. Nevertheless, the relevant qualitative picture of microscopic modes decaying faster then the macroscopic ones, and the long time evolution being described by the hydrodynamic mode is still valid. In other words, the "aging to hydrodynamics" also applies for the description of an inelastic Brownian particle in a low density granular flow.

The theory developed in this Section, i.e. the Brownian limit of the Boltzmann-Lorentz equation and the exact consequences we have derived from the Fokker-Planck equation (51), has also been confirmed on the basis of the DSMC method applied to the Boltzmann-Lorentz equation [31]. In particular, excellent agreement has been found for the approach to a homogeneous cooling state, the temperature of that state, the approach to diffusion as measured, for instance, by the mean square displacement, and the dependence of the diffusion coefficient on the coefficient of restitution. Since the discussion parallels in many points the presentation in the previous section, we do not repeat it here, and refer the reader to the literature for details.

5 Conclusion

We have presented several transport situations in which the transition from a kinetic regime to a hydrodynamic one has been verified by comparing the theoretical results coming from the hydrodynamic equations with the numerical solutions to the kinetic equations. Moreover, in one of the considered situations, self-diffusion, the results have also been shown to agree with Molecular Dynamics simulation data. The agreement extends in all cases over a wide range of values of dissipation and it is by no means restricted to the quasielastic limit. The Brownian motion problem is particularly relevant in this respect since it allows a detailed and exact study of the aging to hydrodynamics and the decay of kinetic or microscopic excitations.

The main physical consequence of the time dependence of the transport coefficients in the context of the validity of a hydrodynamic description, is the introduction of a new relevant time scale, implying that the decay of kinetic excitations in the original time scale t is algebraical rather than exponential. This, in principle, is not inconsistent with a separation of time scales that is the condition required for a hydrodynamic description. In any case and in order to put the results presented in this paper in proper context, let us stress that we have considered values of the restitution coefficient larger than 0.6. It is then possible that for smaller α the hydrodynamic description and the own Boltzmann equation do not provide an accurate description of the evolution of the system.

The energy dissipation in collisions modifies the transport equations in both, a trivial and expected way and also in a subtle and hard to anticipate manner. The former is essentially given by the time dependence of the transport coefficients on time through the temperature and the presence of the energy source term in the equation for the temperature. Examples for the latter are the density gradient contribution to the heat flux and the modification of the fluctuation-dissipation relation in Brownian motion.

In this presentation many relevant aspects of rapid granular flows have not been addressed. In particular, all the discussion has been restricted to near the homogeneous cooling state situations and the stability of a freely evolving granular gas has not been considered. In spite of these limitations we believe it has been clearly established that the combination of analytical kinetic theory, direct simulation Monte Carlo methods, and Molecular Dynamics provides a unique way for approaching the study of rapid granular flows. As already mentioned, this possibility is not restricted to the very low density limit in which the Boltzmann equation applies, but can also be used in the context of the Enskog equation.

Acknowledgements

The material presented in this paper is based on published and unpublished work in collaboration with J. W. Dufty, M. J. Ruiz-Montero, A. Santos, F. Moreno, and R. García-Rojo, to whom I am greatly indebted. This research was partially supported by Grant No. PB98-1124 from the Dirección General de Investigación Científica y Técnica (Spain).

References

1. A classical and useful reference is: C. S. Campbell, Annu. Rev. Fluid Mech. **22**, 57 (1990).
2. J. J. Brey, J. W. Dufty, and A. Santos, J. Stat. Phys. **87**, 1051 (1997).

3. C. K. K. Lun, S. B. Savage, D. J. Jeffrey, and N. Chepurniy, J. Fluid Mech. **140**, 223 (1984).
4. J. T. Jenkins and M. W. Richman, Arch. Ration. Mech. Anal. **87**, 355 (1985); Phys. Fluids **28**, 3485 (1986).
5. N. Sela and I. Goldhirsch, J. Fluid Mech. **361**, 41 (1998).
6. A. Goldshtein and M. Shapiro, J. Fluid Mech. **282**, 75 (1995).
7. J. J. Brey, J. W. Dufty, C. S. Kim, and A. Santos, Phys. Rev. E **58**, 4638 (1998); J. J. Brey and D. Cubero, unpublished.
8. V. Garzó and J. W. Dufty, Phys. Rev. E **59**, 5895 (1999).
9. J. A. McLennan, *Introduction to Nonequilibrium Statistical Mechanics* (Prentice-Hall, New Jersey, 1989).
10. P. Résibois and M. de Leener, *Classical Kinetic Theory of Fluids*, (John Wiley and Sons, New York, 1977).
11. H. Jaeger, S. Nagel, and R. Behringer, Rev. Mod. Phys. **68**, 1250 (1996).
12. V. V. R. Natarajan, M. L. Hunt, and E. D. Taylor, J. Fluid Mech. **304**, 1 (1995), and references therein.
13. O. Zik and J. Stavans, Europhys. Lett. **16**, 255 (1991).
14. R. D. Wildman, J. M. Huntley, and J.-P. Hansen, Phys. Rev. E, **60**, 7066 (1999).
15. C. S. Campbell, J. Fluid Mech. **348**, 85 (1997).
16. R. F. Rodríguez, E. Salinas-Rodríguez. and J. W. Dufty, J. Stat. Phys. **32**, 279 (1983).
17. J. J. Brey, J. W. Dufty, and A. Santos, J. Stat. Phys. **97**, 281 (1999).
18. G. Bird, *Molecular Gas Dynamics and the Direct Simulation of Gas Flows*, (Clrendon Press, Oxford, 1994).
19. M. L. Tan and I. Goldhirsch, Phys. Rev. Lett. **81**, 3022 (1998).
20. L. P. Kadanoff, Rev. Mod. Phys. **71**, 435 (1999).
21. J. W. Dufty and J. J. Brey, Phys. Rev. Lett. **82**, 4566 (1999).
22. J. J. Brey, M. J. Ruiz–Montero, and D. Cubero, Phys. Rev. E **54**, 3664 (1996).
23. H. van Beijeren and M. H. Ernst, Physica A **68**, 437 (1973); **70**, 225 (1973).
24. J. J. Brey, M. J. Ruiz–Montero, and D. Cubero, Europhys. Lett. **48**, 359 (1999)
25. D. M. Gass, J. Chem. Phys. **54**, 1898 (1971).
26. P. Deltour and J. L. Barrat, J. Phys. I **7**, 137 (1997).
27. J. J. Brey, M. J. Ruiz–Montero, D. Cubero, and R. García-Rojo, Phys. Fluids **12** 876 (2000).
28. T. P. C. van Noije and M. H. Ernst, Granular Matter **1**, 57 (1998).
29. I. Goldhirsch and G. Zanetti, Phys. Rev. Lett. **70**, 1619 (1993); S. McNamara and W. R. Young, Phys. Rev. E **50**, R28 (1994).
30. H. Risken, *The Fokker-Planck Equation* (Springer-Verlag, Berlin, 1984).
31. J. J. Brey, M. J. Ruiz-Montero, D. Cubero, and J. W. Dufty, Phys. Rev. E **60**, 7174 (1999).

Granular Gases:
Probing the Boundaries of Hydrodynamics

Isaac Goldhirsch

Department of Fluid Mechanics and Heat Transfer, Faculty of Engineering,
Tel-Aviv University, Ramat-Aviv, Tel-Aviv 69978. ISRAEL
email: isaac@eng.tau.ac.il

Abstract. The dissipative nature of the particle interactions in granular systems renders granular gases mesoscopic and bearing some similarities to regular gases in the "continuum transition regime" (where shear rates and/or thermal gradients are very large). The following properties of granular gases support the above claim: (i) Mean free times are of the same order as macroscopic time scales. (ii) Mean free paths can be macroscopic and comparable to the system's dimensions. (iii) Typical flows are supersonic. (iv) Shear rates are typically "large". (v) Stress fields are scale (resolution) dependent. (vi) Burnett and super-Burnett corrections to both the constitutive relations and the boundary conditions are of importance. It is concluded that while hydrodynamic descriptions of granular gases are relevant, they are probing the boundaries of applicability of hydrodynamics and perhaps slightly beyond.

1 Introduction

Granular materials have been recently defined as a "new state of matter" [1]. The justification of this notion is, of course, the fact that granular materials exhibit a large number of properties and states which differ very significantly from those of their molecular counterparts. Einstein's statement, in his celebrated essay on Brownian motion [2], that the only difference between a Brownian particle and a molecule is one of size and not of principle, does apply, as is well known, to Brownian motion but not to collections of macroscopic particles. The root cause of this difference is the dissipative nature of the grain interactions, or, in other words, the fact that time reversal is broken on the "microscopic" or particle scale. Of course, when the atomic level of the grain interactions is accounted for, time reversal symmetry is restored but this is relevant on time scales which are far too large to be of practical interest.

When studying granular systems one should always bear in mind that practically all states of granular matter are metastable; for instance, the ground state of a sand-pile is one in which all grains are on the ground. All motions of granular materials need to be sustained by external pumping of energy to overcome the loss of energy in the particle interactions.

When granular materials are strongly forced, e.g. by shearing, all frictional bonds can be broken and the material can be fluidized. In this state, known as 'rapid granular flow', the grain interactions are practically instantaneous *inelastic* collisions and as such the state of the system is reminiscent of that of a (classical) molecular gas. One might have hoped that in this gaseous-like state and, in particular, when the gas is dilute, the dynamics is far simpler than in the dense flows (or quasi-static flows) and that kinetic theory in general and the Boltzmann equation, in particular, can be straightforwardly applied. That this is not exactly the case one can conclude e.g. from the observation that granular materials do not possess an equivalent of the state of equilibrium. Indeed, when an 'initial state' of a granular gas is (e.g. computationally) prepared to be statistically homogeneous and have an isotropic velocity distribution function, the kinetic energy of this state decays due to the inelasticity of the collisions until it vanishes. The only steady state of an unforced granular gas is one of zero kinetic energy or zero 'granular temperature'. It turns out that even this picture of a granular gas is an oversimplification since such gases are unstable to density fluctuations which give rise to clusters [3, 4] and destroy the homogeneity of the system. Thus, even one of the simplest possible states of a granular system is far from being truly simple or trivial. One may conclude that intuition gained from the study of e.g. the kinetics of molecular gases cannot be directly applied to the study of granular materials; in particular, many well established and internalized notions and methods of the theory of gases have to be revisited and modified when dealing with granular gases. The significant differences between granular and molecular gases (or materials, in general) have even led to the suspicion that such materials may defy a hydrodynamic description [5]. While the author of this paper does not take such an extreme view, he agrees that, at least on a fundamental level, the notion of a hydrodynamic, or macroscopic description of granular materials is based on unsafe grounds and it requires further study. On the other hand one cannot ignore successes of both phenomenological and rational theories of granular materials in general and granular gases in particular. Therefore, it is perhaps safe to conclude that macroscopic descriptions of granular materials (and gases) are useful, that they capture many of the dynamical features of these materials, but not all.

The main thesis of this paper is that granular gases should be considered to be mesoscopic in the sense that both the microscopic spatial and temporal scales are typically not well separated from the relevant corresponding macroscopic scales and that this property of granular gases is the root cause of many and perhaps most of the peculiar properties of granular gases. One could claim that the lack of scale separation in granular gases is a practical matter as the typical number of grains in a container is far smaller than the Avogadro number and thus one cannot expect as strong a scale separation as in molecular gases. This observation is certainly true for all granular systems that one can study in practice and it should undoubtedly be taken into account when modeling granular dynamics, but this is not the kind of lack of scale separation this paper is referring to. The claim here is that in addition to the above mentioned practical reason for the lack of scale separation there is another absence of scale separation in granular materials which is of fundamental nature and should hold even when a system comprising an Avogadro number of grains is considered. This property is responsible for the significant "normal stress differences" characterizing granular materials [6], for the importance of the Burnett and super-Burnett terms in the hydrodynamic equations of motion for granular gases [6], for the possible non-locality of macroscopic descriptions of granular gases [8] and for the scale dependence of stresses in granular gases [9], to name just a few implications.

Some of the mesoscopic properties of rapid granular flows are not unique to them as they can be partly realized in molecular gases. Indeed, when molecular gases are subject to large shear rates or large thermal gradients (i.e. when the velocity field or the temperature field changes significantly over the scale of a mean free path or the time defined by the mean free time) there is no scale separation between the microscopic and macroscopic scales and the gas can be considered to be mesoscopic. In this case [10] the Burnett and super-Burnett corrections (and perhaps beyond) are of importance and the gas exhibits normal stress differences and other properties characteristic of granular gases. While clusters are not expected in molecular gases, strongly sheared gases do exhibit ordering which violates the molecular-chaos assumption [11]. The class of flows of molecular gases in which all of these and many more phenomena occur due to strong forcing of the gas is known as the Continuum Transition Regime [10], and it is of importance e.g. in the design of high altitude flights. In contrast to molecular gases, in which a mesoscopic state can be obtained by strong forcing, granular gases are *generically* mesoscopic. One may think of granular gases as amplifiers of some (usually weak) properties of atomic or molecular gases, such as normal stress differences; as such they may serve as a laboratory for testing predictions pertaining to the latter. It is important to stress that while quasi-static and static flows are not discussed here, nor are even dense rapid flows, these systems can be considered to be mesoscopic as well: for instance, arches in dense granular systems can span the entire system and thus they create correlations whose typical length is macroscopic.

As mentioned above, some phenomena occurring in granular gases pertain to these systems alone; for instance clustering [3] and collapse [4] have no parallel in the elastic world. In addition, elastically interacting particles do not exhibit many of the properties of avalanches nor do they seem to possess oscillon excitations [12].

As is known from other fields in which mesoscopic systems are of importance, such as solid state physics, such systems exhibit properties which are different from those expected in "the bulk"; among other things fluctuations are expected to be stronger and the ensemble averages of many entities need not be representative of their typical values. Furthermore, like in turbulent systems or systems close to second order phase transitions, in which scale separation is non-existent, one expects the constitutive relations to be scale dependent. This is indeed the case for granular gases.

The structure of this paper is as follows. Section 2 demonstrates, employing the case of a steady sheared granular system, that granular systems are mesoscopic in nature in the sense that both spatial and temporal scale separation are weak or non-existent. It is also shown that dilute rapid granular flows are typically supersonic and that mean free paths can be of macroscopic dimensions (suggesting "non-locality"). Another aspect of the lack of scale separation, which is described in Section 3, is the the scale dependence of the stress field. Section 4 outlines a kinetic approach to granular dynamics and it shows that the lack of scale separation limits the applicability of this approach to near elastic systems; in addition it demonstrates that, unlike in typical atomic gases, the Burnett and super-Burnett terms are sizeable and physically consequential in granular gases. Section 5 presents an outline of a derivation of boundary conditions for granular gases and there again one encounters the the importance of the Burnett and super-Burnett contributions. In particular, it turns out that mass conservation is violated if the super-Burnett terms are not accounted for. Section 6 provides a brief summary and a biased outlook on required future work.

2 Mesoscopic Nature of Granular Flows

The present section is devoted to the demonstration of the mesoscopic nature of rapid granular flows [8]. To this end consider a simply sheared stationary monodisperse granular system, whose collisions are characterized by a fixed coefficient of normal restitution, e. The macroscopic velocity field, $\boldsymbol{V}$, is taken to be $\boldsymbol{V} = \gamma y \hat{\boldsymbol{x}}$ where γ is the shear rate, y is the spanwise coordinate and $\hat{\boldsymbol{x}}$ is a unit vector in the streamwise direction.

First, a simple derivation of the basic equation of state for a sheared granular gas is presented. A more accurate derivation can be obtained by employing a Chapman-Enskog analysis of the pertinent Boltzmann equation [6]. The result of the derivation below, as well as of the more accurate derivation, should be considered as a local relation between granular temperature, shear rate and the degree of inelasticity. The granular temperature, T, is defined here as the (ensemble) average of the square of the fluctuating velocity. Below, a minimal model which contains only the essential physics is employed. A granular gas experiences "heating" by shear (as every gas does) and "cooling" by inelasticity. Let γ denote the (local) shear rate (or its norm, in the general case). Following standard (or granular) hydrodynamics, or just mean free path considerations, the rate of heating due to shear is: $\dot{T}_{\text{heating}} = c_1 \nu \gamma^2$, where ν is the kinematic viscosity and c_1 is an $\mathcal{O}(1)$ (volume fraction dependent) constant. Following kinetic theory and/or mean free path arguments: $\nu \propto l_0 \sqrt{T}$, where l_0 is the mean free path. Hence:

$\dot{T}_{\text{heating}} = c_2 l_0 \gamma^2 \sqrt{T}$, where c_2 is another $\mathcal{O}(1)$ prefactor. Next consider the rate of cooling due to the inelasticity. This rate can be read off the equations of motion but here we prefer to present a mean field derivation, since it is physically more transparent. Consider a monodisperse collection of spheres of radius a, whose collisions are characterized by a single coefficient of normal restitution, e. The binary collision between spheres labeled i and j results in the following velocity transformation:

$$\boldsymbol{v}_i' = \boldsymbol{v}_i - \frac{1+e}{2}(\hat{\boldsymbol{n}} \cdot \boldsymbol{v}_{ij})\hat{\boldsymbol{n}} \ , \tag{1}$$

where $(\boldsymbol{v}_i, \boldsymbol{v}_j)$ are the precollisional velocities, $(\boldsymbol{v}_i', \boldsymbol{v}_j')$ are the corresponding postcollisional velocities, $\boldsymbol{v}_{ij} \equiv \boldsymbol{v}_i - \boldsymbol{v}_j$ and $\hat{\boldsymbol{n}}$ is a unit vector pointing from the center of sphere i to that of sphere j at the moment of contact. Simple geometrical considerations reveal that a collision can occur only if $\boldsymbol{v}_{ij} \cdot \hat{\boldsymbol{n}} < 0$. It is easy to show that: $\boldsymbol{v}_i'^2 + \boldsymbol{v}_j'^2 = \boldsymbol{v}_i^2 + \boldsymbol{v}_j^2 - \frac{\epsilon}{2}(\boldsymbol{v}_{ij} \cdot \hat{\boldsymbol{n}})^2$, where $\epsilon \equiv 1 - e^2$ is the *degree of inelasticity*. Thus, the average change of the mean squared velocity, $\delta\langle v^2\rangle$, per particle, per collision equals: $\delta\langle v^2\rangle = -\frac{\epsilon}{4}\langle(\boldsymbol{v}_{ij} \cdot \hat{\boldsymbol{n}})^2\rangle$. Assuming (for simplicity) isotropy of the probability distribution of $\hat{\boldsymbol{n}}$ (and recalling that $\boldsymbol{v}_{ij} \cdot \hat{\boldsymbol{n}} < 0$ when a collision occurs): $\langle(\boldsymbol{v}_{ij} \cdot \hat{\boldsymbol{n}})^2\rangle = \frac{2}{3}\langle v_{ij}^2\rangle$. It is easy to show that for an isotropic and uncorrelated distribution of velocities: $\langle v_{ij}^2\rangle = 2T$. The average change in the kinetic energy (per unit mass per particle) per collision is also the change in the granular temperature induced by the collision (per particle), hence: $\delta T = -\frac{\epsilon}{3}T$. Next, since $\sqrt{T}$ is the typical relative velocity of nearby particles up to a factor of order unity, one may conclude that the mean free time, τ, i.e. the time between consecutive collisions of a particle, is (up to $\mathcal{O}(1)$ prefactors): $\tau = \frac{l_0}{\sqrt{T}}$. Upon dividing δT by τ one obtains the 'time derivative' of the temperature due to collisions, i.e. the cooling rate: $\dot{T}_{\text{cooling}} = -\frac{\epsilon}{3l_0}T^{\frac{3}{2}}$. All in all, the equation of motion for the temperature field, *ignoring diffusive effects*, such as heat conduction, is:

$$\dot{T} = c_2 l_0 \gamma^2 T^{1/2} - c_3 \frac{\epsilon}{l_0} T^{3/2} \ , \tag{2}$$

where c_3 is an $\mathcal{O}(1)$ prefactor replacing the mean field prefactor in the general case. Since linear stability analysis (as well as the intuitively clear fact that shear modes decay slowly for small wavenumbers - a result of momentum conservation) shows [3] that the decay rate of the shear rate (or the vorticity) is far slower than that of the temperature for fluctuations of small enough wavenumber, it follows that γ^2 is a slowly decaying quantity. This fact can be used to show [3, 13] (by solving (2) with γ^2 taken to be constant) that T converges to a value:

$$T = C\frac{\gamma^2 l_0^2}{\epsilon} \tag{3}$$

on a *finite* time scale [3] of $t_1 \propto (\gamma^2\epsilon)^{-1/2}$. Here C is a volume fraction dependent prefactor, whose value at low volume fractions can be shown to equal approximately 0.6 in two dimensions and 3 in three dimensions [6]. Notice that (3) could have been guessed on the basis of dimensional considerations. Indeed, consider a simply sheared stationary system; when the effect of gravity is neglected (or is physically negligible) the only *input* (or externally imposed) parameter which contains time in its dimensions is the inverse shear rate, γ^{-1}. Since the dimension of granular temperature is L^2/t^2, it must be proportional to $\gamma^2 l_0^2$. Now, since when $\epsilon \to 0$ there is no energy sink in the system, its "steady state" temperature must diverge. This does not prove that T diverges like $1/\epsilon$ but this is a first choice (and, as (3) shows, a correct one). Thus, (3) follows from rather general considerations and its validity must be more general than that of a kinetic theory (or mean field method) used to obtain it.

The above relation between the granular temperature, the shear rate and the degree of inelasticity is the basis for the considerations presented next. The change of the macroscopic velocity over a distance of a mean free path, in the y direction, is given by: γl_0. A shear rate can be considered small if γl_0 is small with respect to the thermal speed, $\sqrt{T}$. Here: $\frac{\gamma l_0}{\sqrt{T}} = \frac{\sqrt{\epsilon}}{\sqrt{C}}$, i.e. the shear rate is not 'small' unless the system is nearly elastic (notice that for e.g. $e = 0.9$: $\sqrt{\epsilon} = 0.44$). Thus, except for very low values of ϵ the shear rate is always 'large'. This result alone implies [10] that Chapman-Enskog (CE) expansion of kinetic theory must be carried out beyond the Navier-Stokes order, the lowest next order being the Burnett and super-Burnett orders. While the resulting hydrodynamic equations are suitable for the description of steady states they are generally ill-posed [14]! Thus, unless a resummation scheme that tames this ill posedness is applied, one cannot obtain useful results from the 'higher orders' (except in steady states). The method developed by Rosenau [14] and recently further developed by Slemrod [14] shows some promise in this direction. Another problem associated with the gradient expansion is that higher orders in the gradients may be non-analytic [15], indicating non-locality (a second argument for non-locality is presented below).

Consider next the mean free time, τ, i.e. the ratio of the mean free path and the thermal speed: $\tau \equiv \frac{l_0}{\sqrt{T}}$. Clearly, τ is the microscopic time scale characterizing the system at hand and γ^{-1} is the macroscopic time scale characterizing this system (the simple sheared state). The ratio $\tau/\gamma^{-1} = \tau\gamma$ is a measure of the temporal scale separation in the system. Since $\tau\gamma = \frac{\sqrt{\epsilon}}{\sqrt{C}}$, it is an $\mathcal{O}(1)$ quantity. It follows that (unless $\epsilon \ll 1$) there is no temporal scale separation in this system, *irrespective of its size or the size of the grains.* Consequently, one cannot a-priori employ the assumption of "fast local equilibration" and/or use local equilibrium as a zeroth order distribution function (both for solving the Boltzmann equation and for the study of generalized hydrodynamics [16] of these systems [17]) unless the system is nearly elastic (in which case, scale separation is restored) and (in unsteady states) the rate of change of the external parameters (e.g. the shear rate) is sufficiently slow. The latter condition severely limits the applicability of the hydrodynamic description. For instance, consider the application of these equations to a stability study. As expected (and is well known [3]) some of the eigenvalues of the granular stability problem (including those corresponding to instabilities) must be of the order of the only "input" inverse time scale (in the absence or irrelevance of gravity), i.e. $1/\gamma$. Since, as explained above, $\tau \propto 1/\gamma$, one obtains instabilities whose characteristic times are comparable with the mean free time. If one adopts the conservative view that hydrodynamics should be valid only on time scales which significantly exceed the mean free time, one encounters the paradoxic situation in which the hydrodynamic equations predict instabilities on time scales which they are not supposed to resolve. A third conclusion from this observation, namely the fact that one cannot distinguish between microscopic and macroscopic spatial fluctuations, is presented below.

The mean free time is usually defined as the time between consecutive collisions of a particle. It is clear that mean free times depend on the *relative velocities* of the particles, hence they are Galilean invariant. A simple, textbook-like (and mean field) derivation of the above expression for the mean free time, τ, proceeds as follows: the flux of particles impinging on a given particle is (proportional to) $n\sqrt{T}$,where n is the number density, hence the typical number of collisions per unit time experienced by a particle is $n\sigma_T\sqrt{T}$, where σ_T is the total collision cross section of two particles, and thus the mean free time is proportional to $\frac{1}{n\sigma_T\sqrt{T}}$, which, following the standard definition of the mean free path, l_0, also equals: $\frac{l_0}{\sqrt{T}}$. During a mean free time a 'typical' particle traverses a distance that is determined by its *absolute speed.* This distance *is* the mean free path. Thus, the mean free path, l, is given by $u^*\tau$, where u^* is the average speed of a particle, a quantity that depends on the frame of reference! In the case of a simple shear flow, the velocity, $\boldsymbol{u}$, of a particle equals : $\boldsymbol{u} = \gamma y\hat{\boldsymbol{x}} + \boldsymbol{v}_{th}$, where $\boldsymbol{v}_{th}$ is the thermal component of the velocity (the average of $\boldsymbol{v}_{th}^2$ being T). Assuming statistical independence of the thermal and average velocities, the steady state average of u^2 is given by: $\gamma^2y^2 + T$, hence the typical speed, u^*, of a particle, can be taken to be: $u^* = \sqrt{\gamma^2y^2 + T}$. It follows that the mean free path, as a function of the spanwise coordinate, y, is given by: $l(y) = \sqrt{\gamma^2y^2 + T}\,\tau = \sqrt{\gamma^2y^2 + T}\,\frac{l_0}{\sqrt{T}}$. At values of y at which the speed is subsonic (following the above considerations this happens when $|y|$ is less than l_0) one can neglect γ^2y^2 with respect to T, in which case $l \approx l_0$. However, when $|y| > |l_0|$, in particular when $|y| \gg l_0$, the thermal speed is far smaller than the average speed (i.e. the flow is supersonic) and in this case: $l(y) \approx l_0\frac{\gamma|y|}{\sqrt{T}} = \frac{\sqrt{\epsilon}}{\sqrt{C}}|y| = \frac{\sqrt{\epsilon}}{\sqrt{C}}\frac{|y|}{l_0}l_0$, i.e. the true mean free path is (much) larger than the equilibrium mean free path. Moreover it is of macroscopic dimensions, being an $\mathcal{O}(1)$ quantity times $|y|$; in particular, if the system is wide enough (in the spanwise direction) the mean free path can exceed the length of the system (in the streamwise direction). This implies that the considered system has long range correlations, unless

ϵ is small enough for a given system size, and this fact may invalidate the hydrodynamic equations, unless additional fields are used. Moreover, following the above considerations, a finite homogeneous sheared system can "tell apart" different values of y since the ratio of the mean free path to the size of the system is y dependent. A physical manifestation of this fact is provided by the strong y dependence of the rms of the fluctuations of the collisional stress (a Galilean invariant quantity) which has been observed in simulations of a simply sheared granular system [8]. Several additional remarks are in order here. One can, in principle, define a local mean free path in a (Lagrangian) frame in which the local average velocity is (instantaneously) zero. Indeed, kinetic derivations (such as the Chapman-Enskog expansion) are performed 'around' the local macroscopic velocity. However, these expansions assume the existence of a state of local equilibrium as a zeroth order distribution function; as mentioned above, local equilibration is a slow process in granular systems and thus it may not occur on the required macroscopic time scale. Furthermore, when finite systems are considered, the fact that a mean free path, as defined here, can be comparable to the system size (this actually defines the Knudsen regime [18]) one expects different physics than when the ratio of these two lengths is far from order one. Also, when free paths are large, perturbations applied at a given point in a system may travel a long distance, thus creating long range correlations. Indeed, long range correlations have been observed (and theorized upon) in granular systems [19]. In strongly sheared elastic systems one expects similar phenomena. One of the known results in this case is the existence of long range atomic ordering [11]; a similar phenomenon seems to exist in sheared flows of granular systems and it is presently under investigation.

It is perhaps important to reiterate that rapid granular flows are typically supersonic. There are two facets to this property: the fact that the velocity field is supersonic with respect to the boundary and the fact that the typical fluctuating speed is small with respect to the change of the macroscopic velocity field over the distance of a mean free path. The source of this property is the inelasticity of the collisions. Consider e.g. two particles moving in the same direction in such a way that they can collide. A collision between these particles reduces the relative velocity but does not reduce the sum of their momenta, by momentum conservation. Thus the fluctuating part of the velocity is reduced i.e. the granular temperature is lowered while the 'average' velocity of these particles remains unaffected.

3 Scale Dependence of Stresses and Fluctuations

In the realm of molecular fluids (when they are not under very strong thermal or velocity gradients) there is a range, or *plateau*, of scales, which are larger than the mean free path and far smaller than the scales characterizing macroscopic gradients, and which can be used to define "scale independent" densities (e.g. mass density, momentum density, energy density or temperature) and fluxes (e.g. stresses, heat fluxes). Such *plateau* is virtually non-existent in systems in which scale separation is weak and therefore these entities can be scale dependent. By way of example, the "eddy viscosity" in turbulent flows is a scale dependent (or resolution dependent) quantity, since in the inertial range of turbulence there is no scale separation. There is a plenitude of "rheological materials" in which the lack of scale separation is associated with scale dependence of stresses and other fields.

The scale dependent entity discussed below is the stress tensor. For simplicity we shall mostly discuss the kinetic part of the stress tensor, $\boldsymbol{\tau}^k$, which dominates at low volume fractions. The kinetic theoretical expression for this tensor is: $\tau^k_{\alpha\beta} = \langle v'_\alpha v'_\beta \rangle$, where $\langle A \rangle$ is the ensemble average of A and $\boldsymbol{v}'$ is the fluctuating part of the velocity.

It can be shown [9, 20] that the stress field can be defined for single realizations (hence one does not need to invoke the notion of an ensemble; the latter may not be known in some systems such as granular systems) in such a way that the standard continuum equation of motion, $\rho \frac{D}{Dt} V_\alpha = -\frac{\partial}{\partial r_\beta} \tau_{\alpha\beta}$, holds (with obvious notation).

Let $\{\boldsymbol{r}_i(t); \boldsymbol{v}_i(t); m_i\}$ be the center of mass coordinates and the corresponding velocities, $\dot{\boldsymbol{r}}_i(t) = \boldsymbol{v}_i(t)$, and masses of a set of N particles, indexed by $\{i; 1 \le i \le N\}$. Let $\phi(\boldsymbol{R})$ be a spatial coarse graining function (or a weight function) which possesses the following properties: (i) It is a scalar positive semidefinite function, (ii) its integral over space is unity (normalization), (iii) it has a single maximum at $\boldsymbol{R} = 0$ and no other extrema, (iv) it has at least one derivative (this includes the possibility of the derivative being a generalized function), and (v) it has a well defined 'width' (e.g. the average of $|\boldsymbol{R}|$), which defines the *spatial coarse graining scale*, w. Let $F(t)$ be a temporal coarse graining function having the properties (i)-(v), with $\boldsymbol{R}$ being replaced by t. It can be shown [9, 20] that when the coarse grained mass and momentum densities, are defined by $\rho(\boldsymbol{r}, t) = \int dt' F(t - t') \sum_i m_i \phi(\boldsymbol{r} - \boldsymbol{r}_i(t'))$ and $\boldsymbol{p}(\boldsymbol{r}, t) = \int dt' F(t - t') \sum_i m_i \boldsymbol{v}_i(t') \phi(\boldsymbol{r} - \boldsymbol{r}_i(t'))$, respectively, and the macroscopic velocity is defined by $\boldsymbol{V}(\boldsymbol{r}, t) \equiv \frac{\boldsymbol{p}(\boldsymbol{r},t)}{\rho(\boldsymbol{r},t)}$, one obtains a closed expression for the stress tensor. The kinetic part of the stress tensor is then given by:

$$\tau^k_{\alpha\beta}(\boldsymbol{r}, t) = \int dt' F(t - t') \sum_i m_i v'_{i\alpha}(\boldsymbol{r}, t, t') v'_{i\beta}(\boldsymbol{r}, t, t') \phi(\boldsymbol{r} - \boldsymbol{r}_i(t')) \,, \quad (4)$$

where the fluctuating velocity of a particle i is defined by: $\boldsymbol{v}'_i(\boldsymbol{r}, t, t') \equiv \boldsymbol{v}_i(t') - \boldsymbol{V}(\boldsymbol{r}, t)$. Notice that the fluctuation of the velocity of a particle i is defined with respect to $\boldsymbol{V}$ at the spatio-temporal "coarse graining center" $\{\boldsymbol{r}, t\}$ and not with respect to $\boldsymbol{V}(\boldsymbol{r}_i(t'), t')$, else the formula for the stress would not have been compatible with the general equations of continuum mechanics. Eq. (4) is also compatible with standard practice in computer simulations: one chooses a coarse graining box, calculates its center of mass velocity and subtracts this velocity from the velocity of every particle in the box to obtain the fluctuating velocities. Consider first spatial coarse graining alone (choosing $F(t) = \delta(t)$). Define: $\boldsymbol{v}_i''(t) \equiv \boldsymbol{v}_i(t) - \boldsymbol{V}(\boldsymbol{r}_i(t), t)$ to be the fluctuation of the velocity of a particle with respect to the average velocity at its instantaneous position and let $\boldsymbol{V}_i'(\boldsymbol{r}_i(t), \boldsymbol{r}, t) = \boldsymbol{V}(\boldsymbol{r}_i(t), t) - \boldsymbol{V}(\boldsymbol{r}, t)$ be the difference between the average velocity at $\boldsymbol{r}_i(t)$ and the coarse graining center $\boldsymbol{r}$. Clearly: $\boldsymbol{v}' = \boldsymbol{v}'' + \boldsymbol{V}'$ and $\boldsymbol{V}' \neq 0$ when the velocity is not uniform. It is easy to see that the above decomposition yields two contributions to the kinetic stress tensor: the first is the (spatio-temporal) avarage of the product of velocity fluctuations (with respect to the local velocity field) and it corresponds to the standard (e.g. kinetic) definition of the kinetic stress tensor, and the second contribution is proportional to a product of gradients of the macroscopic velocity field (multiplied by tthe square of the coarse graining scale, w). As an example of an application of this decomposition consider the case of simple shear flow $\boldsymbol{V} = \gamma y \hat{\boldsymbol{x}}$. Estimating the first contribution by the kinetic result (assuming dilute flow) it follows that the ratio of the contribution of $V_x' V_x'$ to the contribution of $\boldsymbol{v}_x'' \boldsymbol{v}_x''$ to the xx component of the stress tensor is proportional to $\frac{\gamma^2 \rho w^2}{\rho T} = \frac{\gamma^2 w^2}{T}$, where w is the coarse graining scale. In molecular systems there is usually a 'wide' range ('plateau') of values of w for which $\gamma^2 w^2 < T$ and in which the contribution of $\boldsymbol{v}_x'' \boldsymbol{v}_x''$ to τ_{xx}, i.e. ρT, is dominant. For instance, a typical value for air at STP is $T \approx 500$ $\mathrm{m}^2/\mathrm{sec}^2$; taking $\gamma = 1\ \mathrm{sec}^{-1}$ the condition $\gamma^2 w^2 < T$ translates into $w < 500$ m, hence this condition corresponds to a very wide plateau and it poses no practical restrictions. In contrast, in granular gases (due to (3)) this ratio is proportional to $\frac{w^2 \epsilon}{l_0^2}$, a quantity that usually exceeds unity for $w > l_0$, rendering the contribution of $V_x' V_x'$ dominant. In other words, the average velocity changes significantly over the scale of a mean free path and this is a source of 'velocity fluctuations' that contribute to the stress tensor. It is important to reiterate that this result is not a consequence of an arbitrary choice of coarse graining but rather an outcome of the standard definitions of the macroscopic fields. The only freedom in the above definitions is in the choice of the coarse graining functions and different choices (provided they are 'reasonable') yield qualitatively similar results.The above results and related results concerning other components of the stress tensor have been corroborated by numerical simulations [9].

Numerical simulation of sheared (dilute) granular systems reveal [9, 21] that the time series for τ_{yy}^k "looks" intermittent, much like in the experimental result presented in [22]. Though, unlike in the simulations, the cited experiments were performed on dense systems, the agreement with the experiments is due to the fact that the physics underlying this "intermittency" is the same, i.e. the lack of scale separation. In other words, single collisions, which are usually averaged over in molecular systems, appear as "intermittent events" in granular systems as they are separated by macroscopic (and experimentally resolvable) times. A numerical study [9] of the time correlation function of τ_{yy}^k reveals that it decays exponentially, the correlation time being $t_{cor} = \mathcal{O}(1/\gamma)$, which is also a microscopic time, as explained above. Similar behavior is displayed by τ^k_{xx} and τ^k_{xy}. These results hold for a variety of choices of the spatial coarse graining scales. The collisional stress (which dominates at relatively high volume fractions) possesses similar properties (here one needs temporal coarse graining as well); in particular, the corresponding correlation time is $\mathcal{O}(1/\gamma)$. The $1/f^2$ like decay of the spectrum of fluctuations, observed experimentally [22] could be a manifestation of the 'high frequency' tail of the Lorentzian corresponding to the above exponential decay of correlations. One may thus conclude that the "intermittent" stress fluctuations are truly microscopic fluctuations of the kind that exists in every many body system, though their nature may vary from one system to another; due to the lack of scale separation in granular systems these fluctuations are also macroscopic and observable in macroscopic measurements.

4 Remarks on the Construction of a Kinetic Theory of Granular Gases

One of the main problems one encounters when developing a perturbative approach to the kinetics of rapid granular flows is the absence of a finite temperature equilibrium state in free (unforced) systems. Indeed, when a granular gas is left to its own fate, its energy decays (asymptotically) to zero due to the inelasticity, i.e. the only "equilibrium state" is that of vanishing temperature. It is obviously inconvenient to employ such a state as a zeroth order in a perturbation theory for a system at a finite granular temperature. One solution of this problem is to devise a perturbation theory around a decaying granular flow [23]. A different way is based on the observation that in the limit of vanishing gradients (formally, the Knudsen number, K) *and* inelasticity, a granular gas becomes elastic and it possesses an equilibrium state [6]. Therefore, one can employ K and ϵ as small parameters in a perturbation expansion applied to the pertinent Boltzmann equation. One may worry though that the limit of $\epsilon \to 0$ and $K \to 0$ may be singular. The following argument (as well as detailed calculations [6]) indicates that this limit is not singular: consider the physically idealized case in which the degree of inelasticity is so small that the energy decays by less than say 1% in 1000 collisions per particle. Since typical (local) equilibration requires only a few collisions per particle [24], the system "converges" rather rapidly to a near-local-equilibrium state and the effect of inelasticity, if the system is unforced, is to produce a slow decay of the granular temperature. This extreme example should not be taken literally: it does not indicate the the expansion in powers of ϵ is meaningful only for extremely low values of ϵ in the same way that the $K \to 0$ limit in which an elastic system tends to a state of equilibrium does not imply that e.g. the Navier-Stokes equations are limited to infinitesimal gradients only. The above argument has only been presented in order to demonstrate that the limit of vanishing inelasticity is smooth, an observation that enables to expand the single particle distribution function in powers of ϵ. The above argument does not actually hold for the tail of the distribution function [24], i.e. for large, with respect to thermal, fluctuating velocities, as its dynamics is usually much slower than that of the bulk of the distribution function. However, the tail does not affect low order moments (when ϵ is not too large) hence the fact that the tail is not "near equilibrium" is not consequential for the nature of the hydrodynamic equations. It can be shown that the perturbative expansion for the single particle distribution function, $f(\boldsymbol{v})$, in powers of ϵ, breaks down [6, 24] when $v/\sqrt{T} > 1/\sqrt{\epsilon}$ and that (in the unforced case) f crosses over from a near-Gaussian dependence on v to an exponential dependence [24] when $v/\sqrt{T} \approx 1/\epsilon$. Consequently, when ϵ is $\mathcal{O}(1)$ the low moments of f (which determine the hydrodynamic transport coefficients) are affected by the tail of the distribution and the (generalized) Chapman-Enskog expansion is no longer a reliable method for obtaining the transport properties.

The expansion of the solution of the Boltzmann equation in ϵ and K begets constitutive relations which differ, both qualitatively and quantitatively, from those obtained in previous studies [25–28] (see, however [23]). In particular, the Navier-Stokes (order) terms have a different dependence on the degree of inelasticity and number density than in previously derived constitutive relations; for instance the expression for the heat flux contains a term which is proportional to $\epsilon\nabla \log n$, where ϵ is a measure of the degree of inelasticity and n denotes the number density. This contribution to the heat flux is of zeroth order in the density; a similar term, i.e. one that is proportional to $\epsilon\nabla n$, has been previously obtained by using the Enskog correction [28], but this term is $\mathcal{O}(n)$ and it vanishes in the Boltzmann limit. Some minor quantitative differences between our results and previous ones exist as well. These are due to the fact that in our work an isotropic correction to the leading Maxwellian distribution, which has not been considered before, is taken into account and also because the full dependence of the corrections to the Maxwellian distribution on the (fluctuating) speed is computed and employed in the calculations.

In spite of the above explanation, it is rather remarkable that, given the differences between granular and elastic systems, that methods borrowed from the kinetic theory of gases are so useful in the realm of granular gases. The agreement of constitutive relations derived by employing these methods with numerical and physical experiments for which ϵ cannot be considered to be small is truly remarkable; it is possible that this ϵ expansion is an asymptotic expansion (like many perturbative expansions), in which case one expects a larger range of validity than naively expected.

The normal stress difference (between τ_{xx} and τ_{yy}, normalized by their average), calculated by employing the above formulation, equals 0.45 for $e =$ 0.8 and 0.88 for $e = 0.6$, in good agreement with numerical results [30]: 0.42

and 0.86, respectively (for a volume fraction of $\nu = 0.025$). In general the normal stress difference in granular gases is $\mathcal{O}(1)$ [6].

As mentioned, the double expansion, in ϵ and the Knudsen number, K, has been employed [6] to obtain constitutive relations for granular gases up to the Burnett order. The leading order (elastic) viscous contribution to the stress tensor is $\mathcal{O}(K)$ (also known as the Navier-Stokes order) and the leading inelastic correction is $\mathcal{O}(\epsilon K)$. In a steady shear flow, at a given value of the granular temperature, T, it follows from Eq. (3) that: $\epsilon \propto \gamma^2 = \mathcal{O}(K^2)$, thus the leading inelastic correction is also $\mathcal{O}(K^3)$, hence one needs to calculate the super-Burnett, i.e. $\mathcal{O}(K^3)$, contributions alongside the leading order inelastic corrections to render the equations of motion appropriate for the description of steady states. The way Jenkins and Richman [26] implemented the Grad method of moments [29] is directly suitable for steady states, as in their derivation they neglected the time dependence of the mean free time (which depends on the temperature T, a fact which introduces additional ϵ dependence in the constitutive relations in the time dependent case); indeed, when a steady state is considered, the $\mathcal{O}(\epsilon)$ corrections to the elastic result already incorporate the super-Burnett contributions, provided γ^2 (or, in general, K^2) is taken to be $\mathcal{O}(\epsilon)$ in the expansion. Since the result of such a derivation does not describe unsteady states, it cannot be used for stability analyses. As mentioned, a complete hydrodynamic description requires the calculation of the Burnett and super-Burnett coefficients. Similar statements hold for the other transport coefficients and, as shown below, for the boundary conditions, as well.

5 Boundary Conditions

A systematic method for deriving boundary conditions for granular gases (which is also relevant to molecular gases) has been developed [31, 32]. Only the essential ingredients of the method are presented below. The method is based on an expansion in the number of collisions with the boundary, as explained below.

Consider a solid boundary situated at $z = 0$ and a (granular) gas occupying $z > 0$. The solution of the pertinent Boltzmann equation can be written as $f = f_{ce} + f_0\Phi_w = f_0(1 + \Phi_{ce} + \Phi_w)$, where f_0 is a local equilibrium distribution function, Φ_w represents the effect of the wall and $f_0(1 + \Phi_{ce})$ is the Chapman-Enskog, or in principle, an exact solution of the Boltzmann equation far enough from the boundary. The function Φ_w should vanish far away from the boundary (in practice, a few mean free paths away from it). Notice that the value of Φ_{ce} at the boundary is determined by extrapolating the Chapman-Enskog solution to the boundary, since the Chapman-Enskog solution itself is not correct near the boundary. It is convenient to choose a frame of reference in which the solid boundary is stationary (when the boundary moves at constant velocity this involves only a Galilean transformation; below we implicitly assume this to be the case). Inside the domain influenced by the boundary, i.e. the *Knudsen layer*, the macroscopic flow field $\boldsymbol{V}$ (rescaled by the square root of the temperature field, T) is of the order of the Knudsen number (since, by choice of the frame of reference, the boundary is stationary, $\boldsymbol{V}$ should vanish in the absence of gradients), hence it is justified to expand the (extrapolated) Chapman-Enskog solution f_{ce} in the Knudsen layer in powers of the rescaled velocity field.

The boundary conditions are conditions to be satisfied by the hydrodynamic fields extrapolated to the boundary, not by the true values of the these fields there. The role of the boundary conditions is to ensure that the solutions of the hydrodynamic equations outside the Knudsen layer, whose width is a few mean free paths, are compatible with the kinetics near the boundary. In other words, the hydrodynamic equations provide 'outer solutions' that should match the 'inner' kinetic solution next to the boundary. Since Φ_w vanishes when the hydrodynamic fields are space independent it follows that Φ_w is $\mathcal{O}(K)$.

The method we have developed [31] for obtaining boundary conditions [32] is outlined below at $\mathcal{O}(\epsilon^0 K)$; results are presented to $\mathcal{O}(\epsilon K)$. At order $\epsilon^0 K$ it is easy to see that it is sufficient to consider the following linearized Boltzmann equation:

$$v_z \frac{\partial \Phi_w}{\partial z} = \boldsymbol{L}\Phi_w \ , \tag{5}$$

where $\boldsymbol{L}$ is the linearized Boltzmann operator [6, 31]. The solubility conditions (orthogonality relations) for (5) are: $\int d\boldsymbol{v} v_z \psi_i(\boldsymbol{v}) e^{-v^2} \Phi_w(z=0) = 0$, where ψ_i are the invariants of $\boldsymbol{L}$ (i.e. the conserved quantities: $\psi_1 \equiv 1$, $\psi_2 \equiv v_x$, $\psi_3 \equiv v_y$, $\psi_4 \equiv v_z$ and $\psi_5 \equiv v^2$). These relations require that Φ_w does not contribute to the mass, momentum and energy flux in the z direction. The latter fluxes are completely determined by the Chapman-Enskog solution. The physical role of these requirements is to ensure the continuity of the fluxes, i.e. the values of the fluxes in the bulk, given by the Chapman-Enskog expansion, should match the rate of the transfer of the corresponding moments to the boundary. Eq. (5) is not easy to solve since it is a non-trivial integrodifferential equation.

The basis of the method described below is the observation that initial distribution functions converge rapidly (for $\epsilon \ll 1$) to local equilibrium or equilibrium-like distributions (in a matter of a few collisions [24]). Indeed, as the results presented below indicate, this approach is justified since the contributions of multiple particle collisions to the transport coefficients are increasingly smaller. It is convenient to use the Fredholm [33] form of the linearized Boltzmann operator, $\boldsymbol{L}$, i.e. the decomposition: $\boldsymbol{L} = \boldsymbol{A} - q$, where

$$\boldsymbol{A}\Phi = \frac{1}{\pi^{\frac{3}{2}}} \int d\boldsymbol{v}' e^{-v'^2} \left(\frac{2}{R} e^{w^2} - R \right) \Phi(\boldsymbol{v}') \equiv \int d\boldsymbol{v}' K(\boldsymbol{v}, \boldsymbol{v}') \Phi(\boldsymbol{v}') \ , \tag{6}$$

$R = |\boldsymbol{v} - \boldsymbol{v}'|$, $w = \frac{\boldsymbol{v}\times\boldsymbol{v}'}{R}$ and $q(v) = \frac{1}{\sqrt{\pi}}\left(e^{-v^2} + \frac{\sqrt{\pi}}{2}\left(2v + \frac{1}{v}\right)\mathrm{erf}(v)\right)$ is a positive definite function which depends on the speed v alone. The operator $\boldsymbol{A}$ includes the full 'gain term' of $\boldsymbol{L}$ and part of the 'loss term'. The second part of the 'loss term' is q. The function q (approximately) represents the rate of 'loss' of particles, having velocity $\boldsymbol{v}$, due to collisions and it is trivially related to the (velocity dependent) mean free path. The operator $\boldsymbol{A}$ represents the rate of 'creation' of particles with velocity $\boldsymbol{v}$ at a given point in space. Using this decomposition one can transform (5) as follows:

$$\frac{\partial}{\partial z}\left(e^{\frac{q}{v_z}z}\Phi_w\right) = \frac{1}{v_z}e^{\frac{q}{v_z}z}\boldsymbol{A}\Phi_w \ . \tag{7}$$

The solution of (7) can be formally written as follows. When $v_z > 0$:

$$\Phi_w(z) = e^{-\frac{q}{v_z}z}\Phi_w(0) + \frac{1}{v_z}\int_0^z dz' e^{\frac{q}{v_z}(z'-z)}\boldsymbol{A}\Phi_w(z') \ , \tag{8}$$

where only the dependence of Φ_w on z is explicitly spelled out. When $v_z < 0$:

$$\Phi_w(z) = -\frac{1}{v_z}\int_z^\infty dz' e^{\frac{q}{v_z}(z'-z)}\boldsymbol{A}\Phi_w(z') \ . \tag{9}$$

Let $\boldsymbol{P}$ and $\boldsymbol{N}$ be projection operators on the $v_z \geq 0$ and $v_z < 0$ velocity subspaces, respectively (with $\boldsymbol{P} + \boldsymbol{N} = \boldsymbol{I}$). It follows that $\boldsymbol{P}\Phi_w = \boldsymbol{PGP}\Phi_w(0) + \boldsymbol{PQP}\Phi_w + \boldsymbol{PQN}\Phi_w$, and $\boldsymbol{N}\Phi_w = \boldsymbol{NSP}\Phi_w + \boldsymbol{NSN}\Phi_w$, where

$$\boldsymbol{G}\phi = e^{-\frac{q}{v_z}z}\phi \ , \tag{10}$$

$$\begin{aligned}\boldsymbol{Q}\phi &= \frac{1}{v_z}\int_0^z dz' e^{\frac{q}{v_z}(z'-z)}\boldsymbol{A}\phi(z') \\ &= \frac{1}{v_z}\int_0^z dz' \int d\boldsymbol{v}' e^{\frac{q}{v_z}(z'-z)}K(\boldsymbol{v},\boldsymbol{v}')\phi(\boldsymbol{v}',z') \ , \quad \text{and}\end{aligned} \tag{11}$$

$$\begin{aligned}\boldsymbol{S}\phi &= -\frac{1}{v_z}\int_z^\infty dz' e^{\frac{q}{v_z}(z'-z)}\boldsymbol{A}\phi(z') \\ &= -\frac{1}{v_z}\int_z^\infty dz' \int d\boldsymbol{v}' e^{\frac{q}{v_z}(z'-z)}K(\boldsymbol{v},\boldsymbol{v}')\phi(\boldsymbol{v}',z') \ .\end{aligned} \tag{12}$$

The operator $\boldsymbol{NS}$ represents the events in which a particle whose velocity is $\boldsymbol{v}$, with $v_z < 0$, emerges from a collision at a point $z' > z$ and moves to the point z without further collision. Similarly, the operator $\boldsymbol{PQ}$ represents the events in which a particle whose velocity is $\boldsymbol{v}$, with $v_z > 0$, collides at $z' < z$ and proceeds, without further collision, to the point z. The operator $\boldsymbol{G}$ is the propagator corresponding to the motion of a particle from the boundary, $z = 0$, to z, without collision. Straightforward algebra yields:

$$\Phi_w(z) = (\boldsymbol{I} - \boldsymbol{C})^{-1}\boldsymbol{PGP}\Phi_w(0) = (\boldsymbol{I} + \boldsymbol{C} + \boldsymbol{C}^2 + \ldots)\boldsymbol{PGP}\Phi_w(0) \ . \tag{13}$$

where $\boldsymbol{C} \equiv \boldsymbol{NS} + \boldsymbol{PQ}$. The interpretation of (13) is rather simple: the function Φ_w is determined from its value at the boundary, $\Phi_w(0)$, via successive processes of collisions and free motions. For example, the nth order term $\boldsymbol{C}^n \boldsymbol{PGP}\Phi_w(0)$ is the contribution of the particles that come from the boundary (with positive z-component velocity), collide n times, following which their velocity is $\boldsymbol{v}$ and they move without collision to the point z. The next step is to apply the boundary conditions. These can be expressed as follows:

$$(\boldsymbol{v} \cdot \hat{\boldsymbol{n}}(\boldsymbol{x}))f(\boldsymbol{v}, \boldsymbol{x}) = -\int_{\boldsymbol{v}' \cdot \hat{\boldsymbol{n}} < 0} d\boldsymbol{v}'(\boldsymbol{v}' \cdot \hat{\boldsymbol{n}}(\boldsymbol{x}))f(\boldsymbol{v}', \boldsymbol{x})W(\boldsymbol{v}' \to \boldsymbol{v}) \ , \tag{14}$$

where W is the accommodation (or transfer function) which determines the distribution of outgoing velocities of a particle that collides with the boundary with a given velocity and $\hat{\boldsymbol{n}}$ is the vector normal to the boundary (in the case considered here it is the unit vector in the z direction). Upon employing the representation $f = f_0(1 + \Phi_{ce} + \Phi_w)$ one can write the boundary condition in operatorial form as follows:

$$\boldsymbol{P}\Phi_w(0) = \boldsymbol{PRN}\Phi_w(0) + \boldsymbol{PB}f_{ce}(0) \ , \tag{15}$$

where $\boldsymbol{R}$ and $\boldsymbol{B}$ are operators that can be read off the result of the substitution of the decomposition of f in (14). Next, defining the operator Z as the projection on the value of a function at $z = 0$, e.g. $\boldsymbol{Z}\phi(z) = \phi(0)$, one obtains from (13):

$$\boldsymbol{N}\Phi_w(0) = \boldsymbol{ZNS}(\boldsymbol{I} - \boldsymbol{C})^{-1}\boldsymbol{PGP}\Phi_w(0) \ , \tag{16}$$

where use has been made of the identity $\boldsymbol{N}(\boldsymbol{I} - \boldsymbol{C})^{-1}\boldsymbol{P} = \boldsymbol{NS}(\boldsymbol{I} - \boldsymbol{C})^{-1}\boldsymbol{P}$, which follows from $\boldsymbol{NP} = 0$ and $\boldsymbol{NC} = \boldsymbol{NS}$. Substituting the boundary condition (i.e. the expression for $\boldsymbol{P}\Phi_w(0)$ from (15)) in (16) one obtains:

$$\boldsymbol{N}\Phi_w(0) = \boldsymbol{ZNS}(\boldsymbol{I} - \boldsymbol{C})^{-1}\boldsymbol{PGP}(\boldsymbol{PRN}\Phi_w(0) + \boldsymbol{PB}f_{ce}(0)) \ . \tag{17}$$

Equation (17) can now be solved for $\boldsymbol{N}\Phi_w(0)$ to yield:

$$\boldsymbol{N}\Phi_w(0) = (\boldsymbol{I} - \boldsymbol{ZNS}(\boldsymbol{I} - \boldsymbol{C})^{-1}\boldsymbol{PGPR})^{-1}\boldsymbol{ZNS}(\boldsymbol{I} - \boldsymbol{C})^{-1}\boldsymbol{PGPB}f_{ce}(0) \ . \tag{18}$$

Now that an expression for the wall contribution to the distribution function is available once can obtain boundary conditions. To this end recall that the solubility conditions for (5) follow from the fact that the set of invariants $\{\Psi_i\} \equiv \{1, v_x, v_y, v_z, v^2\}$ are eigenfunctions of $\boldsymbol{L}$ corresponding to vanishing eigenvalues. It thus follows that:

$$\int \left(\Psi_i f_0\right)\left(v_z \frac{\partial}{\partial z}\Phi_w\right) d\boldsymbol{v} = \int \left(\Psi_i f_0\right)\left(\boldsymbol{L}\Phi_w\right) = \int \left(\boldsymbol{L}\Psi_i\right) f_0 \Phi_w d\boldsymbol{v} = 0 \ , \tag{19}$$

i.e.

$$\frac{\partial}{\partial z}\int \Psi_i f_0 v_z \Phi_w dv = 0 \; , \tag{20}$$

and hence (since $\Phi_w \overset{z\to\infty}{\longrightarrow} 0$):

$$\int \Psi_i v_z \Phi_w f_0 dv = 0 \; . \tag{21}$$

The physical significance of (21) is that the fluxes are continuous, the value of each flux in the bulk equals its value at the boundary, hence Φ_w cannot contribute to any of the fluxes. The condition (21) are conditions on f_{ce} since Φ_w has been expressed as a functional of f_{ce}. In practice, the orthogonality (or solubility) conditions create relations between the fields and their derivatives. A similar formulation holds for the case of an inelastic gas.

Upon employing the above formulation one obtains for the case of an inelastic gas and a diffusely (but elastically) reflecting boundary that the (slip) velocity parallel to the boundary is given by: $V_x = \alpha l_0 \frac{\partial V_x}{\partial z}$, where $\alpha \approx 0.728 + 0.130\epsilon$, to second order in the collisions. When the boundary is characterized by a degree of inelasticity (for the normal part of the velocity), ϵ_w, and a thermal gradient in the z direction is present (as must be the case for an energy absorbing boundary), one obtains the following result: $V_z = -\zeta l_0 \sqrt{T}\left(\frac{\partial \log n}{\partial z} + \frac{1}{2}\frac{\partial \log T}{\partial z}\right)$ where $\zeta \approx 0.044\epsilon$, to lowest order in the collisions. This boundary condition for V_z is quite surprising. It implies that V_z does not necessarily vanish in the general case, unless a specific relation between the gradients of the number density and granular temperature is satisfied. It is easy to check that the above expression for V_z does not vanish in the case of steady shear, hence mass conservation seems to be violated. This result pertains only to inelastically colliding systems as V_z is predicted to vanish for $\epsilon = 0$. The resolution of this "paradox" can be found by noting (again) that in a steady sheared state the orders $K\epsilon$ (which is the order of the above expression for V_z) and K^3 are the same, hence a correction that is of super-Burnett order should be added to the above expression for V_z. When this is done the value of V_z vanishes in steady states, as it should. In the same situation, the boundary condition for the temperature, T, is $\epsilon_w T = \beta l_0 \frac{\partial T}{\partial z} + \delta l_0 \frac{T}{n}\frac{\partial n}{\partial z}$, where $\beta \approx 2.671 + 1.945\epsilon$ and $\delta \approx 3.810\epsilon$, to second order in the collisions.

6 Conclusion

While some properties of granular gases, such as collapse and clustering are direct consequences of inelasticity, others are indirect consequences as they follow from the lack of scale separation alone. Consequently, they also exist, in a much weakened form in elastic or molecular gases but they can be prominent there only under extreme conditions.

Near elastic systems can be studied by a generalized Chapman-Enskog expansion, yielding a near-standard hydrodynamic description. It is possible to derive Green-Kubo relations for granular gases [17] using methods that are similar to those employed in the derivation of these relations for molecular gases, thus enabling the extension of the constitutive relations to moderate densities in a systematic fashion.

The fact that constitutive relations for granular gases, derived by employing a Boltzmann kinetic theory, seem to agree with Molecular Dynamics measurements [34], even for large values of ϵ, is quite surprising. It is possible that this agreement is due to the fact that these constitutive relations have been studied only for steady states, where, as mentioned, the derivation in [26] seems appropriate, as it indirectly incorporates the effect of the super-Burnett terms. Another reason for this agreement could be the fact that 'higher order' terms in the 'ϵ expansion' have small prefactors. In any case this problem is worth further study; in particular it is important to derive equations of motion which apply both to steady and unsteady situations and, as mentioned, this requires the evaluation of the super-Burnett contributions.

The possible non-local nature of the effective interactions in granular gases is indicated by several observations: (i) The mean free paths can be macroscopic. (ii) Higher orders are required in the gradient expansion of kinetic theory. (iii) In the general (density) case such higher orders may be non-analytic in the gradients (a similar result is obtained when the Chapman-Enskog expansion is "resummed"). (iv) Long range correlations have been shown to exist in free granular gases. (v) Long range atomic ordering has been observed in strongly sheared elastic systems. These observations may imply that a full description of the dynamics of granular gases should include a larger set of fields than standard hydrodynamics does. As a matter of fact, the hydrodynamic description of one dimensional granular systems [35] requires the introduction of an additional field which measures the anisotropy of the distribution function (and is essentially the heat flux); the introduction of such an additional field has been also proposed in studies of atomic systems under large shear or thermal gradients [10].

In summary, quite a few arguments and observations suggest that rapid granular flows are mesoscopic systems and that their theoretical description requires the study of situations which would qualify as extreme in molecular systems. The scale dependence of stresses and the observability of microscopic fluctuations have obvious experimental ramifications. Finally, the problem of strongly inelastic granular gases remains quite open.

Acknowledgements

This work has been partially supported by the National Science Foundation, the Department of Energy, the United-States - Israel Binational Science Foundation (BSF) and the Israel Science Foundation (ISF).

References

1. J. M. Jaeger and S. R. Nagel, Granular Solids, Liquids and Gases, Rev. Mod. Phys. **68**, 1259–1273 (1996).
2. A. Einstein, Investigations on the Theory of the Brownian Movement, edited by R. Fürth, translated into English by A. D. Cowper, Dover (1956).
3. (i) M. A. Hopkins, M. Y. Louge, Inelastic Microstructure in Rapid Granular Flows of Smooth Disks, Phys. Fluids A **3**(1), 47–57 (1991); (ii) I. Goldhirsch, G. Zanetti, Clustering Instability in Dissipative Gases, Phys. Rev. Lett. **70**, 1619–1622 (1993); (iii) I. Goldhirsch, M. L. Tan and G. Zanetti, A Molecular Dynamical Study of Granular Fluids I: The Unforced Granular Gas in Two Dimensions, J. Sci. Comp. **8**(1), 1–40 (1993); (iv) M. L. Tan, I. Goldhirsch, Intercluster Interactions in Rapid Granular Shear Flows, Phys. Fluids **9**(4), 856–869 (1997); (v) I. Goldhirsch, M. L. Tan, The Single Particle Distribution Function for Rapid Granular Shear Flows of Smooth Inelastic Disks, Phys. Fluids **8**(7), 1752–1763 (1996).
4. (i) S. McNamara, W. R. Young, Inelastic Collapse and Clumping in a One-Dimensional Granular Medium, Phys. Fluids **A4**, 496–504 (1992); (ii) S. McNamara, W. R. Young, Kinetics of a One-Dimensional Granular Medium in the Quasielastic Limit, Phys. Fluids **A5**(1), 34–35 (1993); (iii) S. McNamara, W. R. Young, Inelastic Collapse in Two Dimensions. Phys. Rev. E **50**, R28–R31 (1994).
5. L. P. Kadanoff, Built Upon Sand: Theoretical Ideas Inspired by Granular Flows. Rev, Mod. Phys. **71**(1), 435–444 (1999).
6. (i) J. T. Jenkins, M. W. Richman, Plane Simple Shear of Smooth Inelastic Circular Disks: the Anisotropy of the Second Moment in the Dilute and Dense Limits, J. Fluid. Mech. **192**, 313–328 (1988); (ii) I. Goldhirsch, N. Sela, and S. H. Noskowicz, Kinetic Theoretical Study of a Simply Sheared Granular Gas - to Burnett Order, Phys. Fluids **8** (9), 2337–2353 (1996); (iii) I. Goldhirsch, N. Sela Origin of Normal Stress Differences in Rapid Granular Flows, Phys. Rev. E **54**(4), 4458–4461 (1996); (iv) N. Sela, I. Goldhirsch, Hydrodynamic Equations for Rapid Flows of Smooth Inelastic Spheres – to Burnett Order, J. Fluid Mech. **361**, 41–74 (1998), and refs. therein.
7. (i) D. Burnett The Distribution of Molecular Velocities and the Mean Motion in a Nonuniform Gas, Proc. Lond. Math. Soc. **40**, 382–435 (1935); (ii) Textbook on kinetic theory: M. K. Kogan, Rarefied Gas Dynamics, Plenum Press, New-York (1969); (iii) S. Chapman, T. G. Cowling, The Mathematical Theory of Nonuniform Gases, Cambridge University Press, Cambridge (1970).
8. M. L. Tan, I. Goldhirsch, Rapid Granular Flows as Mesoscopic Systems, Phys. Rev. Lett. **81**(14), 3022–3025 (1998).
9. B. J. Glasser, I. Goldhirsch, Scale Dependence, Correlations and Fluctuations of Stresses in Rapid Granular Flows, Unpublished (1999).
10. e.g.: (i) L. C. Woods, Transport Processes in Dilute Gases Over the Whole Range of Knudsen Numbers. Part I: General Theory, J. Fluid. Mech. **93**, 585–607 (1979); (ii) K. Fiscko, D. Chapman, Comparison of Burnett, Super-Burnett and Monte-Carlo Solutions for Hypersonic Shock Structure, Prog. Aerounau. Astronaut. **118**, 374–395 (1989).
11. J. F. Lutsko, Molecular Chaos, Pair Correlations and Shear-Induced Ordering of Hard Spheres, Phys. Rev. Lett. **77**, 2225–2228 (1996).

12. P. B. Umbanhowar, F. Melo, H. L. Swinney, Localized Excitations in a Vertically Vibrated Layer, Nature **382**, 793–796 (1996).
13. P. K. Haff, Grain Flow as a Fluid Mechanical Phenomenon, J. Fluid Mech. **134**, 401–948 (1983).
14. (i) A. V. Bobylev, Exact Solutions of the Nonlinear Boltzmann Equation and the Theory of Maxwell Gas Relaxation, Theor. Math. Phys. **60**(2) 280–310 (1984); (ii) A. N. Gorban, and I. V. Karlin, Transport Theory and Statistical Physics **21**, 101–117 (1992). (iii) P. Rosenau, Extending Hydrodynamics via the Regularization of the Chapman-Enskog Expansion, Phys. Rev. A **40**, 7193–7196 (1989); (iv) M. Slemrod, Renormalization of the Chapman-Enskog Expansion: Isothremal Fluid Flow and Rosenau Saturation, J. Stat. Phys. **91**(1-2) 285–305 (1998).
15. M. H. Ernst, J. R. Dorfman, Nonanalytic Dispersion Relations for Classical Fluids, J. Stat. Phys. **12**(4), 311–361 (1975).
16. (i) I. Oppenheim, Nonlinear Response Theory, in "Correlation Functions and Quasiparticle Interactions in Condensed Matter, ed. J. Woods Halley, Plenum Press, 235–258 (1978); (ii) H. Mori, Time-Correlation Functions in the Statistical Mechanics of Transport Processes, Phys. Rev. **111**, 694–706 (1958); (iii) H. Mori, Statistical Mechanical Theory of Transport in Fluids, Phys. Rev. **112**, 1829–1842 (1958).
17. I. Goldhirsch, T. P. C. van Noije, Green-Kubo Relations for Granular Fluids, Phys. Rev. E **61**, 3241–3244 (1999).
18. cf. e.g. (i) S. Harris, Introduction to the Theory of the Boltzmann Equation. Holt, Reinhart and Winston, N.Y. (1971); (ii) C. Cercignani, Theory and Application of the Boltzmann Equation, Scottish Acad. Press, Edinburgh and London (1975).
19. T. P. C. van Noije, M. H. Ernst, R. Brito, and J. A. G. Orza, Mesoscopic Theory of Granular Fluids, Phys. Rev. Lett. **79**, 411–414 (1997).
20. (i) A. I. Murdoch, On Effecting Averages and Changes of Scale via Weighting Functions, Arch. Mech. **50**, 531–539 (1998); (ii) M. Babic, Average Balance Equations for Granular Materials, Int. J. of Eng. Sci. **35**, 523–548 (1997).
21. S. B. Savage, Disorder, Diffusion and Structure Formation in Granular Flows, In: Disorder and Granular Media, D. Bideau and A. Hansen eds., Random Materials and Processes Ser., 255–285, North Holland (1993).
22. B. Miller, C. O'Hern, and R. P. Behringer, Stress Fluctuations for Continuously Sheared Granular Materials, Phys. Rev. Lett. **77**, 3110–3113 (1996).
23. (i) J. J. Brey, F. Moreno, and J. W. Dufty, Model Kinetic Equations for Low Density Granular Flow, Phys. Rev. E **54**(1), 445–456 (1996); (ii) J. J. Brey, J. W. Dufty, C. S. Kim, and A. Santos, Hydrodynamics for Granular Flow at Low Density, Phys. Rev. E **58**, 4638–4653 (1998).
24. (i) I. Kuscer, and M. M. R. Williams, Relaxation Constants of a Uniform Hard-Sphere Gas, Phys. Fluids. **10**, 1922–1927 (1967); (ii) F. Schurrer, and G. Kugerl, The Relaxation of Single Hard Sphere Gases, Phys. Fluids. **A 2**, 609–618 (1990); (iii) S. E. Esipov, and T. Pöschel, The Granular Phase Diagram, J. Stat. Phys. **86**, 1385–1395 (1997); (iv) J. J. Brey, D. Cubero, and M. J. Ruiz-Montero, High Energy Tail in the Velocity Distribution of a Granular Gas, Phys. Rev. E **59** 1256–1258 (1999).
25. (i) J. T. Jenkins, S. B. Savage, A Theory for Rapid Granular Flow of Identical, Smooth, Nearly Elastic, Spherical Particles, J. Fluid Mech. **130**, 187–202 (1983);

(ii) C. K. K. Lun, S. B. Savage, D. J. Jeffrey, and N. Chepurniy, Kinetic Theories of Granular Flow: Inelastic Particles in a Couette Flow and Slightly Inelastic Particles in a General Flow Field, J. Fluid Mech. **140** 223–256 (1984); (iii) J. T. Jenkins, M. W. Richman, Grad's 13-Moment System for a Dense Gas of Inelastic Particles, Phys. Fluids **28**, 3485–3494 (1985); (iv) C. K. K. Lun, and S. B. Savage, A Simple Kinetic Theory for Granular Flow of Rough Inelastic Spheres, J. Appl. Mech. **154**, 47–53 (1987).

26. J. T. Jenkins, and M. W. Richman, Plane Simple Shear of Smooth Inelastic Circular Disks: the Anisotropy of the Second Moment in the Dilute and Dense Limits, J. Fluid. Mech. **192**, 313–328 (1988).
27. C. K. K. Lun, Kinetic Theory for Granular Flow of Dense, Slightly Inelastic, Slightly Rough Spheres, J. Fluid Mech. **223**, 539–559 (1991).
28. E. J. Boyle, M. Massoudi, A Theory for Granular Materials Exhibiting Normal Stress Effects Based on Enskog's Dense Gas Theory, Int. J. Engng. Sci. **28**, 1261–1275 (1990).
29. H. Grad, On the Kinetic Theory of Rarefied Gases, Comm. Pure and Appl. Math. **2**, 331–407 (1949).
30. (i) O. R. Walton, R. L. Braun, Stress Calculations for Assemblies of Inelastic Spheres in Uniform Shear, Acta Mech. **63**, 73–86 (1986); (ii) O. R. Walton, R. L. Braun, Viscosity and Temperature Calculations for Shearing Assemblies of Inelastic, Frictional Disks, J. Rheol. **30**, 949–980 (1986).
31. N. Sela, and I. Goldhirsch, Boundary Conditions for Granular and Molecular Gases, Unpublished (1999).
32. Some previous work on boundary conditions for granular gases: (i) J. T. Jenkins, and M. W. Richman, Boundary Conditions for Plane Flows of Smooth, Nearly Elastic, Circular Disks, J. Fluid Mech. **171**, 53–69 (1986); (ii) M. W. Richman, Boundary Conditions Based Upon a Modified Maxwellian Velocity Distribution for Flows of Identical, Smooth, Nearly Elastic Spheres, Acta Mech. **75**, 227–240 (1988); (iii) J. T. Jenkins, Boundary Conditions for Rapid Granular Flow: Flat, Frictional Walls, J. Appl. Mech. **114**, 120–127 (1992); (iv) J. T. Jenkins, and E. Askari, Boundary Conditions for Rapid Granular Flows. J. Fluid Mech. **223**, 497–508 (1991). (v) J. T. Jenkins, *Boundary conditions for collisional grain flows at bumpy, frictional walls*, (in this volume, page 125).
33. C. L. Pekeris, Solution of the Boltzmann-Hilbert Integral Equation, Proc. N. A. S. **41** 661–664 (1955).
34. M. Hopkins, and H. H. Shen, A Monte-Carlo Solution for Rapidly Shearing Granular Flows Based on the Kinetic Theory of Dense Gases, J. Fluid Mech. **244**, 477–491 (1992).
35. N. Sela, and I. Goldhirsch, Hydrodynamics of a One-Dimensional Granular Medium, Phys. Fluids **7**(3), 507–525 (1995).

Granular Gases with Impact-Velocity-Dependent Restitution Coefficient

Nikolai V. Brilliantov[1,2] and Thorsten Pöschel[2]

[1] Moscow State University, Physics Department, Moscow 119899, Russia. e-mail: nbrillia@physik.hu-berlin.de
[2] Humboldt-Universität, Institut für Physik, Invalidenstr. 110, D-10115 Berlin, Germany. e-mail: thorsten@physik.hu-berlin.de, http://summa.physik.hu-berlin.de/~thorsten

Abstract. We consider collisional models for granular particles and analyze the conditions under which the restitution coefficient might be a constant. We show that these conditions are not consistent with known collision laws. From the generalization of the Hertz contact law for viscoelastic particles we obtain the coefficient of normal restitution ϵ as a function of the normal component of the impact velocity v_{imp}. Using $\epsilon(v_{\text{imp}})$ we describe the time evolution of temperature and of the velocity distribution function of a granular gas in the homogeneous cooling regime, where the particles collide according to the viscoelastic law. We show that for the studied systems the simple scaling hypothesis for the velocity distribution function is violated, i.e. that its evolution is not determined only by the time dependence of the thermal velocity. We observe, that the deviation from the Maxwellian distribution, which we quantify by the value of the second coefficient of the Sonine polynomial expansion of the velocity distribution function, does not depend on time monotonously. At first stage of the evolution it increases on the mean-collision time-scale up to a maximum value and then decays to zero at the second stage, on the time scale corresponding to the evolution of the granular gas temperature. For granular gas in the homogeneous cooling regime we also evaluate the time-dependent self-diffusion coefficient of granular particles. We analyze the time dependence of the mean-square displacement and discuss its impact on clustering. Finally, we discuss the problem of the relevant internal time for the systems of interest.

1 Introduction

Granular gases, i.e. systems of inelastically colliding particles, are widely spread in nature. They may be exemplified by industrial dust or interterrestrial dust; the behavior of matter in planetary rings is also described in terms of the granular gas dynamics. As compared with common molecular gases, the steady removal of kinetic energy in these systems due to dissipative collisions causes a variety of nonequilibrium phenomena, which have been very intensively studied (e.g. [1–11]). In most of these studies the coefficient of restitution, which characterizes the energy lost in the collisions, was assumed to be constant. This approximation, although providing a considerable simplification, and allowing to understand the main effects in granular gas dynamics, is not always justified (see also the paper by Thornton in this book [12]). Moreover, sometimes it occurs to be too crude to describe even qualitatively the features of granular gases. Here we discuss the properties of granular gases consisting of viscoelastically colliding particles which implies an impact-velocity dependent restitution coefficient. The results are compared with results for gases consisting of particles which interact via a constant restitution coefficient and we see that the natural assumption of viscoelasticity leads to *qualitative* modifications of the gas properties.

The following problems will be addressed:

- Why does the restitution coefficient ϵ depend on the impact velocity $v_{\rm imp}$?
- How does it depend on the impact velocity?
- What are the consequences of the dependence of ϵ on $v_{\rm imp}$ on the collective behavior of particles in granular gases? In particular how does $\epsilon = \epsilon\left(v_{\rm imp}\right)$ influence:
  - the evolution of temperature with time?
  - the evolution of the velocity distribution function with time?
  - the self-diffusion in granular gases?

In what follows we will show that the dependence of the restitution coefficient on the impact velocity is a very basic property of dissipative particle collisions, whereas the assumption of a constant restitution coefficient for the collision of three-dimensional spheres may lead to a physically incorrect dependence of the dissipative force on the compression rate of the colliding particles. From the Hertz collision law and the general relation between the elastic and dissipative forces we deduce the dependence of the restitution coefficient on the impact velocity, which follows purely from scaling considerations. We also give the corresponding relation obtained from rigorous theory. Using the dependence $\epsilon(v_{\rm imp})$ we derive the time dependence of the temperature, the time-evolution of the velocity distribution function and describe self-diffusion in granular gases in the homogeneous cooling regime.

2 Dependence of the Restitution Coefficient on the Impact Velocity

The collision of two particles may be characterized by the compression ξ and by the compression rate $\dot{\xi}$, as shown on Fig. 1. The compression gives rise to the elastic force $F_{\rm el}(\xi)$, while the dissipative force $F_{\rm diss}(\xi,\dot{\xi})$ appears due to the compression rate.

If the compression and the compression rate are not very large, one can assume the dependence of the elastic and dissipative force on ξ and $\dot{\xi}$

$$F_{\rm el}(\xi) \sim \xi^{\alpha} \tag{1}$$

$$F_{\rm diss}(\xi,\dot{\xi}) \sim \dot{\xi}^{\beta}\xi^{\gamma}\,. \tag{2}$$

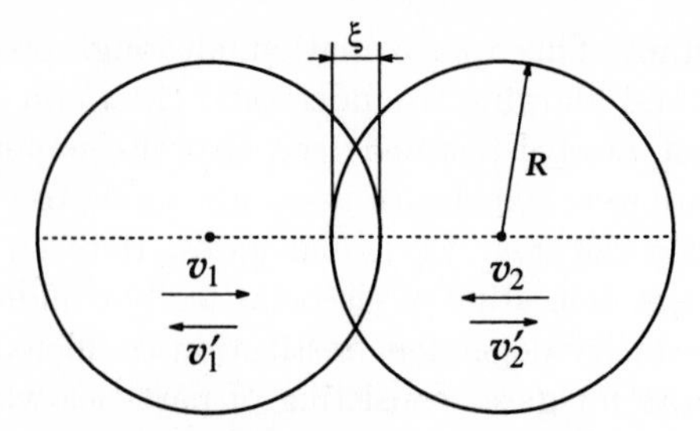

Fig. 1. Sketches of two colliding spheres. The compression ξ is equal to $2R-|\boldsymbol{r}_1-\boldsymbol{r}_2|$, with $\boldsymbol{r}_{1/2}$ being the particles positions. The compression rate is $\dot{\xi} = v_1 - v_2$. For simplicity the head on collision of identical spheres is shown.

The dimension analysis yields the following functional form for the dependence of the restitution coefficient on the impact velocity [13]:

$$\epsilon(v_{\text{imp}}) = \epsilon\left(v_{\text{imp}}^{\frac{2(\gamma-\alpha)}{1+\alpha}+\beta}\right) \tag{3}$$

Therefore, the condition for a constant restitution coefficient imposes the relation between the exponents α, β and γ [14, 15]:

$$2\,(\gamma-\alpha)+\beta\;(1+\alpha)=0\,. \tag{4}$$

For compressions which do not exceed the plasticity threshold, the particle's material behaves as a viscoelastic medium. Then it may be generally shown [16–18] that the relation

$$F_{\text{diss}} = A\,\dot{\xi}\frac{\partial}{\partial\xi}F_{\text{el}}(\xi) \tag{5}$$

between the elastic and dissipative force holds, independently on the shape of the bodies in contact, provided three conditions are met [19]:

(i) The elastic components of the stress tensor σ_{el}^{ik} depend linearly on the components of the deformation tensor u_{ik} [20].
(ii) The dissipative components of the stress tensor $\sigma_{\text{diss}}^{ik}$ depend linearly on the components of the deformation rate tensor $\dot{u}_{ik}$ [20].
(iii) The conditions of quasistatic motion are provided, i.e. $\dot{\xi} \ll c$, $\tau_{\text{vis}} \ll \tau_c$ [16, 17] (here c is the speed of sound in the material of particles and τ_{vis} is the relaxation time of viscous processes in its bulk).

The constant A in Eq. (5) reads [16, 17]

$$A = \frac{1}{3}\frac{(3\eta_2-\eta_1)^2}{(3\eta_2+2\eta_1)}\left[\frac{(1-\nu^2)(1-2\nu)}{Y\nu^2}\right]\,. \tag{6}$$

where Y and ν are respectively the Young modulus and the Poisson ratio of the particle material and the viscous constants η_1, η_2 relate (linearly) the dissipative stress tensor $\sigma_{\rm diss}^{ik}$ to the deformation rate tensor $\dot{u}_{ik}$ [16, 17, 20].

From Eq. (5) follows that

$$\beta = 1 \qquad \gamma = \alpha - 1 \,. \tag{7}$$

Consider now a collision of three-dimensional spherical particles of radii R_1 and R_2. The Hertz contact contact law gives for the elastic force [21]

$$F_{\rm el} = \rho \xi^{3/2} \,, \qquad \rho \equiv \frac{2Y}{3(1-\nu^2)} \sqrt{R^{\rm eff}} \,, \tag{8}$$

where $R^{\rm eff} \equiv R_1 R_2/(R_1 + R_2)$. With the set of exponents, $\alpha = 3/2$, $\beta = 1$ and $\gamma = 1/2$, which generally follows from the basic laws of the viscoelastic collision, the condition for the constant restitution coefficient, Eq. (4), is obviously not satisfied. For spherical particles the restitution coefficient could be constant only for $\gamma = 1/4$; this, however, is not consistent with the collision laws. Instead one obtains the functional dependence

$$\epsilon = \epsilon \left(v_{\rm imp}^{1/5} \right) \,. \tag{9}$$

Note that this conclusion comes from the general analysis of viscoelastic collisions with no other assumptions needed. Therefore, the dependence of the restitution coefficient on the impact velocity, Eq. (9), is a natural property, provided the assumption on viscoelasticity holds true which is the case in a wide range of impact velocities (see discussion on page 102). We want to mention that the functional dependence Eq. (9) was already given in [22] using heuristic arguments.

Rigorous calculations [23] yield for the dependence of the restitution coefficient on the impact velocity:

$$\epsilon = 1 - C_1 A \kappa^{2/5} v_{\rm imp}^{1/5} + C_2 A^2 \kappa^{4/5} v_{\rm imp}^{2/5} \mp \cdots \tag{10}$$

with

$$\kappa = \left(\frac{3}{2} \right)^{3/2} \frac{Y \sqrt{R^{\rm eff}}}{m^{\rm eff} (1-\nu^2)} \tag{11}$$

where $m^{\rm eff} = m_1 m_2/(m_1 + m_2)$ ($m_{1/2}$ are the masses of the colliding particles). Numerical values for the constants C_1 and C_2 obtained in [23] may be also written in a more convenient form [13]:

$$C_1 = \frac{\Gamma(3/5)\sqrt{\pi}}{2^{1/5} 5^{2/5} \Gamma(21/10)} = 1.15344, \qquad C_2 = \frac{3}{5} C_1^2 \,. \tag{12}$$

Although the next-order coefficients of the above expansion $C_3 = -0.483582$, $C_4 = 0.285279$, are now available [13], we assume that the dissipative constant A is small enough to ignore these high-order terms. (For large A a very accurate Padé approximation for $\epsilon(v_{\rm imp})$ has been proposed recently [13]).

3 Time-Evolution of Temperature and of the Velocity Distribution Function

We consider a granular gas composed of N identical particles confined in a volume Ω. The particles are assumed to be smooth spheres, so that the collision properties are determined by the normal component of the relative motion only. The gas is supposed to be dilute enough so that one can assume binary collisions (i.e. neglect multiple collisions) and ignore the collision duration as compared with the mean free time in between successive collisions. We assume that the initial velocities of the particles (more precisely the temperature, which we define below) are not very large to assure viscoelastic properties of the collisions, i.e. to avoid plastic deformations and fragmentation. The final velocities are assumed not to be very small which allows to neglect surface forces as adhesion and others. Under these restrictions one can apply the viscoelastic collision model. Furthermore, we assume that dissipation is not large, so that the second-order expansion (10) for $\epsilon(v_{\rm imp})$ describes the collisions accurately. We analyze the granular gas in the regime of homogeneous cooling, i.e. in the pre-clustering regime, when the gas is homogeneously distributed in space.

The impact velocity $v_{\rm imp}$ of colliding smooth spheres, which determines the value of the restitution coefficient according to Eq. (10), is given by the normal component of the relative velocity

$$v_{\rm imp} = |\boldsymbol{v}_{12} \cdot \boldsymbol{e}| \qquad \text{with} \qquad \boldsymbol{v}_{12} = \boldsymbol{v}_1 - \boldsymbol{v}_2 \,. \tag{13}$$

The unit vector $\boldsymbol{e} = \boldsymbol{r}_{12}/|\boldsymbol{r}_{12}|$ gives the direction of the intercenter vector $\boldsymbol{r}_{12} = \boldsymbol{r}_1 - \boldsymbol{r}_2$ at the instant of the collision.

The evolution of the granular gas proceeds by elementary collision events in which the pre-collisional velocities of colliding particles $\boldsymbol{v}_1$, $\boldsymbol{v}_2$, are converted into after-collisional ones, $\boldsymbol{v}_1^*$, $\boldsymbol{v}_2^*$, according to the rules

$$\begin{aligned} \boldsymbol{v}_1^* &= \boldsymbol{v}_1 - \frac{1}{2}\left[1+\epsilon(|\boldsymbol{v}_{12}\cdot\boldsymbol{e}|)\right](\boldsymbol{v}_{12}\cdot\boldsymbol{e})\boldsymbol{e} \\ \boldsymbol{v}_2^* &= \boldsymbol{v}_2 + \frac{1}{2}\left[1+\epsilon(|\boldsymbol{v}_{12}\cdot\boldsymbol{e}|)\right](\boldsymbol{v}_{12}\cdot\boldsymbol{e})\boldsymbol{e}\,, \end{aligned} \tag{14}$$

where ϵ depends on the impact velocity $v_{\rm imp} = |\boldsymbol{v}_{12} \cdot \boldsymbol{e}|$. Due to the *direct* collision (14) the population in the velocity phase-space near the points $\boldsymbol{v}_1$, $\boldsymbol{v}_2$ decreases, while near the points $\boldsymbol{v}_1^*$, $\boldsymbol{v}_2^*$ it increases. The decrease of the population near $\boldsymbol{v}_1$, $\boldsymbol{v}_2$ caused by the direct collision, is (partly) counterbalanced by its increase in the *inverse* collision, where the after-collisional velocities are $\boldsymbol{v}_1$, $\boldsymbol{v}_2$ with the pre-collisional ones $\boldsymbol{v}_1^{**}$, $\boldsymbol{v}_2^{**}$. The rules for the inverse collision read

$$\begin{aligned} \boldsymbol{v}_1 &= \boldsymbol{v}_1^{**} - \frac{1}{2}\left[1+\epsilon(|\boldsymbol{v}_{12}^{**}\cdot\boldsymbol{e}|)\right](\boldsymbol{v}_{12}^{**}\cdot\boldsymbol{e})\,\boldsymbol{e} \\ \boldsymbol{v}_2 &= \boldsymbol{v}_2^{**} + \frac{1}{2}\left[1+\epsilon(|\boldsymbol{v}_{12}^{**}\cdot\boldsymbol{e}|)\right](\boldsymbol{v}_{12}^{**}\cdot\boldsymbol{e})\,\boldsymbol{e}\,. \end{aligned} \tag{15}$$

Note that in contrast to the case of $\epsilon = \text{const}$, the restitution coefficients in the inverse and in the direct collisions are different.

The Enskog-Boltzmann equation describes the evolution of the population of particles in the phase space on the mean-field level. The evolution is characterized by the distribution function $f(\boldsymbol{r}, \boldsymbol{v}, t)$, which for the force-free case does not depend on $\boldsymbol{r}$ and obeys the equation [4, 24]

$$\frac{\partial}{\partial t} f(\boldsymbol{v}_1, t) = g_2(\sigma)\sigma^2 \int d\boldsymbol{v}_2 \int d\boldsymbol{e}\Theta(-\boldsymbol{v}_{12}\cdot\boldsymbol{e})|\boldsymbol{v}_{12}\cdot\boldsymbol{e}| \\ \times \left\{\chi f(\boldsymbol{v}_1^{**}, t) f(\boldsymbol{v}_2^{**}, t) - f(\boldsymbol{v}_1, t) f(\boldsymbol{v}_2, t)\right\} \equiv g_2(\sigma) I(f,f) \quad (16)$$

where σ is the diameter of the particles. The contact value of the pair distribution function [25]

$$g_2(\sigma) = (2-\eta)/2(1-\eta)^3 \,, \quad (17)$$

accounts on the mean-field level for the increasing frequency of collisions due to excluded volume effects with $\eta = \frac{1}{6}\pi n \sigma^3$ being the volume fraction.

The first term in the curled brackets in the right-hand side of Eq. (16) refers to the "gain" term for the population in the phase-space near the point $\boldsymbol{v}_1$, while the second one is the "loss" term. The Heaviside function $\Theta(-\boldsymbol{v}_{12}\cdot\boldsymbol{e})$ discriminates approaching particles (which do collide) from separating particles (which do not collide), and $|\boldsymbol{v}_{12}\cdot\boldsymbol{e}|$ gives the length of the collision cylinder. Integration in Eq. (16) is performed over all velocities $\boldsymbol{v}_2$ and interparticle vectors $\boldsymbol{e}$ in the direct collision. Equation (16) accounts also for the inverse collisions via the factor χ, which appears due to the Jacobian of the transformation $\boldsymbol{v}_1^{**}, \boldsymbol{v}_2^{**} \to \boldsymbol{v}_1, \boldsymbol{v}_2$, and due to the difference between the lengths of the collision cylinders of the direct and the inverse collision:

$$\chi = \frac{\mathcal{D}(\boldsymbol{v}_1^{**}, \boldsymbol{v}_2^{**})}{\mathcal{D}(\boldsymbol{v}_1, \boldsymbol{v}_2)} \frac{|\boldsymbol{v}_{12}^{**}\cdot\boldsymbol{e}|}{|\boldsymbol{v}_{12}\cdot\boldsymbol{e}|} \,. \quad (18)$$

For constant restitution coefficient the factor χ is a constant

$$\chi = \frac{1}{\epsilon^2} = \text{const} \,, \quad (19)$$

while for $\epsilon = \epsilon\,(v_{\text{imp}})$, as given in Eq. (10), it reads [26]

$$\chi = 1 + \frac{11}{5} C_1 A \kappa^{2/5} |\boldsymbol{v}_{12}\cdot\boldsymbol{e}|^{1/5} + \frac{66}{25} C_1^2 A^2 \kappa^{4/5} |\boldsymbol{v}_{12}\cdot\boldsymbol{e}|^{2/5} + \cdots \quad (20)$$

From Eq. (20) it follows that $\chi = \chi(|\boldsymbol{v}_{12}\cdot\boldsymbol{e}|)$. Since the average velocity in granular gases changes with time, such a dependence of χ means, as we will show below, that χ and, therefore, the velocity distribution function itself depend explicitly on time. The time dependence of χ changes drastically the properties of the collision integral and destroys the simple scaling form of the velocity distribution function, which holds for the case of the constant restitution coefficient (e.g. [4,5]).

Nevertheless, some important properties of the collision integral are preserved. Namely, it may be shown that the relation

$$\frac{d}{dt}\langle\psi(t)\rangle = \int d\boldsymbol{v}_1\psi(\boldsymbol{v}_1)\frac{\partial}{\partial t}f(\boldsymbol{v}_1,t) = g_2(\sigma)\int d\boldsymbol{v}_1\psi(\boldsymbol{v}_1)I(f,f) = \tag{21}$$

$$\frac{g_2(\sigma)\sigma^2}{2}\int d\boldsymbol{v}_1 d\boldsymbol{v}_2\int d\boldsymbol{e}\,\Theta(-\boldsymbol{v}_{12}\cdot\boldsymbol{e})|\boldsymbol{v}_{12}\cdot\boldsymbol{e}|f(\boldsymbol{v}_1,t)f(\boldsymbol{v}_2,t)\Delta\left[\psi(\boldsymbol{v}_1)+\psi(\boldsymbol{v}_2)\right]$$

holds true, where

$$\langle\psi(t)\rangle \equiv \int d\boldsymbol{v}\psi(\boldsymbol{v})f(\boldsymbol{v},t) \tag{22}$$

is the average of some function $\psi(\boldsymbol{v})$, and

$$\Delta\psi(\boldsymbol{v}_i) \equiv [\psi(\boldsymbol{v}_i^*)-\psi(\boldsymbol{v}_i)] \tag{23}$$

denotes the change of $\psi(\boldsymbol{v}_i)$ in a direct collision.

Now we introduce the temperature of the three-dimensional granular gas,

$$\frac{3}{2}nT(t) = \int d\boldsymbol{v}\frac{mv^2}{2}f(\boldsymbol{v},t)\,, \tag{24}$$

where n is the number density of granular particles ($n = N/\Omega$), and the characteristic velocity $v_0^2(t)$ is related to temperature via

$$T(t) = \frac{1}{2}mv_0^2(t)\,. \tag{25}$$

First we try the scaling ansatz

$$f(\boldsymbol{v},t) = \frac{n}{v_0^3(t)}\tilde{f}(\boldsymbol{c}) \tag{26}$$

where $\boldsymbol{c} \equiv \boldsymbol{v}/v_0(t)$ and following [4,7] assume that deviations from the Maxwellian distribution are not large, so that $\tilde{f}(\boldsymbol{c})$ may be expanded into a convergent series with the leading term being the Maxwellian distribution $\phi(c) \equiv \pi^{-3/2}\exp(-c^2)$. It is convenient to use the Sonine polynomials expansion [4,7]

$$\tilde{f}(\boldsymbol{c}) = \phi(c)\left\{1+\sum_{p=1}^{\infty}a_pS_p\left(c^2\right)\right\}\,. \tag{27}$$

These polynomials are orthogonal, i.e.

$$\int d\boldsymbol{c}\,\phi(c)S_p(c^2)S_{p'}(c^2) = \delta_{pp'}\mathcal{N}_p\,, \tag{28}$$

where $\delta_{pp'}$ is the Kronecker delta and $\mathcal{N}_p$ is the normalization constant. The first few polynomials read

$$\begin{aligned} S_0(x) &= 1 \\ S_1(x) &= -x^2 + \frac{3}{2} \\ S_2(x) &= \frac{x^2}{2} - \frac{5x}{2} + \frac{15}{8} \,. \end{aligned} \tag{29}$$

Writing the Enskog-Boltzmann equation in terms of the scaling variable $\boldsymbol{c}_1$, one observes that the factor χ may not be expressed only in terms of the scaling variable, but it depends also on the characteristic velocity $v_0(t)$, and thus depends on time. Therefore, the collision integral also occurs to be time-dependent. As a result, it is not possible to reduce the Enskog-Boltzmann equation to a pair of equations, one for the time evolution of the temperature and another for the time-independent scaling function, whereas for ϵ =const. the Boltzmann-Enskog equation is separable, e.g. [4, 5, 7]. Formally adopting the approach of Refs. [4, 7] for ϵ =const., one would obtain time-dependent coefficients a_p of the Sonine polynomials expansion. This means that the simple scaling ansatz (26) is violated for the case of the impact-velocity dependent restitution coefficient.

Thus, it seems natural to write the distribution function in the following general form

$$f(\boldsymbol{v}, t) = \frac{n}{v_0^3(t)} \tilde{f}(\boldsymbol{c}, t) \tag{30}$$

with

$$\tilde{f}(\boldsymbol{c}) = \phi(c) \left\{ 1 + \sum_{p=1}^{\infty} a_p(t) S_p(c^2) \right\} \tag{31}$$

and find then equations for the *time-dependent* coefficients $a_p(t)$. Substituting (30) into the Boltzmann equation (16) we obtain

$$\frac{\mu_2}{3} \left(3 + c_1 \frac{\partial}{\partial c_1} \right) \tilde{f}(\boldsymbol{c}, t) + B^{-1} \frac{\partial}{\partial t} \tilde{f}(\boldsymbol{c}, t) = \tilde{I} \left(\tilde{f}, \tilde{f} \right) \tag{32}$$

with

$$B = B(t) \equiv v_0(t) g_2(\sigma) \sigma^2 n \,. \tag{33}$$

We define the dimensionless collision integral:

$$\tilde{I} \left(\tilde{f}, \tilde{f} \right) = \\ \int d\boldsymbol{c}_2 \int d\boldsymbol{e} \Theta(-\boldsymbol{c}_{12} \cdot \boldsymbol{e}) |\boldsymbol{c}_{12} \cdot \boldsymbol{e}| \left\{ \tilde{\chi} \tilde{f}(\boldsymbol{c}_1^{**}, t) \tilde{f}(\boldsymbol{c}_2^{**}, t) - \tilde{f}(\boldsymbol{c}_1, t) \tilde{f}(\boldsymbol{c}_2, t) \right\} \tag{34}$$

with the reduced factor $\tilde{\chi}$

$$\tilde{\chi} = 1 + \frac{11}{5} C_1 \delta' |\boldsymbol{c}_{12} \cdot \boldsymbol{e}|^{1/5} + \frac{66}{25} C_1^2 \delta'^2 |\boldsymbol{c}_{12} \cdot \boldsymbol{e}|^{2/5} + \cdots \tag{35}$$

which depends now on time via a quantity

$$\delta'(t) \equiv A\kappa^{2/5} \left[2T(t)\right]^{1/10} \equiv \delta \left[2T(t)/T_0\right]^{1/10} . \tag{36}$$

Here $\delta \equiv A\kappa^{2/5}[T_0]^{1/10}$, T_0 is the initial temperature, and for simplicity we assume the unit mass, $m = 1$. We also define the moments of the dimensionless collision integral

$$\mu_p \equiv - \int d\boldsymbol{c}_1 c_1^p \tilde{I}\left(\tilde{f}, \tilde{f}\right) , \tag{37}$$

so that the second moment describes the rate of the temperature change:

$$\frac{dT}{dt} = -\frac{2}{3} BT\mu_2 . \tag{38}$$

Equation (38) follows from the definitions of the temperature and of the moment μ_2. Note that these moments depend on time, in contrast to the case of the constant restitution coefficient, where these moments are time-independent [4].

Multiplying both sides of Eq. (32) with c_1^p and integrating over $d\boldsymbol{c}_1$, we obtain

$$\frac{\mu_2}{3} p \left\langle c^p \right\rangle - B^{-1} \sum_{k=1}^{\infty} \dot{a}_k \nu_{kp} = \mu_p \tag{39}$$

where integration by parts has been performed and we define

$$\nu_{kp} \equiv \int \phi(c) c^p S_k(c^2) d\boldsymbol{c} \tag{40}$$

$$\left\langle c^p \right\rangle \equiv \int c^p \tilde{f}(\boldsymbol{c}, t) d\boldsymbol{c} . \tag{41}$$

The calculation of ν_{kp} is straightforward; the first few of these read: $\nu_{22} = 0$, $\nu_{24} = \frac{15}{4}$. The odd moments $\left\langle c^{2n+1} \right\rangle$ vanish, while the even ones $\left\langle c^{2n} \right\rangle$ may be expressed in terms of a_k with $0 \le k \le n$, namely, $\left\langle c^2 \right\rangle = \frac{3}{2} - \frac{3}{2} a_1$. On the other hand, from the definition of temperature and of the thermal velocity in Eqs. (25) and (24) follows that $\left\langle c^2 \right\rangle = \frac{3}{2}$ and thus, $a_1 = 0$. Similar considerations yield $\left\langle c^4 \right\rangle = \frac{15}{4}(1 + a_2)$. The moments μ_p may be expressed in terms of coefficients $a_2, a_3, \cdots$ too; therefore, the system Eq. (39) is an infinite (but closed) set of equations for these coefficients.

It is not possible to get a general solution of the problem. However, since the dissipative parameter δ is supposed to be small, the deviations from the

Maxwellian distribution are presumably small too. Thus, we assume that one can neglect all high-order terms with $p > 2$ in the expansion (31). Then Eq. (39) is an equation for the coefficient a_2. For $p = 2$ Eq. (39) converts into an identity, since $\langle c^2 \rangle = \frac{3}{2}$, $a_1 = 0$, $\nu_{22} = 0$ and $\nu_{24} = \frac{15}{4}$. For $p = 4$ we obtain

$$\dot{a}_2 - \frac{4}{3} B\mu_2 (1 + a_2) + \frac{4}{15} B\mu_4 = 0 . \tag{42}$$

In Eq. (42) B depends on time as

$$B(t) = (8\pi)^{-1/2} \tau_c(0)^{-1} [T(t)/T_0]^{1/2} , \tag{43}$$

where $\tau_c(0)$ is related to the initial mean-collision time,

$$\tau_c(0)^{-1} = 4\pi^{1/2} g_2(\sigma) \sigma^2 n T_0^{1/2} . \tag{44}$$

The time evolution of the temperature is determined by Eq. (38), i.e. by the time dependence of μ_2.

The time-dependent coefficients $\mu_p(t)$ may be expressed in terms of a_2 owing to their definition Eq. (37) and the approximation $\tilde{f} = \phi(c)[1 + a_2(t) S_2(c^2)]$. We finally obtain:

$$\begin{aligned} \mu_p = -\frac{1}{2} \int d\mathbf{c}_1 \int d\mathbf{c}_2 \int d\mathbf{e} \Theta(-\mathbf{c}_{12} \cdot \mathbf{e}) |\mathbf{c}_{12} \cdot \mathbf{e}| \phi(c_1) \phi(c_2) \\ \times \left\{ 1 + a_2 \left[S_2(c_1^2) + S_2(c_2^2) \right] + a_2^2 \, S_2(c_1^2) S_2(c_2^2) \right\} \Delta(c_1^p + c_2^p) \end{aligned} \tag{45}$$

with the definition of $\Delta(c_1^p + c_2^p)$ given above. Calculations performed up to the second order in terms of the dissipative parameter δ yield [26]:

$$\mu_2 = \sum_{k=0}^{2} \sum_{n=0}^{2} \mathcal{A}_{kn} \delta'^k a_2^n \tag{46}$$

where

$$\begin{array}{lll} \mathcal{A}_{00} = 0; & \mathcal{A}_{01} = 0; & \mathcal{A}_{02} = 0 \\ \mathcal{A}_{10} = \omega_0; & \mathcal{A}_{11} = \frac{6}{25}\omega_0; & \mathcal{A}_{12} = \frac{21}{2500}\omega_0 \\ \mathcal{A}_{20} = -\omega_1; & \mathcal{A}_{21} = -\frac{119}{400}\omega_1; & \mathcal{A}_{22} = -\frac{4641}{640000}\omega_1 \end{array} \tag{47}$$

with

$$\omega_0 \equiv 2\sqrt{2\pi} 2^{1/10} \Gamma\left(\frac{21}{10}\right) C_1 = 6.48562\ldots \tag{48}$$

$$\omega_1 \equiv \sqrt{2\pi} 2^{1/5} \Gamma\left(\frac{16}{5}\right) C_1^2 = 9.28569\ldots \tag{49}$$

Similarly

$$\mu_4 = \sum_{k=0}^{2}\sum_{n=0}^{2} \mathcal{B}_{kn}\delta'^k a_2^n \tag{50}$$

with

$$\begin{aligned}
&\mathcal{B}_{00} = 0; && \mathcal{B}_{01} = 4\sqrt{2\pi}; && \mathcal{B}_{02} = \frac{1}{8}\sqrt{2\pi} \\
&\mathcal{B}_{10} = \frac{56}{10}\omega_0; && \mathcal{B}_{11} = \frac{1806}{250}\omega_0; && \mathcal{B}_{12} = \frac{567}{12500}\omega_0 \\
&\mathcal{B}_{20} = -\frac{77}{10}\omega_1; && \mathcal{B}_{21} = -\frac{149054}{13750}\omega_1; && \mathcal{B}_{22} = -\frac{348424}{5500000}\omega_1
\end{aligned} \tag{51}$$

Thus, Eqs. (38) and (42), together with Eqs. (46) and (50) form a closed set to find the time evolution of the temperature and the coefficient a_2. We want to stress an important difference for the time evolution of temperature for the case of the impact-velocity dependent restitution coefficient, as compared to that of a constant restitution coefficient. In the former case it is coupled to the time evolution of the coefficient a_2, while in the latter case there is no such coupling since $a_2 = \text{const}$. This coupling may lead in to a rather peculiar time-dependence of the temperature.

Introducing the reduced temperature $u(t) \equiv T(t)/T_0$ we recast the set of equations (38) and (42) into the following form:

$$\begin{aligned}
\dot{u} + \tau_0^{-1} u^{8/5}\left(\frac{5}{3} + \frac{2}{5}a_2 + \frac{7}{500}a_2^2\right) \\
- \tau_0^{-1} q_1 \delta\, u^{17/10}\left(\frac{5}{3} + \frac{119}{240}a_2 + \frac{1547}{128000}a_2^2\right) = 0
\end{aligned} \tag{52}$$

$$\dot{a}_2 - r_0 u^{1/2}\mu_2\,(1 + a_2) + \frac{1}{5} r_0 u^{1/2}\mu_4 = 0\,, \tag{53}$$

where we introduce the characteristic time

$$\tau_0^{-1} = \frac{16}{5} q_0 \delta \cdot \tau_c(0)^{-1} \tag{54}$$

with

$$q_0 = 2^{1/5}\Gamma(21/10)C_1/8 = 5^{-2/5}\sqrt{\pi}\Gamma(3/5)/8 = 0.173318\ldots \tag{55}$$

$$r_0 \equiv \frac{2}{3\sqrt{2\pi}}\tau_c(0)^{-1} \tag{56}$$

$$q_1 \equiv 2^{1/10}(\omega_1/\omega_0) = 1.53445\ldots \tag{57}$$

As shown below the characteristic time τ_0 describes the time evolution of the temperature. To obtain these equations we use the expressions for $\mu_2(t)$, $B(t)$, and for the coefficients $\mathcal{A}_{nk}$. Note that the characteristic time τ_0 is $\delta^{-1} \gg 1$ times larger than the collision time $\sim \tau_c(0)$.

We will find the solution to these equations as expansions in terms of the small dissipative parameter δ ($\delta'(t) = \delta \cdot 2^{1/10} u^{1/10}(t)$):

$$u = u_0 + \delta \cdot u_1 + \delta^2 \cdot u_2 + \cdots \tag{58}$$

$$a_2 = a_{20} + \delta \cdot a_{21} + \delta^2 \cdot a_{22} + \cdots \tag{59}$$

Substituting Eqs. (46,50,58,59) into Eqs. (52,53), one can solve these equations perturbatively, for each order of δ. The solution of the order of $\mathcal{O}(1)$ reads for the coefficient $a_2(t)$ [26]:

$$a_{20}(t) \approx a_{20}(0) e^{-4t/(5\tau_{\rm E}(0))} \,, \tag{60}$$

where $\tau_{\rm E} = \frac{3}{2}\tau_c$ is the Enskog relaxation time, so that $a_{20}(t)$ vanishes for $t \sim \tau_0$. This refers to the relaxation of an initially non-Maxwellian velocity distribution to the Maxwellian distribution. Note that the relaxation occurs within few collisions per particle, similarly to the relaxation of common molecular gases.

We now assume that the initial distribution is Maxwellian, i.e., that $a_{20}(0) = 0$ for $t = 0$. Then the deviation from the Maxwellian distribution originates from the inelasticity of the interparticle collisions. For the case of $a_{20}(0) = 0$ the solution of the order of $\mathcal{O}(1)$ for the reduced temperature reads

$$\frac{T(t)}{T_0} = u_0(t) = \left(1 + \frac{t}{\tau_0}\right)^{-5/3} \,, \tag{61}$$

which coincides with the time-dependence of the temperature obtained previously using scaling arguments [23] (up to a constant τ_0 which may not be determined by scaling arguments).

The solution for $a_2(t)$ in linear approximation with respect to δ reads

$$a_2(t) = \delta \cdot a_{21}(t) = -\frac{12}{5} w(t)^{-1} \left\{ \mathrm{Li}\left[w(t)\right] - \mathrm{Li}\left[w(0)\right] \right\} \tag{62}$$

where

$$w(t) \equiv \exp\left[(q_0 \delta)^{-1} \left(1 + t/\tau_0\right)^{1/6} \right] . \tag{63}$$

and with the logarithmic Integral

$$\mathrm{Li}(x) = \int_0^x \frac{1}{\ln(t)} dt \,. \tag{64}$$

For $t \ll \tau_0$ the coefficient $a_2(t)$ (62) reduces to

$$a_2(t) = -\delta \cdot h \left(1 - e^{-4t/(5\tau_{\rm E}(0))}\right) \tag{65}$$

where

$$h \equiv 2^{1/10} \left(\mathcal{B}_{10} - 5\mathcal{A}_{10}\right)/16\pi = (3/10)\Gamma(21/10)2^{1/5}C_1 = 0.415964 \,. \tag{66}$$

As it follows from Eq. (65), after a transient time of the order of few collisions per particle, i.e. for $\tau_{\rm E}(0) < t \ll \tau_0$, $a_2(t)$ saturates at the "steady-state"-value $-h\,\delta = -0.415964\,\delta$, i.e. it changes only slowly on the time-scale $\sim \tau_c(0)$. On the other hand, for $t \gg \tau_0$ one obtains

$$a_2(t) \simeq -\delta \cdot h\,(t/\tau_0)^{-1/6} \tag{67}$$

so that $a_2(t)$ decays to zero on a time-scale $\sim \tau_0$, i.e. slowly in the collisional time-scale $\sim \tau_c(0) \ll \tau_0$. The velocity distribution thus tends asymptotically to the Maxwellian distribution. For this regime the first-order correction for the reduced temperature, $u_1(t)$, reads [26]:

$$u_1(t) = \left(\frac{12}{25}h + 2q_1\right)(t/\tau_0)^{-11/6} = 3.26856\,(t/\tau_0)^{-11/6} \,, \tag{68}$$

where we used the above results for the constants h and q_1. From the last equation one can see how the coupling between the temperature evolution and the evolution of the velocity distribution influences the evolution of temperature. Indeed, if there were no such coupling, there would be no coupling term in Eq. (52), and thus no contribution from $\frac{12}{25}h$ to the prefactor of $u_1(t)$ in Eq. (68). This would noticeably change the time behavior of $u_1(t)$. On the other hand, the leading term in the time dependence of temperature, $u_0(t)$, is not affected by this kind of coupling.

In Fig. 2 and Fig. 3 we show the time dependence of the coefficient $a_2(t)$ of the Sonine polynomial expansion and of the temperature of the granular gas. The analytical findings are compared with the numerical solution of the system (52,53). As one can see from the figures the analytical theory reproduces fairly well the numerical solution for the case of small δ.

As it follows from Fig. 2 (where the time is given in collisional units), for small δ the following scenario of evolution of the velocity distribution takes place for a force-free granular gas. The initial Maxwellian distribution evolves to a non-Maxwellian distribution, with the discrepancy between these two characterized by the second coefficient of the Sonine polynomials expansion a_2. The deviation from the Maxwellian distribution (described by a_2) quickly grows, until it saturates after a few collisions per particle at a "steady-state" value. At this instant the deviation from the Maxwellian distribution is maximal, with the value $a_2 \approx -0.4\delta$ (Fig. 2a). This refers to the first "fast" stage of the evolution, which takes place on a mean-collision

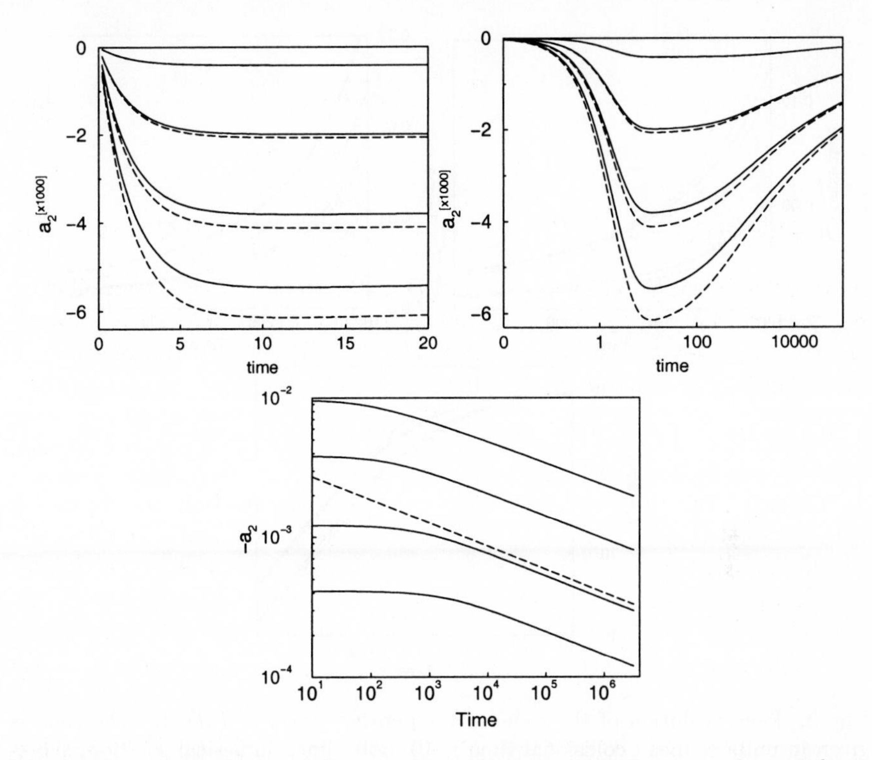

Fig. 2. Time dependence of the second coefficient of the Sonine polynomial expansion $a_2(t)$. Time is given in units of the mean collisional time $\tau_c(0)$. (Left): $a_2 \times 1000$ (solid lines) for $\delta = 0.001, 0.005, 0.01, 0.015$ (top to bottom) together with the linear approximation (dashed lines); (Right): the same as (left) but for larger times; (Middle): $-a_2(t)$ over time (log-scale) for $\delta = 0.03, 0.01, 0.003, 0.001$ (top to bottom) together with the power-law asymptotics $\sim t^{-1/6}$.

time-scale $\sim \tau_c(0)$. After this maximal deviation is reached, the second "slow" stage of the evolution starts. At this stage a_2 decays to zero on the "slow" time scale $\tau_0 \sim \delta^{-1}\tau_c(0) \gg \tau_0(0)$, which corresponds to the time scale of the temperature evolution (Fig. 2b); the decay of the coefficient $a_2(t)$ in this regime occurs according to a power law $\sim t^{-1/6}$ (Fig. 2c). Asymptotically the Maxwellian distribution would be achieved, if the clustering process did not occur.

Fig. 3 illustrates the significance of the first-order correction $u_1(t)$ in the time-evolution of temperature. This becomes more important as the dissipation parameter δ grows (Figs. 3a,b). At large times the results of the first-order theory (with $u_1(t)$ included) practically coincide with the numerical results, while zero-order theory (without $u_1(t)$) demonstrates noticeable deviations (Fig. 3c).

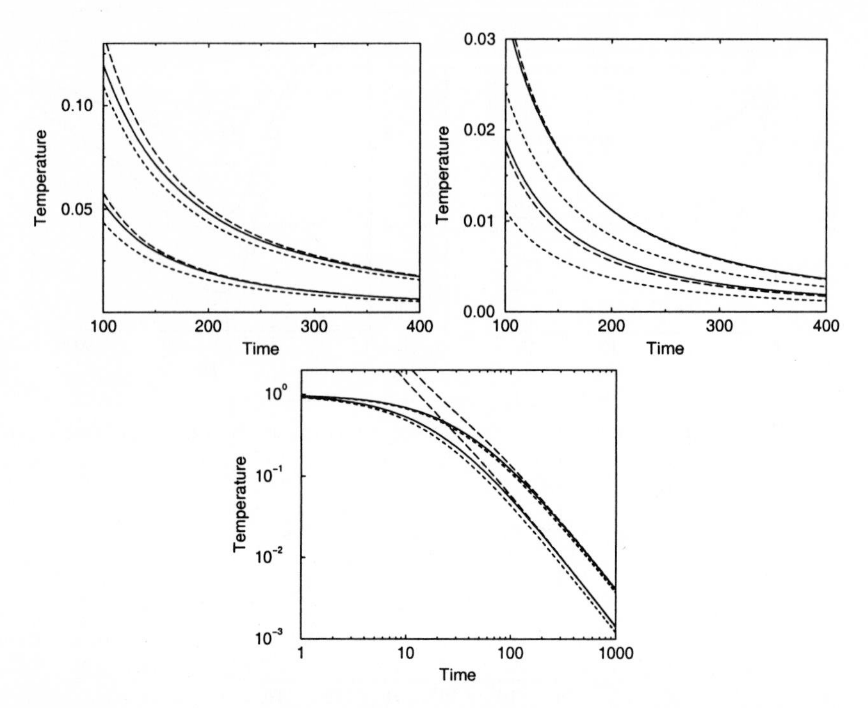

Fig. 3. Time-evolution of the reduced temperature, $u(t) = T(t)/T_0$. The time is given in units of mean collisional time $\tau_c(0)$. Solid line: numerical solution, short-dashed: $u_0(t) = (1 + t/\tau_0)^{-5/3}$ (zero-order theory), long-dashed: $u(t) = u_0(t) + \delta\, u_1(t)$ (first-order theory). (Left): for $\delta = 0.05, 0.1$ (top to bottom); (Right): $\delta = 0.15, 0.25$ (top to bottom); (Middle): the same as (Left) but log-scale and larger ranges.

For larger values of δ the linear theory breaks down. Unfortunately, the equations obtained for the second order approximation $\mathcal{O}(\delta^2)$ are too complicated to be treated analytically. Hence, we studied them only numerically (see Fig. 4). As compared to the case of small δ, an additional intermediate regime in the time-evolution of the velocity distribution is observed. The first "fast" stage of evolution takes place, as before, on the time scale of few collisions per particle, where maximal deviation from the Maxwellian distribution is achieved (Fig. 4). For $\delta \geq 0.15$ these maximal values of a_2 are positive. Then, on the second stage (intermediate regime), which continues $10 - 100$ collisions, a_2 changes its sign and reaches a maximal negative deviation. Finally, on the third, slow stage, $a_2(t)$ relaxes to zero on the slow time-scale $\sim \tau_0$, just as for small δ. In Fig. 4 we show the first stage of the time evolution of $a_2(t)$ for systems with large δ. At a certain value of the dissipative parameter δ the behavior changes qualitatively, i.e. the system then reveals another time scale as discussed above.

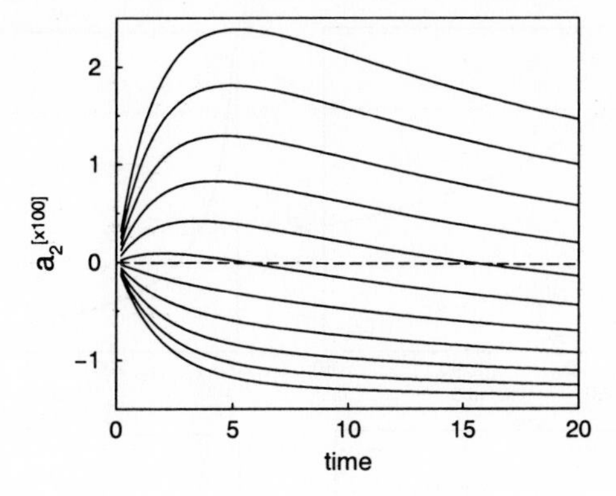

Fig. 4. Time dependence of the second coefficient of the Sonine polynomial expansion $a_2(t) \times 100$. Time is given in units of mean collisional time $\tau_c(0)$. $\delta = 0.1, 0.11, 0.12, \ldots, 0.20$ (bottom to top).

Figure 5 shows the numerical solution of Eqs. (52) and (53) for the second Sonine coefficient $a_2(t)$ as a function of time. One can clearly distinguish the different stages of evolution of the velocity distribution function. A more detailed investigation of the evolution of the distribution function for larger dissipation is subject of present research [26].

The interesting property of the granular gases in the regime of homogeneous cooling is the overpopulation of the high-velocity tails in the velocity distribution [5], which has been shown for granular gases consisting of particles which interact via a constant restitution coefficient, $\epsilon = \text{const}$. How does the velocity dependence of the restitution coefficient as it appears for viscoelastic spheres influence this effect? We observe, that for the case of $\epsilon = \epsilon(v_{\text{imp}})$ the functional form (i.e. the exponential overpopulation [5]) persists, but it decreases with time on the "slow" time-scale $\sim \tau_0$. Namely we obtain for the velocity distribution for $c \gg 1$ [26]:

$$\tilde{f}(c,t) \sim \exp\left[-\frac{b}{\delta}c\left(1+\frac{t}{\tau_0}\right)^{1/6}\right]. \tag{69}$$

where $b = \sqrt{\pi/2}\,(16q_0/5)^{-1} = 2.25978\ldots$, which holds for $t \gg \tau_c(0)$. Again we see that the distribution tends asymptotically to the Maxwellian distribution, since the overpopulation vanishes as $t \to \infty$.

Using the temperature and the velocity distribution of a granular gas as were derived in this section, one can calculate the kinetic coefficients. In the next section we consider the simplest one – the self-diffusion coefficient.

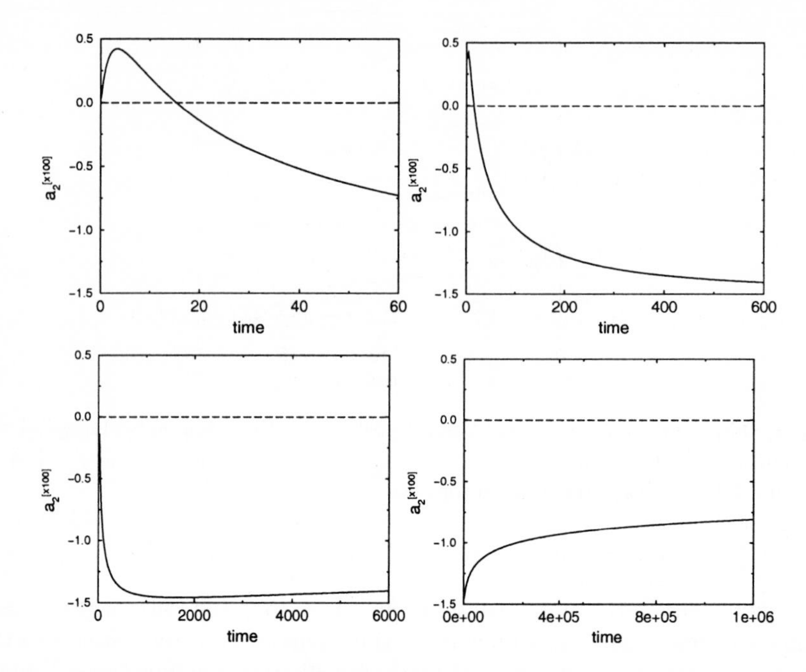

Fig. 5. The second Sonine coefficient a_2 for $\delta = 0.16$ over time. The numerical solutions of Eqs. (52) and (53) show all stages of evolution discussed in the text.

4 Self-Diffusion in Granular Gases of Viscoelastic Particles

In the simplest case diffusion of particles occurs when there are density gradients in the system. The diffusion coefficient D relates the flux of particles $\boldsymbol{J}$ to the density gradient ∇n according to a linear relation, provided the gradients are not too large:

$$\boldsymbol{J} = -D\nabla n\,. \tag{70}$$

The coefficient D also describes the statistical average of the migration of a single particle. For *equilibrium* 3D-systems the mean-square displacement of a particle reads

$$\left\langle (\Delta r(t))^2 \right\rangle_{\rm eq} = 6\,D\,t\,, \tag{71}$$

where $\langle \cdots \rangle_{\rm eq}$ denotes the *equilibrium* ensemble averaging. For *nonequilibrium* systems, such as granular gases, one should consider the time-dependent diffusion coefficient $D(t)$ and the corresponding generalization of Eq. (71):

$$\left\langle (\Delta r(t))^2 \right\rangle = 6 \int^t D(t')dt'\,, \tag{72}$$

where $\langle \cdots \rangle$ denotes averaging over the nonequilibrium ensemble. If the migration of a particle occurs in a uniform system composed of particles of the same kind, this process is called "self-diffusion". Correspondingly, the kinetic coefficient D is called self-diffusion coefficient.

To find the mean-square displacement, one writes

$$\left\langle (\Delta r(t))^2 \right\rangle = \left\langle \int_0^t \boldsymbol{v}(t')dt' \int_0^t \boldsymbol{v}(t'')dt'' \right\rangle \tag{73}$$

and encounters then with the velocity autocorrelation function

$$K_v(t',t) \equiv \langle \boldsymbol{v}(t')\boldsymbol{v}(t'') \rangle$$

which should be evaluated in order to obtain the mean-square displacement and the self-diffusion coefficient.

To calculate $K_v(t',t)$ we use the approximation of uncorrelated successive binary collision, which is valid for moderately dense systems, and an approach based on the formalism of the pseudo-Liouville operator $\mathcal{L}$ [27]. The pseudo-Liouville operator is defined as

$$i\mathcal{L} = \sum_j \boldsymbol{v}_j \cdot \frac{\partial}{\partial \boldsymbol{r}_j} + \sum_{i<j} \hat{T}_{ij} . \tag{74}$$

The first sum in (74) refers to the free streaming of the particles (the ideal part) while the second sum refers to the particle interactions which are described by the binary collision operators [28]

$$\hat{T}_{ij} = \sigma^2 \int d\boldsymbol{e}\, \Theta\left(-\boldsymbol{v}_{ij}\cdot\boldsymbol{e}\right)|\boldsymbol{v}_{ij}\cdot\boldsymbol{e}|\delta\left(\boldsymbol{r}_{ij}-\sigma\boldsymbol{e}\right)\left(\hat{b}_{ij}^{\boldsymbol{e}}-1\right) , \tag{75}$$

where $\Theta(x)$ is the Heaviside function. The operator $\hat{b}_{ij}^{\boldsymbol{e}}$ is defined as

$$\hat{b}_{ij}^{\boldsymbol{e}} f\left(\boldsymbol{r}_i, \boldsymbol{r}_j, \boldsymbol{v}_i, \boldsymbol{v}_j \cdots\right) = f\left(\boldsymbol{r}_i, \boldsymbol{r}_j, \boldsymbol{v}_i^*, \boldsymbol{v}_j^* \cdots\right) , \tag{76}$$

where f is some function of the dynamical variables and $\boldsymbol{v}_i^*$ and $\boldsymbol{v}_j^*$ are the postcollisional velocities from Eq. (14). The pseudo-Liouville operator gives the time derivative of any dynamical variable B (e.g. [24]):

$$\frac{d}{dt} B\left(\{\boldsymbol{r}_i, \boldsymbol{v}_i\}, t\right) = i\mathcal{L}\, B\left(\{\boldsymbol{r}_i, \boldsymbol{v}_i\}, t\right) . \tag{77}$$

Therefore, the time evolution of B reads $(t > t')$

$$B\left(\{\boldsymbol{r}_i, \boldsymbol{v}_i\}, t\right) = e^{i\mathcal{L}(t-t')} B\left(\{\boldsymbol{r}_i, \boldsymbol{v}_i\}, t'\right) . \tag{78}$$

With Eq. (78) the time-correlation function reads

$$\langle B(t')B(t) \rangle = \int d\Gamma \rho(t') B(t') e^{i\mathcal{L}(t-t')} B(t') , \tag{79}$$

where $\int d\Gamma$ denotes integration over all degrees of freedom and $\rho(t')$ depends on temperature T, density n, etc., which change on a time-scale $t \gg \tau_c$.

Now we assume that

(i) the coordinate part and the velocity part of the distribution function $\rho(t)$ factorize, and
(ii) the molecular chaos hypothesis is valid.

This suggests the following form of the distribution function:

$$\rho(t) = \rho(\boldsymbol{r}_1, \ldots, \boldsymbol{r}_N) \cdot f(\boldsymbol{v}_1, t) \ldots f(\boldsymbol{v}_N, t) . \tag{80}$$

In accordance with the molecular chaos assumption the sequence of the successive collisions occurs without correlations. If the variable B does not depend on the positions of the particles, its time-correlation function decays exponentially [29]:

$$\langle B(t')B(t)\rangle = \left\langle B^2\right\rangle_{t'} e^{-\left|t-t'\right|/\tau_B(t')} \qquad (t > t') . \tag{81}$$

where $\langle\cdots\rangle_{t'}$ denotes the averaging with the distribution function taken at time t'. The relaxation time τ_B is inverse to the initial slope of the autocorrelation function [29], as it may be found from the time derivative of $\langle B(t')B(t)\rangle$ taken at $t = t'$. Equations (79) and (81) then yield

$$-\tau_B^{-1}(t') = \int d\Gamma \rho(t') B i\mathcal{L} B / \left\langle B^2\right\rangle_{t'} = \frac{\langle B i\mathcal{L} B\rangle_{t'}}{\langle B^2\rangle_{t'}} . \tag{82}$$

The relaxation time $\tau_B^{-1}(t')$ depends on time via the distribution function $\rho(t')$ and varies on the time-scale $t \gg \tau_c$.

Let $B(t)$ be the velocity of some particle, say $\boldsymbol{v}_1(t)$. Then with $3T(t) = \left\langle v^2\right\rangle_{t'}$ Eqs. (81) and (82) (with Eqs. (74) and (75)) read [10]

$$\langle \boldsymbol{v}_1(t') \cdot \boldsymbol{v}_1(t)\rangle = 3T(t') e^{-|t-t'|/\tau_v(t')} \tag{83}$$

$$-\tau_v^{-1}(t') = (N-1) \frac{\left\langle \boldsymbol{v}_1 \cdot \hat{T}_{12} \boldsymbol{v}_1 \right\rangle_{t'}}{\langle \boldsymbol{v}_1 \cdot \boldsymbol{v}_1\rangle_{t'}} . \tag{84}$$

To obtain Eq. (84) we take into account that $\mathcal{L}_0 \boldsymbol{v}_1 = 0$, $\hat{T}_{ij} \boldsymbol{v}_1 = 0$ (for $i \neq 1$) and the identity of the particles.

Straightforward calculation yields for the case of a constant restitution coefficient:

$$\tau_v^{-1}(t) = \frac{\epsilon+1}{2} \frac{8}{3} n \sigma^2 g_2(\sigma) \sqrt{\pi T(t)} = \frac{\epsilon+1}{2} \tau_E^{-1}(t) , \tag{85}$$

where $\tau_E(t) = \frac{3}{2}\tau_c(t)$ is the Enskog relaxation time [24]. Note that according to Eq. (85), $\tau_v = \frac{2}{1+\epsilon}\tau_E > \tau_E$, i.e., the velocity correlation time for inelastic collisions exceeds that of elastic collisions. This follows from partial suppression of the backscattering of particles due to inelastic losses in their normal relative motion, which, thus, leads to more stretched particle trajectories, as compared to the elastic case.

Similar (although somewhat more complicated) computations may be performed for the system of viscoelastic particles yielding

$$\tau_v^{-1}(t) = \tau_E^{-1}(t)\left[1 + \frac{3}{16}a_2(t) - 4q_0\,\delta\,u^{1/10}(t)\right], \tag{86}$$

where $q_0 = 0.173318$ has been already introduced and $a_2(t)$, $u(t)$ are the same as defined above. To obtain Eq. (86) we neglect terms of the order of $\mathcal{O}(a_2^2)$, $\mathcal{O}(\delta^2)$ and $\mathcal{O}(a_2\,\delta)$.

Using the velocity correlation function one writes

$$\left\langle (\Delta r(t))^2 \right\rangle = 2\int_0^t dt' 3T(t') \int_{t'}^t dt'' e^{-|t''-t'|/\tau_v(t')}\,. \tag{87}$$

On the short-time scale $t \sim \tau_c$, $T(t')$ and $\tau_v(t')$ may be considered as constants. Integrating in (87) over t'' and equating with (72) yields for $t \gg \tau_c \sim \tau_v$ the time-dependent self-diffusion coefficient

$$D(t) = T(t)\tau_v(t)\,. \tag{88}$$

Substituting the dependencies for $u(t) = T(t)/T_0$ and $a_2(t)$ as functions of time, which has been derived in the previous section, we obtain the time dependence of the coefficient of self-diffusion $D(t)$. For $t \gg \tau_0$ this may be given in an explicit form:

$$\frac{D(t)}{D_0} \simeq \left(\frac{t}{\tau_0}\right)^{-5/6} + \delta\left(4q_0 + q_1 + \frac{21}{400}h\right)\left(\frac{t}{\tau_0}\right)^{-1}, \tag{89}$$

where the constants q_0, q_1 and h are given above. Hence, the prefactor in the term proportional to δ reads $\left(4q_0 + q_1 + \frac{21}{400}h\right) = 2.24956$, and D_0 is the initial Enskog value of the self-diffusion coefficient

$$D_0^{-1} = \frac{8}{3}\pi^{1/2} n g_2(\sigma)\sigma^2 T_0^{-1/2}\,. \tag{90}$$

Correspondingly, the mean-square displacement reads asymptotically for $t \gg \tau_0$:

$$\left\langle (\Delta r(t))^2 \right\rangle \sim t^{1/6} + b\,\delta\,\log t + \dots\,, \tag{91}$$

where b is some constant. This dependence holds true for times

$$\tau_c(0)\,\delta^{-1} \ll t \ll \tau_c(0)\,\delta^{-11/5}\,. \tag{92}$$

The first inequality in Eq. (92) follows from the condition $\tau_0 \ll t$, while the second one follows from the condition $\tau_c(t) \ll \tau_0$, which means that temperature changes are slow on the collisional time-scale. For the constant restitution coefficient one obtains

$$T(t)/T_0 = [1 + \gamma_0 t/\tau_c(0)]^{-2} \,, \tag{93}$$

where $\gamma_0 \equiv \left(1 - \epsilon^2\right)/6$ [3, 9]. Thus, using Eqs. (85) and (88) one obtains for the mean-square displacement in this case

$$\left\langle (\Delta r(t))^2 \right\rangle \sim \log t \,. \tag{94}$$

As it follows from Eqs. (91) and (94) the impact-velocity dependent restitution coefficient, Eq. (10), leads to a significant change of the long-time behavior of the mean-square displacement of particles in the laboratory-time. Compared to its logarithmically weak dependence for the constant restitution coefficient, the impact-velocity dependence of the restitution coefficient gives rise to a considerably faster increase of this quantity with time, according to a power law.

One can also compare the dynamics of the system in its inherent-time scale. First we consider the average cumulative number of collisions per particle $\mathcal{N}(t)$ as an inherent measure for time (e.g. [2, 8]). It may be found by integrating $d\mathcal{N} = \tau_c(t)^{-1}dt$ [9]. For a constant restitution coefficient ϵ one obtains $\mathcal{N}(t) \sim \log t$, while for the impact-velocity dependent $\epsilon\left(v_{\text{imp}}\right)$ one has $\mathcal{N}(t) \sim t^{1/6}$. Therefore, the temperature and the mean-square displacement behave in these cases as

$\epsilon = \text{const}$	$\epsilon = \epsilon\left(v_{\text{imp}}\right)$
$T(\mathcal{N}) \sim e^{-2(1-\epsilon^2)\mathcal{N}}$	$T(\mathcal{N}) \sim \mathcal{N}^{-10}$
$\left\langle (\Delta r(\mathcal{N}))^2 \right\rangle \sim \mathcal{N}$	$\left\langle (\Delta r(\mathcal{N}))^2 \right\rangle \sim \mathcal{N}$

If the number of collisions per particle $\mathcal{N}(t)$ would be the relevant quantity specifying the stage of the granular gas evolution, one would conjecture that the dynamical behavior of a granular gas with a constant ϵ and velocity-dependent ϵ are identical, provided an $\mathcal{N}$-based time-scale is used. Whereas in equilibrium systems the number of collisions is certainly an appropriate measure of time, in nonequilibrium systems this value has to be treated with more care. As a trivial example may serve a particle bouncing back and forth between two walls, each time it hits a wall it loses part of its energy: If one describes this system using a $\mathcal{N}$-based time, one would come to the conclusion that the system conserves its energy, which is certainly not the proper description of the system. According to our understanding, therefore, the number of collision is not an appropriate time scale to describe physical reality. Sometimes, it may be even misleading.

Indeed, as it was shown in Ref. [8], the value of $\mathcal{N}_c$, corresponding to a crossover from the linear regime of evolution (which refers to the homogeneous cooling state) to the nonlinear regime (when clustering starts) may differ by orders of magnitude, depending on the restitution coefficient and on the density of the granular gas. Therefore, to analyze the behavior of a granular gas, one can try an alternative inherent time-scale, $\mathcal{T}^{-1} \equiv T(t)/T_0$ which is based on the gas temperature. Given two systems of granular particles at the same density and the same initial temperature T_0, consisting of particles colliding with constant and velocity-dependent restitution coefficient, respectively, the time $\mathcal{T}$ allows to compare directly their evolution. A strong argument to use a temperature-based time has been given by Goldhirsch and Zanetti [3] who have shown that as the temperature decays, the evolution of the system changes from a linear regime to a nonlinear one. Recent numerical results of Ref. [8] also support our assumption: It was shown that while $\mathcal{N}_c$ differs by more than a factor of three for two different systems, the values of $\mathcal{T}_c$, (defined, as $\mathcal{T}_c = T(\mathcal{N}_c)/T_0$) are very close [8]. These arguments show that one could consider $\mathcal{T}$ as a relevant time-scale to analyze the granular gas evolution.

With the temperature decay $T(\mathcal{N})/T_0 \sim e^{-2\gamma_0\mathcal{N}}$ for a constant restitution coefficient and $T(\mathcal{N})/T_0 \sim \mathcal{N}^{-10}$ for the impact-velocity dependent one, we obtain the following dependencies:

$\epsilon = \text{const}$	$\epsilon = \epsilon\left(v_{\text{imp}}\right)$
$T \sim \frac{1}{\mathcal{T}}$	$T \sim \frac{1}{\mathcal{T}}$
$\left\langle (\Delta r(\mathcal{T}))^2 \right\rangle \sim \log \mathcal{T}$	$\left\langle (\Delta r(\mathcal{T}))^2 \right\rangle \sim \mathcal{T}^{1/10}$

This shows that in the temperature-based time-scale, in which the cooling of both systems is synchronized, the mean-square displacement grows logarithmically slow for the case of constant restitution coefficient and much faster, as a power law, for the system of viscoelastic particles with $\epsilon = \epsilon(v_{\text{imp}})$. Thus, we conclude that clustering may be retarded for the latter system.

5 Conclusion

In conclusion, we considered kinetic properties of granular gases composed of viscoelastic particles, which implies the impact-velocity dependence of the restitution coefficient. We found that such dependence gives rise to some new effects in granular gas dynamics: (i) complicated, non-monotonous time-dependence of the coefficient a_2 of the Sonine polynomial expansion, which describes the deviation of the velocity distribution from the Maxwellian and (ii) enhanced spreading of particles, which depends on time as a power law, compared to a logarithmically weak dependence for the systems with a constant ϵ.

The Table below compares the properties of granular gases consisting of particles interacting via a constant coefficient of restitution ϵ = const and consisting of viscoelastic particles where the collisions are described using an impact velocity dependent restitution coefficient $\epsilon = \epsilon(v_{\rm imp})$:

$\epsilon = \text{const}$	$\epsilon = \epsilon\left(v_{\rm imp}\right)$
ϵ is a model parameter	$\epsilon = 1 - C_1 A \kappa^{2/5} v_{\rm imp}^{1/5} + \cdots$ $C_1 = 1.15396$, $C_2 = \frac{3}{5} C_1^2$, ... $\kappa = \kappa(Y, \nu, m, R)$ $A = A(\eta_1, \eta_2, Y, \nu)$ all quantities are defined via parameters of the particle material Y, ν, $\eta_{1/2}$ and their mass and radius.
Small parameter	
$1 - \epsilon^2$ – does not depend on the state of the system	$\delta = A \kappa^{2/5} T_0^{1/10}$ – depends on the initial temperature T_0.
Temperature	
$T = T_0 \left(1 + t/\tau_0'\right)^{-2}$	$T = T_0 \left(1 + t/\tau_0\right)^{-5/3}$
Velocity distribution	
$f(\boldsymbol{v}, t) = \frac{n}{v_0^3(t)} \tilde{f}(\boldsymbol{c})$ $\tilde{f}(\boldsymbol{c}, t) = \phi(c) \left\{1 + \sum_{p=1}^{\infty} a_p S_p(c^2)\right\}$ $a_2 = \text{const.}$	$f(\boldsymbol{v}, t) = \frac{n}{v_0^3(t)} \tilde{f}(\boldsymbol{c}, t)$ $\tilde{f}(\boldsymbol{c}) = \phi(c) \left\{1 + \sum_{p=1}^{\infty} a_p(t) S_p(c^2)\right\}$ $a_2 = a_2(t)$ – is a (complicated) function of time.
Self-diffusion	
$\left\langle \left(\Delta r(t)\right)^2 \right\rangle \sim \log t$	$\left\langle \left(\Delta r(t)\right)^2 \right\rangle \sim t^{1/6}$

Acknowledgements

We thank M. H. Ernst and I. Goldhirsch for valuable discussions.

References

1. Y. Du, H. Li, and L. P. Kadanoff, Phys. Rev. Lett. **74**, 1268 (1995); T. Zhou and L. P. Kadanoff, Phys. Rev. E **54**, 623 (1996); A. Goldshtein, M. Shapiro, and C. Gutfinger, J. Fluid Mech. **316**, 29 (1996); A. Goldshtein, V. N. Poturaev, and I. A. Shulyak, Izvestiya Akademii Nauk SSSR, Mechanika Zhidkosti i Gaza **2**, 166 (1990); J. T. Jenkins and M. W. Richman, Arch. Part. Mech. Materials **87**, 355 (1985); V. Buchholtz and T. Pöschel, Granular Matter **1**, 33 (1998); E. L. Grossman, T. Zhou, and E. Ben-Naim, Phys. Rev. E **55**, 4200 (1997); F. Spahn, U. Schwarz, and J. Kurths, Phys. Rev. Lett. **78**, 1596 (1997); S. Luding and H. J. Herrmann, Chaos **9**, 673 (1999); S. Luding and S. McNamara, Gran. Matter **1**, 113 (1998); T. P. C. van Noije, M. H. Ernst, and R. Brito, Phys. Rev. E **57**, R4891 (1998); J. A. C. Orza, R. Brito, T. P. C. van Noije, and M. H. Ernst, Int. J. Mod. Phys. C **8**, 953 (1997); N. Sela and I. Goldhirsch, J. Fluid. Mech., **361**, 41 (1998); N. Sela, I. Goldhirsch, and H. Noskowicz, Phys. Fluids **8**, 2337 (1996); I. Goldhirsch, M.-L. Tan, and G. Zanetti, J. Sci. Comp. **8**, 1 (1993); T. Aspelmeier, G. Giese, and A. Zippelius, Phys. Rev. E **57**, 857 (1997); J. J. Brey and D. Cubero, Phys. Rev. E **57**, 2019 (1998); J. J. Brey, J. W. Dufty, C. S. Kim, and A. Santos, Phys. Rev. E **58**, 4648 (1998); J. J. Brey, M. J. Ruiz-Montero, and F. Moreno, Phys. Rev. E **55**, 2846 (1997); J. J. Brey, D. Cubero, and M. J. Ruiz-Montero, Phys. Rev. E **59**, 1256 (1999); V. Kumaran, Phys.Rev.E, **59**, 4188, (1999); V. Garzo and J. W. Dufty, Phys. Rev. E, **59**, 5895, (1999).
2. S. McNamara and W. R. Young, Phys. Rev. E **53**, 5089 (1996).
3. I. Goldhirsch and G. Zanetti, *Phys. Rev. Lett.*, **70**, 1619 (1993).
4. T. P. C. van Noije and M. H. Ernst, Granular Matter, **1**, 57 (1998).
5. S. E. Esipov and T. Pöschel, J. Stat. Phys., **86**, 1385 (1997).
6. T. P. C. van Noije, M. H. Ernst and R.Brito, Physica A **251**, 266 (1998).
7. A. Goldshtein and M. Shapiro, J. Fluid. Mech., **282**, 75 (1995).
8. R. Brito and M. H. Ernst, Europhys. Lett. **43**, 497 (1998).
9. T. P. C. van Noije and M. H. Ernst, R. Brito and J. A. G. Orza, Phys. Rev. Lett. **79**, 411 (1997).
10. N. V. Brilliantov and T. Pöschel, Phys. Rev. E **61**, 1716 (2000).
11. M. Huthmann and A. Zippelius, Phys. Rev. E. **56**, R6275 (1997); S. Luding, M. Huthmann, S. McNamara and A. Zippelius, Phys. Rev. E. **58**, 3416 (1998).
12. C. Thornton, *Contact mechanics and coefficients of restitution*, (in this volume, page 184).
13. R. Ramírez, T. Pöschel, N. V. Brilliantov, and T. Schwager, Phys. Rev. E **60**, 4465 (1999).
14. Y. Taguchi, Europhys. Lett. **24**, 203 (1993).
15. S. Luding, *Collisions and Contacts between two particles*, in: H. J. Herrmann, J. -P. Hovi, and S. Luding (eds.), *Physics of dry granular media - NATO ASI Series E350*, Kluwer (Dordrecht, 1998), p. 285.
16. N. V. Brilliantov, F. Spahn, J.-M. Hertzsch, and T. Pöschel, Phys. Rev. E **53**, 5382 (1996).
17. J.-M. Hertzsch, F. Spahn, and N. V. Brilliantov, J. Phys. II (France), **5**, 1725 (1995).
18. Derivation of the dissipative force given in [16, 17] for colliding spheres may be straightforwardly generalized to obtain the relation (5) (or (A17) in [16, 17])

for colliding bodies of any shape, provided that displacement field in the bulk of the material of bodies in contact is a one-valued function of the compression (see also [19]).
19. N. V. Brilliantov and T. Pöschel, in preparation.
20. L. D. Landau and E. M. Lifschitz, *Theory of Elasticity*, Oxford University Press (Oxford, 1965).
21. H. Hertz, J. f. reine u. angewandte Math. **92**, 156 (1882).
22. G. Kuwabara and K. Kono, Jpn. J. Appl. Phys. **26**, 1230 (1987).
23. T. Schwager and T. Pöschel, Phys. Rev. E **57**, 650 (1998).
24. P. Resibois and M. de Leener, *Classical Kinetic Theory of Fluids* (Wiley, New York, 1977).
25. N. F. Carnahan, and K. E. Starling, J. Chem. Phys., **51**, 635 (1969).
26. N. V. Brilliantov and T. Pöschel, Phys. Rev. E, **61**, 5573 (2000)
27. The term "pseudo" was initially referred to the dynamics of systems with singular hard-core potential [24, 30].
28. For application to "ordinary" fluids, see [29] and to granular systems [6, 11]. A rigorous definition of $\mathcal{L}$ includes a pre-factor, preventing successive collisions of the same pair of particles [24, 30] which, however, does not affect the present analysis.
29. D. Chandler, J. Chem. Phys. **60**, 3500, 3508 (1974); B. J. Berne, *ibid* **66**, 2821, (1977); N. V. Brilliantov and O. P. Revokatov, Chem. Phys. Lett. **104**, 444 (1984).
30. M. H. Ernst, J. R. Dorfman, W. R. Hoegy and J. M. J. van Leeuwen, Physica A **45**, 127 (1969).

Boundary Conditions for Collisional Grain Flows at Bumpy, Frictional Walls

James T. Jenkins

Department of Theoretical and Applied Mechanics
Cornell University, Ithaca, NY 14853 USA. e-mail: jtj2@cornell.edu

Abstract. We outline the derivation of conditions that determine the slip velocity and flux of fluctuation energy for collisional flows of frictional spheres at bumpy, frictional boundaries with either cylindrical or spherical bumps. We illustrate their use by solving for the profiles of fluctuation velocity, mean velocity, and volume fraction in steady, fully developed shearing flows between boundaries with frictional, cylindrical features.

1 Introduction

Collisional grain flows are often driven by and inevitably influenced by the surfaces that bound them. Because the thickness of such flows is often limited to tens of particles, the influence of the boundaries can be extremely pervasive. Experiments carried out in shear cells with the boundaries roughened in different ways provide a clear indication of this [1–3].

Boundary conditions for collisional flows of smooth disks and spheres have been derived by calculating the average collisional transfer of momentum and energy from flat walls made bumpy by attaching to them frictionless disks or spheres [4–7]. These boundary conditions all incorporate the fact that slip occurs at a bumpy boundary. They differ in regard to the distribution function employed to do the averaging. Because of the collisions that result from the slip at the boundary, the energy of the mean flow may be converted to fluctuation energy there. The possibility that fluctuation energy may be supplied to or removed from the flow at a bumpy boundary appears to be crucial to the understanding of many collisional grain flows [8]. Such boundary conditions have also been tested in numerical simulations of collisional shear flows of frictionless disks [9] and spheres [10, 11]. The agreement between the predictions and quantities measured in the simulations is, in general, good.

Friction is also important to the transfer of momentum and energy at boundaries. It has been incorporated in boundary conditions in a heuristic way by assuming that some fraction of the particles are colliding and the remainder are sliding, resisted by Coulomb friction [12]. Also, boundary conditions at a flat, frictional wall have been derived [13] based upon a model of a collision that distinguishes between sticking and sliding collisions [14, 15]. This explicit calculation of the average rate transfer of momentum and energy has been tested in numerical simulations [16] and improved upon [17], making use of the information provided by the numerical simulations.

In this paper, we first use existing expressions for the collisional shear stress and rate of dissipation at a frictionless boundary with cylindrical bumps [7] and at a flat, frictional wall [17] to obtain boundary conditions for a collisional flow of frictional spheres at a frictional boundary with cylindrical bumps. This is done by the simple expedient of assuming that in each collision, the total tangential impulse is the sum of that due to the bumpiness of a frictionless boundary and that due to the friction at a flat wall. This ignores the influence of the geometry of the bumps on the frictional transfer of momentum, but it provides a simple method of incorporating both the geometry and friction using existing results.

We use these boundary conditions and the continuum equations that result from the kinetic theory in the limit of dense flows to phrase and solve boundary value problems for two steady, fully developed shearing flows between boundaries with frictional cylindrical bumps. In the first problem, the boundaries are identical; in the second, the upper boundary is bumpier than the lower boundary. Analytical solutions are obtained for the profiles of fluctuation velocity, mean velocity, and volume fraction. Also, the values of the shear stress and pressure required to maintain a flow with a given gap width, boundary velocity, and filling are predicted.

Finally, we provide the formula that lead to the analogous boundary conditions for boundaries with frictional spherical bumps and, in an Appendix, give an indication of how the bumpiness of such boundaries is characterized.

2 Boundaries with Frictional, Cylindrical Features

2.1 Preliminaries

The boundary geometry is defined by the bumpiness θ and the mean diameter $\bar{\sigma}$. These are given in terms of the diameter d of the boundary cylinders, the diameter σ of the flow spheres, and the spacing s between the edges of the boundary cylinders by [4]

$$\sin\theta \equiv \frac{d+s}{d+\sigma} \text{ and } \bar{\sigma} \equiv \frac{1}{2}\left(\sigma + d\right).$$

The geometry of the boundary is sketched in Fig. 1.

In the flow, the particle pressure p may be expressed in terms of two functions of the solid volume fraction ν and the mass density ρ_s of the material of the spheres by [5]

$$p = 4\rho_s \nu G F T, \tag{1}$$

where

$$G \equiv \frac{\nu\left(2-\nu\right)}{2\left(1-\nu\right)^3}, \quad F \equiv 1 + \frac{1}{4G},$$

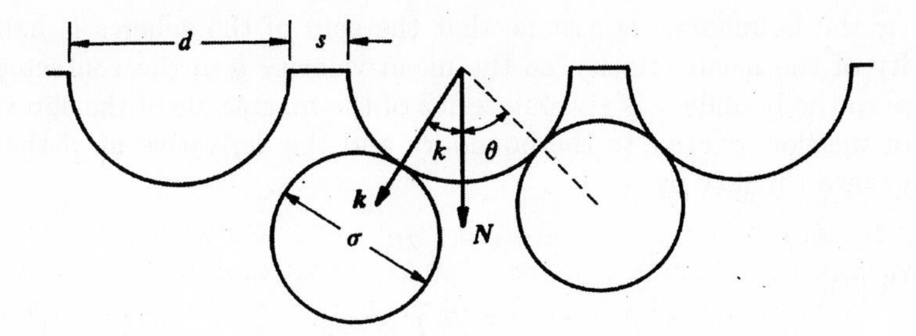

Fig. 1. The boundary geometry

and $\sqrt{3T}$ is the r.m.s value of the velocity fluctuations. When the flow is rectilinear and fully developed, the shear stress S is given by [5]

$$S = \frac{2}{5\sqrt{\pi}} \frac{J}{F} \frac{\sigma p}{\sqrt{T}} u' , \tag{2}$$

where

$$J \equiv 1 + \frac{\pi}{12}\left(1 + \frac{5}{8G}\right)^2 ,$$

u is the mean velocity of the flow, and the prime denotes a derivative across the flow. Similarly, the flux Q of fluctuation energy across the flow is given by [5]

$$Q = -\frac{1}{\sqrt{\pi}} \frac{M}{F} \frac{\sigma p}{\sqrt{T}} T', \tag{3}$$

where

$$M \equiv 1 + \frac{9\pi}{32}\left(1 + \frac{5}{12G}\right)^2 .$$

Collisions are of two types: sliding and sticking [13]. In a sliding collision between a sphere and the boundary, the coefficient of friction is μ; the corresponding coefficient in a collision between two spheres is μ_f. In a sticking collision between a sphere and the boundary, the coefficient of tangential restitution is β. In a collision between a sphere and the boundary, the coefficient of normal restitution is e; the corresponding coefficient in a collision between two spheres is e_f. We assume that e and e_f are near one and that the coefficients of sliding friction μ and μ_f are small, so that collisions are nearly elastic. We introduce the quantity

$$\overline{\mu} \equiv \frac{7}{2} \frac{(1+e)}{(1+\beta)} \mu \approx 7 \frac{\mu}{(1+\beta)}$$

that is important in distinguishing between sliding and sticking collisions at the wall [13].

Near the boundary, we assume that the spin of the spheres is half the vorticity of the mean velocity, so the mean velocity g of the contact point relative to the boundary is given in terms of the magnitude of the slip velocity v of the flow relative to the boundary and the derivative u' of the flow velocity into the flow by

$$g = v - \frac{1}{4}\bar{\sigma}u',$$

or, with (2),

$$g = v - \frac{5\sqrt{\pi}}{8}\frac{F}{J}\frac{S}{p}\sqrt{T}.$$

At the boundary, the collisional rates of production of momentum and energy are calculated by averaging the change in momentum and energy in a single collision between a flow sphere and the boundary over all possible collisions. When the boundary is bumpy but frictionless, this involves summing the collisions over the surface of the bumps [5]; when the boundary is flat but frictional, this involves distinguishing between sticking and sliding collisions [13]. In the former case, the velocity distribution function used to carry out the average is

$$f(\boldsymbol{C},\boldsymbol{r}) = \frac{1}{(2\pi T)^{3/2}}\frac{6\nu}{\pi\sigma^3}\left[1 - \left(\frac{2}{\pi}\right)^{1/2}\frac{\sigma B}{T^{3/2}}C_i u_{i,j} C_j\right]\exp\left(-\frac{C^2}{2T}\right),$$

where $\boldsymbol{C}$ is the particle velocity fluctuation, the mean fields are evaluated at a position $\boldsymbol{r}$ a distance $\overline{\sigma}/2$ from the flat part of the boundary and the coefficient B, associated with the perturbation to a Maxwellian distribution, is given by

$$B \equiv \frac{\pi}{12\sqrt{2}}\left(1 + \frac{5}{8G}\right).$$

In the latter case, a Maxwellian velocity distribution or something simpler is employed.

2.2 Shear Stress

As mentioned earlier, we assume that in each collision between a flow sphere and the boundary, a portion of the tangential impulse is due to the bumpiness of a frictionless boundary and a portion is due to friction at a flat wall. The frictional contribution is calculated in two limits. In the first, the points of contact of all colliding particles are assumed to stick; in the second, the points of contact of all colliding particles are assumed to slide. Then the shear stress S at the boundary is the sum of the tangential components of the average collisional rates of production of momentum per unit area S^B in frictionless collisions with cylindrical bumps and S^F in frictional collisions with a flat boundary:

$$S = S^B + S^F,$$

where [7]

$$
\begin{aligned}
S^B &= \left(\frac{2}{\pi}\right)^{1/2} \frac{p}{\sqrt{T}} \{ v \left(\theta \csc\theta - \cos\theta\right) \\
&\quad - \bar{\sigma} u' \left[\left(\theta \csc\theta - \cos\theta\right) - \frac{2}{3}\left(1 + \frac{\sigma}{\bar{\sigma}} B\right) \sin^2\theta \right] \} \\
&= \left(\frac{2}{\pi}\right)^{1/2} \frac{v}{\sqrt{T}} p \left(\theta \csc\theta - \cos\theta\right) \\
&\quad - \frac{5}{\sqrt{2}} \frac{\bar{\sigma}}{\sigma} \frac{F}{J} S \left[\left(\theta \csc\theta - \cos\theta\right) - \frac{2}{3}\left(1 + \frac{\sigma}{\bar{\sigma}} B\right) \sin^2\theta \right];
\end{aligned}
$$

and, for sticking collisions, $S^F \leq \mu p$, [13]

$$
\begin{aligned}
S^F &= \frac{3}{14}\left(1+\beta\right) \frac{g}{\sqrt{3T}} p \\
&= \frac{\sqrt{3}}{14}\left(1+\beta\right) \left(\frac{v}{\sqrt{T}} - \frac{5\sqrt{\pi}}{8} \frac{F}{J} \frac{S}{p} \right) p,
\end{aligned}
$$

while, for sliding collisions, [13]

$$
S^F = \mu p.
$$

The expressions for the shear stress S that result in the two regimes of frictional behavior are linear in the ratio $v/\sqrt{T}$, because in both the derivation of S^B and S^F, terms of higher order have been ignored.

2.3 Slip Velocity

In the regime of sticking collisions, the shear stress may be expressed in terms of the slip velocity and inverted to yield

$$
\begin{aligned}
\frac{v}{\sqrt{T}} &= \left(\frac{\pi}{2}\right)^{1/2} \frac{1}{\left(\theta \csc\theta - \cos\theta\right) + \left(\frac{\pi}{2}\right)^{1/2} \frac{\sqrt{3}}{14}\left(1+\beta\right)} \\
&\quad \times \left\{ 1 + \frac{15}{112}\left(\frac{\pi}{3}\right)^{1/2} \left(1+\beta\right) \frac{F}{J} \right. \\
&\quad \left. + \frac{\bar{\sigma}}{\sigma} \frac{5}{\sqrt{2}} \frac{F}{J} \left[\left(\theta \csc\theta - \cos\theta\right) - \frac{2}{3}\left(1 + \frac{\sigma}{\bar{\sigma}} B\right) \sin^2\theta \right] \right\} \frac{S}{p}.
\end{aligned}
$$

Similarly, for sliding collisions

$$
\begin{aligned}
\frac{v}{\sqrt{T}} &= \left(\frac{\pi}{2}\right)^{1/2} \frac{1}{\left(\theta \csc\theta - \cos\theta\right)} \left\{ \left(\frac{S}{p} - \mu \right) \right. \\
&\quad \left. + \frac{\bar{\sigma}}{\sigma} \frac{5}{\sqrt{2}} \frac{F}{J} \left[\left(\theta \csc\theta - \cos\theta\right) - \frac{2}{3}\left(1 + \frac{\sigma}{\bar{\sigma}} B\right) \sin^2\theta \right] \frac{S}{p} \right\}.
\end{aligned}
$$

Note that the characterization of the region of sticking collisions may be phrased in terms of the total stress S as

$$S - S^B \leq \mu p$$

with $v/\sqrt{T}$ given in terms of S in S^B by its expression for sticking collisions..

2.4 Energy Flux

In a similar fashion, the flux Q of fluctuation energy into the flow at the boundary is the sum of the bumpy, frictionless flux Q^B and the additional flat, frictional flux ΔQ^F:

$$Q = Q^B + \Delta Q^F,$$

where

$$Q^B = S^B v - D^B,$$

with [4, 7],

$$D^B = \left(\frac{2}{\pi}\right)^{1/2} p\sqrt{T}(1-e)\theta \csc\theta .$$

The additional frictional flux ΔQ^F is due to the working of the frictional shear stress and the additional frictional dissipation ΔD^F:

$$\Delta Q^F = S^F g - \Delta D^F,$$

where, for sticking collisions, up to an error of order $\mu\overline{\mu}^3$ [17],

$$\Delta D^F \equiv \left\{\left(\frac{\pi}{2}\right)^{1/2}\left[\frac{\pi}{2}\mu\left(1-\frac{g^2}{3T}\right) - 7\mu^2\right] + \frac{\sqrt{3}}{14}(1+\beta)\frac{g^2}{T}\right\} p\sqrt{T},$$

and, for sliding collisions [17],

$$\Delta D^F \equiv \left(\frac{2}{\pi}\right)^{1/2}\left[\left(\frac{\pi}{2}\right)^{1/2}\mu\frac{g}{\sqrt{T}} - \frac{7}{3}\mu^2\right] p\sqrt{T}.$$

Consequently, for sticking collisions,

$$\Delta Q^F = -\left(\frac{\pi}{2}\right)^{1/2}\left[\frac{\pi}{2}\mu\left(1-\frac{g^2}{3T}\right) - 7\mu^2 + \frac{3}{14}(1+\beta)\frac{g}{\sqrt{3T}}\left(\frac{2}{\pi}\right)^{1/2}\frac{v}{\sqrt{T}}\right] p\sqrt{T};$$

while, for sliding collisions,

$$\Delta Q^F = -\left(\frac{2}{\pi}\right)^{1/2}\left[-\frac{7}{3}\mu^2 + \left(\frac{\pi}{2}\right)^{1/2}\frac{v}{\sqrt{T}}\mu\right] p\sqrt{T}.$$

3 A Boundary-Value Problem

3.1 Boundary Conditions

We consider a dense, steady, fully-developed shearing flow of identical spheres of diameter σ that are made of a material of density ρ_s. In the limit of dense flows, G is relatively large and (e.g. [8, 18])

$$F \approx 1, \ J \approx 1 + \frac{\pi}{12}, \ M \approx 1 + \frac{9\pi}{32}, \text{ and } B \approx \frac{\pi}{12\sqrt{2}}.$$

The flow is maintained by the relative velocity U of two parallel bumpy frictional boundaries separated by a given distance L. The boundary conditions are applied at the position of the centers of the flow spheres that are in contact with the tips of the boundary cylinders. The subscripts 0 and 1 denote the lower and upper boundaries, respectively. We first determine the profile of the fluctuation velocity $w \equiv \sqrt{T}$.

In this simple flow, the balance of fluctuation energy has the form (e.g. [8, 18])

$$-Q' + Su' - \gamma = 0,$$

where the rate of dissipation γ is given by

$$\gamma = \frac{6}{\sqrt{\pi}}(1 - e_{\text{eff}})\frac{pw}{\sigma}, \tag{4}$$

with [19]

$$e_{\text{eff}} \equiv e_f - \frac{\pi}{2}\mu_f.$$

With (3), (2), and (4), the balance of fluctuation energy within the cell may be written as

$$\sigma^2 w'' + K^2 w = 0, \tag{5}$$

where

$$K^2 \equiv \frac{1}{M}\left[\frac{5\pi}{4J}\left(\frac{S}{p}\right)^2 - 3(1 - e_{\text{eff}})\right].$$

We write the solution of this equation as

$$w(y) = A \sin\left(\frac{Ky}{\sigma}\right) + w_0 \cos\left(\frac{Ky}{\sigma}\right),$$

where A is a constant to be determined and $w_0 \equiv w(0)$.

The boundary condition at the upper boundary is

$$Q_1 = -\left(\frac{2}{\pi}\right)^{1/2}\left[\frac{\pi}{2}f_1\left(\frac{S}{p}\right)^2 - h_1(1 - e_1) + \Delta b_1^F\right] pw_1$$

$$\equiv -b_1\left(\frac{2}{\pi}\right)^{1/2} M\sqrt{2}pw_1 \ ;$$

while that at the lower boundary is

$$Q_0 = \left(\frac{2}{\pi}\right)^{1/2} \left[\frac{\pi}{2} f_0 \left(\frac{S}{p}\right)^2 - h_0 (1 - e_0) + \Delta b_0^F\right] p w_0$$

$$\equiv b_0 \left(\frac{2}{\pi}\right)^{1/2} M \sqrt{2} p w_0 \ .$$

Here f and h are given in terms of the bumpiness θ of the boundary and the parameters e, μ and β that characterize a collision by

$$h = \theta \csc \theta,$$

and, for sticking collisions,

$$f \equiv \frac{1}{(\theta \csc \theta - \cos \theta) + \left(\frac{\pi}{2}\right)^{1/2} \frac{\sqrt{3}}{14} (1 + \beta)} \times$$
$$\left\{1 + \frac{15}{112} \left(\frac{\pi}{3}\right)^{1/2} (1 + \beta) \frac{1}{J} + \frac{5}{\sqrt{2}J} \left[(\theta \csc \theta - \cos \theta) - \frac{2}{3} (1 + B) \sin^2 \theta\right]\right\}$$

and

$$\Delta b^F \equiv -\left(\frac{\pi}{2}\right) \left[\frac{\pi}{2} \mu \left(1 - \frac{g^2}{3T}\right) - 7\mu^2\right]$$
$$- \frac{\pi}{2} f \frac{S}{p} \frac{\sqrt{3}}{14} (1 + \beta) \frac{g}{\sqrt{T}};$$

while, for sliding collisions,

$$f \equiv \frac{1}{(\theta \csc \theta - \cos \theta)} \left\{\left(1 - \mu \frac{p}{S}\right)\right.$$
$$\left. + \frac{5}{\sqrt{2}J} \left[(\theta \csc \theta - \cos \theta) - \frac{2}{3} (1 + B) \sin^2 \theta\right]\right\}$$

and

$$\Delta b^F \equiv \frac{7}{3} \mu^2 - \frac{\pi}{2} f \frac{S}{p} \mu.$$

The boundary condition at $y = 0$ requires that

$$A = -\frac{b_0}{K} w_0.$$

The corresponding condition at $y = L$ provides the relation between L and S/p that must be satisfied for the flow to be steady and fully-developed:

$$\tan\left(\frac{KL}{\sigma}\right) = \frac{(b_1 + b_0) K}{(b_0 b_1 - K^2)}.$$

When K is imaginary, we write $K = -ik$ and express the solutions in terms of the corresponding hyperbolic functions.
As a consequence of the above relation, the $w_1 \equiv w(L)$ is related to w_0 by

$$w_1 = w_0 \left(\frac{b_0^2 + K^2}{b_1^2 + K^2}\right)^{1/2} .$$

3.2 The Mean Velocity

The derivative of the mean velocity u across the flow is given in terms of the fluctuation velocity and the stress ratio though the constitutive relation (2) for the shear stress:

$$\frac{du}{dy} = \frac{5\pi^{1/2}}{2J}\frac{w}{\sigma}\frac{S}{p} .$$

Upon integrating this in the dense limit, we obtain

$$u(y) = \frac{5\pi^{1/2}}{2J}\frac{S}{p}\frac{1}{K}w_0\left[\frac{b_0}{K}\cos\left(\frac{Ky}{\sigma}\right) + \sin\left(\frac{Ky}{\sigma}\right)\right] + B ,$$

where B is a constant to be determined by the boundary condition on the slip velocity.

The slip velocity is given in terms of the stress ratio and fluctuation velocity by

$$v = \left(\frac{\pi}{2}\right)^{1/2} f\frac{S}{p}w .$$

At the lower boundary, $u_0 = v_0$; so

$$B = \left(\frac{\pi}{2}\right)^{1/2}\left(f_0 - \frac{5}{\sqrt{2}J}\frac{b_0}{K^2}\right)\frac{S}{p}w_0 .$$

The velocity of the upper boundary is $U = u_1 + v_1$, so

$$U = \left(\frac{\pi}{2}\right)^{1/2}\left\{\frac{5}{\sqrt{2}J}\frac{b_0}{K^2}\left[\cos\left(\frac{KL}{\sigma}\right) + \frac{K}{b_0}\sin\left(\frac{KL}{\sigma}\right)\right]\right.$$
$$\left. + f_0 - \frac{5}{\sqrt{2}J}\frac{b_0}{K^2} + f_1\left(\frac{b_0^2+K^2}{b_1^2+K^2}\right)^{1/2}\right\}\frac{S}{p}w_0 .$$

The trigonometric terms involving L/σ may be eliminated by integrating the differential equation (5) across the gap and using the boundary conditions and the solution to obtain

$$b_0\cos\left(\frac{KL}{\sigma}\right) + K\sin\left(\frac{KL}{\sigma}\right) = -b_1\frac{w_1}{w_0} .$$

Then

$$\frac{U}{w_0} = \left(\frac{\pi}{2}\right)^{1/2}\frac{1}{K^2}\left[-\frac{\alpha_0}{M} - \frac{5}{\sqrt{2}J}b_1\left(\frac{b_0^2+K^2}{b_1^2+K^2}\right)^{1/2} + f_1\left(\frac{b_0^2+K^2}{b_1^2+K^2}\right)^{1/2}\right]\frac{S}{p},$$

where

$$\alpha_0 \equiv 3(1-e_{\text{eff}})f_0 - \frac{5}{2J}\left[(1-e_0)h_0 - \Delta b_0^F\right] .$$

Given S/p, this determines w_0 in terms of U.

3.3 Volume Fraction

The distribution of volume fraction is obtained from the constitutive relation (1) for the pressure, using the solution for the fluctuation velocity:

$$\frac{2\nu^2(2-\nu)}{(1-\nu)^3} = \frac{p}{\rho_s U^2}\left(\frac{U}{w}\right)^2 .$$

Given $p/\rho_s U^2$, this cubic equation may be solved to obtain $\nu(y)$.

4 Graphical Results

In Fig. 2 we show profiles of fluctuation velocity, mean velocity, and solid volume fraction for two shear cells with the same thickness, $L/\sigma = 8.20$, and filling, $\nu_{avg} = 0.50$. In both cells we employ typical values for the collision parameters [15]: $e_f = 0.95$, $\mu_f = 0.10$, $e_0 = e_1 = 0.80$, $\mu_0 = \mu_1 = 0.10$, $\beta_0 = \beta_1 = 0.40$; and we take $\theta_0 = \pi/5$. The cells differ in the bumpiness of their upper boundary. The first cell is symmetric with $\theta_1 = \pi/5$; in the second, non-symmetric cell, $\theta_1 = \pi/4$.

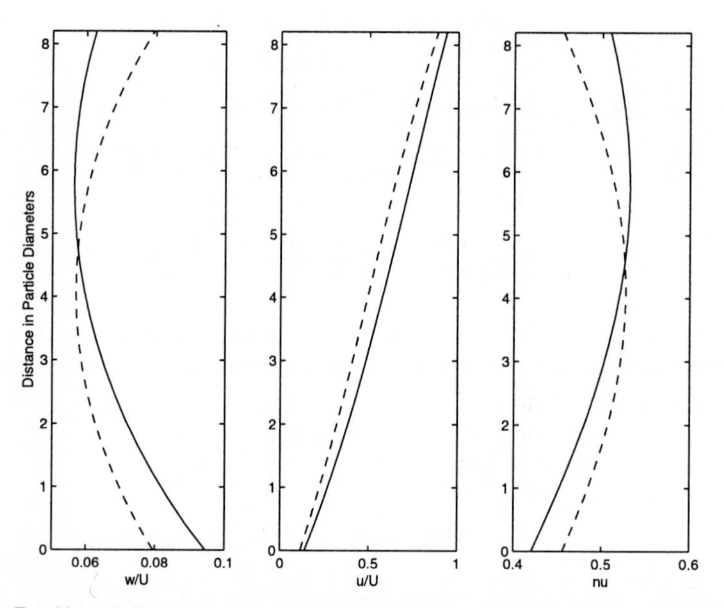

Fig. 2. Profiles of dimensionless fluctuation velocity, dimensionless mean velocity, and solid volume fraction in shear cells with the same thickness and filling, but with different upper boundaries

In obtaining the solutions, we first iterate to determine the stress ratio that results in the given gap width. With this, we calculate the profiles of w/U

and u/U. Then, we iterate to determine the value of the normalized pressure that results in the given average volume fraction. For the symmetric solution, $S/p = 0.41572$ and $p/\rho_s U^2 = 0.0251$; for the non-symmetric solution, $S/p = 0.42107$ and $p/\rho_s U^2 = 0.0257$. The number of significant figures retained is an indication of the sensitivity of the stress ratio and the normalized pressure on the gap width and average volume fraction, respectively.

As indicated in Fig. 2, the slip velocity is decreased when the upper boundary is made bumpier. As a consequence, the strength of the velocity fluctuations is lower there and, because the pressure is constant in the cell, the volume fraction is higher. In general, in a cell of fixed gap width, changes in the boundary influence the fluctuation energy in the cell in two ways: first, by changing the stress ratio and the rate of production of fluctuation energy in the interior and, second, by changing the flux of fluctuation energy at the boundary. Changes in the profiles of fluctuation velocity in the cell result from a competion between these two mechanisms. Finally, we note that for the symmetric solution, $v_0/w_0 = 1.4491$ and, for the asymmetric solution, $v_0/w_0 = 1.4740$. Numerical simulations [20] indicate that these values of the slip velocity are very near the largest for which a theory that is linear in v/w applies.

Acknowledgments

This research was sponsored by the Life Sciences and Microgravity Research Program of the National Aeronautics and Space Administration.

Appenidx

A Bumpiness for Spheres

We suppose that identical spheres are attached to a flat wall in a periodic way. Let σ be the diameter of a sphere in the flow, d be the diameter of a sphere on the wall, and s be the closest distance between the surfaces of neighboring wall spheres. We define

$$\varepsilon \equiv \frac{\sigma + s}{\sigma + d},$$

and assume that s/d and d/σ are such that a flow sphere never penetrates deeply enough to touch the flat wall. Then ε is less than one. For example, for equal diameter flow spheres and wall spheres, at the greatest permitted separation, ε is 0.85361.

We want to characterize how, on average, a sphere of the flow can penetrate between two spheres on the wall. Consider a spherical polar coordinate system centered on a wall sphere. The line normal to the wall is the polar axis, θ is the angle between the polar axis and a line joining the centers of a wall sphere and a flow sphere, and ϕ is the circumferential angle, assumed to have a value of $\pi/2$ when the distance between the surfaces of the wall spheres is s. Then the relation between θ and the circumferential angle ϕ that must obtain for the flow sphere to be in contact with at least two wall spheres is

$$\theta = \arcsin\left(\frac{\varepsilon}{\sin\phi}\right).$$

The smallest interval of ϕ over which θ ranges from its largest to smallest values depends upon the type of periodic packing. We take $\phi = \pi/2$ to be the endpoint of this interval that corresponds to the smallest value and denote by ϕ_0 the endpoint that corresponds to the largest. Then, for a square packing, $\phi_0 = \pi/4$; while, for an hexagonal packing, $\phi_0 = \pi/3$.

The average angle of penetration $\overline{\theta}$ is obtained by integrating over this interval and dividing by the difference in angle:

$$\overline{\theta} = \frac{1}{(\pi/2) - \phi_0} \int_{\phi_0}^{\frac{\pi}{2}} \arcsin\left(\frac{\varepsilon}{\sin\phi}\right) d\phi.$$

We expand the integrand in powers of ε:

$$\arcsin\left(\frac{\varepsilon}{\sin\phi}\right) = \frac{1}{\sin\phi}\varepsilon + \frac{1}{6\sin^3\phi}\varepsilon^3 + O\left(\varepsilon^5\right)$$

and integrate

$$\begin{aligned}
&\int_{\phi_0}^{\frac{\pi}{2}} \left(\frac{1}{\sin\phi}\varepsilon + \frac{1}{6\sin^3\phi}\varepsilon^3\right) d\phi \\
&= -\left[\ln\left(\sin\phi_0\right) - \ln\left(\cos\phi_0 + 1\right)\right]\varepsilon \\
&+ \frac{1}{12}\left[\cos\phi_0 \csc^2\phi_0 - \ln\left(\sin\phi_0\right) + \ln\left(\cos\phi_0 + 1\right)\right]\varepsilon^3.
\end{aligned}$$

For a square packing,

$$\overline{\theta} = 0.1061\varepsilon\left(10.576 + 2.2956\varepsilon^2\right);$$

while for a hexagonal packing,

$$\overline{\theta} = 2.6526 \times 10^{-2}\varepsilon\left(39.55 + 7.2958\varepsilon^2\right).$$

For the tightest such packing of equal spheres, $\varepsilon = 0.5$; so $\overline{\theta} = 0.5915$ and 0.5487, respectively. For comparison, $\pi/5 = 0.6283$ and $\pi/6 = 0.5236$.

B Boundaries with Spherical, Frictional Features

We note that for boundaries with spherical features, formulas similar to the preceding apply with θ interpreted as the average angle of penetration (Appendix A) and [5]

$$S^B \equiv \left(\frac{2}{\pi}\right)^{1/2} \frac{p}{\sqrt{T}} \left\{ v\frac{2}{3}\left[2\csc^2\theta\,(1-\cos\theta)-\cos\theta\right] \right.$$
$$\left. -\bar{\sigma}u' \left[\frac{2}{3}\left[2\csc^2\theta\,(1-\cos\theta)-\cos\theta\right] - \frac{1}{2}\left(1+\frac{\sigma}{\bar{\sigma}}B\right)\sin^2\theta\right]\right\},$$
$$= \left(\frac{2}{\pi}\right)^{1/2} \frac{v}{\sqrt{T}} p \frac{2}{3}\left[2\csc^2\theta\,(1-\cos\theta)-\cos\theta\right]$$
$$-\frac{5}{\sqrt{2}}\frac{\bar{\sigma}}{\sigma}\frac{F}{J} S \left[\frac{2}{3}\left[2\csc^2\theta\,(1-\cos\theta)-\cos\theta\right] - \frac{1}{2}\left(1+\frac{\sigma}{\bar{\sigma}}B\right)\sin^2\theta\right];$$

$$D^B = \left(\frac{2}{\pi}\right)^{1/2} 2p\sqrt{T}(1-e)(1-\cos\theta)\csc^2\theta;$$

and

$$h = 2(1-\cos\theta)\csc^2\theta,$$

and, for small sliding,

$$\frac{v}{\sqrt{T}} = \left(\frac{\pi}{2}\right)^{1/2} \frac{1}{\frac{2}{3}\left[2\csc^2\theta\,(1-\cos\theta)-\cos\theta\right] + \left(\frac{\pi}{2}\right)^{1/2}\frac{\sqrt{3}}{14}(1+\beta)}$$
$$\times \left\{ 1 + \frac{15}{56}\left(\frac{\pi}{3}\right)^{1/2}(1+\beta)\frac{F}{J} \right.$$
$$\left. + \frac{\bar{\sigma}}{\sigma}\frac{5}{\sqrt{2}}\frac{F}{J}\left[\frac{2}{3}\left[2\csc^2\theta\,(1-\cos\theta)-\cos\theta\right] - \frac{1}{2}\left(1+\frac{\sigma}{\bar{\sigma}}B\right)\sin^2\theta\right]\right\}\frac{S}{p}$$

$$f \equiv \frac{1}{\frac{2}{3}\left[2\csc^2\theta\,(1-\cos\theta)-\cos\theta\right] + \left(\frac{\pi}{2}\right)^{1/2}\frac{\sqrt{3}}{14}(1+\beta)}$$
$$\times \left\{ 1 + \frac{15}{56}\left(\frac{\pi}{3}\right)^{1/2}(1+\beta)\frac{1}{J} \right.$$
$$\left. + \frac{\bar{\sigma}}{\sigma}\frac{5}{\sqrt{2}J}\left[\frac{2}{3}\left[2\csc^2\theta\,(1-\cos\theta)-\cos\theta\right] - \frac{1}{2}\left(1+\frac{\sigma}{\bar{\sigma}}B\right)\sin^2\theta\right]\right\};$$

while for large sliding,

$$\frac{v}{\sqrt{T}} = \left(\frac{\pi}{2}\right)^{1/2} \frac{1}{\frac{2}{3}\left[2\csc^2\theta\,(1-\cos\theta) - \cos\theta\right]} \left\{ \left(\frac{S}{p} - \mu\right) + \frac{\bar{\sigma}}{\sigma}\frac{5}{\sqrt{2}}\frac{F}{J}\left[\frac{2}{3}\left[2\csc^2\theta\,(1-\cos\theta) - \cos\theta\right] - \frac{1}{2}\left(1+\frac{\sigma}{\bar{\sigma}}B\right)\sin^2\theta\right]\frac{S}{p}\right\}.$$

$$f = \frac{1}{\frac{2}{3}\left[2\csc^2\theta\,(1-\cos\theta) - \cos\theta\right]} \left\{ \left(1 - \mu\frac{p}{S}\right) + \frac{\bar{\sigma}}{\sigma}\frac{5}{\sqrt{2}J}\left[\frac{2}{3}\left[2\csc^2\theta\,(1-\cos\theta) - \cos\theta\right] - \frac{1}{2}\left(1+\frac{\sigma}{\bar{\sigma}}B\right)\sin^2\theta\right]\right\}.$$

References

1. S. B. Savage and M. Sayed. Stresses developed by dry cohesionless granular materials sheared in an annular shear cell. *Journal of Fluid Mechanics* **142**, 391–430 (1984).
2. D. M. Hanes and D. L. Inman. Observations of rapidly flowing granular-fluid materials. *Journal of Fluid Mechanics* **150**, 357–380, 1985.
3. K. Craig, R. H. Buckholz, and G. Domoto. Effect of shear surface boundaries on stress for shearing flow of dry metal powders - an experimental study. *Journal of Tribology* **109**, 232–237 (1987).
4. J. T. Jenkins and M. W. Richman. Boundary conditions for plane flows of smooth, nearly elastic, circular disks. *Journal of Fluid Mechanics* **171**, 313–328 (1986).
5. M. W. Richman. Boundary conditions based upon a modified Maxwellian velocity distribution function for flows of identical, smooth, nearly elastic spheres. *Acta Mechanica* **75**, 227–240 (1988).
6. G. C. Pasquarell and N. L. Ackermann. Boundary conditions for plane granular flow. *ASCE Journal of Engineering Mechanics* **115**, 1283–1302 (1989).
7. M. W. Richman and C. S. Chou. Boundary effects on granular shear flows of smooth disks. *Journal of Applied Mechanics and Physics (ZAMP)* **39**, 885–901 (1988).
8. J. T. Jenkins and E. Askari. Rapid granular shear flows driven by identical, bumpy, frictionless boundaries. In C. Thornton, editor, *Powders and Grains 93*, pages 295–300, Balkema, Rotterdam (1993).
9. M. Y. Louge, J. T. Jenkins, and M. A. Hopkins. Computer simulations of rapid granular flows between parallel bumpy boundaries. *Physics of Fluids A* **2**, 1042–1044 (1990).
10. H. Kim and A. D. Rosato. Particle simulations of the flow of smooth spheres between bumpy boundaries. In et al. H. H. Shen, editor, *Advances in the Micromechanics of Granular Materials*, pages 91–100, Elsevier, Amsterdam, (1992).
11. S. B. Savage and R. Dai. Studies of granular shear flow. *Mechanics of Materials* **16**, 225–238 (1993).

12. P. C. Johnson and R. Jackson. Frictional-collisional constitutive relations for granular materials and their applications to plane shearing. *Journal of Fluid Mechanics* **176**, 67–93 (1987).
13. J. T. Jenkins. Boundary conditions for rapid granular flows: Flat, frictional walls. *Journal of Applied Mechanics* **59**, 120–127 (1992).
14. O. R. Walton. Numerical simulation of inelastic, frictional particle-particle interactions. *In Particulate Two-Phase Flow Butterworth-Heinemann, (M. C. Roco, editor)*, pages 1249–1253 (1992).
15. S. F. Foerster, M. Y. Louge, H. Chang, and K. Allia. Measurements of the collision properties of small spheres. *Physics of Fluids* **6**, 1108–1115 (1994).
16. M. Y. Louge. Computer simulations of rapid granular flows of spheres interacting with a flat, frictional boundary. *Physics of Fluids* **6**, 2253–2269 (1994).
17. J. T. Jenkins and M. Y. Louge. On the flux of fluctuation energy in a collisional grain flow at a flat, frictional wall. *Physics of Fluids* **9**, 2835–2840 (1997).
18. J. T. Jenkins and E. Askari. Boundary conditions for granular flows: phase interfaces. *Journal of Fluid Mechanics* **223**, 497–508 (1991).
19. J. T. Jenkins and C. Zhang. Kinetic theory for rapid shearing flows of slightly frictional, nearly elastic spheres. *Physics of Fluids* (under review, 2000).
20. J. T. Jenkins, S. V. Myagchilov, and Haitao Xu. Nonlinear boundary conditions for collisional grain flows. (in preparation, 2000).

Diffusion Process in Two-Dimensional Granular Gases

Christian Henrique[1], George Batrouni[2], and Daniel Bideau[1]

[1] Groupe matière condensée et Matériaux (UMR 6626 CNRS) Université de Rennes 1, Bât 11A, 263 avenue du général Leclerc. CS 74205, 35042 Rennes Cedex, France. e-mail: christian.henrique@univ-rennes1.fr

[2] Institut Non-Linéaire de Nice, Université de Nice-Sophia Antipolis, 1361 route des Lucioles, 06560 Valbonne, France

Abstract. We study the diffusion process in a granular gas. We first show that for finite size systems the choice of boundary conditions is of crucial importance. With periodic boundary conditions, the coefficient of diffusion is found to depend on the system size and does not saturate for large systems, which is of course not physical. The problem is solved by using reflecting boundaries. In that case, we find good agreement between numerical results and the Langevin theory. We also study the influence of an external random force on the diffusion process for a forced system. In particular, we analyze differences in the mean square velocity and displacement between the elastic and inelastic case.

1 Introduction

We are interested in the diffusion process in a granular gas. As granular gases are dissipative, it is necessary to bring energy into the system to keep the particles agitated (see section 4). Our goal is to study the dependence of the dynamic properties of the granular gas on the mode used to force the system. This work constitutes the first step to understand the diffusion process in a binary system composed of two grain sizes. The system considered here is composed of one particle, **s**, of radius R_s in a sea of particles of radius R_b. The particles are spheres constrained to move in a plane so that the system is two dimensional. The system considered here is dilute with a surface packing fraction of 30 %. The simulations are conducted with the molecular dynamics algorithms (*time step driven* [1] and *event driven* [2]) and the results are compared to theoretical predictions. It is well known that for a 2D gas, the integral of the auto-correlation function does not converge [3]. This means that the mean square displacement does not vary linearly with time. Therefore, we can not define a coefficient of diffusion in 2D. However, we show that in a limited range of time, in the stationary state, the mean square displacement can be approximated by the linear function:

$$\langle (\boldsymbol{r}(t+t_0) - \boldsymbol{r}(t_0))^2 \rangle \propto 4Dt \tag{1}$$

where D can be interpreted as a diffusion coefficient. All the measures are expressed in arbitrary units.

2 Choice of Boundary Conditions

In this section we show that periodic boundary conditions introduce strong correlations and therefore alter the diffusion process.

2.1 Periodic Boundary Conditions

Consistent with common practice, we have used periodic boundary conditions to simulate a system of identical spheres $R_s = R_b$. Initially the particles are placed randomly in a square box of length L. The periodic boundary conditions are applied in both directions. In this case, for elastic or forced gases (section 4), we have observed a dependence of D on the system size.

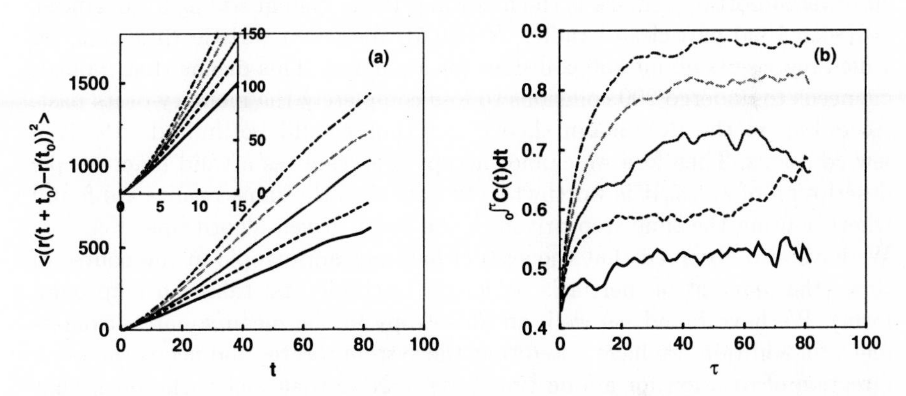

Fig. 1. Dependence of the mean square displacement on the system size. **(a)**: $\langle(\boldsymbol{r}(t+t_0) - \boldsymbol{r}(t_0))^2\rangle$ as a function of t. From bottom to top the system size is 20, 40, 30, 50, 60. **(b)**: Integral of $C(t)$ as a function of t for the same system. From bottom to top the system size is 20, 40, 30, 50, 60, respectively

In Fig. 1, we show both the mean square displacement $\langle(\boldsymbol{r}(t+t_0)-\boldsymbol{r}(t_0))^2\rangle$ and $\int C(t)dt$, both calculated in the stationary state, as function of t. $C(t)$ corresponds to the normalized autocorrelation function:

$$C(t) = \frac{\langle \boldsymbol{v}(t_0+t)\boldsymbol{v}(t_0)\rangle - \langle \boldsymbol{v}(t_0)\rangle^2}{\langle \boldsymbol{v}(t_0)^2\rangle - \langle \boldsymbol{v}(t_0)\rangle^2} \tag{2}$$

First we note that the mean square displacement varies linearly with time as expected but the slope of the curve, i.e. the diffusion coefficient, increases with system size. This feature can be also observed in $\int_0^{t\sim\infty} C(t)dt$, which is also proportional to the diffusion coefficient. Similary, we observe that the relaxation time τ_r (i.e. $C(\tau_r) \simeq 0$) increases with size. In summary, the bigger the system is, the longer is the characteristic time τ_r and the larger is the diffusion coefficient D. We recall that such dependence has been observed by Alder *et al.* [4]. They proposed the following law for the dependence of D on the number of particles, n,

$$D(n) = D(\infty)(1 - 2/n) \tag{3}$$

However [5] their numerical simulations do not support this conjecture since they fail to observe any saturation of D for large systems. In addition, they found strong correlations in the velocity field characterized by the presence of vortex flow pattern at the microscopic scale.
Our results confirm the lack of convergence for D with system size. In addition, this variation of D with L is also observed in the case of inelastic collisions.
Another important remark is the following. If the system size is, for example, 60 (with 1400 particles of radius $R = 0.5$), the characteristic time $\tau_r \simeq 20$ which represents about 200 collisions for a particle. This means that a particle needs to undergo 200 collisions to lose completely the memory of its past. According to the Boltzmann theory this time should be limited to only a few collisions. Therefore we cannot accept this result as a valid macroscopic description of a gas. It is worthwhile to note that the same results are found whether using the time step driven or the event driven algorithms.
We have also observed that the system becomes anisotropic in the course of time (the moment of inertia is no longer isotropic and takes an ellipsoidal form). We have found, as well, an anisotropy in the mean square displacement. In addition, we have noted that the system rotates and keeps the same direction of rotation for a long time ($\sim \tau_r$). Note that we set the initial angular momentum to zero in our numerical simulations. We suspect that this rotation induces an anomalous temporal correlation of velocities. One should point out that this rotation phenomenon seems similar to that observed by Alder *et al.* in their simulations with similar periodic boundary conditions. We strongly believe that the use of periodic boundary conditions is responsible for this anomalous correlation. These boundary conditions present another inconvenience. Because the distance between particles is not conserved by rotation, the interaction potential used in the algorithm, which depends only on the relative positions of particles, is itself not invariant by rotation. The angular momentum of the system is not conserved. The use of periodic boundary conditions amounts to replicating the system in a square lattice. There are, therefore, several identical systems which interact through the boundaries. The rotation observed in our system is then extended to all these systems and can create some shear stress, due to frustration of rotation, between neighboring systems. These boundary conditions can have other consequences on the dynamics of granular systems. For example, during simulation of a cooling state the system evolves towards clusters [6] whose orientation depends on the type of boundary conditions [7].

2.2 Reflecting Boundaries

In the case of reflecting boundaries, the mean square displacement is limited at long time by the system size. To circumvent this problem, we proceed as follows

- the test particle **s** is put at the center of system at $t = 0$
- We calculate its position, velocity etc., until it reaches the walls delimiting the system (we will refer to t_w as the time needed by the particle **s** to the reach the system boundaries).
- We make several simulations modifying the initial conditions.

The mean square displacement is calculated over 500 trajectories and limited to time smaller than the smallest t_w. In this case, we do not find any dependence of the mean square displacement on the system size, see Fig. 2. We

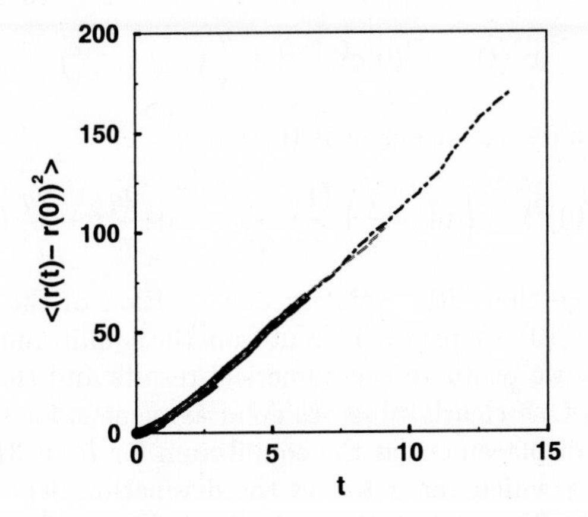

Fig. 2. Variation of $\langle(\boldsymbol{r}(t) - \boldsymbol{r}(0))^2\rangle$ as function of t using reflecting boundaries. Are superposed (like on the Fig. 1) the results for system size: 20, 30, 40, 50, 60

should recall that the integral of the velocity correlation function does not converge in 2D. However, in a limited range of time, see Fig. 2, the quantity $\langle(\boldsymbol{r}(t + t_0) - \boldsymbol{r}(t_0))^2\rangle$ can be approximated by a straight line and D calculated according to Eq. (1). We should keep in mind that the estimation of D according to this method is an approximation.

3 Diffusion in an Elastic Gas

We first validate our algorithm using reflecting boundaries for elastic gas (i.e., where the collision between particles are elastic). In the case elastic case, one can compare the numerical results with those given by the Langevin equation.

Indeed, near equilibrium, the dynamics of s can be described approximately, by a Langevin equation:

$$\dot{\boldsymbol{v}}(t) = -\gamma \boldsymbol{v}(t) + \Gamma(t), \tag{4a}$$

$$\langle \Gamma_i(t)\Gamma_j(t')\rangle = q\delta_{i,j}\delta(t-t'). \tag{4b}$$

Knowing the total kinetic energy of the system E_k^{tot}, which is given by the initial momentum of each particle, we can easily calculate the square velocity in the equilibrium state $\langle v^2(\infty)\rangle = q/\gamma$, where $1/\gamma$ corresponds to the relaxation time and the coefficient of diffusion D in this case is equal to $\langle v^2(\infty)\rangle/(2\gamma)$.
For example, if $R_s \gg R_b$ or equivalently $m_s \gg m_b$ ($m_{s,b}$ is the mass of the particle of radius $R_{s,b}$), $1/\gamma$ is larger: the collision of s with a light particle b will not affect strongly s. Thus a great number of collisions is needed before s reaches its equilibrium state. The dependence of the square velocity on time is simply given by integrating Eq. (4a)

$$v^2(t) = v^2(0)e^{(-2\gamma t)} + \frac{q}{\gamma}\left(1-e^{-2\gamma t}\right) \tag{5}$$

and the mean square displacement is therefore:

$$\langle(\boldsymbol{r}(t)-\boldsymbol{r}(0))^2\rangle = \left(v_0^2 - \frac{q}{\gamma}\right)\frac{(1-e^{-\gamma t})^2}{\gamma^2} + \frac{2q}{\gamma^2}t - \frac{2q}{\gamma^3}\left(1-e^{-\gamma t}\right) \tag{6}$$

$v_s(0)$ and $\boldsymbol{r}(0)$ are the initial velocity and position of the particle s. One should note that $v_s^2(0)$ is a priori different from the equilibrium velocity square $v_s^2(\infty)$. In Fig. 3, we compare the numerical results and those given by the Langevin theory. One clearly observes good agreement for the mean square velocity and the displacement at the equilibrium for $R_s = 3R_b$.

The value of γ which characterizes the dissipation depends on both velocities, V_s and V_b. To estimate theoretically the value of γ, we consider the deviation due to a collision of the particle s moving at v_s in the x-direction. The dissipative term $-\gamma \boldsymbol{v}$ appearing in (4a) can therefore be formally written as

$$-\gamma \boldsymbol{v}_s = \left\langle \frac{\boldsymbol{v}_s' \cdot \boldsymbol{x} - v_s}{v_s} \right\rangle \omega_c \boldsymbol{v}_s \tag{7}$$

where $\boldsymbol{v}_s'$ is the velocity after the collision and ω_c is the rate of collision. To calculate the different terms, we proceed as follows. We consider the collision of s with a particle b which is moving at a velocity $\boldsymbol{v}_b$. The collision is characterized by two angles: α the angle between $\boldsymbol{r}_s - \boldsymbol{r}_b$ and the x axis, and β the angle between $\boldsymbol{v}_b$ and the x axis. We then calculate theoretically $\boldsymbol{v}_s'(\alpha,\beta)$. Integrating over α and β, one can have an estimate of γ thanks to Eq. (7), see also [8]. The collision frequency which also depends on both velocities can be calculated analytically [9].

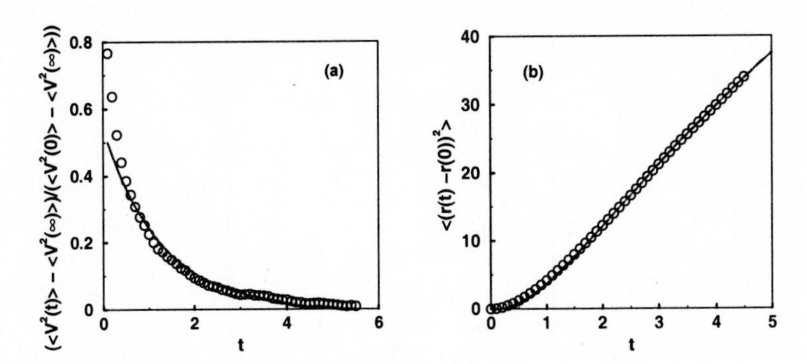

Fig. 3. Comparison between numerical results and Langevin approximation. $R_s = 1.5$ and $R_b = 0.5$. **(a)** dependence of the mean squared velocity on t; (*circle*): numerical result, (*line*): approximation in $e^{-2\gamma t}$ according to 5. **(b)** the mean squared displacement, (*circle*): numerical result, (*line*): theoretical prediction according to 6

In Fig. 4, we compare for different values of R_s the diffusion coefficient found from the simulation with the theoretical value predicted by the Langevin equation combined with our analytical calculation of γ. The theoretical calculation of γ is rather good. We recall that $1/\gamma$ corresponds to the characteristic time for the diffusive behavior. It is important to notice that γ can be approximated by ω_c only when $R_s \ll R_b$. The comparison with a random walk is possible but the typical time of the walk should not be taken equal to τ_c (which is the mean time between two collisions) but to τ_r. Even for the monodisperse case, the velocities of **s** after and before a collision are correlated. The agreement between numerical results and theoretical prediction allows us to validate our numerical algorithm.

4 Forced Systems

In a real granular system dissipation occurs through collisions. The collision laws used in the simulations and the mechanical properties of grains (restitution and friction coefficients) are the same as in Ref. [10].
Due to dissipation, we need to feed energy into the system to maintain the particles agitated. We choose here, like in [11, 12], a random heating. At every time step δt we give each particle a random acceleration. The equation of motion can now be written formally as:

$$m_i \dot{\boldsymbol{v}}_i = F_i^c + F_i^t \ , \tag{8a}$$

$$\langle F_{i,x}^t(t) F_{j,y}^t(t') \rangle = m_i m_j \delta_{i,j} \delta(t - t') \eta_0^2 \ . \tag{8b}$$

F_i^c corresponds to the total force due to the collisions on the particle i of mass m_i. The random acceleration has been chosen independent of the mass

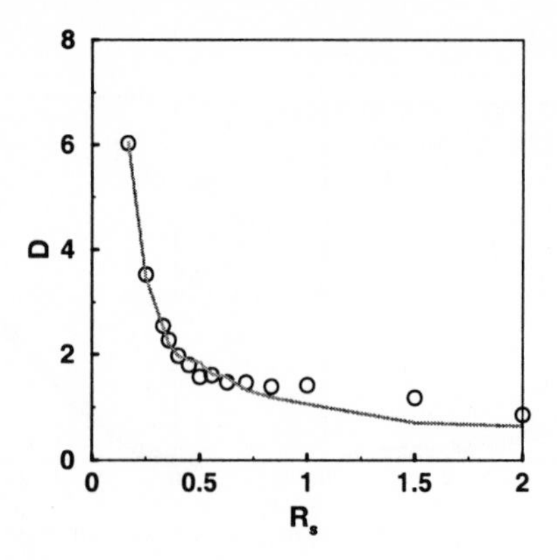

Fig. 4. Coefficient of diffusion for different values of R_s. $R_b = 0.5$. (*circle*): numerical values obtained by simulation. (*line*): theoretical values calculated from $\langle v^2(\infty)\rangle/(2\gamma)$

of the particle and corresponds to a Gaussian noise of variance η_0^2. This choice is of course arbitrary. At long time, the loss of energy due to collision and the gain due to F_i^t balance each other such that the system reaches a steady state out of equilibrium. It can be shown [13] that the velocity distribution in the steady state can be well described by a Maxwellian.

4.1 Stationary State

We focus first on the stationary state for a monodisperse gas ($R = R_s = R_b$). In the stationary state the loss and the gain of energy balance in average. The loss of energy Δ per unit time can be expressed [11] as:

$$\Delta \propto \omega_c m \langle v^2 \rangle \ . \tag{9}$$

On the other hand, the gain of energy due the stochastic force should be given by:

$$\frac{1}{2} m[\langle v^2(t+\delta t)\rangle - \langle v^2(t)\rangle] = m\eta_0^2 \delta t \ . \tag{10}$$

In the steady state, which is characterized by a constant mean quadratic velocity $\langle v^2(\infty)\rangle$, we find (with $\omega_c \propto \sqrt{v^2}$ [9]) the following scaling for the mean quadratic velocity:

$$\langle v^2(\infty)\rangle \propto (\eta_0^2)^{2/3} \ , \tag{11a}$$

$$\langle v^2(\infty)\rangle \propto \tau_c \ . \tag{11b}$$

These two scaling laws have been checked numerically, see Fig. 5. We obtain the exponent 2/3 for various coefficients of restitution used in the contact

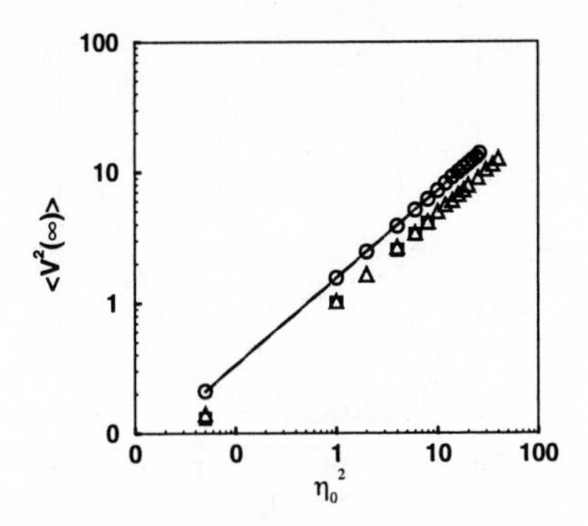

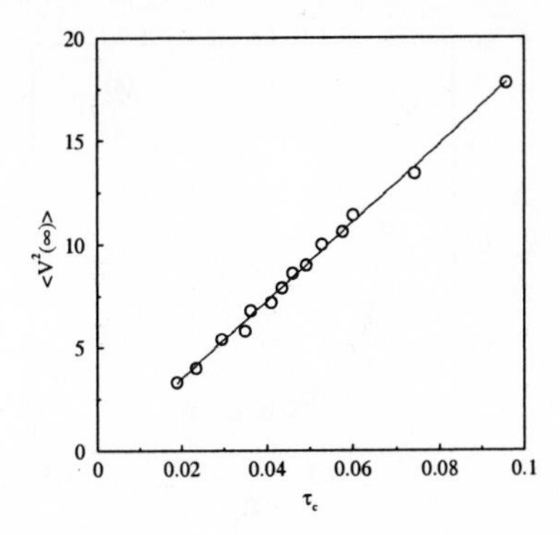

Fig. 5. (a) $\langle v^2(\infty)\rangle$ versus η_0^2 for different coefficients of restitution. (*circle*): $e_n = 0.87$, $e_s = 0.4$, $\mu = 0.25$; (*triangle*): $e_n = 0.4$, $e_s = 0.4$, $\mu = 0.25$. **(b)** $\langle v^2(\infty)\rangle$ versus the mean time between collisions τ_c.

laws. We have verified the predicted dependence with τ_c for different values of R.

In summary, in the case of a monodisperse system, we can predict the dependence of the granular temperature on the various parameters and consequently characterize the stationary state.

4.2 Diffusion of One Particle

We consider here the bidisperse case (a single particle of radius R_s in a sea of particles of radius R_b). The difference of mass between the two types of particles will alter the character of the collisions. The loss of energy due to collisions will now depend on the two masses m_s and m_b and on both velocities. This loss of energy can be formally written as

$$\Delta = P(m_s, m_b)\, \omega_c m_s \langle v_s^2\rangle \tag{12}$$

$P(m_s, m_b)$ is a function of both masses m_s and m_b. The frequency ω_c depends on both velocities (v_s and v_b) and on the collision cross-section between the particle s and a particle b. We have not yet calculated the expression of $P(m_s, m_b)$ theoretically. We have only results from the simulations which are presented in Fig. 6. We show the mean squared velocity $\langle v^2(\infty)\rangle$, the diffusion coefficient D, and the relaxation time τ_r as functions of R_s. Note that η_0^2 has been chosen such that the temperature of the gas is the same as in the previous section. We see that $\langle v^2\rangle$ first decreases with R_s for $R_s < R_b$ but then increases when $R_s > R_b$. Because of dissipation, the repartition of the energy with the mass is no longer proportional to $1/m_s$. The behavior of $\langle v^2(\infty)\rangle$ strongly modifies the curve of D versus R_s compared to the elastic case. The relaxation time represented in Fig. 6(b) clearly increases as R_s increases. We have seen in the elastic case that $D = \langle v^2\rangle/(2\gamma)$. In Fig. 6, we show the numerical variation of D as a function of R_s and the corresponding

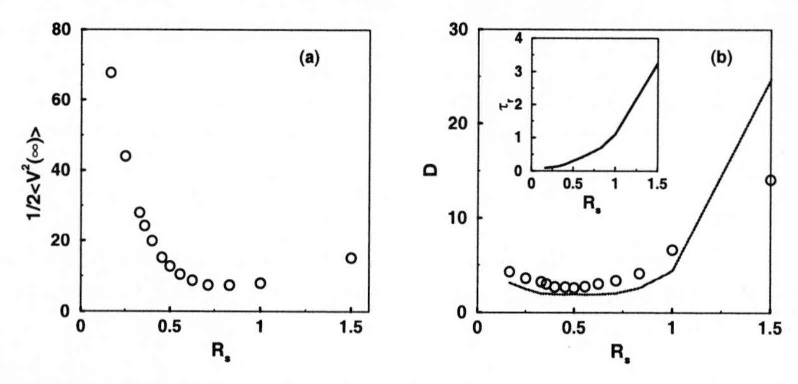

Fig. 6. **(a)** $\langle v^2(\infty)\rangle$ vs R_s ($R_b = 0.5$). **(b)** Coefficient of diffusion D as a function of R_s. (*circle*): numerical values obtained from simulation. (*line*): corresponding values given by $\langle v^2(\infty)\rangle/(2\gamma)$. The inset shows τ_r versus R_s

values given by $\langle v^2\rangle/(2\gamma)$. One can see clearly that the external noise modifies the dynamics of the granular gas and in particular the diffusion coefficient D. The numerical value of D is found to be larger than that obtained by the corresponding random walk. Indeed, at short time, due to the random force, $\langle v^2\rangle$ is not constant. Between two collisions, i.e. in the absence of dissipation, $\langle v^2(t)\rangle$ increases linearly with t. Starting from the dynamic equation of the particle **s** without dissipation, i.e. at short time, $\dot{v}(t) = \eta(t)$, we find that the mean quadratic displacement scales as :

$$\langle x^2(t)\rangle = \frac{\eta_0^2}{6}t^3 . \tag{13}$$

On the other hand, in the case of a random walk the mean square displacement at short time scales as t^2. This difference explains the disagreement between D and $\langle v^2\rangle/(2\gamma)$. For the same granular temperature, the external heating chosen here influences at short time the particle dynamics and therefore alters the value of the diffusion coefficient.

5 Conclusion

We have presented some general results about the diffusion process in an agitated granular gas. We have first shown that the boundary conditions used in the simulations are of crucial importance. In particular, periodic boundary conditions introduce artificially strong temporal correlations which alter the macroscopic properties of the gas.
If we ensure that no correlations are induced by the algorithm (for example by using reflecting boundaries), the numerical results obtained for an elastic gas can be described very well by a Langevin equation. In particular, the theoretical calculation of the relaxation time allows us to predict the diffusion

coefficient in all cases studied. This was not a priori intuitive since the radius of the particles is here of the order of the mean free path.
Finally we have analyzed the influence of uniform heating (a random acceleration) on dissipative gases. We have shown that heating influences the dynamic at short times. This is perceptible through the value of the diffusion coefficient which is different from that expected from the Langevin description.
We are now applying with success these results to the diffusion process in a granular mixture consisting of two type of grains (differing by the mass or the size) in equal proportion.

Acknowledgements

We thank the CNRS and the NFR for support though the Franco-Norwegian *PICS*. C. H. thanks Alexandre Valance for his support during this work and his help for the writing of this paper.

References

1. M. P. Allen and D. J. Tildesley, Computer simulations of liquids. Oxford University Press (1987).
2. J. M. Haile, Molecular dynamics simulation, elementary methods. Wiley-Interscience (1997).
3. M. H. Ernst, E. H. Hauge, and J. M. J. van Leeuwen, Asymptotic Behavior of Correlation function. I. Kinetic Terms. Phys. Rev. A **4**, 2055 (1971).
4. B. J. Alder and T. E. Wainwright, Velocity Autocorrelations for hard Spheres. Phys. Rev. Lett. **18**, 988 (1967).
5. B. J. Alder and T. E. Wainwright, Decay of the Velocity autocorrelation Function. Phys. Rev. A **1**, 18 (1970).
6. I. Goldhirsh and G. Zanetti, Clustering Instability in Dissipative Gases. Phys. Rev. Lett. **70**, 1619 (1993).
7. I. Goldhirsh, Private communication.
8. C. Henrique, Diffusion et mélange dans un milieux granulaire modèle à deux dimensions. PhD thesis Université de Rennes I (1999).
9. S. Chapmann and T. G. Cowling, The mathematical theory of non-uniform gases. Cambridge University Press, (1970).
10. S. F. Foerster, M. Y. Louge, H. Chang and K. Allia, Measurements of the collision properties of small spheres. Phys. Fluids, **6**, 1108 (1994).
11. T. P. C. van Noije, M. H. Ernst, E. Trizac and I. Pagonarabarraga, Randomly Driven granular Fluids: Large scale structure. Phys. Rev. E **59**, 4326 (1999).
12. D. R. Williams and F. C. MacKintosh Phys. Rev. E **54**, R9 (1969).
13. G. Peng and T. Ohta, Steady state of a driven granular medium. Phys. Rev. E **58**, 4737 (1998).

II

Collisions and One-Dimensional Models

Energy Loss and Aggregation Processes in Low Speed Collisions of Ice Particles Coated with Frosts or Methanol/Water Mixtures

Frank Bridges[1], Kimberley Supulver[1,2], and Douglas N.C. Lin[1]

[1] University of California, Santa Cruz CA 95064,USA
[2] Malin Space Science Systems, P. O. Box 910148, San Diego, CA 92191-0148, USA.

Abstract. At low speeds, the energy loss in collisions and the adhesion (sticking) between particles depends strongly on the surface layer composition and morphology. In this paper we review our measurements of the coefficient of restitution in low speed collisions for a variety of surface coatings, for speeds in the range 0.01–2 cm/s, and also the investigations of sticking forces (under both dynamic and static conditions), including the conditions for which sticking contacts have been observed. These results are compared with a few other similar measurements, and with theoretical models for energy loss in collisions. Specific applications to the dynamics of Saturn's rings and to the formation of planetesimals in the early solar nebula, which motivated these studies, are also discussed.

1 Introduction

Recent detection of protostellar disks or flattened structures with properties similar to those of the primordial solar nebula surrounding the protosun [1, 2] suggests that the necessary environment for planetary-system formation is commonly found around young stellar objects [3–5]. The discoveries of planetary companions around nearby stars [6, 7] indicate that they are ubiquitous and their formation process is robust. This progress in observational astronomy plus the recent theoretical advancements in the understanding of mechanisms of star formation and nebular disk evolution [1, 8–11] have spurred on the investigation of the development of our (and other) solar systems. Similar to ring systems such as Saturn's rings, the primordial solar nebula is an excellent example of dilute granular gases, in which collisions and aggregation processes play crucial roles in the dynamical evolution.

In this paper we focus on the formation and dynamical evolution of small objects in the solar system. These entities may also be the progenitors of comets. Some appear to be clusters of smaller particles weakly bonded together. For example, comets Shoemaker-Levy 9 (1992) and 16P/Brooks 2 (1886) both broke into many pieces following close passage near Jupiter, most likely as a result of relatively gentle tidal forces. Estimates of the average tensile strengths for these objects, based on tidal force disruption, range from 100–1000 dyn/cm ^{2} [12–16]. Several investigators have suggested that these objects are "rubble piles" of weakly bound small particles; however, such models are not universally accepted.

There are many outstanding issues to be addressed. How do small particles, many cm in size, form and how do they aggregate into larger planetesimals within the short time-span available? How is energy lost in collisions and what are the sticking mechanisms that enable aggregates to form? What is the "glue" that holds them together after formation? Are they roughly homogeneous (like snow balls) or quite inhomogeneous objects with significant internal voids as expected for a cluster of small objects (i.e., the rubble pile model)? What are the dynamics of collisions between such objects? It is likely that the answer(s) will vary depending on the distance from the sun and the local temperature. Our investigations focus on the cooler regions of the solar system where the temperatures were < 150 K and most of the volatile components of the solar system were in the form of ices or frosts, deposited on a variety of refractory dust grains and particle surfaces.

Of the many experimental issues that need to be addressed to gain a full understanding of the solar system and its formation, we consider only two — the energy losses in slow speed collisions for cold surfaces, coated with various frosts and ices, and the magnitudes of the sticking forces that can be obtained at these low temperatures. For aggregation to occur, the relative speed must first be reduced until the operative sticking mechanism(s) can hold the particles together in a collision. To model such dynamics the coefficient of restitution is needed as a function of the impact velocity for a variety of temperatures and surface coatings.

Relatively little work has been done to investigate sticking mechanisms for small particles, with sizes for which self-gravity is not significant. Recent experiments by Würm and Blum [17] show that significant clustering occurs in collisions of micron-sized particles, for which electrostatic forces should dominate. However, collisions of small fluffy dust balls (roughly 1 mm in size) at speeds from 14 $\mathrm{cm\,s^{-1}}$ to $\sim 1\ \mathrm{m\,s^{-1}}$ did not produce any sticking in ~ 500 collisions [18]. In an interesting experimental simulation, Pinter et al. and Blum [19, 20] observed coagulation among sub-cm-sized glass spheres coated with a thin layer of hydrocarbons when collided at speeds up to $\sim 10^2\ \mathrm{cm\,s^{-1}}$. Although the conditions under which these experiments were carried out are far removed from those expected in the solar nebula, they are important in that they exhibit sticking at quite large impact speeds, which means the sticking force per unit area for the organic surface layer must also be large. Consequently, frosts or surface layers containing organic material may result in higher sticking forces than for water frost.

The outline of the rest of the paper is as follows: In section 2 we describe briefly the experimental apparatus and then review our experimental results in sections 3 and 4. In section 5 we consider various theoretical models for understanding the coefficient of restitution results and in section 6 discuss some of the applications of these results to planetary systems. A brief summary is given in section 7.

2 Experimental Apparatus and Procedures

Two important issues that need to be addressed for any models of the solar system are the energy loss in low speed collisions as a function of impact speed (parameterized by the coefficient of restitution, ϵ) and contact sticking mechanisms (the sticking parameter). In the outer regions of the solar system, including Saturn's rings, the particle surfaces are very cold (likely covered with frosts of water and other volatile components) and the experiments must be carried out in a low temperature environment. Two different pieces of equipment have been designed for these purposes (see the next section). The first is a compound pendulum apparatus for collision experiments, at speeds in the range 0.005–10 cm/s; it is primarily for energy loss experiments although some contact sticking measurements have also been carried out [21]. The second is an apparatus for measuring sticking forces under static conditions using well defined contact areas [22].

For each experiment, a surface layer of a volatile material such as water, CO_2, methanol, or water-methanol mixtures, is deposited onto the surface of a spherical particle and/or a flat plate of ice or other material. The vapor is transported using nitrogen as a carrier gas, through an insulated tube (3 mm diameter) into the cold environment; a heater, wrapped around the small tube (under the insulation), keeps the tube temperature above 5° C. The frosty layer is usually deposited in an atmosphere of 25–100 torr, at temperatures between 70-200 K. The low-density atmosphere environment serves two purposes. First it rapidly cools and thermalizes the vapor; second, it provides nucleation sites for micro-snow-flakes to form, thereby simulating frost-coated dust grains in the outer solar system. The morphology of the frosty surface depends on several factors including the rate of deposition, the ambient gas pressure, and the temperature. For temperatures in the range 100 to 200 K and moderate deposition rates (1 mm/hr), water frost forms a dendritic-like porous surface for thin surface layers < 0.5 mm. For thicker layers, the frost becomes denser and smoother in this temperature range. At significantly lower temperatures (70 K), the surface frost has quite a different morphology; very small, roughly spherical particles form which do not adhere well to the plates. These particles flow easily much like grains of sand.

Surface layers of methanol, or water/methanol mixtures, form a smoother more homogeneous surface that remains slightly viscous to low temperatures. It is this viscosity that likely accounts for the sticking of these surfaces, rather than the mechanical interlocking mechanism proposed for water frost.

2.1 Compound Pendulum Apparatus

The measurement of energy loss in low speed collisions, using contact surfaces that may be slightly irregular, poses particular problems for experiments that are carried out on earth. The major problem is to cancel out the effects of earth gravity. If the surface has irregularities on the size of a hundred microns or more, the displacement needed for measuring the velocity must be significantly larger — typically at least 0.5 mm. Consequently to achieve velocities as slow as 0.01 cm/s, for distance of order 0.5 mm, the acceleration must be of order 10^{-3} cm/s^2, or a micro-g. This low value for the acceleration of the particle is achieved using a long-period, compound pendulum, with the center of mass very close to the rotation axis. The apparatus is shown in Fig. 1. An ice particle is mounted on one side of the aluminum disk pendulum and a counter-weight is located on the opposite side for balancing. A split spherical mold is used to freeze an ice ball (5 cm diameter) onto an aluminum shaft which can be bolted to the pendulum. A small weight at the top of the pendulum allows for an adjustment of the position of the center of mass, and hence the period of the pendulum. Periods as long as 30 seconds can be achieved although typically a period of 10 seconds is used. For these long periods, the center of mass is located only about 100 micron below the rotation axis. Consequently it often takes over an hour to balance the pendulum; it is so sensitive, that removing the air from the cryostat in which the pendulum is located, changes the balance condition because of the difference in buoyancy forces on the ice particle and the counter weight. Differential thermal contraction of the ice and aluminum disk when the apparatus is cooled, also changes the balance condition, and the pendulum must be re-balanced once low temperatures are achieved.

The compound pendulum is suspended on agate knife edges as shown in the figure. The ice particle on the side of the pendulum moves vertically and impacts a flat ice surface mounted on a lead brick to provide inertia. For this arrangement, the reaction force is normal to the flat surfaces on which the knife edges sit, and there is little energy loss in the knife-edge suspension during a collision.

Most of the experiments presented here used a 5 cm diameter ice particle. However a few measurements were done with different radii of curvature; the desired radius of curvature, was molded onto the lower section of the 5 cm diameter particle using another mold [24]. In addition, the effective mass of the particle is greatly increased by the inertia of the pendulum to which it is attached. In typical experiments this effective mass is about 1500 g.

The pendulum is displaced using a small "magnetic kicker"; it consists of a small magnet, mounted at the bottom of the pendulum, which moves inside a small, stationary coil of wire. Applying a voltage to this coil will deflect the pendulum and thus provide a means of controlling its motion. The position of the pendulum is measured capacitively (Capacitive Displacement Device, CADD), using plates mounted at the top of the pendulum (see Fig. 1) [23]. Square wave signals, 180° out of phase, are applied to the two outer sets of electrodes, and a signal proportional to the deflection of the pendulum is detected at the central electrodes, using a lock-in amplifier. (The capacitive coupling is proportional to the overlap of the plates attached to the moving disk, with the outer set of fixed plates.) The displacement of the particle can

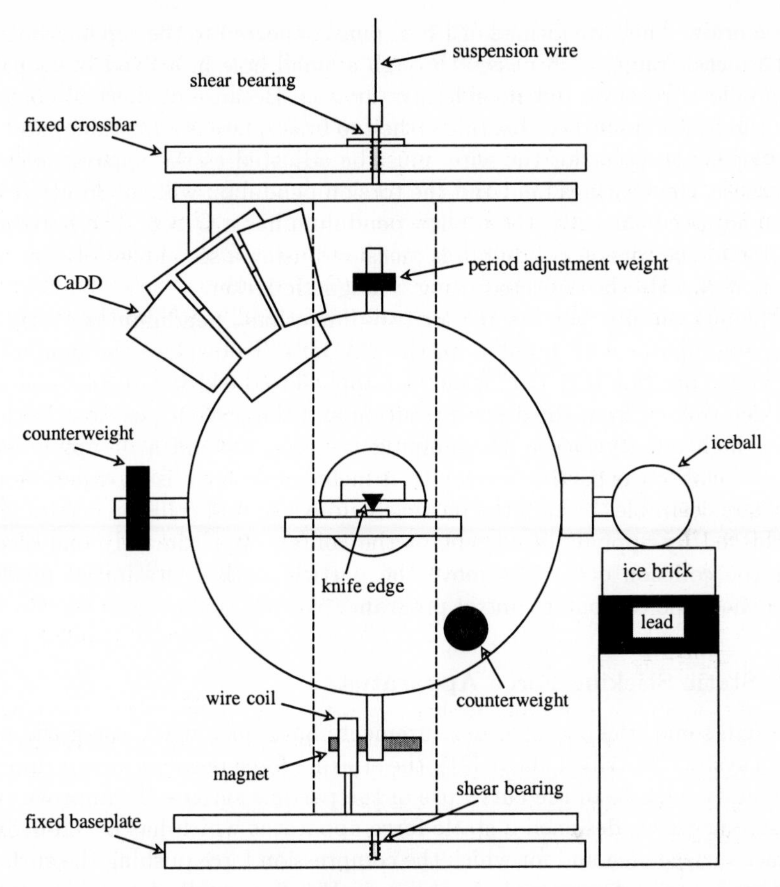

Fig. 1. A side view of the compound pendulum system for 2-D collisions. For the 1-D version, the support wire is removed and the frame is fastened to the bottom of the cryostat.(From Ref. [23], Fig. 2.)

be measured to about 1 micron, and is calibrated using a deflected laser beam from a mirror attached to the pendulum. The calibration of the displacement force for the kicker magnet is carried out at the same time, by measuring the pendulum displacement as a function of current in the coil. The losses for such a pendulum are very low, and a quality factor, $Q > 10$ is possible when the pendulum has a long period (10 seconds). The Q is proportional to the ratio of the energy stored in the oscillator to the energy disipated per cycle.

For two-dimensional, glancing collisions, the compound pendulum described above is mounted in a metal frame suspended from a fine stainless steel wire. This forms a torsion pendulum; the length and gauge of the wire are chosen to make its period approximately 10 seconds. In addition to the torsion motion, a rocking motion of the metal frame, about its center-of-mass can also be excited in a collision and shear bearings are needed to eliminate such motion. They are formed of 1 mm pins connected to the top and bottom of the metal frame; each passes through a small hole in a fixed brass plate, which allows rotation but no sideways motion. Because of thermal contraction, the positions of these bearings (the top brass plate is adjustable) and the top suspension point for the wire, must be adjusted as the apparatus cools. Otherwise, the bearings bind and the torsion pendulum will not rotate freely. When adjusted correctly, the torsion pendulum has a $Q > 5$. The horizontal position of the torsion pendulum is measured using a second set of capacitor plates; it can also be deflected using a magnetic kicker.

The motion of either the one- or two-dimensional pendulum is controlled using a computer with input from the CADD's. To displace the pendulum, the desired position is given; the current applied to the kicker magnet coil(s) is then determined from the desired position and the present position. Because of the high quality factor, Q, damping is necessary and a damping term, proportional to the particle's velocity, is included. In a collision experiment it is usually desirable to catch the particle before a second collision occurs. This is achieved by applying a current to the coil(s), approximately one second after the collision occurs, to move the particle back to its initial position using the same computer control program.

2.2 Static Sticking Force Apparatus

Although some sticking force measurements have been made using the pendulum system described above [21], the sticking force per unit area is difficult to quantify because of the curvature of the particle surface. To improve such measurements we designed a static force apparatus which has a well-defined contact surface area and for which the compression force pushing the surfaces together is easily determined. As shown in Fig. 2, a small plate is suspended from a load cell located at the top of the cryostat. The upper part of the cryostat can be moved vertically using either a coarse or fine control, so that the suspended plate can be raised or lowered onto a base plate. The load cell is calibrated in grams and can measure weights from 0.02 to 150 g, corresponding to forces from 20–150,000 dyne.

Prior to a measurement, the suspended plate is raised and a frost layer is deposited onto the two surfaces; then the upper plate is lowered very slowly until it rests on the base plate. The compression force is set by the predetermined weight placed on the upper plate. The upper plate (and load-cell) is then raised slowly using a micrometer (resolution of 1.5 micron) until the surfaces separate. Sticking occurs when the force measured by the load cell exceeds the total weight (bare plate plus added weight) of the upper plate.

The stretch of the frost layer can also be measured with this apparatus by measuring the displacement of the load-cell as a function of force. However because the support wire or thread, from which the small plate is suspended, can also stretch slightly, the spring constant of the thread or wire must first be determined. This is accomplished by connecting the support thread (wire) to

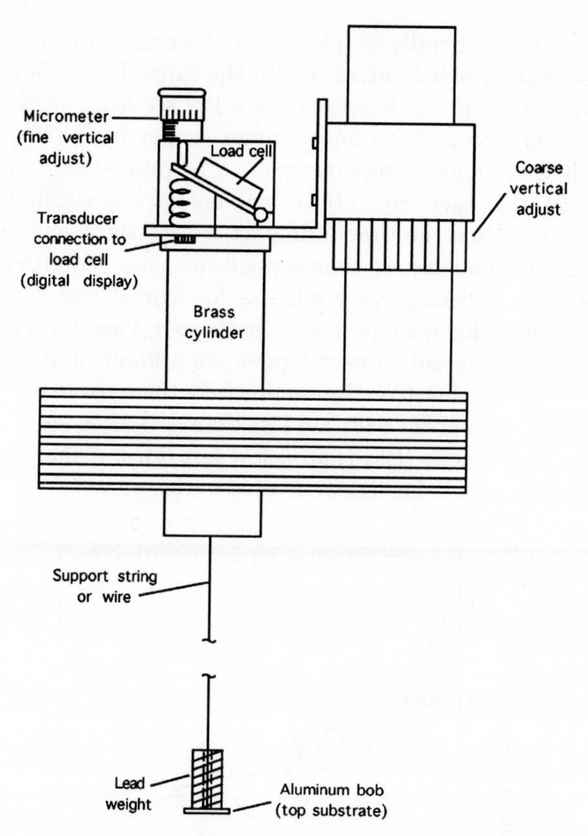

Fig. 2. A sketch of the central section of the static sticking apparatus (From Ref. [22], Fig. 2.)

the base plate and measuring the displacement of the load-cell as a function of the applied force. It should also be noted that the entire upper suspension of the apparatus is under stress from atmospheric pressure; the tiny additional forces associated with the frost sticking force will not deflect it significantly.

3 Energy Loss in Collisions

3.1 Normal Incidence Collision Experiments, Smooth Surfaces

The value for the coefficient of restitution, ϵ, varies considerably with the surface morphology, the temperature, and the impact speed of the ice particles. For a smooth, frost-free surface of poly-crystalline ice, ϵ can be large, approaching 1.0 at low impact speeds (less than 1 mm/s) and remaining quite high at speeds of a few cm/s. In Fig. 3, ϵ is plotted as a function of velocity up to 2 cm/s, for a temperature of 105 K [24]. Two different radii of curvature were used for the ice particles in these measurements and gave essentially the same results, with ϵ generally > 0.8. A few other measurements with a very smooth ice surfaces yielded values of ϵ in the range 0.85–0.95 at 2 cm/s. In each case however, ϵ approaches 1.0 at very low speeds. For comparison, ϵ is typically between 0.85 and 0.9 for a rubber "super ball" [23]. Thus, at low impact speeds and at low temperatures, ice is quite elastic in collisions. It should be noted, that there could be a small energy loss within the pendulum itself; particularly when the time of impact is very short and larger impulse forces are present. (The impact time depends on the surface morphology and is short when the surface is very stiff, as for a frost-free surface; however it can be quite long for heavily frosted surfaces.) Therefore the values of ϵ plotted in Fig. 3 and in subsequent figures are a lower limit for a particular measurement. It is difficult to estimate precisely the error in ϵ. Relative errors (i.e. repeatability on a given day) are less than $\pm$ 0.02; systematic errors are comparable at low speeds (less than a few mm/s), but may be as much as 0.05 low at higher speeds for lightly frosted surfaces.

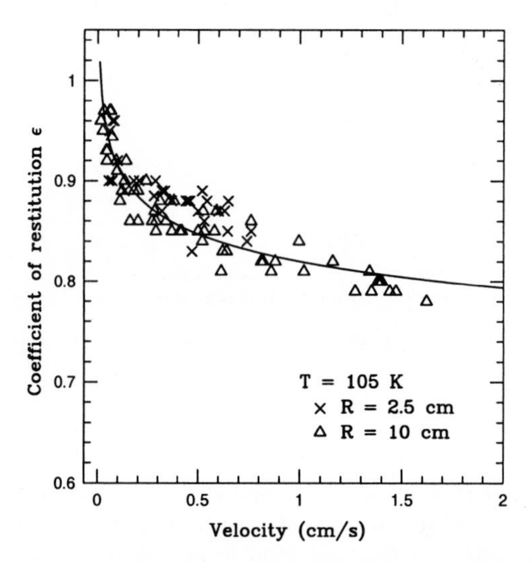

Fig. 3. A plot of the coefficient of restitution as a function of impact velocity (normal incidence) for smooth ice surfaces; particle radii = 2.5 and 10 cm.

In these measurements, the contact surface remains the same throughout the experiment. After each series of measurements the surfaces were examined and always found to be slightly fractured (the clear surface became white) over a region of two to four mm^2. This indicates that a small amount of brittle fracture occurs at these low temperatures, which must contribute to the energy loss and values of ϵ less than 1.0. Other recent measurements by Higa et al. [25, 26] for temperatures near 270 K, show that ϵ is close to 1 at significantly higher speeds than used in the present measurements, and that ϵ remains close to 1 at lower temperatures (113 K) for a range of masses. The slightly larger value for ϵ that they obtain, may be the result of a difference in the size of the crystallites forming the ice-particle, a less brittle surface at high temperatures, or possibly that the contact surfaces are different for each successive collision in their measurements. In contrast, Dilley and Crawford [27] observed a large change with particle mass at similar speeds.

3.2 Normal Incidence Collision Experiments, Frost-Coated Surfaces

For ice particles coated with a layer of frost the energy loss in a collision is much higher. In Fig. 4 we compare the values of ϵ for different surface layers as a function of velocity (T = 120 K, surface radius = 20 cm) [24]. The sublimated surface is achieved by raising the temperature to 150–180 K, evacuating the chamber for roughly an hour which speeds up the sublimation, and then lowering the temperature to that required for the measurement. Because of the poly-crystalline nature of the ice particle surface, different ice-facets are exposed at various points on the surface (which sublimate at different rates). This leads to a low density layer with voids and micro-bridges on a micron scale.

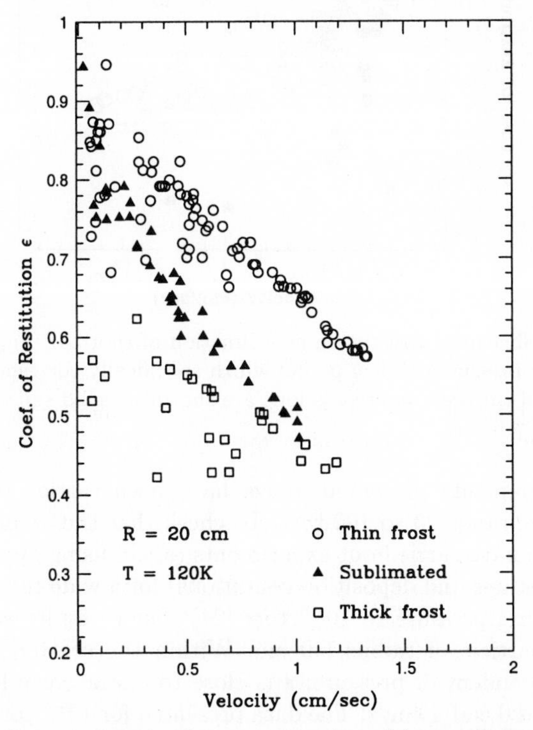

Fig. 4. The coefficient of restitution as a function of (normal) impact velocity for compacted frosted and sublimated surfaces; surface radius = 20 cm.

The slightly frosted surface shows more loss than the frost-free surfaces discussed above; adding a thick layer of frost always reduces ϵ further. For a large number of measurements using heavily frosted surfaces, and for temperatures in the 100–150 K range, ϵ is usually less than 0.5 and sometimes less than 0.4 at 2 cm/s. Sublimated surfaces show a similar high energy loss in collisions, although ϵ is somewhat higher than for heavily frosted surfaces. The fragile micro-bridge structure of the sublimated surface is easily damaged and the low value of ϵ is attributed to fracturing of this brittle surface. After many collisions, the compacted, sublimated surface behaves similarly to that for compacted frost.

For water-frosted surfaces (with the same frost thicknesses and deposition conditions), ϵ has also been investigated as a function of the contact surface radius, R. A weak dependence is found with ϵ increasing slowly with R.

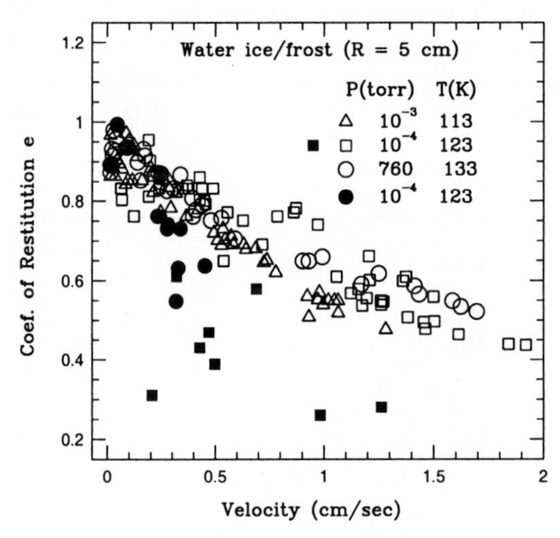

Fig. 5. The coefficient of restitution as a function of (normal) impact velocity for various ambient pressures. A few points which specifically correspond to collisions of freshly frosted surfaces, where ϵ is low, are shown by solid squares (only).

The measurements presented above have been carried out in ambient pressures in the range 25 to 100 torr. To check that the ambient pressure is not a factor we have carried out experiments (again using approximately the same frost thickness and deposition conditions) for a wide range of pressures, from atmospheric pressure to 10^{-5} torr [24]. These results are displayed in Fig. 5 for a particle of radius 5.0 cm. Within the scatter the results are the same independent of pressure; ϵ is close to one at very low speeds and between 0.4 and 0.5 at 2 cm/s. The data presented for 10^{-4} torr show another interesting feature of these collision experiments. For freshly frosted surfaces, i.e., surfaces that have not yet been collided, the coefficient of restitution is always very low as shown by the solid squares in Fig. 5. As the frost layer becomes compacted, ϵ increases and becomes nearly constant after 5–10 collisions. This is shown explicitly in Fig. 6 where ϵ is plotted versus the collision number at a constant speed of 0.3 cm/s [24]. This figure also shows that after the initial rapid rise, ϵ continues to increase slowly above 10 collisions, and after a very large number of collisions can be in the range 0.8 to 0.9 for a speed of 3 mm/s. Consequently, frost surface layers that are highly compacted are relatively elastic in low speed collisions, and do not remove kinetic energy quickly from the system. Fluffy, porous frost on the other hand, can have a very high loss, often over 90%, in each collision.

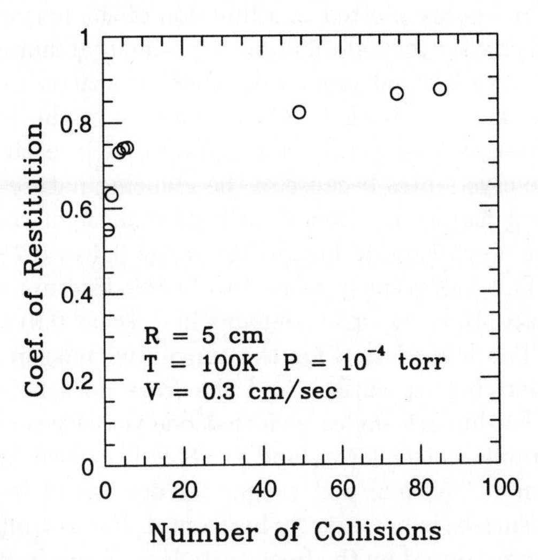

Fig. 6. The change of the coefficient of restitution with the number of impacts at the same impact speed.

We have made similar measurements for surface layers of other volatile materials. Frost layers formed of CO_2 [28] show comparable energy losses in low speed collisions as observed for water frost. A few measurements were also carried out using a surface layer of methanol. Such collisions had high loss but often resulted in sticking at low speeds, which limited our ability to collect data.

3.3 Energy Loss in Glancing Collisions

A few experiments with glancing collisions were also carried out using the 2-D apparatus [23]. This apparatus is much more difficult to balance at low temperatures because of the two independent degrees of freedom, and consequently measurements are time-consuming. For such measurements two quantities are determined, the decrease in magnitude of the normal component of the velocity after the collision and any change in the tangential velocity. e_n is defined as

$$e_n = v_{nf}/v_{ni} \tag{1}$$

where v_{nf} and v_{ni} are the final and initial normal velocity components respectively. e_t is defined similarly,

$$e_t = v_{tf}/v_{ti} \tag{2}$$

with v_{tf} and v_{ti} the final and initial tangential velocities.

In Fig. 7 e_n and e_t are plotted as a function of the normal and tangential velocity components for several different experimental runs [23] with lightly frosted surfaces. As observed previously, there is scatter in the data which is attributed in this case both to small changes in the local morphology of the surface frost as ice-crystallites are fractured in each collision and to movement of the crystallites because of the glancing nature of the collisions. The most striking feature of these data is that most of the loss occurs for normal incidence — e_n is quite low, in the range 0.4 to 0.7 over most of the velocity range. This loss is likely related to brittle fracture of ice crystallites in the frost surface. In contrast, e_t remains high (near 0.9) over most of this velocity range. The low friction for the tangential motion is attributed to the smooth underlying ice surface and the frost does not appear to have much influence. For impact angles such that one velocity component is small, there is considerable scatter for e_n and e_t. This is caused in part by a weak coupling between the normal and tangential degrees of freedom when the effective impact surface is not exactly horizontal. For low impact speeds, the slight irregularity produced by the frost particles can produce such a coupling and transform vertical (normal) motion into tangential motion or vice versa. This coupling is particularly important when one component is much smaller than the other. Thus the large values of e_n and the low values of e_t (and the large scatter in each quantity), for component velocities below 0.5 mm/s may not be real features of these glancing collisions. Note that in most of these measurements if v_{ni} is small then v_{ti} is not, or vica versa.

3.4 Dynamical Sticking Measurements in Normal Incidence Collision Experiments

Sticking collisions were first observed for very low speed collisions with surfaces freshly coated with water frost [21]. Usually sticking is only observed for speeds below a few mm/s — the highest speed for which impact sticking has been observed is roughly 7 mm/s. A surprising feature of such collisions is that the frost layer behaves much like a linear spring for small displacements. After sticking, the particle motion is well described by damped harmonic motion, with a frequency much faster than the pendulum frequency. Such

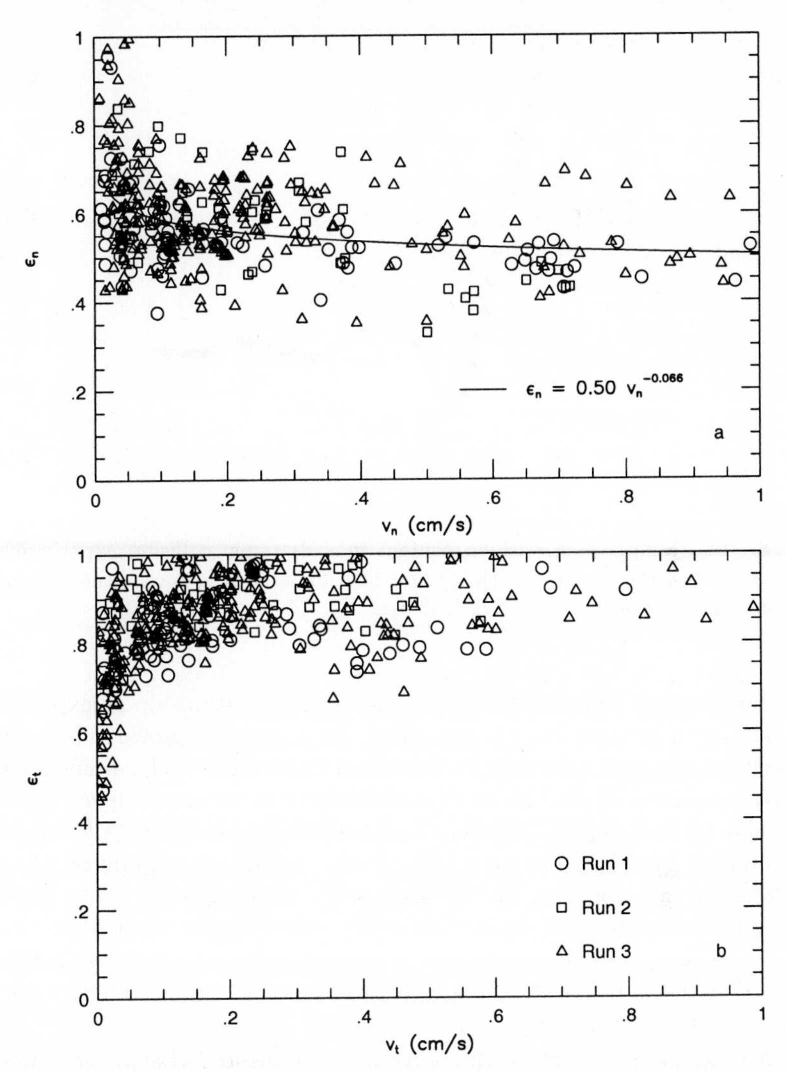

Fig. 7. Measurements of the coefficient of restitution in glancing collisions as a function of the tangential and normal components of the incident velocity. (From Ref. [23], Fig. 5.)

motion is shown explicitly in Fig. 8 [21]. For these data, the oscillation period for the damped motion is about 0.53 seconds, much shorter than the 10 second period of the pendulum. Using the moment of inertia for the pendulum and the period, an effective spring constant for the frost contact can be determined.

When sticking has occurred, a force can be applied to the particle via the magnetic kicker to separate the surfaces; measurements of the displacement

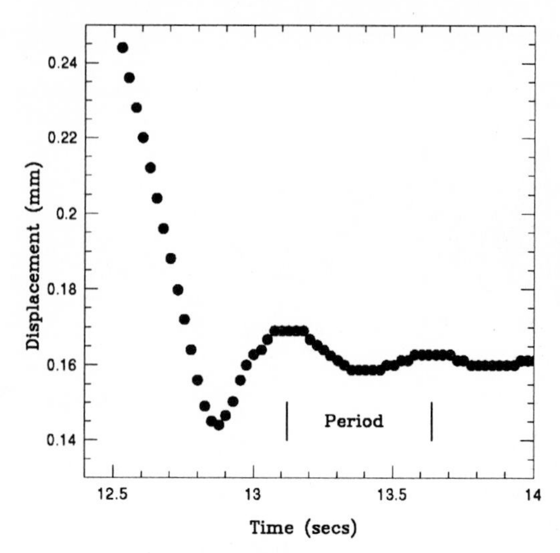

Fig. 8. An example of a sticking contact in a dynamic collision experiment. In this case, after the surfaces stick, the particle undergoes low-amplitude, damped, harmonic-motion with a period of 0.53 sec.

as a function of the applied force yields a linear relationship as expected, if the frost layer behaves like a linear spring. An example is shown in Fig. 9b. At some point the frost layer breaks; the value of the force at this point is called the sticking force, F_s. In Fig. 10 F_s is plotted as a function of impact speed for several experiments [28]. In these and other similar experiments F_s increases with impact velocity up to some critical value and then drops quickly to zero for higher speed impacts. The increase of F_s with impact velocity suggests that the sticking force is, at least in part, caused by an interpenetration of the two surface frost layers. Once a few "high speed" (> 8mm/s) collisions have taken place, the frost is compacted and generally sticking will no longer occur even at very low impact speeds.

Once sticking contacts were observed, the unusual behavior of some low speed collisions at normal incidence, such as that shown in Fig. 9a, could be understood. For this collision, the rebound appears to start at a speed comparable to the incoming speed (i.e., the magnitudes of the displacement as a function of time are comparable and nearly symmetric about the time of impact). However the rebound speed becomes lower than the incident speed as the particle separates, and the final speed once separation is complete, is significantly lower than the incident speed. For this collision, the surfaces do stick together initially, but the sticking force is not sufficient to stop the rebound. As the surfaces separate, the sticking force of the frost surface retards the outward motion of the particle (reducing its speed) until the frost surface breaks. Thus a part of the energy loss in this collision is due to the energy required to stretch the frost, up to the breaking point.

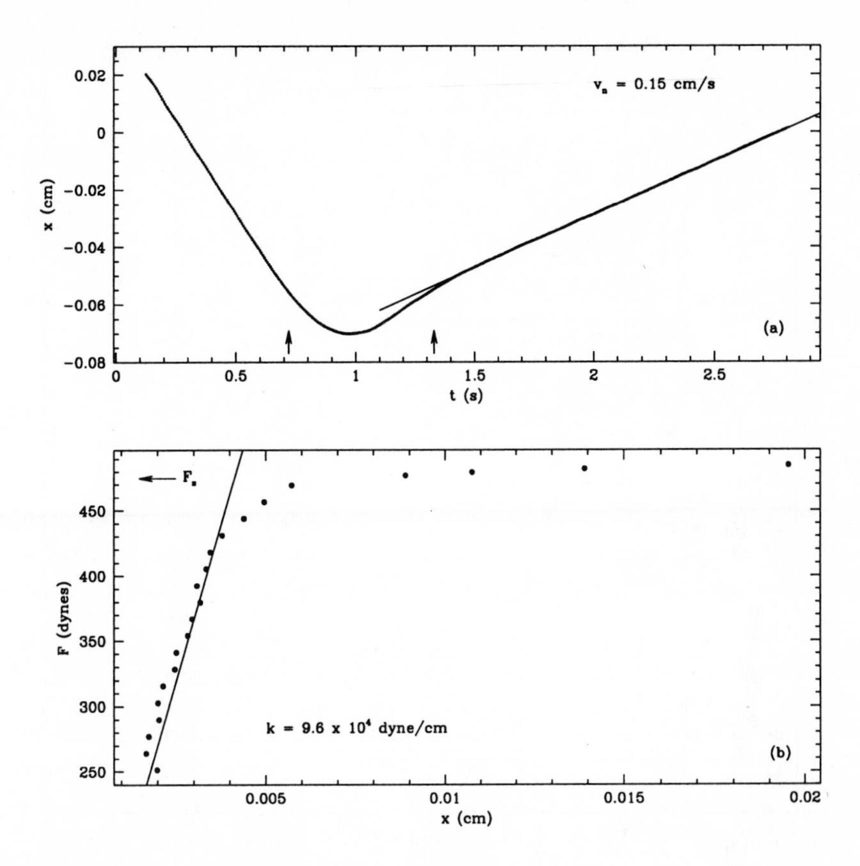

Fig. 9. (**a**) A low speed collision for which the speed slows down as the particle rebounds away from the surface as a result of a sticky contact. (**b**) The change of the displacement of the particle as the force applied to it is increased linearly with time. The force for which separation occurs is called the sticking force F_s. (From Ref. [28], Fig. 2.)

4 Static Sticking Measurements

A large number of static sticking force measurements have been carried out using a well-defined surface contact area of 0.78 cm^2 [22]. In these measurements, after the surfaces are brought together very slowly as described in Sec. 2.2, a force is applied to the upper plate via the load cell. As the load cell is raised the force increases; at a critical displacement the frost breaks, similar to the behavior described above for dynamical sticking measurements. An example is shown in Fig. 11 [22]; the applied force increases linearly with displacement and then suddenly drops back to roughly 17,300 dyne, which is the total weight of the upper plate. The magnitude of the step-drop gives the

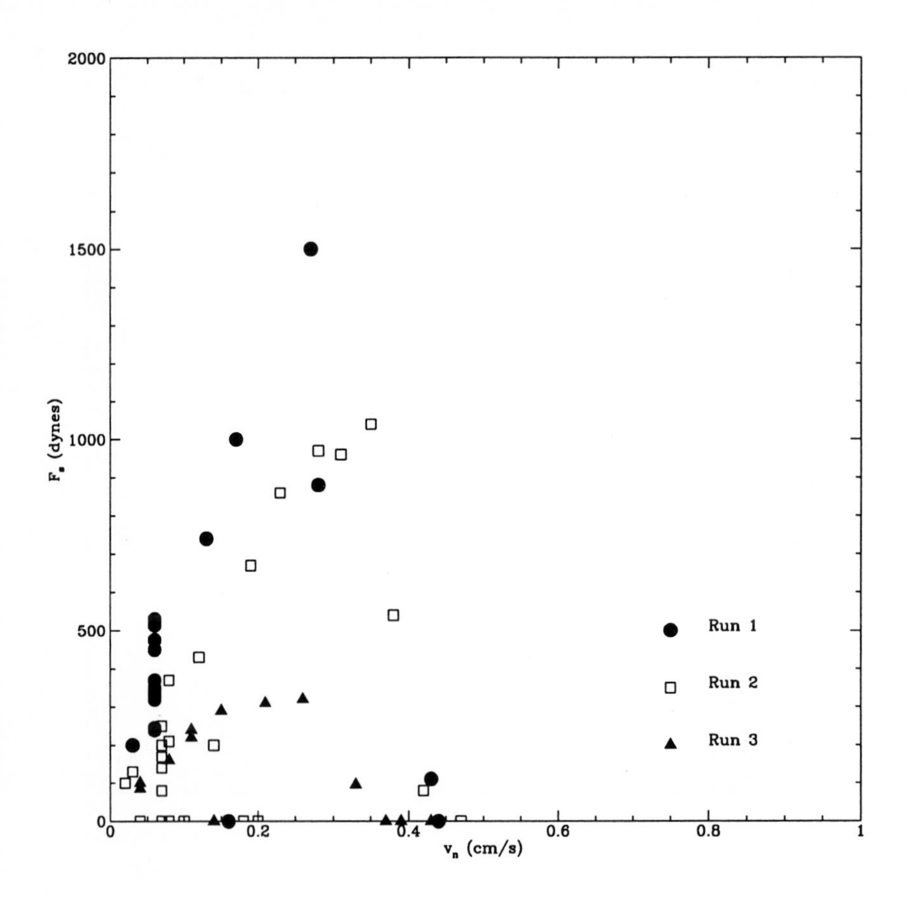

Fig. 10. A plot of the sticking force as a function of impact speed (beginning at very low speeds), for several experimental runs. Generally the sticking increases with speed to some critical value (different for each frost layer) and then goes quickly to zero for higher speed impacts; in subsequent collisions, sticking does not occur, even at very low speeds (From Ref. [28], Fig. 3.)

sticking force, in this case 1520 dynes. Note that there is no plastic regime for these measurements — the measured force does not change significantly with time at each new position and the break point is very abrupt.

The slope of the data up to the breaking point gives an effective spring constant for the system which is a combination of the spring constants for the frost layer and the support thread. Using the known spring constant for the support thread (or wire) the spring constant for the frost can then be extracted.

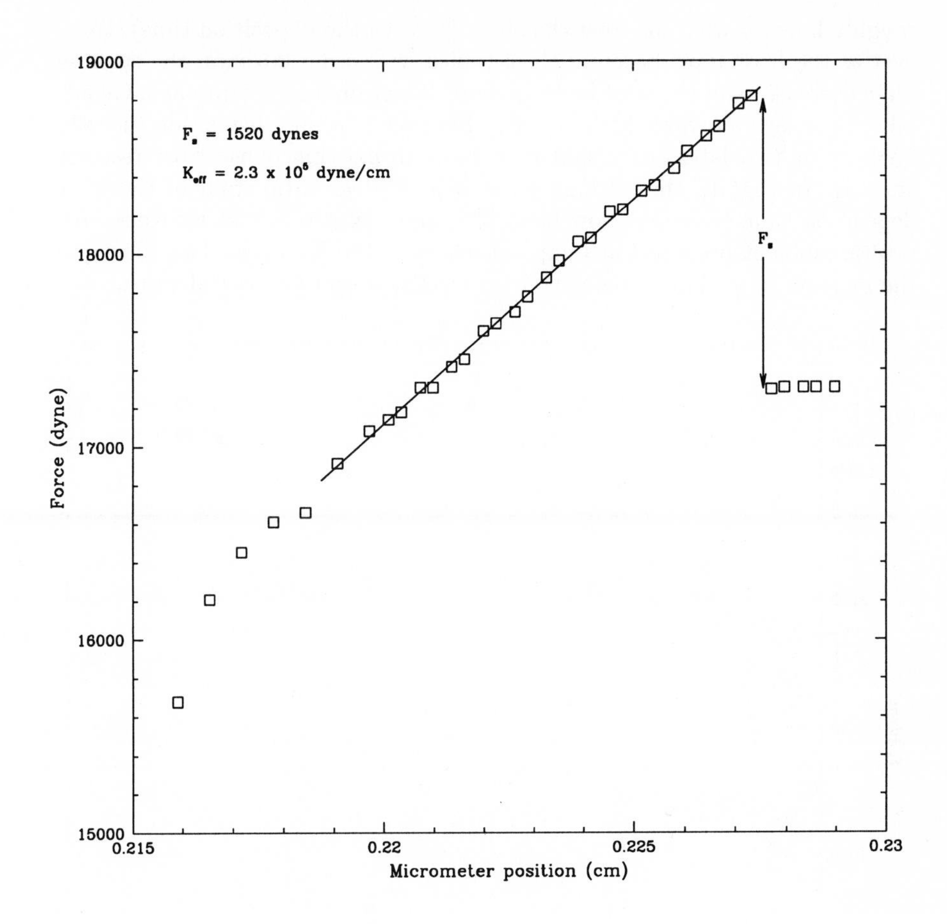

Fig. 11. An example of the linear force vs dispacement plot that is observed when the surface frost layer is stretched. Beyond a certain displacement the frost layer breaks and the measured force is the weight of the upper plate (plus attached weight). The difference between the highest force measured and the weight after the frost breaks is called the sticking force, F_s. (From Ref. [22], Fig. 4.)

4.1 Static Sticking Measurements for Water Frost Layers

The sticking force is strongly dependent upon the morphology of the frost layer, particularly on the dendritic-like structure and the thickness of the frost. In these experiments the structure of the frost layer can be varied in several ways — by changing the ambient pressure in the chamber (which changes the nucleation of small crystallites in the gas before they deposit on the plates), the frost layer thickness, the deposition rate, or the deposition temperature. The thickness of the frost increases with the deposition time for given deposition conditions. For thin frosts, the sticking force increases roughly linearly with the frost thickness (or with the deposition time). However as the frost layer becomes thick typically of order 0.5 mm, the sticking force decreases and the frost layer appears denser under a microscope. Generally, dense layers exhibit little sticking; however a second deposition (usually an hour or two later) of a thin frost onto an existing dense frost restores sticking. In Fig. 12 the sticking force is plotted as a function of the frost deposition time (constant flow rate) [22], and hence the frost thickness, for several ambient pressures at temperatures near 100 K. These data illustrate the increase in sticking force with frost thickness up to a critical value.

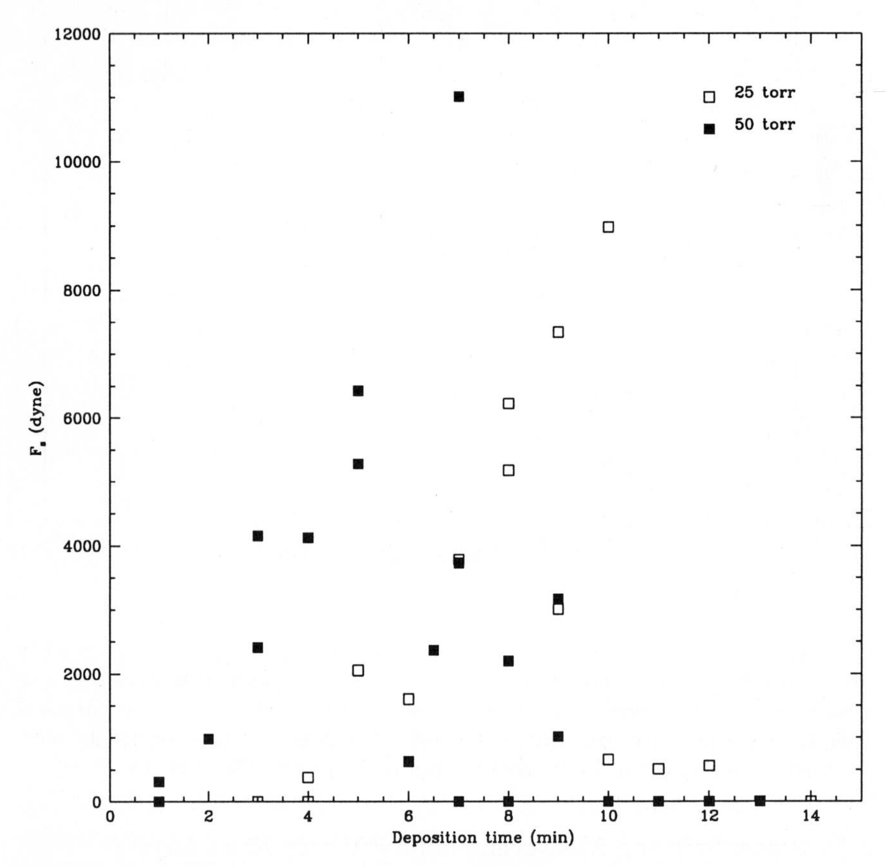

Fig. 12. The sticking force as a function of deposition time, and hence of frost thickness, for a constant deposition rate. The sticking force increases with thickness to some critical thickness and then drops to nearly zero; in this regime, the frost has fewer voids and becomes much smoother. However, subsequent depositions can make the surface-sticking force large again. (From Ref. [22], Fig. 5.)

We have also carried out experiments with different weights set onto the upper plate to change the compression force on the frost surface. For moderate weights the sticking force increases roughly linearly with the applied weight, but for large weights (100 g) sticking no longer occurs — likely because the dendritic-like structure of the frost has been damaged and inter-penetration of the two frost layers no longer occurs. This is essentially the same result as in the dynamic experiments — that above some impact speed (and hence some maximum contact force) there is no sticking.

The effective spring constant and the total stretch of the frost were also determined in each static measurement. For water frost, this provides a means of estimating the energy stored in the frost layer. For the upper plate area of 0.78 cm^2, the average spring constant is roughly $K_f = 10^5$ dyne/cm for a large number of measurements. This can be seen in Fig. 13 where the energy, $1/2K_f x^2$, is plotted as a function of the maximum frost stretch, at the breaking point of the frost [22]. The data cluster about a straight line with this value for K_f. The maximum force per unit area is about 10^4 dyne/cm^2, and the stored energy in a frost layer 0.5 mm thick can be as large as 10^4 erg/cm^3.

For water frost layers there is another type of sticking mechanism that is produced by an annealing of the frost interface when the temperature is raised. After the frost layers have been brought into contact several times, the sticking force usually drops to zero. If the cryostat temperature is raised from roughly 100 K to a value in the range 150 to 180 K for 15 minutes and then cooled back to the measurement temperature while the surfaces are left in contact, sticking is again observed and can be as large as that observed for contact sticking. This effect is only observed if the temperature is raised above approximately 140 K; consequently it is likely related to a partial transition of amorphous frost to crystalline frost which occurs near 135 K. During such a transition, crystallites that are formed at the interface region will bond the two surfaces together. A few examples of such sticking forces are included in Fig. 13.

4.2 Other Surface Layers

A few experiments were also carried out using other volatile components of the outer solar system including layers of CO_2 frost and methanol. Sticking contacts were observed for CO_2 frosts, but the magnitude of the sticking force was roughly a factor of 5 lower than for water frost surfaces. Consequently, such frosts likely play a minor role compared to water frost in particle aggregation. In contrast, pure methanol "frosts" have very high sticking forces for temperatures in the range 110–180 K. In the first series of experiments, the surfaces could not be separated because the required force exceeded the range of the force gauge. In the next series of measurements the area was reduced by factors of 5 and 10 (to 0.16 and 0.08 cm^2), using a ring annulus on the upper plate with a low surface area. Again the force gauge was saturated for some measurements (even for the smallest contact area), corresponding to a minimum sticking force per unit area of 2.4×10^6 dyne/cm^2. This is more

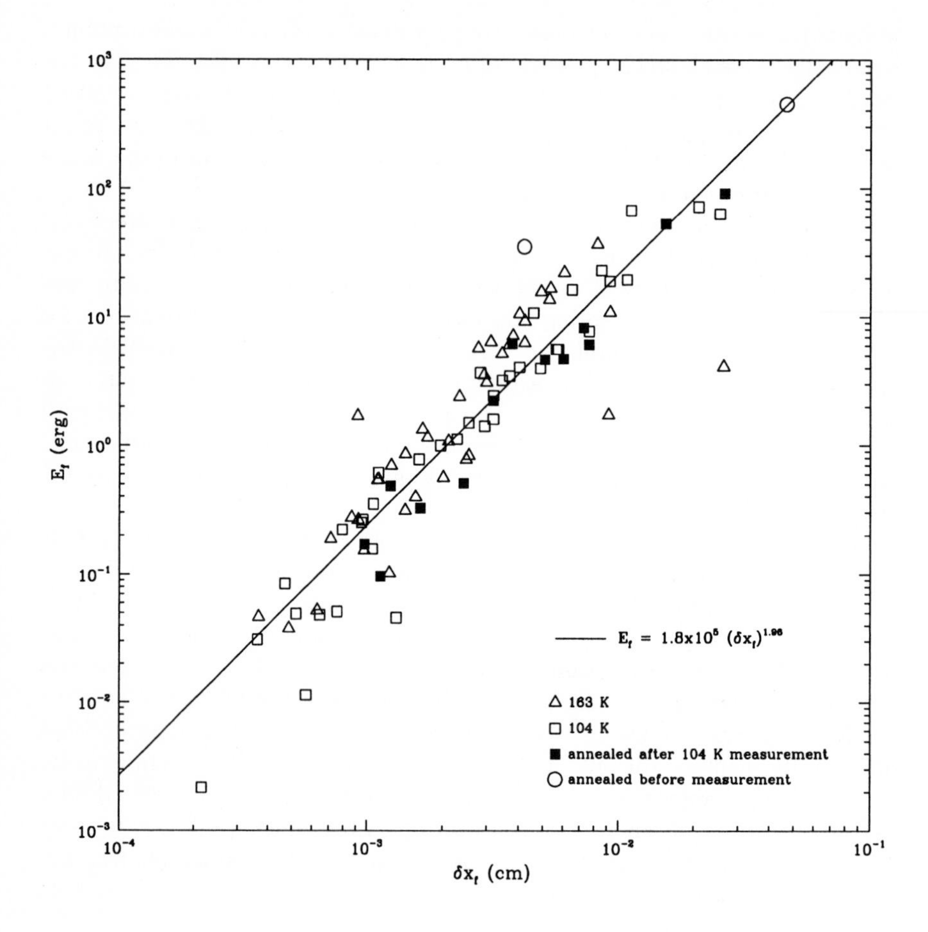

Fig. 13. A plot of the energy stored in the frost spring layer ($0.5K_f x^2$), at the breaking point of the layer. This plot shows that for a fixed contact area (0.78 cm^2) the average, effective spring constant of the frost is roughly constant. (From Ref. [22], Fig. 9.)

than a factor of 100 times larger than the largest forces obtained for water frost, yet this is only a lower limit for sticking with pure methanol.

Several examples of recent measurements by J. Donev [29] are shown in Fig. 14a. The time of contact before separation of the plates was varied and the results lumped into two groups — measurements taken after less than 1 min of contact and those for longer contact times. One feature seemed to emerge but would require more extensive measurements to verify — that the sticking force was often larger for the longer contact times, particularly at the lowest temperatures. This would be consistent with a highly viscous material whose viscosity increases as T is decreased. Longer contact times under a weak compressive force would allow the surfaces to flow together and thus increase the sticking contact force.

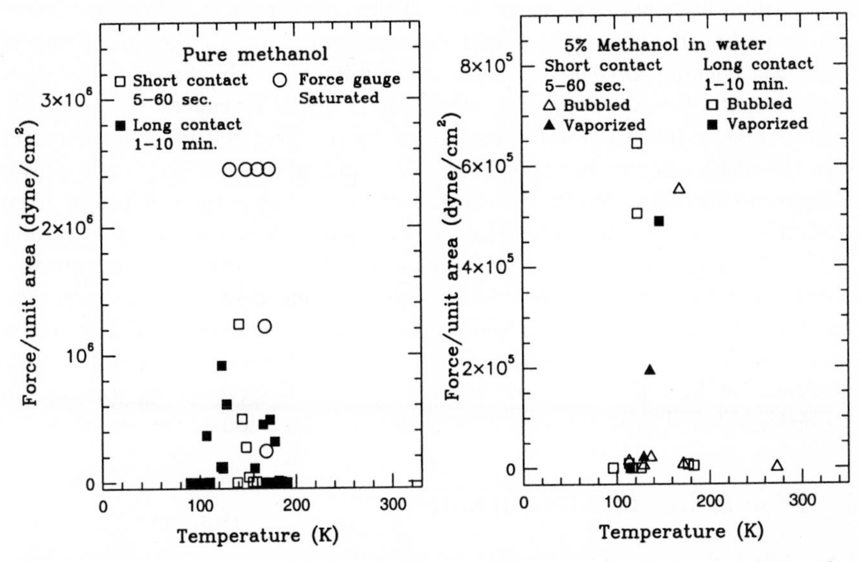

Fig. 14. (**a**) Examples of the measured sticking force per unit area, for surface layers of methanol for various temperatures and different contact times. The circles (long contact times) indicate lower limits for the three different areas used (0.78, 0.16 and 0.08 cm^2 [29]), for which the strain gauge load cell saturates at a weight corresponding to 200g. (**b**) Examples of the measured sticking force per unit area, for 5% methanol/water mixtures. For some measurements nitrogen gas was bubbled through the mixture before passing into the cryostat; in others the mixture was sprayed directly into the flow of nitrogen

Pure methanol does not exist in outer space; however, many comets have methanol concentrations of order 5% and several series of measurements using a 5% methanol/water solution were carried out, mostly between 100 and 200 K. For these measurements, two methods were used to introduce the vapor; nitrogen gas was bubbled through the methanol/water mixture or the mixture was vaporized directly into the carrier gas chamber. Large sticking forces were obtained but not as large as observed for pure methanol, with the largest sticking force about 6×10^5 dyne/cm^2 (still more than 20 times larger than for water frost). Again there is a tendency for larger sticking forces (at lower temperatures) when the contact time is long as shown in Fig. 14b. In nearly all cases, the data represented by squares (long contact times) is at or above the data for short contact times shown by triangles.

It is well known that water/organic or organic/organic mixtures form glasses at low temperatures; such glasses are used to provide a structureless matrix in which to suspend small particles or molecules, often for low temperature spectroscopy [30]. However the viscosity of these mixtures at low temperatures has generally not been studied and applications to Saturn's rings or the outer solar system, from the perspective of sticking and aggregation, have not been considered. The viscous nature of both methanol and water/methanol mixtures becomes very apparent in measurements of the static sticking force for these layers. Unlike the behavior of water frost layers, when a force is applied to these glassy layers by raising the load cell, there is plastic flow. The force registered by the load cell initially increases with the displacement, but then over a long period of time, the surface layer relaxes and the force slowly decreases. For short measurement times of order a minute, a plot of force vs. displacement can be obtained that is similar to that for water frost; however each measurement must be taken quickly. In experiments for which the time between each successive displacement was long (over ten minutes) the surface layer continued to stretch, and the overall stretch of the layer could be very large, of order of 1cm. Therefore for these materials we do not have a good estimate of the effective spring constant or the energy stored in the surface layer prior to separation of the surfaces.

5 Discussion of Measurements

5.1 Modeling the Coefficient of Restitution

The coefficient of restitution, ϵ, for water ice remains high over a significant range of impact speeds for clean surfaces [24–26]. However, it develops a significant velocity dependence for low impact speeds when the surfaces are coated with a frost layer, with the velocity dependence determined by the thickness and morphology of the frost layer. Several attempts have been made to model the energy loss in these collisions. All calculations begin with Hertz's elastic theory [31] and extend it by including an energy loss mechanism. Early studies by Goldreich and Tremaine [32] applied the model developed by Andrew [33] for plastic deformation of metals, but the resulting velocity dependence is too slow to explain the data for frosted surfaces. Dilley [34] added a viscous dissipation term (with loss primarily in the surface layer) and obtained an equation of the form

$$\epsilon = \exp(-\pi\xi\sqrt{1-\xi^2}). \tag{3}$$

For a particle of radius R_p impacting an infinite surface;

$$\xi = AR_p^{-\beta}v^{\alpha} \tag{4}$$

where A, β and α are constants to be determined. This model can describe the data for frost coated surfaces quite well, but the significance of these parameters is not clear.

Another approach has been developed by Pöschel, Spahn and co-workers [35–38]. Viscoelastic stress is added to the Hertz elastic model through a term proportional to $v_s^{3/2}$ (more generally v_s^α), where v_s is the velocity of compression at the contact surface. In this model the material properties are assumed uniform - that is, there is no explicit consideration of a soft surface layer over a hard surface. This general approach should be applicable to collisions of solid ice particles for which a local radius of curvature can be defined. It should also apply for low speed collisions of a particle coated with a surface layer, but the collision must be soft enough that the surface layer is not highly compressed. For locally spherical surfaces, the coefficient of restitution has the general form

$$\epsilon = 1 - \gamma_1 v_o^{1/5} + \gamma_2 v_o^{2/5} + ... \tag{5}$$

where γ_1 and γ_2 are constants and v_o is the incident velocity. The constants depend both on material properties (masses, radii, viscous constants etc.) and some universal constants - the latter have been evaluated analytically and numerically [37, 38]. The velocity dependence of ϵ for frost-coated surfaces can be fit well using this model; the fitting parameters, γ_1 and γ_2, required to fit the data for frost-coated ice particles indicate a low surface viscosity and hence a soft surface, consistent with the presence of a frost layer. However, this model does not appear to work well for describing the coefficient of restitution for clean surfaces [35].

Finally Higa et al. [25, 26] (see next section) have obtained an empirical expression for ϵ in high speed collisions of clean ice surfaces:

$$\epsilon = \epsilon_{qe}\left(\frac{v_i}{v_c}\right)^{-\log(v_i/v_c)}, \tag{6}$$

where ϵ_{qe} is the low speed value for ϵ, v_i is the incident velocity and v_c is the critical speed above which cracks form. This may be important in high speed collisions if some particles are formed of solid ice in the outer regions of the solar nebula.

Some of the experimental data indicate that there is more than one mechanism that contributes to the total energy loss in low speed collisions, particularly for freshly frosted surfaces. Much of the energy loss must occur through brittle fracture of the ice crystallites on the surface as well as re-arrangements of the surface particles (these are modeled by the viscous terms described above). However for low speed collisions and for surfaces for which some sticking will occur, another mechanism is also present. When the surfaces stick together, energy is required to stretch the surface layer before it breaks. This reduces the rebound velocity as shown in Fig. 9a and for very low speed collisions can provide the dominant energy loss interaction. However it would probably not play a significant role for compacted surfaces.

5.2 Other Collision Experiments

There are very few investigations of slow speed collisions of ice particles reported in the literature. Higa and coworkers [25, 26] have measured ϵ for a range of speeds in free fall conditions with speeds up to 10 m/s. They have relatively few measurements below 5 cm/s and for these very slow speeds, there is a potential problem because the bounce amplitude becomes less than 100 μ, comparable to possible surface roughness or surface dust. They find that for clean ice surfaces, ϵ is close to 1 (between 0.8 and 1) for a wide range of temperatures and particle sizes. For such collisions, there is no significant fracturing of the ice. Above a critical speed, which changes with temperature, fracturing occurs and ϵ drops quickly to zero. These experiments are consistent with our low speed measurements for clean surfaces and show that this quasi-elastic regime can extend to quite high impact speeds of order 1 m/s.

Most of the results described in this review are for rough surfaces — compacted frost layers or surfaces that have partially sublimated. These surfaces are very lossy even at speeds as low as 2 cm/s. Higa et al. also find greatly increased loss for surfaces coated with frost. Such surfaces efficiently remove kinetic energy from a system of particles. Interestly however, in our experiments ϵ again approaches 1 for such surfaces at low impact speeds, below 1 mm/s.

In another set of experiments, Dilley and Crawford [27] used a long (10 m) pendulum to investigate ϵ as a function of mass for low speeds of order 1 cm/s. They obtained very low values of ϵ for particle masses below 100 g. However, in these experiments the ice particles likely had some frost on the surface from exposure to air. In view of the high values of ϵ in the experiments of Higa et al and also for our experiments using clean ice surfaces, we attribute the low values of ϵ in the experiments of Dilley and Crawford to a layer of frost on the surface.

5.3 Sticking Mechanisms

The measurements presented in Sec. 4 indicate that three different types of sticking mechanisms may be present. The first is a mechanical contact sticking mechanism for water frost surfaces. In this case the surface has a dendritic-like structure, and the sticking force is assumed to arise from inter-penetration of the two surface layers, with a mechanical locking of the ice filaments. As the impact force is increased the inter-penetration and the sticking force are also increased. This mechanism may also be applicable to CO_2 frosts.

The second mechanism observed for water frost surfaces is thermally induced. It is caused by molecular rearrangement, most likely a partial transition from an amorphous to crystalline phase, that occurs above 140 K, even for short annealing times. This mechanism is applicable both for freshly frosted surfaces which would show contact sticking, and for surfaces which are compacted and no longer exhibit any sticking interaction. For cases where contact sticking has produced aggregation, this mechanism may generate stronger surface bonds if the particles cycle through regions in the solar system or a ring system where the temperature is higher.

The third sticking mechanism applies to viscous, glassy materials such as water/methanol mixtures. In this case the surfaces will flow together under a weak contact force, thereby generating a sticking contact. As the temperature is lowered and the viscosity increases, more time is required for the surfaces to flow together. Consequently there is little sticking at very low temperatures — for example, for the water/methanol mixtures there is essentially no sticking below roughly 110 K for short contact times.

6 Application to Planetary Systems

The experimental data presented here have many planetary and astrophysical applications, ranging from planetary rings and planetesimal disks to the internal structure of meteorites, comets, and asteroids. We briefly discuss below some of the implications of the data presented here.

6.1 Planetary Rings

The structure and evolution of planetary rings are determined by the velocity distribution of the constituent particles [32, 39]. Energy is transferred through gravitational scatterings and direct collisions, from the particles' mean differential Keplerian motion about the planet into dispersive random motion. Kinetic energy is also dissipated through inelastic collisions. If ϵ is a decreasing function of the impact velocity, an energy equilibrium would be established when these two processes are balanced against each other, which leads to an average relative velocity called the dispersion velocity, σ. The velocity dispersion also determined the ring thickness $H \sim \sigma/\Omega$, where Ω is the Keplerian angular frequency. The inferred ring thickness based on the data in Figs. 3-8 varies from a few meters [40] to nearly 100 m depending on the surface structure of the ring particles and the decrease in ϵ with σ. The current upper limit of the ring thickness inferred from radio occultation is 100 m [41], and that from damping of the density waves is an order of magnitude smaller [42]. From these observations and our experimental results, we can rule out the possibility that ring particles have regular crystalline surfaces.

In the opaque regions where the particles' collision frequency is $> \Omega$, the collisional mean free path is comparable to the particle size, S, the dynamics of the rings are better described by granular flow [43–45]. In these regions, collisional damping of the velocity dispersion results in $\sigma \sim S\Omega$ [46] and the effect of self gravity leads to the formation of transient gravitational wakes [47–50]. Reflection of sunlight by these wakes produces the azimuthal brightness variation which has been observed in the rings [51, 52].

Due to the large cross section of the rings and the strong gravitational influence of Saturn on interplanetary particles in the outer solar system, impacts by cometary bodies occur frequently [53, 54]. Since comets carry a concentration of methanol and other organic molecules [55], at least some ring particles may be covered with frost and organic molecules. The results in Sec. 4 suggest that slow speed collisions would cause these particles to adhere with a range of sticking strengths. These particles are also being pulled apart by the planet's differential tidal force, the magnitude of which increases with S. Using the sticking strength for frost-covered particles, the size of the largest particles inferred from the balance between the tidal tearing and the sticking strength is $\sim$ 10 m [21] and possibly much larger if surface layers of water/methanol mixtures are present. This upper limit for the size is consistent with that inferred from Voyager data [56]. Since $S \sim H$, at the largest particle size, the rings may be regarded as essentially a monolayer [32, 48, 49, 57].

6.2 Planetesimal Growth

The growth of solid particles is an essential step in the formation of planetesimals which are the building blocks of planets. Depending on their sizes, particles acquire dispersive motion through their interaction with the disk gas [11, 58, 59] and other planetesimals [60–62]. Collisions between particles with relatively large impact speeds are likely to be damped repeatedly [63, 64] until they become cohesive. This process eventually lead to the formation of loose "rubble piles" as the end product of multiple collisions. The spectacular images of Shoemaker-Levy 9 show that it fragmented into more than twenty identifiable pieces due to planetary tides before it crashed into Jupiter [65]. Numerical simulations indicate that the parent body of Shoemaker-Levy-9 needs to be very weakly bound as a loose rubble pile in order for it to be tidally disrupted. The miniscule sticking strength which best reproduces the morphology of this comet [13] is comparable to that obtained in our experiments.

The gaseous, dusty solar nebula from which the planets formed was very likely turbulent [8, 66, 67]. This turbulence lofts small particles above the warmer, denser midplane regions of the nebula to cooler, more rarefied regions where frost layers can grow [68]. Over such a temperature range, frost layers accumulate on particle surfaces [69]. Collisions between small particles occur frequently, some of which could lead to sticking and growth of aggregates of cm-sized particles. Experiments show that the sticking strength decreases with the thickness and compactness the frost layer. Although, particle collisions could lead to compaction of the frost layer and thus to less-sticky frost, such a layer (or a rough, sublimated ice surface) also dissipate more energy upon collision than a bare ice or rock surface (Section 3.1). Dissipation decreases relative particle velocities in the nebula and the chances that an existing aggregate would be disrupted upon collision with another particle.

The sticking experiments showed that frost sticking forces are significant, high enough to hold centimeter-sized aggregates together in collisions at typical relative speeds in a turbulent nebula, particularly if the aggregates undergo temperature cycling [28]. A strong sticking force is needed for bodies to grow beyond centimeter sizes in a turbulent nebula, since centimeter-sized particles are too small for their mutual gravitational attraction to hold them together in collisions, and they are too large for van der Waals attraction to bind them. Frosts of volatiles such as water, methanol, and carbon dioxide could provide this sticking force.

7 Summary

In this review we summarize the energy losses that we have observed in low speed collisions of ice particles coated with various surface layers, and compared these results with other recent measurements. Clean ice is very elastic (ϵ is close to 1) while rough surfaces and surfaces coated with frosts of water or CO_2 show significant loss and a strong velocity dependence of ϵ. Surfaces coated with thick frosts provide an efficient means of reducing the relative particle speeds, if the speeds are quite high — many cm/s; however, such surfaces can be quite elastic at very low speeds and may not be able to reduce the relative speeds to zero. Consequently, the cooling of such granular gases depends strongly on the velocity dependence of the coefficient of restitution. These experimental results have important consequences for improving our understanding of the dynamics of ring systems such as Saturn's rings and for the dynamics of planetesimals in the early solar system.

Theoretical models of collisions have been extended to include viscous dissipation forces and can model the observations for frost coated surfaces quite well. These calculations provide functional forms for the dependence of ϵ on the impact velocity, which should be helpful in future investigations. For collisions on rough surfaces, where there are effectively two or more points of contact, these results will need to be generalized.

Another feature of the surface coatings is that they provide a sticking mechanism that may have been important in the initial aggregation processes by which planetesimals and comets formed. Contact sticking forces may also be important in the dynamics of Saturn's rings if the distribution of particle sizes is nearly static because formation and disintegration processes are in quasi-equilibrium. Low density frost produces a mechanical sticking force for slow speed collisions of at most a few mm/s. An interpenetration of dendritic-like structures on each surface may account for this sticking interaction. In addition, this frost layer acts like a linear spring for small displacements and translational kinetic energy can be stored as elastic energy in the spring. If this stretched spring breaks, energy is lost; this may be an important contribution to energy loss in low speed collisions.

Layers of other materials such as methanol/water mixture appear more uniform and have a different behavior; they provide larger sticking forces in the range 100–150 K and show plastic deformation when stretched. It is well known in other fields that mixtures of organics or mixtures of water and organics form glasses at low temperatures; the higher contact sticking for such surfaces is attributed to the slightly viscous nature of such materials at relatively low temperatures. Since many of the volatile constituents of the solar system are found both in comets and in the ices of the outer solar system we have suggested that contact sticking produced by surfaces composed of water/organic mixtures, may dominate the aggregation processes.

These measurements provide important information needed for modeling a range of phenomena in the solar system, including the formation of planetesimals, the weak structural strength of comets and the dynamics of ring systems. For Saturn's rings, the measured values of ϵ provide constraints on the thickness of the rings which are consistent with other measurements. A number of observations indicate that many comets are weakly bound clusters of denser objects. Such an aggregate will break up in a tidal force into its constituent pieces as observed for Shoemaker-Levy 9 prior to its impacting Jupiter. Estimates of the tidal forces indicate that the tensile strength is very weak, but consistent with the sticking force measurements reported here.

Computer simulations of planetesimal formation in the outer regions of the solar system show that small ($\sim$ mm- to cm-sized) icy particles experience wide ranges of ambient temperature in a turbulent solar nebula. At the lower temperatures, frost layers grow and can provide damping of relative velocities as well as a sticking mechanism. As the particle moves into warmer regions, any frost bonds that have formed could be cemented further by annealing at higher temperatures; such annealing has been observed in our experiments with water frost bonds.

A strong sticking force is needed to hold aggregates together in the turbulent solar nebula, since relative speeds between bodies several centimeters in size were high enough to break apart weakly-bonded aggregates. Frost layers of water alone are not as strongly adhesive as layers of methanol/water mixtures, as discussed above; in either case, the frost layers would provide an important mechanism to damp the relative velocities and decrease the probability that aggregates would break upon collision. A freshly-formed frost layer would be less dense and more dendritic in nature, and therefore more likely to stick upon collision with another frosty particle, than a layer which had experienced many collisions and become compacted. Thus frost plays two important roles in planetesimal formation: decreasing particle relative velocities; and providing a significant sticking force.

Acknowledgements

The authors thank a large number of students who have worked on this project over more than a decade — A. Hatzes, R. Knight, J. Lievore, S. Sachtjen, S. Tiscareno, M. Zafra and most recently J. Donev who collected most of the data for methanol and the methanol/water mixtures. This project was supported at various times by NASA, CalSpace, and Faculty research grants.

References

1. Cameron A. G. W. (1978) The primitive solar accretion disk and the formation of the planets. In *The Origin of the Solar System* (S. F. Dermott, Ed.) 49–75. J. Wiley & Sons, New York
2. Cameron A. G. W. (1978a) Physics of the primitive solar accretion disc. Moon and Planets **18** 5–40
3. Sargent A. I., Beckwith S. (1987) Kinematics of the circumstellar gas of HL Tauri and R Monocerotis. Astrophys. J. **323** 294–305
4. Sargent A. I., Beckwith S. V. W. (1991) The molecular structure around HL Tauri. Astrophys. J. Lett. **382** 31–35
5. O'Dell C. R., Wen Z. (1994) Post-refurbishment mission Hubble Space Telescope images of the core of the Orion Nebula: Proplyds, Herbig-Haro objects, and measurements of a circumstellar disk. Astrophys. J. **436** 194–202
6. Marcy G. W., Butler R. P. (1996) A planetary companion to 70 Virginis. Astrophys. J. **464** L147–L151
7. Mayor M., Queloz D. (1995) A Jupiter-mass companion to a solar-type star. Nature **378** 355–359
8. Lin D. N. C., Papaloizou J. (1985) On the dynamical origin of the solar system. In *Protostars and Planets II* (D. Black, M. Matthews, Ed.) 981–1072. U. of Arizona Press, Tucson
9. Lin D. N. C., Papaloizou J. (1993) Growth of planets from planetesimals. In *Protostars and Planets III* (E. Levy, J. Lunine, Ed.) 749–835. U. of Arizona Press, Tucson
10. Shu F. H., Johnstone D., Hollenbach D. (1993) Photoevaporation of the solar nebula and the formation of the giant planets. Icarus **106** 92–101
11. Weidenschilling S. J., Cuzzi J. N. (1993) Formation of planetesimals in the solar nebula. In *Protostars and Planets III* (E. H. Levy, J. I. Lunine, Ed.) 1031–1060. U. of Arizona Press, Tucson
12. Asphaug E., Benz W. (1994) Density of Comet Shoemaker-Levy deduced by modeling breakup of the parent rubble pile. Nature **370** 120–124
13. Asphaug E., Benz W. (1996) Size, density, and structure of comet Shoemaker-Levy 9 inferred from the physics of tidal breakup. Icarus **121** 225–248
14. Sekanina Z. (1993) Disintegration phenomena expected during collision of Comet Shoemaker-Levy 9 with Jupiter. Science **262** 382–387
15. Sekanina Z., Chodas P. W., Yeomans D. K. (1994) Tidal disruption and structure of periodic comet Shoemaker-Levy 9. Astron. Astrophys. **289** 607–636
16. Sekanina Z., Yeomans D. K. (1985) Orbital motion, nucleus precession, and splitting of Periodic Comet Brooks 2. Astron. J. **90** 2335–2352
17. Würm G., Blum J. (1998) Experiments on preplanetary dust aggregation. Icarus **132** 125–136
18. Blum J., Münch M. (1993) Experimental investigations on aggregate-aggregate collisions in the early solar nebula. Icarus **106** 151–167
19. Pinter S., Blum J., Grün E. (1989) Mechanical properties of "fluffy" agglomerates consisting of core-mantle particles. In *Proc. Intl. Workshop on the Physics and Mechanics of Cometary Materials* (J. Hunt, T. D. Guyenne, Ed.) 215–219. ESA SP-302
20. Blum J. (1989) Coagulation of protoplanetary dust. Bull. Amer. Astr. Soc. **22** 1082 (abstract)

21. Hatzes A. P., Bridges F., Lin D. N. C., Sachtjen S. (1991) Coagulation of particles in Saturn's rings: Measurements of the cohesive force of water frost. Icarus **89** 113–121
22. Supulver K. D., Bridges F. G., Tiscareno S., Levoire J., Lin D. N. C. (1997) The sticking properties of water frost produced under various ambient conditions. Icarus **129** 539–554
23. Supulver K. D., Bridges F. G., Lin D. N. C. (1995) The coefficient of restitution of ice particles in glancing collisions: Experimental results for unfrosted surfaces. Icarus **113** 188–199
24. Hatzes A. P., Bridges F. G., Lin D. N. C. (1988) Collisional properties of ice spheres at low impact velocities. M. N. R. A. S. **231** 1091–1115
25. Higa M., Arakawa M., Maeno N. (1996) Measurements of restitution coefficients of ice at low temperatures. Planet. Space Sci. **44** 917–925
26. Higa M., Arakawa M., Maeno N. (1998) Size dependence of restitution coefficients of ice in relation to collision strength. Icarus **133** 310–320
27. Dilley J., Crawford D. (1996) Mass dependence of energy loss in collisions of icy spheres: An experimental study. J. Geophys. Res. **101** 9267–9270
28. Bridges F. G., Supulver K. D., Lin D. N. C., Knight R., Zafra M. (1996) Energy loss and sticking mechanisms in particle aggregation in planetesimal formation. Icarus **123** 422–435
29. Donev J. (1999) Low temperature methanol sticking measurements. Senior thesis, unpublished
30. Myer B. (1971) Spectroscopy. p. 203. American Elsevier Pub. Co., New York
31. Hertz H., Reine J. (1882) Angewandte Phys. **92** 156
32. Goldreich P., Tremaine S. (1978) Icarus **34** 227–239
33. Andrew J. P. (1930) Phil. Mag. **9** 593–610
34. Dilley J. P. (1993) Energy loss in collisions of icy spheres: Loss mechanisms and mass-size dependence. Icarus **105** 225–234
35. Hertzsch J. M., Spahn F., Brilliantov N. V. (1995) On low-velocity collisions of viscoelastic particles. J. de Physique II **5** 1725–1738
36. Brilliantov N. V., Spahn, F., Hertzsch J.-M., Pöschel, T. (1996) A model for collisions in granular gases. Phys. Rev. E **53** 5382–5392
37. Schwager T., Pöschel T. (1998) Coefficient of resitution of viscous particles and cooling rate of granular gases. Phys. Rev. E **57** 650–654
38. Ramirez R., Pöschel T., Brilliantov N. V., Schwager T. (1999) Coefficient of restitution of colliding viscoelastic spheres. Phys. Rev. E **60** 4465–4472
39. Stewart G. R., Lin D. N. C., Bodenheimer P. H. (1984) In *Planetary Rings* (A. Brahic, R. Greenberg, Ed.) 447. U. of Arizona Press, Tucson.
40. Bridges F. G., Hatzes A. P., Lin D. N. C. (1984) Nature **309** 333
41. Zebker H. A., Tyler G. L. (1984) Science **223** 396
42. Shu F. H., Dones L., Lissuaer J. J., Yuan C., Cuzzi J. N. (1985) Ap. J. **299** 542
43. Hämeen-Anttila, K. A. (1982) Moon Planets **26** 171
44. Araki S., Tremaine S. D. (1986) Icarus **65** 83
45. Araki S. (1991) Icarus **90** 139
46. Wisdom J., Tremaine S. D. (1988) Icarus **95** 925
47. Salo H. (1992) Nature **359** 619
48. Salo H. (1995) Icarus **117** 287
49. Richardson D. (1993) M. N. R. A. S. **261** 396

50. Toomre A. (1990) In *Dynamics and Interactions of Galaxies* (R. Wielen, Ed) 292. Springer, Berlin.
51. Franklin F. A., Cook A. F., Barrey R. T. F., Roof C. A., Hunt G. E., De Rueta H. B. (1987) Icarus **69** 280
52. Dones L., Cuzzi J. N., Showalter M. R. (1993) Icarus **105** 184
53. Dones L. (1991) Icarus **92** 194
54. Cuzzi J. N., Estrada P. R. (1998) Icarus **132** 1
55. Whittet D. C. B., Schutte W. A., Tielens A. G. G. M., Boogert A. C. A., De Graauw T., Ehrenfreund P., Gerakines P. A., Helmich F. P., Prusti T., Van Dishoeck E. F. (1996) Astr. Ap. **315** 357
56. Showalter M. R., Nicholson P. D. (1990) Icarus **87** 285
57. Petit J.-M., Henon M. (1988) Astr. Ap. **199** 343
58. Morfill G. E., Tscharnuter W., Volk H. J. (1985) In *Protostars and Planets II* (D. Black, M. Matthews, Ed.) 493. U. of Arizona Press, Tucson
59. Weidenschilling S. (1997) Icarus **127** 290
60. Kokubo E., Ida S. (1996) Icarus **123** 180
61. Safronov V. S. (1969) Evolution of the Protoplanetary Cloud and Formation of the Earth and the Planets. Nauka Press, Moscow. Also NASA-TT-F-677 (1972)
62. Wetherill G. W. (1980) Formation of the terrestrial planets. Ann. Rev. Astr. Ap. **18** 77–113
63. Petit J.-M., Henon M. (1987) Astr. Ap. **188** 198
64. Aarseth S. J., Lin D. N. C., Palmer P. L. (1993) Evolution of planetesimals II: Numerical simulations. Ap. J. **403** 351–376
65. Noll K. S., McGrath M. A., Trafton L. M., Atreya S. K., Caldwell J. J., Weaver H. A., Yelle R. V., Barnet C., Edgington S. (1995) Science **267** 1307
66. Dubrulle B. (1993) Differential rotation as a source of angular momentum transfer in the solar nebula. Icarus **106** 59–76
67. Cuzzi J. N., Dobrovolskis A. R., Champney J. M. (1993) Particle-gas dynamics in the midplane of a protoplanetary nebula. Icarus **106** 102–134
68. Supulver K. D., Lin D. N. C. (1999) Formation of icy planetesimals in a turbulent solar nebula. Icarus submitted
69. Supulver K. D. (1997) Planetesimal formation in the outer solar nebula. Ph. D. dissertation, U. of Cal. Santa Cruz.

Contact Mechanics and Coefficients of Restitution

Colin Thornton[1], Zemin Ning[2], Chuan-yu Wu[1], Mohammed Nasrullah[1], and Long-yuan Li[1]

[1] School of Engineering & Applied Science Aston University, UK.
e-mail: c.thornton@aston.ac.uk
[2] Department of Physics, Cavendish Laboratory Cambridge University, UK

Abstract. We assume that the energy dissipated during collisions between solid bodies is due to plastic deformation at the contact. The normal and tangential interactions between elastic and elastic-perfectly plastic spheres are discussed. For elastic-perfectly plastic spheres an analytical solution for the normal coefficient of resitution is presented which is velocity dependent. The predicted rebound kinematics for oblique impacts of elastic spheres are shown to be in excellent agreement with experimental data. Comparisons between predictions and experimental data for oblique impacts of elastoplastic spheres are only in agreement if sliding occurs throughout the impact duration.

1 Introduction

In numerical simulations of rapid granular flows, particles collide with each other and energy is dissipated as a result of the collisions. In order to dissipate energy many researchers use two collisional operators, a constant coefficient of restitution and a friction coefficient, with the implication that the coefficient of restitution is a material property. Although the restitution coefficient is dependent on certain material properties it is more significantly dependent of the relative impact velocity, see Goldsmith [1]. Assuming that the interaction between colliding particles can be modelled by a lumped parameter system consisting of a spring and dashpot in parallel, Brilliantov et al. [2] developed an analytical solution for a velocity dependent coefficient of restitution which was reasonably fitted to experimental data reported by Bridges et al. [3]. In contrast to this viscoelastic model, we assume that the collisional energy dissipation is due to plastic deformation of the colliding bodies. In the paper we consider the normal and tangential contact interactions between (i) elastic spheres and (ii) elastic-perfectly plastic spheres and show the corresponding rebound kinematics in terms of the normal and tangential coefficients of restitution and imparted particle rotation resulting from the collisional interaction.

2 Normal Impact of Elastoplastic Spheres

The normal impact of two elastic spheres of radii R_i $(i = 1, 2)$ is described by the theory of Hertz which provides the following equations for the normal contact pressure distribution, $p(r)$, the contact radius, a, and the force-displacement relationship.

$$p(r) = p_0\sqrt{1-\left(\frac{r}{a}\right)^2}\,, \qquad \text{where} \qquad p_0 = \frac{3F}{2\pi a^2} = \frac{2E^* a}{\pi R^*} \tag{1}$$

$$a = \left(\frac{3FR^*}{4E^*}\right)^{1/3}, \qquad \text{and} \qquad F = \frac{4}{3}E^*\sqrt{R^*}\alpha^{3/2} \tag{2}$$

$$\text{with} \qquad a^2 = R^*\alpha\ , \quad \frac{1}{R^*} = \frac{1}{R_1} + \frac{1}{R_2} \quad \text{and} \quad \frac{1}{E^*} = \frac{1-\nu_1^2}{E_1} + \frac{1-\nu_2^2}{E_2}\,. \tag{3}$$

In the above equations, F is the normal contact force, α is the relative approach of the sphere centres, R_i, E_i and ν_i are the radius, Young's modulus and Poisson's ratio of the spheres respectively.

Hardy et al. [4] reported results of a finite element analysis of a rigid sphere indenting an elastoplastic half-space. They showed that the Hertzian pressure distribution was valid until $p_0 = 1.6\sigma_y$ when the yield stress σ_y was reached below the centre of the contact area. Further indentation resulted in a spreading of the plastic deformation zone below the surface and a slight modification of the shape of the contact pressure distribution as p_0 increased further. When the contact force was ca. six times the contact force at initial yield the plastic deformation zone in the substrate reached the contact surface at the perimeter of the contact area. Beyond this point, further indentation resulted in a dramatic change in the pressure distribution: over an enlarging central portion of the contact area the contact pressure was constant with only a slight increase in p_0.

Thornton [5] suggested that the evolution of the contact pressure distribution, described above, could be approximated by an "elastic" phase during which the pressure distribution was Hertzian followed by a "plastic" phase during which the pressure distribution was described by a truncated Hertzian pressure distribution by defining a "limiting contact pressure" p_y, see Thornton and Ning [6]. The normal contact pressure distribution during "plastic" loading was described by

$$P(r) = \begin{cases} p_y & 0 \le r \le a_p \\ \frac{3F_e}{2\pi a^2}\sqrt{1-\left(\frac{r}{a}\right)^2} & a_p \le r \le a \end{cases} \tag{4}$$

where a_p defines the area over which the contact pressure is constant and F_e is the equivalent Hertzian force for the contact radius a, given by

$$F_e = \frac{4E^* a^3}{3R^*} \quad \text{and} \quad \left(\frac{a_p}{a}\right)^2 = 1 - \left(\frac{\pi R^* p_y}{2E^* a}\right)^2 . \tag{5}$$

Based on the above simplifying assumptions it was shown that the force-displacement relationship during "plastic" loading is given by

$$F = F_y + \pi R^* p_y \left(\alpha - \alpha_y\right) , \tag{6}$$

where F_y and α_y are the contact normal force and the relative approach when the Hertzian contact pressure distribution satisfies the condition $p_0 = p_y$. Equation (6) indicates that for $F > F_y$ the force displacement relationship is linear with a normal contact stiffness $k_p = \pi R^* p_y$. Unloading was assumed to be elastic for which equations (1) to (3) apply but, in order to account for the change in contact curvature due to irrecoverable plastic deformation, R^* is replaced by $R_p^* > R^*$ and

$$R_p^* = \frac{4E^* \left(a^*\right)^3}{3F^*} \tag{7}$$

where a^* and F^* are the values of the contact radius and contact force at the point of unloading. The above theoretical model has been incorporated into

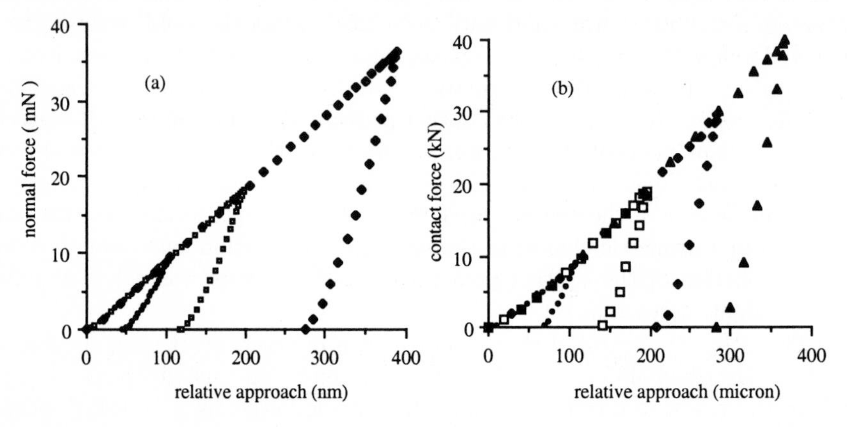

Fig. 1. Normal force-displacement relationship for elastoplastic impact (a) DEM simulations, (b) FEM modelling

the Aston discrete element method (DEM) code and single particle impacts with a plane target wall have been simulated. Figure 1(a) shows the force-displacement curves obtained for a 20 μm diameter sphere ($E = 215$ GPa, $p_y = 1.9$ GPa, $\nu = 0.3$) impacting a wall ($E = 215$ GPa, $\nu = 0.3$) at 5, 10 and 20 ms^{-1}. For comparison, Fig. 1(b) shows finite element method

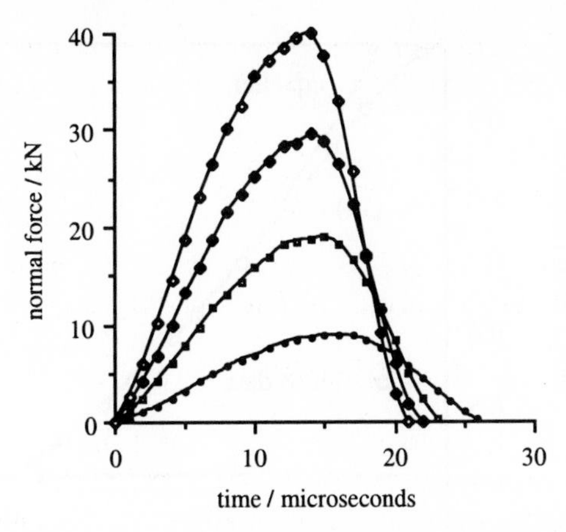

Fig. 2. Normal force-displacement relationship for elastoplastic impact (FEM data)

(FEM) solutions for a 12.6 mm diameter elastic sphere (E = 210 GPa, ν = 0.286) impacting an elastoplastic substrate (E = 204 GPa, σ_y = 0.94 GPa, ν = 0.286) at 10, 20, 30 and 40 ms^{-1}. The corresponding time evolutions of the normal contact force obtained from the FEM modelling are shown in Fig. 2 which shows that the impact duration reduces as the impact velocity increases, primarily due to the shorter unloading period. The small decrease in the total loading time is due to the reduced elastic loading period at higher velocities since the plastic loading period is independent of the impact velocity [7]. Ignoring the initial elastic loading period, the impact duration can be approximated as

$$t_c = \frac{\pi}{2\omega_p}\left(1+\sqrt{\frac{5}{4}}e_n\right) = 1.118\left(\frac{(m^*)^2}{R^*E^*V_y}\right)^{1/5}\left(1+\sqrt{\frac{5}{4}}\,e_n\right) . \tag{8}$$

If $e_n = 1$ the above equation reduces to

$$t_c = 2.368\left(\frac{(m^*)^2}{R^*\left(E^*\right)^2 V}\right)^{1/5} , \tag{9}$$

which may be compared with the elastic solution obtained by Deresiewicz [8]

$$t_c = 2.87\left(\frac{(m^*)^2}{R^*\left(E^*\right)^2 V}\right)^{1/5} . \tag{10}$$

It is worth noting that the elastoplastic model of impact predicts that the unloading period is shorter than the loading period, in contrast to the predictions of viscoelastic impact models.

Based on the above simplified model of elastoplastic impact, Thornton [5] obtained the following analytical solution for the coefficient of restitution e_n as a function of the relative impact velocity V, which is illustrated in Fig. 3.

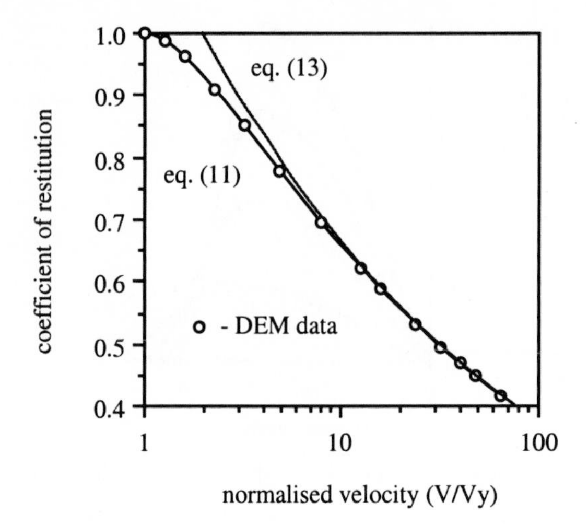

Fig. 3. Normal coefficient of restitution

$$e_n = \sqrt{\frac{6\sqrt{3}}{5}} \sqrt{1 - \frac{1}{6}\left(\frac{V_y}{V}\right)^2} \left[\frac{(V_y/V)}{(V_y/V) + 2\sqrt{\frac{6}{5} - \frac{1}{5}(V_y/V)^2}} \right]^{1/4} , \tag{11}$$

$$\text{where} \quad V_y = 3.194 \sqrt{\frac{p_y^5 (R^*)^3}{(E^*)^4 m^*}} \quad \text{with} \quad \frac{1}{m^*} = \frac{1}{m_1} + \frac{1}{m_2} . \tag{12}$$

V_y is the relative impact velocity below which the collisional interaction is assumed to be elastic, m_i are the particle masses and, for $V \gg V_y$, the asymptotic solution is

$$e_n = 1.185 \left(\frac{V_y}{V}\right)^{1/4} . \tag{13}$$

It should be noted that, in the above, the subscript y does not imply initial yield (when $p_0 = 1.6\sigma_y$) but corresponds to the instant when $p_0 = p_y \sim 2.8\sigma_y$ when, according to the data provided by Hardy et al. [4], significant flattening of the contact pressure distribution commences.

3 Oblique Impact of Elastic Spheres

A theoretical investigation of the behaviour of elastic spheres in contact under varying oblique forces was presented by Mindlin and Deresiewicz [9]. Solutions were presented in the form of instantaneous compliances which, due to the dependence on both the current state and the previous loading history, could not be integrated *a priori*. However, several loading sequences involving variations of both normal and tangential forces were examined from which general procedural rules were identified. Adopting an incremental approach, the procedure is to update the normal force and contact area radius followed by calculating ΔT using the new values of P and a. By reanalysing the loading cases considered by Mindlin and Deresiewicz [9], it can be shown [10–12] that the tangential incremental displacement may be expressed as

$$\Delta\delta = \frac{1}{8G^*a}\left[\pm\mu\Delta P + \frac{\Delta T \mp \mu\Delta P}{\Theta}\right] \tag{14}$$

except when, for $\Delta P > 0$,

$$|\Delta\delta| < \frac{\mu\Delta P}{8G^*a}\,. \tag{15}$$

Rearrangement of (14) defines the tangential stiffness as

$$k_T = \frac{\Delta T}{\Delta\delta} = 8G^*a\Theta \pm \mu\,(1-\Theta)\,\frac{\Delta P}{\Delta\delta} \tag{16}$$

where

$$\Theta^3 = \begin{cases} 1 - \frac{T+\mu\Delta P}{\mu P} & \Delta\delta > 0 \quad \text{(loading)} \\ 1 - \frac{T^*-T+\mu\Delta P}{2\mu P} & \Delta\delta < 0 \quad \text{(unloading)} \\ 1 - \frac{T-T^{**}+\mu\Delta P}{2\mu P} & \Delta\delta > 0 \quad \text{(reloading)} \end{cases} \tag{17}$$

and the negative sign in (16) is only invoked during unloading. The parameters T^* and T^{**} define the load reversal points and need to be continuously updated

$$T^* = T^* + \mu\Delta P \qquad \text{and} \qquad T^{**} = T^{**} - \mu\Delta P \tag{18}$$

to allow for the effect of varying normal force. If (15) is true then (16) is used with $\Theta = 1$ until the following condition is satisfied

$$8G^*a\sum|\Delta\delta| > \mu\sum\Delta P\,. \tag{19}$$

The above theoretical model was implemented in DEM simulations of oblique impacts.

Figure 4 illustrates the linear kinematics of an in-plane oblique impact of a sphere with a plane target wall. The velocity of the centre of the sphere is represented by V and the surface velocity at the contact is represented by v. Subscripts n and t indicate normal and tangential components; and i and r indicate initial and rebound conditions. The normal, tangential and rotational impulses developed at the contact patch are given by

$$F_n = m\,(V_{ni} - V_{nr})\,, \quad F_t = m\,(V_{ti} - V_{tr}) \quad \text{and} \quad F_\omega = mk^2\,(\omega_i - \omega_r) \tag{20}$$

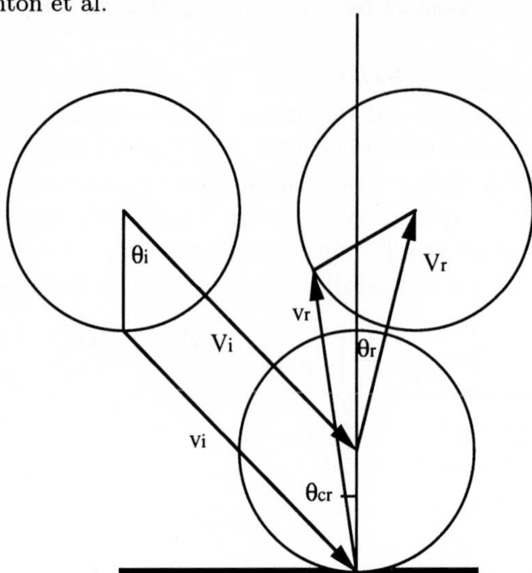

Fig. 4. Oblique impact kinematics

where k is the radius of gyration of the sphere and ω is the rotational velocity. Therefore the change in rotation due to the collision is

$$\omega_i - \omega_r = \frac{R}{k^2}\left(V_{ti} - V_{tr}\right) . \tag{21}$$

The relationship between the surface velocity and the velocity of the sphere centre is $v_t = V_t + R\omega$ and $v_n = V_n$. For a solid sphere $R^2/k^2 = 5/2$ and so the tangential surface velocity of the contact patch at the end of the impact is given by

$$v_{tr} = \frac{7}{2}V_{tr} + R\omega_i - \frac{5}{2}V_{ti} \tag{22}$$

from which we obtain an equation for the tangential coefficient of restitution

$$e_t = \frac{V_{tr}}{V_{ti}} = \frac{5}{7} + \frac{2}{7}\left(\frac{v_{tr} - R\omega_i}{V_{ti}}\right) = \frac{5}{7} + \frac{2}{7}\left(\frac{e_n \tan\Theta_{cr} - R\omega_i/V_{ni}}{\tan\Theta_i}\right) \tag{23}$$

Results of DEM simulations of oblique impacts of elastic and elastic, perfectly plastic spheres show that the complete rebound kinematics are described by the following analytical solutions if sliding occurs throughout the impact. It was demonstrated that sliding throughout the impact occurs ($T = \mu P$) if

$$\frac{v_{ti}}{V_{ni}} = \frac{V_{ti} + R\omega_i}{V_{ni}} \geq \mu\left(\frac{1+e_n}{2}\right)(7 - e_n) \tag{24}$$

where the LHS of (24) defines the effective angle of impact. If (24) is true then

$$e_t = 1 - \frac{\mu\,(1+e_n)}{\tan\Theta_i}\,, \quad \omega_r = \omega_i - \frac{5\mu\,(1+e_n)\,V_{ni}}{2R}\,, \quad \tan\Theta_r = \frac{e_t}{e_n}\tan\Theta_i\,. \tag{25}$$

Gorham and Kharaz [13] report experimental results obtained from impacting 5 mm diameter aluminium oxide spheres against a glass target at different impact angles using a constant impact speed of 3.89 ms^{-1}. Figure 5 shows the normal and tangential coefficients of resitution obtained. The average value of $e_n = 0.982$ which is consistent with energy dissipation due to elastic wave propagation, indicating that the collisional interactions may be assumed to be elastic. From (23) $e_t = 5/7$ if $v_{tr} = 0$ and $\omega_i = 0$. In the experiments $\omega_i = 0$ and $v_{tr} = 0$ when $\tan\Theta_i = 7\mu$ corresponding to $\Theta_i = 33.5^o$, see Fig. 5. Hence, we deduce that $\mu = 0.094$ and using this value in (25) obtain the theoretical prediction for e_t which is superimposed on the figure showing good agreement with the experimental data for the range $\Theta_i > 29.5^o$ $(= 6\mu)$ when sliding occurs throughout the impact as given by (24).

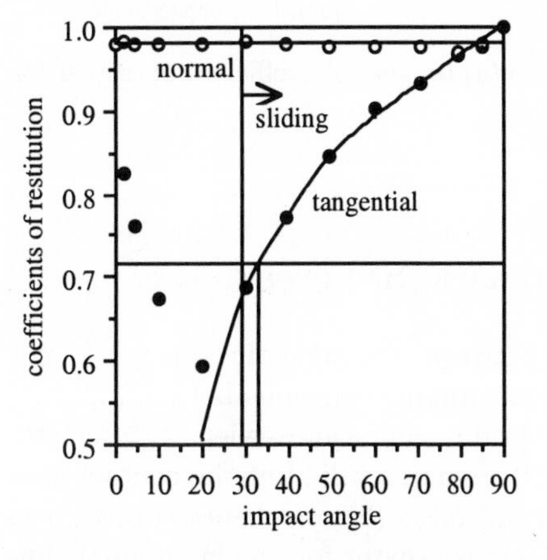

Fig. 5. Normal and tangential coefficients of restitution (elastic impacts)

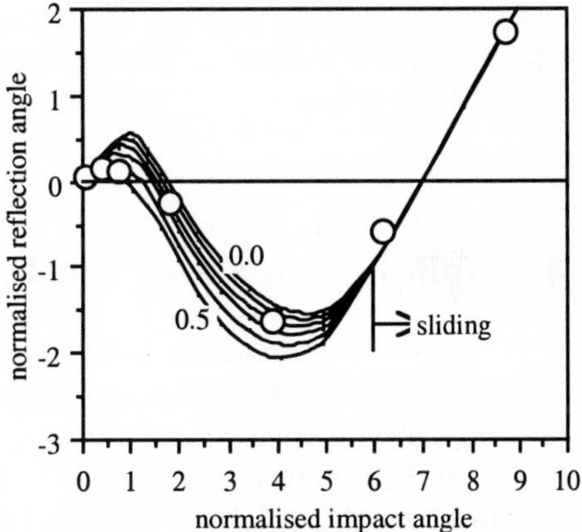

Fig. 6. Reflection angle of the contact patch (elastic impacts)

Figure 6 shows the results of DEM simulations of oblique elastic impacts plotted in terms of the normalised reflection angle $(v_{tr}/\mu V_{ni})$ against the normalised impact angle $(V_{ti}/\mu V_{ni})$. It can be seen that, if sliding does not occur thoughout the impact, the rebound surface velocity is significantly dependent on the value of Poisson's ratio, because ν affects the ratio of the tangential and normal stiffnesses. In the experiments [13] both linear and rotational rebound velocities were measured and the surface velocities were calculated. The results are superimposed on Fig. 6 and fall within the range predicted by the DEM simulations. The experimentally obtained values of e_t are compared with DEM results for $\nu = 0.3$ in Fig. 7(a). Figure 7(b) shows experimental and DEM results plotted in terms of the normalised rotation $(R\omega_r/\mu V_{ni})$ against normalised impact angle. In both figures there is excellent agreement between the experimental and simulated data.

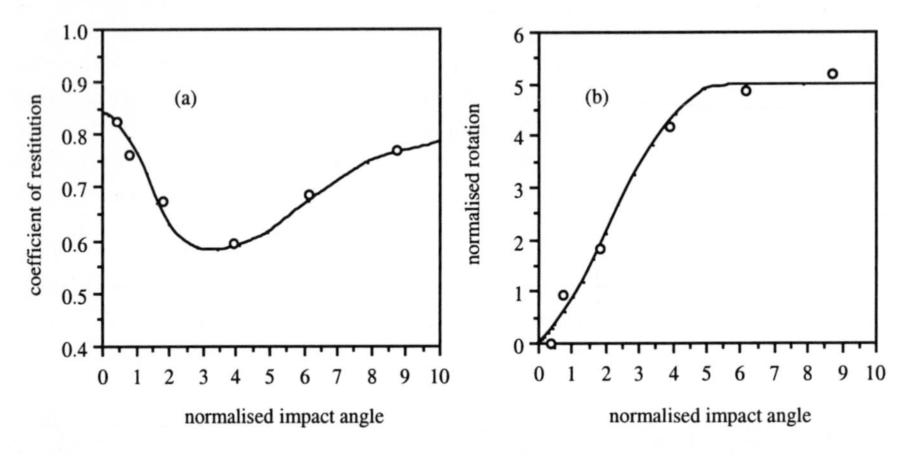

Fig. 7. Rebound kinematics (elastic impacts) (a) tangential coefficient of restitution (b) particle rotation

4 Oblique Impact of Elastoplastic Spheres

The tangential interaction behaviour between elastoplastic spheres is not known. For simplicity, we assume that, for oblique collisions between elastic-perfectly plastic spheres, the form of Mindlin and Deresewiecz's [9] elasto-frictional theory given by Eqs. (14) to (19) remains valid but the contact area is defined by the effective Hertzian normal force (5) and hence, for a given actual normal force, the tangential stiffness is greater for an elastoplastic impact than for the elastic case. In this section we present the results of DEM simulations which assume this very simple model and compare these with experimental data.

Figure 8 shows experimental results obtained by Gorham and Kharaz [13] for 5 mm diameter aluminium oxide spheres impacting an aluminium target at different angles using a constant impact speed of 3.89 ms^{-1}. Assuming a constant value of $e_n = 0.61$ the theoretical predictions obtained from (24) and (25) are superimposed on the figure, with $\mu = 0.163$, showing good agreement with the experimental data for e_t when sliding occurs throughout the impact duration. DEM simulations were performed in which the normal impact velocity was constant and the impact angle was varied.

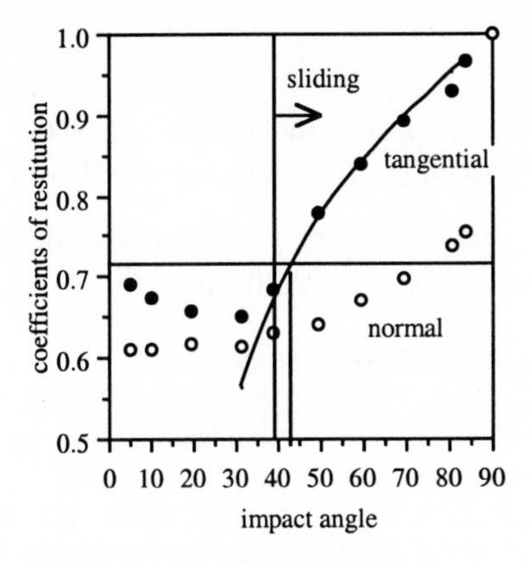

Fig. 8. Normal and tangential coefficients of restitution (plastic impacts)

Figure 9 shows the results plotted in terms of the normalised reflection angle ($v_{tr}/\mu V_{ni}$) against the normalised impact angle ($V_{ti}/\mu V_{ni}$) for different normal impact velocities. (In our simple model, e_n depends only on the normal impact velocity and so is independent of impact angle.) Superimposed on the figure are the experimental data points [13]. It is clear that the pattern of behaviour suggested by our simple tangential interaction model is not supported by the experimental data when sliding does not occur throughout the impact. It is also noted that the experimental data could be reasonably represented by the bilinear approximation suggested by Foerster et al. [14]. Currently, we are performing finite element simulations of oblique impacts in order to obtain an improved theoretical model for the tangential interaction between elastic-perfectly plastic spheres.

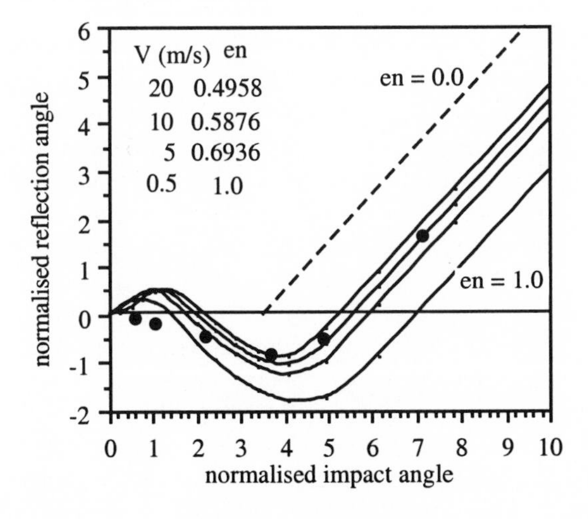

Fig. 9. Reflection angle of the contact patch (plastic impacts)

References

1. W. Goldsmith, *Impact* Edward Arnold (London, 1960).
2. N. Brilliantov, T. Pöschel, F. Spahn and J.-M. Hertzsch, Phys. Rev. E **53**, 5382 (1996).
3. F. G. Bridges, A. P. Hatzes and D. N. C. Lin, Nature **309**, 333 (1984).
4. C. Hardy, C. N. Baronet and G. V. Tordion, Int. J. Num. Methods Engng **3**, 451 (1971).
5. C. Thornton, J. Appl. Mech. **64**, 383 (1997).
6. C. Thornton and Z. Ning, Powder Technology **99**, 154 (1998).
7. K. L. Johnson, *Contact Mechanics*, Cambridge University Press (1985).
8. H. Deresiewicz, Acta Mechanica **6**, 110 (1968).
9. R. D. Mindlin and H. Deresiewicz, J. Appl. Mech. **20**, 327 (1953).
10. C. Thornton and C. W. Randall, In: M. Satake and J. T. Jenkins (eds.), *Micromechanics of Granular Materials*, p. 133 Elsevier (1988).
11. C. Thornton and K. K. Yin, Powder Technology **65**, 153 (1991).
12. C. Thornton, In: M. Oda and K. Iwashita (eds.), *Mechanics of Granular Materials*, p. 187 Balkema (1999).
13. D. A. Gorham and A. Kharaz, Proc. Conf. on Pneumatic and Hydraulic Conveying Systems II , Davos, June (1999).
14. S. F. Foerster, M. Y. Louge, H. Chang and K. Allia, Phys. Fluids **6**, 1108 (1994).

Kinetic Theory for 1D Granular Gases

Rosa Ramírez and Patricio Cordero

Departamento de Física, Facultad de Ciencias Físicas y Matemáticas
Universidad de Chile, Santiago 3, Chile. e-mail: rosa@cecam.fr

Abstract. The Boltzmann like equation for a dissipative 1D granular gas, can be regarded, in the thermodynamic limit, as that of a simple particle inside a viscous medium - viscosity produced by the dissipative collisions with the rest of the particles. Analytical perturbative solutions can be found for this equation. We find that for low dissipative regimes, there is excellent agreement between our theoretical predictions for the macroscopic fields, and the measurementes from molecular dynamics simulations.

1 Introduction

Even if one dimensional granular systems are not totally realistic, they may reproduce several of the phenomena observed in higher dimensions [1–5]. They have also the advantage that both theory and molecular dynamic simulations are considerably easier to understand.

Our aim is to study within the frame of kinetic theory, a system of N inelastic point particles of mass $m = 1$ restricted to move under the action of gravity g in a 1D box of height L and bouncing on a base which we have chosen to behave as a thermal wall. The interaction between these particles is modeled by the inelastic collision rule $c_1' = q\,c_1 + (1-q)\,c_2$ and $c_2' = (1-q)\,c_1 + q\,c_2$ where the constant restitution coefficient is $r = 1 - 2q$. To describe this quasielastic system we write down a Boltzmann's like kinetic equation . The low dissipation thermodynamic limit ($N \to \infty$, $q \to 0$ with qN finite) leads to a highly simplified equation. Further assuming that qN is a small parameter, analytical perturbative solutions arround the elastic case can be found for this equation, that is, the stationary state for the quasielastic system is simply a distortion of the elastic stationary state. In other words, we study the regime in which the trajectories of the particles are only slightly modified by the dissipative collisions as shown in fig. 1. The elastic case—our reference system—deserves some comments.

A 1D system of elastic point particles is tantamount to a non-interacting system. In fact, relabeling the particles involved after each collision corresponds to a system of particles passing through each other without ever interacting. This leads to some important consequences. First, Liouville's distribution function for the N particle system is simply the product of the N one-particle distribution functions $F^{(N)}(z_1 \cdots z_N) = F^{(1)}(z_1) \cdots F^{(1)}(z_N)$ hence Liouville's equation exactly reduces to Boltzmann's equation: there are

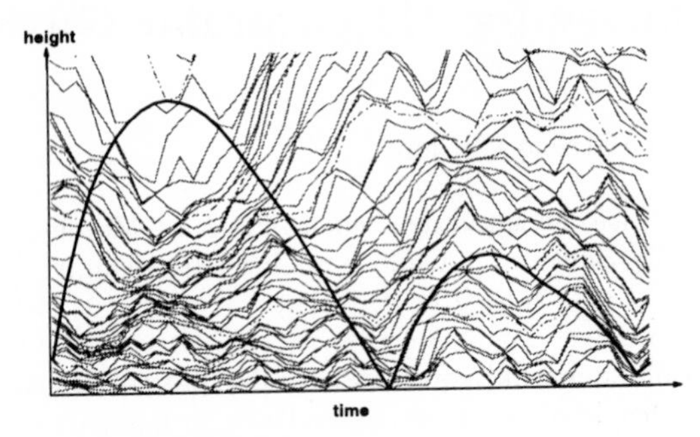

Fig. 1. Movement of one particle in the background of the rest of them for a quasielastic open system on a hot wall.

no particle-particle correlations at all. Second, due to the absence of any interaction between particles, the only way for the velocity distribution to evolve from a given initial condition is by the effect of the boundaries. Stochastic boundary conditions are absolutely necessary to define a stationary velocity distribution independent of the initial condition.

For this reason we will choose at the bottom a stochastic boundary defined in the same way as a thermal wall: any particle hitting the bottom wall will forget its velocity and will take a new one with a probability

$$W(c) = 2\,\frac{c}{\sqrt{2T_0}}\,e^{-c^2/2T_0} \tag{1}$$

T_0 being a temperature with dimensions of velocity squared. This boundary condition does not create any correlation and it destroys any one that the inelastic dynamics could create.

A conservative 1D system with such boundary condition will reach a Maxwellian distribution function in the stationary state—the trajectories of the test particles (relabeled after each collision) are just parabolas with different energies taken from the base and they do not see each other. In our quasielastic case, these parabolas are slightly modified as shown in Fig. 1, so that a perturbative expansion around the Maxwellian distribution is justified.

2 Kinetic Equation and Boundary Conditions

To describe the inelastic system we define the velocity $v_T = \sqrt{2\,T_0/m}$ and the length $L_T = \frac{T_0}{m\,g}$ which corresponds to the length scale of significant variations of the density in the elastic case. Besides we use the Froude number $Fr = L/L_T = \frac{m\,g\,L}{T_0}$ to describe the adimesional height of the system. If L is much larger than L_T, namely $Fr \to \infty$ the system will resemble an open system. On the contrary, if $Fr \ll 1$ the system will behave almost as if there was no gravity and the density and other fields will be nearly homogeneous. We will see both limits in Sec. 4.

We define a dimensionless distribution function for the stationary state $F(\hat{x},\hat{c}) = L_T\, v_T\, f(x,c)$ normalized as

$$\int_0^{Fr} d\hat{x} \int_{-\infty}^{\infty} d\hat{c}\, F(\hat{x},\hat{c}) = 1 \tag{2}$$

where the dimensionless position $\hat{x}$ and velocity $\hat{c}$ are $\hat{x} = x/L_T$ $\hat{c} = c/v_T$. Using these variables the collision term in Boltzmann's equation can be written as an expansion in the small parameter q [3] leading to

$$\left(\hat{c}\,\frac{\partial}{\partial\hat{x}} - \frac{1}{2}\,\frac{\partial}{\partial\hat{c}}\right) F(\hat{x},\hat{c}) = N \sum_{k=1}^{\infty} \frac{q^k}{k!}\,\frac{\partial^k}{\partial\hat{c}^k}\left[M_k(\hat{x},\hat{c},ht)\, F(\hat{x},\hat{c})\right] \tag{3}$$

where

$$M_k(\hat{x},\hat{c}) = \int_{-\infty}^{\infty} |\hat{c}-\hat{c}'|\, (\hat{c}-\hat{c}')^k\; F(\hat{x},\hat{c})\, d\hat{c}' \tag{4}$$

Multiplying and dividing this right hand side by N^k—so that there is a factor $(qN)^k/(k!\,N^{k-1})$—and taking the limit $N \to \infty$ keeping qN fixed, (this is what we call the *hydrodynamic limit*), the kinetic equation becomes simply

$$\left(\hat{c}\,\frac{\partial}{\partial\hat{x}} - \frac{1}{2}\,\frac{\partial}{\partial C}\right) F(\hat{x},\hat{c}) = qN\,\frac{\partial}{\partial\hat{c}}\left[M(\hat{x},\hat{c})\, F(\hat{x},\hat{c})\right] \tag{5}$$

where $M(\hat{x},\hat{c}) \equiv M_1(\hat{x},\hat{c})$. Equation (5) is the *hydrodynamic limit of Boltzmann's equation for a granular system in 1D.* In principle this equation is valid for any finite value of the parameter qN provided N is large enough. The collision term for a test particle in the original Boltzmann equation is replaced by an effective friction produced by the collisions with the rest of the particles. In this sense equation (5) represents a particle passing through a viscous fluid—viscosity produced by dissipative collisions—and suffering the corresponding acceleration, as observed for example in Fig. 1. Piasecki has already dealt with the idea of a particle moving inside a viscous medium [7]. We claim that Eq.(5) is the one which represents exactly the viscosity of this granular medium in the hydrodynamic limit.

As we can see the adimensional problem in the hydrodynamic limit depends solely on the Froude number, Fr, via the normalization condition and on qN via the kinetic equation. The boundary conditions we are about to use do not depend either on Fr nor on qN. If a boundary condition with two different temperatures at the bottom and top walls is used, then a third dimensionless parameter would come in. Since we are not going to deal with such case the only two dimensionless parameters that determine the stationary state of the system are Fr and qN.

3 The Quasielastic Regime

Although the kinetic equation is valid for any finite value of qN the following formalism is valid only in the *quasielastic regime* characterized by $qN \ll 1$. In this limit we look for solutions of the form

$$F(\hat{x},\hat{c}) = F^{(0)}(\hat{x},\hat{c}) + qN\,F^{(1)}(\hat{x},\hat{c}) + (qN)^2 F^{(2)}(\hat{x},\hat{c}) + \dots \tag{6}$$

where $F^{(0)}$ is the solution for the elastic case.

When $F(\hat{x},\hat{c})$ is replaced back in (5) we find a set of equations for each order of the distribution function. Each $F^{(s)}(\hat{x},\hat{c})$ follows an equation of the form

$$\left(c\frac{\partial}{\partial x} - \frac{1}{2}\frac{\partial}{\partial c}\right) F^{(s)}(\hat{x},\hat{c}) = J_{(s)}\left[F^{(s-1)}(\hat{x},\hat{c}), F^{(s-2)}(\hat{x},\hat{c}), \dots, F^{(0)}(\hat{x},\hat{c})\right] \tag{7}$$

where $J_{(s)}\left[F^{(s-1)}(\hat{x},\hat{c}), \dots, F^{(0)}(\hat{x},\hat{c})\right]$ is a function representing the collision term at order s and it only contains lower order functions.

We impose the normalization condition (2) so that at each order

$$\int_{-\infty}^{\infty} d\hat{c} \int_0^{Fr} d\hat{x}\; F^{(s)}(\hat{x},\hat{c}) = \delta_{0s}. \tag{8}$$

Each $F^{(s)}$ must also satisfy a boundary condition at the bottom $|\hat{c}|\,F^{(S)}(\hat{x} = 0, \hat{c} > 0) = w(\hat{c})\,R_s$, where $w(\hat{c}) = 2\,\hat{c}\,e^{-\hat{c}^2}$ is the dimensionless version of W in (1), while the condition at the elastic top wall is $F^{(s)}(\hat{x} = Fr, \hat{c}) = F^{(s)}(\hat{x} = Fr, -\hat{c})$.

With this scheme the set of equations (7) can be solved exactly starting from $F^{(0)}$—this is the elastic solution –and following recursively up to the desired order [5].

This perturbative solution is valid at any position near or far from the walls since it is the kinetic equation with its corresponding microscopic boundary that is being solved.

4 Results

The open system: The stationary state of the open system ($Fr = \infty$) is determined by qN alone. The function corresponding to the elastic case is just a Maxwellian $F^{(0)} = \frac{1}{\sqrt{\pi}} e^{-(\hat{x}+\hat{c}^2)}$ and the first and second order corrections modify this function as shown in Fig. 2 where the theoretical distribution functions up to zeroth, first and second order evaluated at the bottom $\hat{x} = 0$ with $qN = 0.2$ are compared with the measured distribution function. Interpreting the system as one in which particles pass through each other loosing some energy in the process it can be said that particles coming out

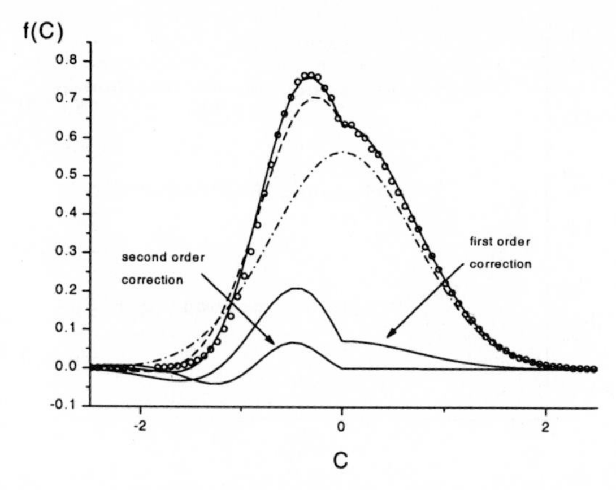

Fig. 2. The distribution function normalized to 1 at $\hat{x} = 0$ evaluated to zeroth (dot-dashed line), first (dashed line) and second order (solid line) are seen in this figure for $qN = 0.2$ and $N = 200$. Circles are the simulational results.

from the base have a vanishing most probable velocity, namely $F(C > 0)$ is maximum at $C = 0$, just like a Boltzmann distribution, while particles coming down are in principle accelerated, but the friction with the background (the rest of the particles) produces an effect similar to a "limit velocity" and $F(C < 0)$ has a maximum away from the origin. Moving away from the base the history of all particles tends to be comparable regardless of the sign of their velocity and the distribution function tends to be more and more symmetric.

The effects of dissipation over density: The zeroth order density $n_0(\hat{x})$ decreases exponentially, and since $qN \leq 0.1$ deviations from this behavior should be small. Dissipation prevents particles from reaching the heights they would in the conservative case implying that the system has a smaller effective height. Consequently the density tends to be higher near the base although corrections due to dissipation have not a maximum on the base but at some distance over the it. This suggests that for higher values of qN (too high for our theoretical description to be valid) a drop floating on a vapor would be formed in the system as, in fact, we have seen in simulations.

Effects on the granular temperature and the heat flux: In Fig. 3 the temperature profile is shown for three values of qN. The zeroth order temperature profile is the straight line $T = 1/4$ in all three cases. The effect of dissipation is to produce a $T(\hat{x})$ with negative gradient and this is already predicted by the negative first order correction. Figure 3 also shows that the temperature reaches an asymptotic value that our formalism predicts to be $T(\hat{x} \sim \infty) \approx 1/4 - qN(1 - qN)/2$ which coincides with what we observe. At $\hat{x} = 0$ there is a temperature gap: the temperature of the system does not coincide with the imposed value $T_0 = 1/4$.

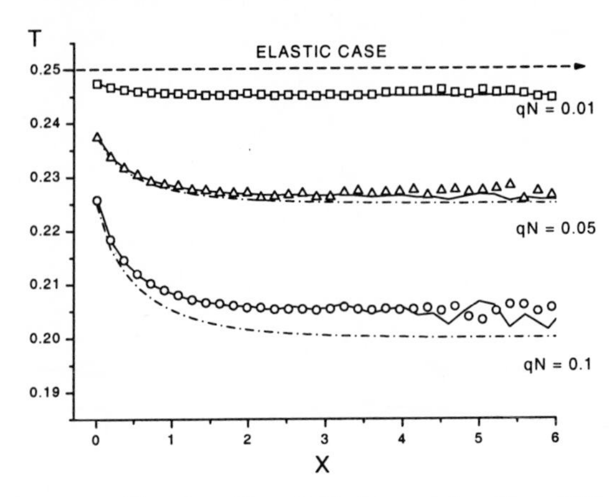

Fig. 3. Temperature profiles for $qN = 0.01$ (squares), $qN = 0.05$ (triangles) and $qN = 0.1$ (circles). The solid (dot-dashed) line is the predicted dimensionless temperature up to second (first) order.

In our case this *temperature slip* at the base is $\delta T \approx qN(1 - qN/2)/4$ which is also the observed gap. This thermal slip at the wall is a well known effect when the system has an externally imposed temperature gradient, but in granular systems δT is due to the dissipative collisions, namely, it is an intrinsic property of the system and it does not vanish with increasing density, but it rather increases.

The heat flux Q—the flux of energy entering through the base and being dissipated in the bulk—has an *analytic* expression to first order in q,

$$Q(\hat{x}) = qN \int_{-\infty}^{\infty} dC \, \frac{C^3}{2} \, F(\hat{x}, C) = \frac{qN}{\sqrt{2\pi}} \, e^{-2\hat{x}} \, . \tag{9}$$

An elastically closed system: The perturbative solution for our system with an elastic wall at height L (dimensionless height $\hat{x} = Fr$) has the form

$$F(\hat{x}, \hat{c}) = \lambda \left(F^{(0)}(\hat{x}, \hat{c}) + qN\lambda \, F^{(1)}(\hat{x}, \hat{c}) + (qN\lambda)^2 \, F^{(2)}(\hat{x}, \hat{c}) + \cdots \right) \tag{10}$$

where the prefactor $\lambda = 1/(1 - e^{-Fr})$ determines the system's density scale. The method to recursively construct the solution is the same one seen in Sec. 3 except that in this case the two boundary conditions have to be imposed [6]. The quasielastic condition for the system is in this case $qN\lambda \ll 1$, namely, not only qN but also λ will determine the behavior of the stationary state. We can see at left in Fig. 4 the corrections to the density profile for a fixed qN value and different values for λ (or Fr). The agreement with theory is quite good even though the shape of the corrections vary widely for the different values of the Froude number conisdered. For small Fr the correction is almost symmetric about $\hat{x} = \frac{1}{2} Fr$ where it has a maximum. As Fr gets larger the maximum of $n_1(\hat{x})$ moves down until it reaches the base and disappears. The correction develops a minimum high up and it moves down as Fr grows.

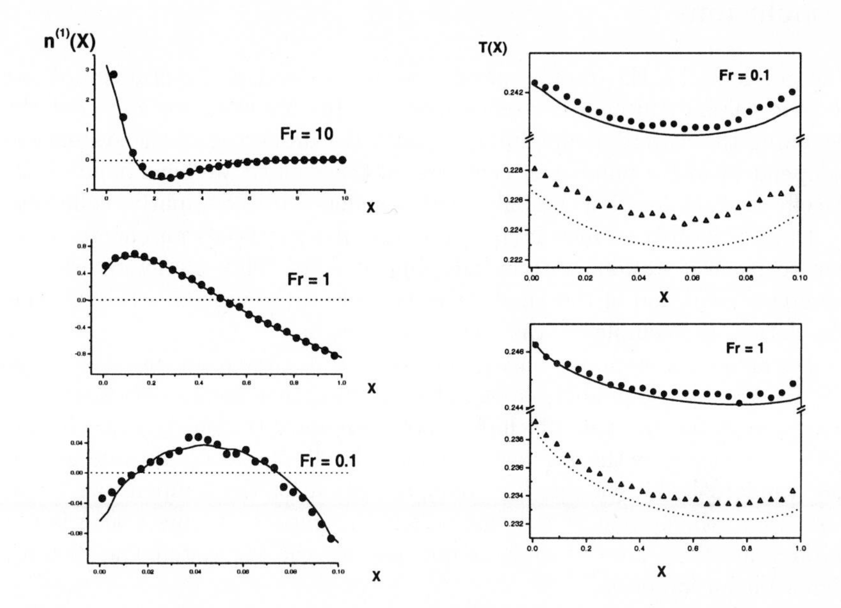

Fig. 4. At left, density corrections due to dissipation for different Froude numbers. At right, the temperature profile for two values of the Froude number with $qN = 0.01$ (dots) and $qN = 0.03$ (triangles). If Fr is small enough T shows a minimum even though the only source of energy is at the bottom. In both cases lines are the theoretical predictions while symbols are the measured data from MD simulations.

The temperature profile: When $Fr \gg 1$ few particles reach the top wall and the behavior of the profile tends to that of the open system observed in Fig. 3. Taking smaller values of Fr the temperature profile eventually develops a minimum near the top. The position of this minimum approaches the mid height as Fr gets smaller, being approximately in the middle of the box when $Fr = 0.1$ as shown in Fig. 4.

In spite of this minimum, notice that the heat flux, which at first order reads,

$$Q(\hat{x}) = \frac{qN}{\sqrt{2\pi}\,(1 - e^{-Fr})^2}\left(e^{-2\hat{x}} - e^{-2Fr}\right) \tag{11}$$

decreases monotonically with height, being zero at $\hat{x} = Fr$, as it should be due to the elastic wall that closes the system.

Namely, the energy coming from the base heats the system near the top more than it heats the region immediately underneath. This remarkable effect is a boundary effect, in the sense that the upper wall is needed to observe such a quasi-elastic regime. For an open system, in the quasi-elastic limit $qN \ll 1$ this kind of temperature profile does not exist.

Conclusions

The problem of a 1D granular gas can be understood as the problem of *one particle* passing through a viscous medium. In this work we analyzed the system in the hydrodynamic limit in which the properties of the system are independent of the number of particles. In this context we have pointed out the effective acceleration this only particle suffers due to dissipative collisions.

Due to dissipative interactions, the granular gas needs an energy injection to reach a stationary time independent state. This energy enters as a boundary condition in the kinetic theory and, in the case of a 1D granular gas, determines completely the quasi-elastic regime.

For an open system, the only parameter entering the problem is the factor qN. In this case we predict, among other results, that there exists a temperature slip at the hot wall which does not decrease with density and which is of the same order as the gradients produced by dissipation. This implies that the standard hydrodynamic boundary condition for the temperature (temperature of the system at the wall equals the imposed temperature) is not suitable for these kind of systems not only for the 1D system but also for higher dimension ones.

Any other boundary in the system plays a role as important as the energetic boundary. We have predicted and observed an inverse heat flux (from colder to hotter zones) just by including an elastic upper wall. Although there have been observed inverse heat fluxes in very low density systems in higher dimensions [8] which could be due to a dissipative bulk effect, our case corresponds mainly to a boundary effect.

Although some of properties of 1D systems are generalizable to two or three dimensional systems, we have to point out that there exists a crucial fact which makes the 1D case special: the velocity distribution function is completely determined by the boundary conditions. Then, some of the 1D results we present here could qualitatively be extended to higher dimensions in the case of Knudsen gases, or even standard gases but just near the boundaries.

References

1. H.M Jaeger and S.R. Nagel, Science, **255**, 1523 (1992).
2. S. Warr, J. M. Huntley and G. T. H. Jaques, Phys. Rev. E **52**, 5583 (1995).
3. S. McNamara and W. R. Young, Phys. Fluids A **5**, 1 (1993).
4. E. L. Grossman and B. Roman, Phys. Fluids **8**, 12 (1996).
5. R. Ramírez and P. Cordero, Phys. Rev. E **59**, 1 (1999).
6. R. Ramírez and P. Cordero, in preparation (1999).
7. J. Piasecki, in Proceedings of the NATO Advanced Study Institute on Methods for Many Body Systems (in press) (1999).
8. R. Soto, M. Mareschal and D. Risso, Phys. Rev. Lett. **83**, 5003 (1999).

Chains of Viscoelastic Spheres

Thorsten Pöschel[1] and Nikolai V. Brilliantov[2,1]

[1] Humboldt-Universität, Institut für Physik, Invalidenstr. 110, D-10115 Berlin, Germany. email thorsten@physik.hu-berlin.de, http://summa.physik.hu-berlin.de/~thorsten

[2] Moscow State University, Physics Department, Moscow 119899, Russia. email nbrillia@physik.hu-berlin.de

Abstract. Given a chain of viscoelastic spheres with fixed masses of the first and last particles. We raise the question: How to chose the masses of the other particles of the chain to assure maximal energy transfer? The results are compared with a chain of particles for which a constant coefficient of restitution is assumed. Our simple example shows that the assumption of viscoelastic particle properties has not only important consequences for very large systems (see [1]) but leads also to qualitative changes in small systems as compared with particles interacting via a constant restitution coefficient.

1 Introduction

We consider a linear chain of inelastically colliding particles of masses m_i, radii R_i $(i = 0 \ldots n)$, with initial velocities $v_0 = v > 0$ and $v_i = 0$ $(i = 1 \ldots n)$ at initial positions $x_i > x_j$ for $i > j$ with $x_{i+1} - x_i > R_{i+1} + R_i$. The masses of the first and last particles m_0 and m_n are given and we address the questions: How have the masses of the particles in between to be chosen to maximize the energy transfer, i.e., to maximize the after-collisional velocity $v_n^\prime$ of the last particle. If n is variable, how should n be chosen to maximize $v_n^\prime$? Throughout this paper we assume that the initial distance of the particles is large enough to neglect "multiple collisions", i.e., only the first impact of each particle influences the final velocity of the Nth sphere of the chain.

Recent investigations show that the properties of very large systems of viscoelastic particles differ significantly from those of particles interacting with constant coefficient of restitution [1–5]. The system considered here may serve as an example of a *small* system which properties change qualitatively when the viscoelastic properties of the particles are considered.

The coefficient of restitution is defined via $\epsilon = \left|\left(v_{i+1}^\prime - v_i^\prime\right) / \left(v_{i+1} - v_i\right)\right|$, which relates the relative velocity of the particles after the collision to the pre-collisional quantity. The elastic collision corresponds to $\epsilon =$ const. For this case, basic mechanics yields:

$$v_1^\prime = \frac{1+\epsilon}{1+\frac{m_1}{m_0}} v_0 , \qquad v_n^\prime = (1+\epsilon)^n \prod_{k=0}^{n-1} \left(1 + \frac{m_{k+1}}{m_k}\right)^{-1} v_0 \,. \tag{1}$$

The final velocity v_n' is maximized by

$$m_k = \left(\frac{m_n}{m_0}\right)^{k/n} m_0 \,, \quad \text{yielding} \quad v_n' = \left[\frac{1+\epsilon}{1+\left(\frac{m_n}{m_0}\right)^{1/n}}\right]^n v_0 \,. \tag{2}$$

The optimal mass distribution for the case of a constant coefficient of restitution ϵ does not depend on the value of ϵ and is, therefore, the same as the optimal mass distribution in a chain of elastic particles. Figure 1 (left) shows the optimal mass distribution for different chain lengths n. The mass of the first particle is $m_0 = 1$ and of the last particle $m_n = 0.1$.

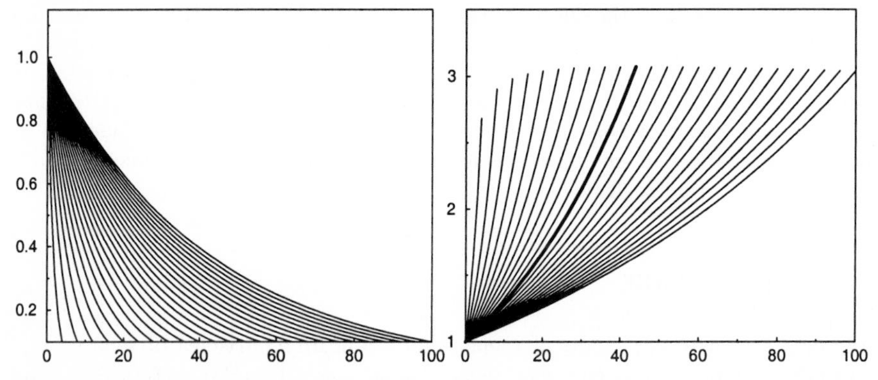

Fig. 1. Left: Optimal mass distribution m_i, $i = 1 \ldots n$, for the case of constant ϵ. Each of the lines shows the mass m_i over the index i for a specified chain length n. The masses of the first and last particles are fixed at $m_0 = 1$ and $m_n = 0.1$. **Right:** Velocity distributions of the particles in chains of length n with the optimal mass distribution according to (2) as a function of the chain length. The dissipative constant is $b \equiv 1 - \epsilon = 5 \cdot 10^{-4}$. The last particle reaches its maximal velocity for chain length $n^* = 44$ (bold drawn). The velocity of the first particle of the chain is $v_0 = 1$.

In contrast to the mass distribution the corresponding velocity distributions do depend on the value of the restitution coefficient ϵ. Figure 1 (right) shows the velocity distribution for $b \equiv 1 - \epsilon = 5 \cdot 10^{-4}$. For the case of dissipative collisions the ratio $R_v = v_n'/v_0$ does not monotonously increase with n as for elastic particles ($\epsilon = 1$), but rather it has an extremum which shifts to smaller chain lengths with increasing dissipative parameter b. The optimal value of n, which maximizes R_v reads

$$n^* = \log\left(m_0/m_n\right) / \log\left(x_0\right) \tag{3}$$

where x_0 is the solution of the equation

$$(1 + x_0) = (1+\epsilon) x_0^{x_0/(1+x_0)} \,. \tag{4}$$

Correspondingly, the extremal value of the R_v reads

$$R_v^* = \left[\frac{1+\epsilon}{1+x_0}\right]^{n^*} . \tag{5}$$

2 Chains of Viscoelastic Particles

It has been shown that for colliding viscoelastic spheres the restitution coefficient depends on the masses of the colliding particles and also on their relative velocity v_{ij} [6]. An explicit expression for the coefficient of restitution is given by the series [4, 7] (see also [1])

$$\epsilon = 1 - C_1 \left(\frac{3A}{2}\right) \alpha^{2/5} v_{ij}^{1/5} + C_2 \left(\frac{3A}{2}\right)^2 \alpha^{4/5} v_{ij}^{2/5} \mp \cdots \tag{6}$$

with

$$\alpha = \frac{2\, Y \sqrt{R^{\text{eff}}}}{3\, m^{\text{eff}} \left(1-\nu^2\right)} \tag{7}$$

with Y and ν being the Young modulus and the Poisson ratio, respectively and $R^{\text{eff}} = R_i R_j/(R_i + R_j)$, $m^{\text{eff}} = m_i m_j/(m_i + m_j)$. The material constant A describes the dissipative properties of the spheres (for details see [6]). The constants $C_1 = 1.15344$ and $C_2 = 0.79826$ were obtained analytically in Ref. [4] and then confirmed by numerical simulations.

In the following calculation we neglect terms $\mathcal{O}\left(v^{2/5}\right)$ and higher and assume for simplicity that all particles are of the same radius R, but have different masses. We abbreviate

$$\epsilon = 1 - b\, v^{1/5} \left(m^{\text{eff}}\right)^{-2/5} \quad \text{with} \quad b = C_1 \left(\frac{3A}{2}\right) \left(\frac{2}{3} \frac{Y\sqrt{R/2}}{1-\nu^2}\right)^{\frac{2}{5}} . \tag{8}$$

Hence, for viscoelastic particles the velocities of the $k+1$-rst particle after colliding with the k-th reads

$$v_{k+1}' = \frac{2 - b\left(\frac{m_{k+1}+m_k}{m_{k+1} m_k}\right)^{2/5} v_k^{1/5}}{1 + \frac{m_{k+1}}{m_k}} v_k . \tag{9}$$

The masses m_k, $k = 1 \ldots n-1$ which maximize v_n' can be determined numerically and the results are shown in Fig. 2 for two different values of the dissipative constant b.

For small chain length or small b, respectively, the optimal mass distribution is very close to that for the elastic chain as shown in Fig. 1. Again we find a monotonously decaying function for the masses. For larger chain length n or larger dissipation b, however, the mass distribution is a non-monotonous function. The according velocities of the particles in chains of spheres of optimal masses are drawn in Fig 3.

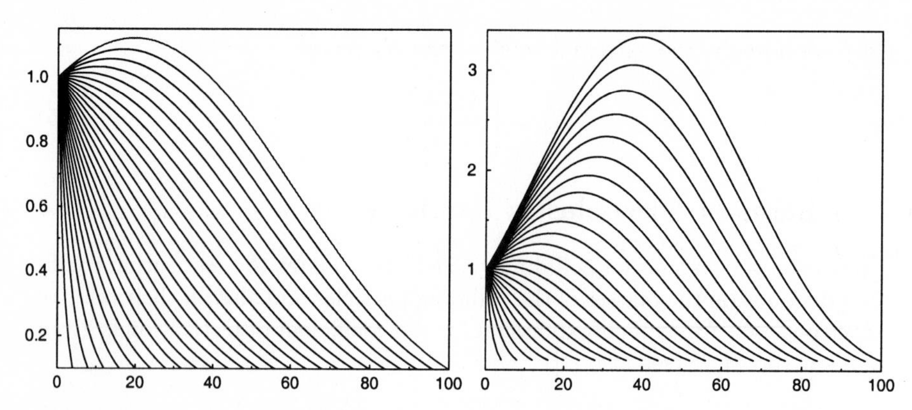

Fig. 2. Optimal mass distribution of collision chains over the chain length n. The dissipative constant was $b = 5 \cdot 10^{-4}$ (left) and $b = 2 \cdot 10^{-3}$ (right).

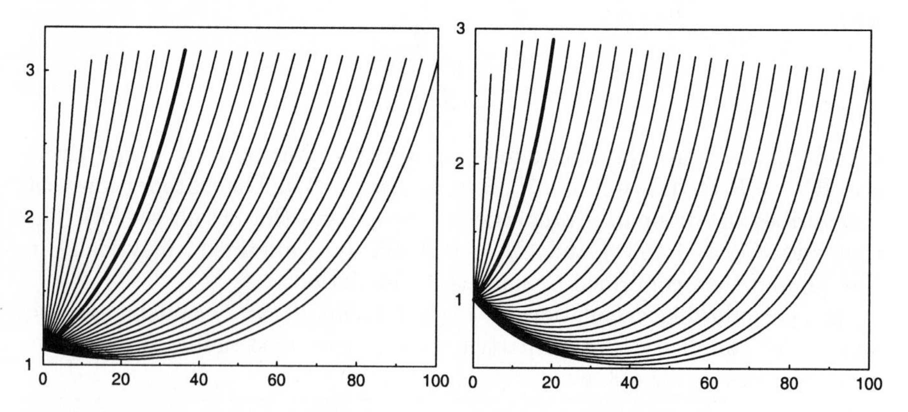

Fig. 3. The velocities of particles in optimal chains according to Fig. 2.

2.1 Optimal Mass-Distribution

The "loss" of energy, i.e., the amount of kinetic energy which is not transferred from the first particle of the chain to the last one may be subdivided into "inertial" and "viscous" losses. Inertial losses occur due to mismatch of subsequent masses, which causes incomplete transfer of momentum even for elastic collisions if the masses differ. Viscous losses are caused by the dissipative nature of collisions. The inertial loss is, thus, given by the energy which remains in the $i-1$rst particle after the collision with the ith:

$$\Delta E_{in}^{(i)} = \frac{m_{i-1}}{2}\left(v_{i-1}^{\prime}\right)^2 = \frac{m_{i-1}}{2}\left(\frac{m_i - m_{i-1}}{m_i + m_{i-1}}\right)^2 v_{i-1}^2 \,. \tag{10}$$

We describe the chain in continuum approximation $m(x)$ with $m_i \approx m_{i-1} + \frac{dm(x)}{dx} \cdot 1$, where we assume that particles are separated on a line by unit distance. Discarding high-order mass gradients within the continuum picture

$\Delta E_{in}^{(i)} \to \frac{dE_{in}}{dx} \cdot 1$ we write for the "line-density" of the inertial losses

$$\frac{dE_{in}}{dx} \approx \frac{\left(\frac{dm(x)}{dx}\right)^2}{8m(x)} v(x)^2 \,. \tag{11}$$

Viscous losses may be quantified as the difference of the kinetic energy of a particle after an *elastic* collision and that of after a *dissipative* collision:

$$\begin{aligned}
\Delta E_{vis}^{(i)} &= \left.\frac{m_i v_i^2}{2}\right|_{\epsilon=1} - \left.\frac{m_i v_i^2}{2}\right|_{\epsilon=\epsilon(v_i)} = \\
&= \frac{m_i}{2}\left(\frac{2}{1+\frac{m_i}{m_{i-1}}}\right)^2 v_{i-1}^2 - \frac{m_i}{2}\left(\frac{1+\epsilon\left(v_{i-1}\right)}{1+\frac{m_i}{m_{i-1}}}\right)^2 v_{i-1}^2 \\
&= \frac{2m_i v_{i-1}^2}{\left(1+\frac{m_i}{m_{i-1}}\right)^2}\left\{1-\left[1-\frac{b}{2}\left(\frac{m_i+m_{i-1}}{m_i m_{i-1}}\right)^{2/5} v_{i-1}^{1/5}\right]^2\right\} .
\end{aligned} \tag{12}$$

Expanding (12) up to linear order in the dissipative parameter b which is assumed to be small and neglecting products of b and mass gradients (which are supposed to be small too), the continuum transition of Eq. (12) yields

$$\frac{dE_{vis}}{dx} \approx \frac{b}{2^{3/5}} m^{3/5} v^{11/5} . \tag{13}$$

Thus, the total energy loss in the entire chain reads

$$E_{tot} = \int_0^n \left[\frac{m_x^2}{8m} v^2 + \frac{b}{2^{3/5}} m^{3/5} v^{11/5}\right] dx = \int_0^n \left[\frac{m_x^2}{8m^2} + \frac{b}{2^{3/5}} \frac{1}{m^{1/2}}\right] dx \,. \tag{14}$$

with $m_x \equiv dm/dx$. For the second part of Eq. (14) we assume in zero-order approximation the "ideal chain Ansatz" for the velocity distribution $v(x)$, which refers to the velocity distribution $v(x)$ in an idealized chain, where the kinetic energy completely transforms through the chain, i.e., where $\frac{1}{2}m(x)v^2(x) = \text{const} = \frac{1}{2}m_0 v_0^2$. With $m_0 = 1$, $v_0 = 1$, i.e., $v(x) = 1/\sqrt{m(x)}$, the right hand side of Eq. (14) follows.

The mass distribution which minimizes E_{tot} satisfies the Euler-equation applied to the integrand in (14):

$$\frac{d}{dx}\frac{2m_x}{8m^2} - \frac{\partial}{\partial m}\left[\frac{m_x^2}{8m^2} + \frac{b}{2^{3/5}}\frac{1}{m^{1/2}}\right] = 0 \tag{15}$$

which leads to an equation for the mass distribution of the optimal chain, written for $y(x) \equiv 1/m(x)$:

$$\frac{d^2y}{dx^2} - \frac{1}{y}\left(\frac{dy}{dx}\right)^2 - 2^{2/5} b y^{3/2} = 0 \,. \tag{16}$$

Figure 4 (lines) shows the numerical solution of (16), i.e., the optimal mass distribution, for a chain of length $n = 40$ for different damping parameters b. The points show the results of a discrete numerical optimization of the full chain problem (see Eq. (9)) applying a steepest descent method to optimize the masses m_k of all particles. For small dissipation b both results agree.

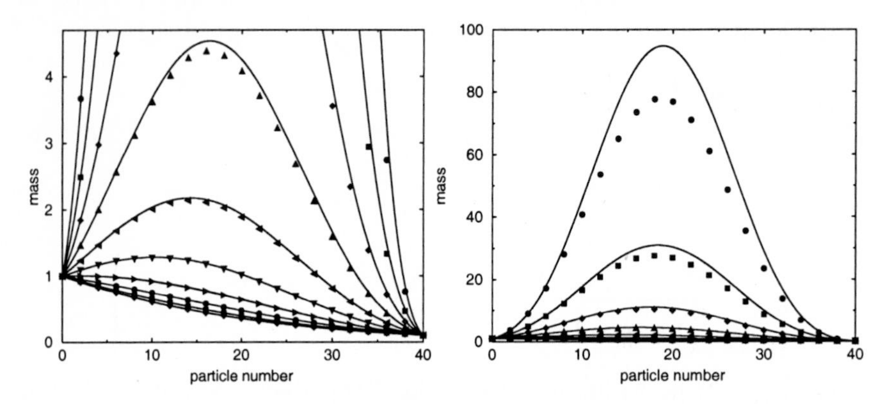

Fig. 4. Left: Mass distribution of the optimal chain of length $n = 40$ for different values of the dissipative parameter b. Lines: numerical solution of Eq. (16), Points: discrete numerical optimization (from top to bottom: • : $b = 0.128$, ■ : $b = 0.064$, ♦ : $b = 0.032$, ▲ : $b = 0.016$, ◄: $b = 0.008$, ▼ : $b = 0.004$, ►: $b = 0.002$, etc.). **Right:** Same data and symbols as left but plotted in larger scale.

For larger values of b the solution of Eq. (16) deviates from the discrete optimization which is understandable since in our approximation we assumed the gradients of the mass distribution to be small which is violated for larger b. While the absolute values of masses deviates from the discrete calculation, Eq. (16) still predicts well the position of the maximum of $m(x)$.

Figure 5 displays the corresponding distribution of velocities for the optimal chains shown in Fig. 4. According to the maximum in the mass distribution, the velocity distribution reveals for larger b a pronounced minimum.

One can give a simple physical explanation of the appearance of a maximum in the mass distribution (and the corresponding minimum in the velocity distribution): As it is seen from Eq. (8) the restitution coefficient increases with decreasing impact velocity and increasing masses of colliding particles; this reduces the viscous losses. Thus, slowing down particles, by increasing their masses in the inner part of the chain, leads to decrease of the viscous losses of the energy transfer. The larger the masses in the middle and the smaller their velocities, the less energy is lost due to dissipation. On the other hand, since the masses m_0 and m_n are fixed, very large masses in the middle of the chain will cause large mass mismatch of the subsequent masses and, thus, large inertial losses [see Eq. (10)]. The optimal mass distribution,

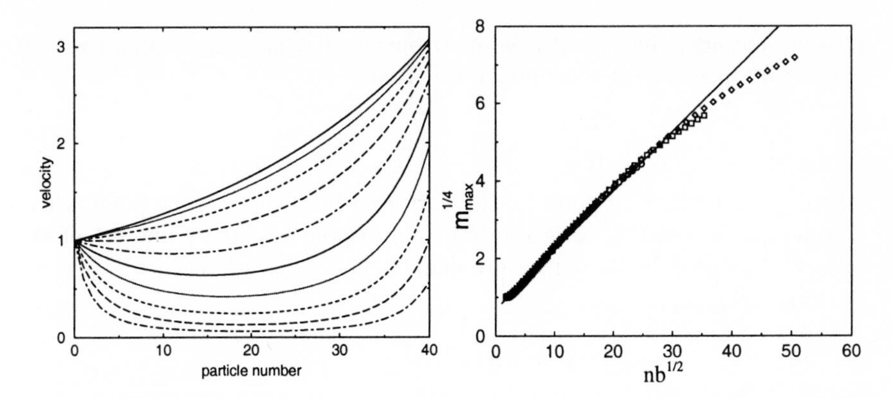

Fig. 5. Left: The velocity distribution along the optimal chain shown in Fig. 4 Lines from top to bottom: $b = 2.5 \cdot 10^{-4}$, $5 \cdot 10^{-4}$, 0.001, 0.002, 0.004, 0.008, 0.016, 0.032, 0.064, 0.128.

Fig. 6. Right: The mass of the heaviest sphere m^* in an optimal chain depends on the dissipative parameter b and on the chain length n. In the figure we plotted $(m^*)^{1/4}$ over $n\sqrt{b}$ for about 3000 different combinations of b and n ($n = 2 \ldots 300$, $b = 0.0001 \ldots 0.256$) including all data presented in Figs. 2, 4. Without any adjustable parameters the data from the numerical optimization of chains agrees well with the analytical expression Eq. (26).

minimizing the *total* losses, compromises (dictated by b) between these two opposite tendencies. For the case of a constant coefficient of restitution the relative part of the kinetic energy, which is lost due to dissipation does not depend on the impact velocity. This means that only minimization of the inertial losses, caused by mass gradient, may play a role in the optimization of the mass distribution. Thus, only a monotonous mass distribution with minimal mass gradients along the chain may be observed as an optimal one for the case of the constant restitution coefficient.

2.2 The Maximum of the Optimal Mass-Distribution

The mass m^* of the heaviest sphere in the optimal mass distribution can be expressed as a function of the chain length n and the dissipative parameter b. With the term $y^{-1}\left(dy/dx\right)^2$ discarded, Eq. (16) describes formally scattering of a particle of unit mass by the potential

$$U(y) = -\frac{1}{2} d\, b y^{5/2} \qquad \text{with} \qquad d \equiv \frac{4}{5} 2^{2/5} . \tag{17}$$

Formally changing notations $x \to t$ ("time") to emphasize the mechanical analogy, we write the equation of motion

$$\ddot{y} = -\frac{dU}{dy}, \quad y_0 = y(t=0) = \frac{1}{m_0}, \quad y_n = y(t=n) = \frac{1}{m_n} . \tag{18}$$

Here we consider the case of mass distributions having a maximum; the generalization, however, is straightforward. Hence,

$$\frac{1}{2}\dot{y}^2 + U(y) = \text{const} = U(y^*) \tag{19}$$

where y^* is the turning point in the scattering problem, i.e., the point where the particle's "velocity" $\dot{y}$ is zero (this corresponds to the point m^* of the mass distribution in the initial problem). The "particle" reaches this point at "time" t^*, i.e.,

$$\dot{y}^2 = d\,b\left(y^{5/2} - y^{*\,5/2}\right) . \tag{20}$$

Solving Eq. (20) with respect to $\dot{y}$ yields

$$\frac{dy}{dt} = \pm\sqrt{d\,b}\,y^{*\,5/4}\sqrt{(y/y^*)^{5/2} - 1}\,. \tag{21}$$

Integration over "time" from $t = 0$ to $t = n$ in Eq. (21), therefore, leads to (with correct choice of signs)

$$\frac{y^{*\,-\frac{5}{4}}}{\sqrt{d\,b}}\left[\int_{y^*}^{y_0}\frac{dy}{\sqrt{\left(\frac{y}{y^*}\right)^{\frac{5}{2}} - 1}} + \int_{y^*}^{y_n}\frac{dy}{\sqrt{\left(\frac{y}{y^*}\right)^{\frac{5}{2}} - 1}}\right] = n\,. \tag{22}$$

Using the substitute $z = (y^*/y)^{5/2}$, the integrals in Eq. (22) may be recasted into the form

$$\frac{2}{5}\,y^* \int\limits_{\left(\frac{y^*}{y_k}\right)^{\frac{5}{2}}}^{1} z^{-\frac{9}{10}}\,(1-z)^{-\frac{1}{2}}\,dz = \mathcal{B}\left(\frac{1}{10},\frac{1}{2}\right) - \mathcal{B}\left[\frac{1}{10},\frac{1}{2},\left(\frac{y^*}{y_k}\right)^{\frac{5}{2}}\right] \tag{23}$$

with $k = 0$ for the first integral in the the left-hand side of Eq. (22) and with $k = n$ for the second integral. $\mathcal{B}(x, y)$ is the Beta-function and $\mathcal{B}(x, y,\, a)$ is the incomplete Beta-function (which has an upper limit a instead of 1 in its integral representation). If we assume the pronounced maximum in the optimal mass-distribution, so that $a \equiv (y^*/y_k)^{5/2} = (m_k/m^*)^{5/2}$ is small, one can approximate the incomplete Beta-function as

$$\mathcal{B}\left(\frac{1}{10},\frac{1}{2},\,a\right) \equiv \int_0^a z^{-\frac{9}{10}}\,(1-z)^{-\frac{1}{2}}\,dz \approx \int_0^a z^{-\frac{9}{10}} = 10\,a\,. \tag{24}$$

With the use of Eqs. (23) and (24), Eq. (22) reads

$$\frac{2y^{*\,-\frac{1}{4}}}{5\sqrt{d\,b}}\left\{2\mathcal{B}\left(\frac{1}{10},\frac{1}{2}\right) - 10\left[\left(\frac{y^*}{y_0}\right)^{\frac{1}{4}} + \left(\frac{y^*}{y_n}\right)^{\frac{1}{4}}\right]\right\} = n\,. \tag{25}$$

In the original variables, which refer to the mass distribution, we obtain a *scaling* relation, connecting the heaviest mass m^*, the chain length n and the dissipative parameter b:

$$(m^*)^{1/4} = p\sqrt{b}\,n + q \tag{26}$$

with the constants

$$p \equiv \frac{5\sqrt{d}}{4\mathcal{B}\left(\frac{1}{10},\frac{1}{2}\right)} \qquad q \equiv \frac{5}{\mathcal{B}\left(\frac{1}{10},\frac{1}{2}\right)}\left[m_0^{\frac{1}{4}} + m_n^{\frac{1}{4}}\right] . \tag{27}$$

So far we considered the solution of the variational Eq. (16) with the term $y^{-1}\left(dy/dx\right)^2$ discarded. For this case the constant d, which has been given above reads: $d = \frac{4}{5}\,2^{2/5}$. It may be shown, however, that perturbative (thus approximate) account of this omitted term leads to an equation of the same form as Eq. (21), but with the *renormalized* coefficient $d \to \frac{9}{5}\,d = \frac{36}{25}\,2^{2/5}$; the details are given in [8]. Using numerical values for $\mathcal{B}\left(\frac{1}{10},\frac{1}{2}\right)$ (see [9]), the renormalized coefficient d and $m_0 = 1$, $m_n = 0.1$ for the first and the last masses of the chain, yields for p and q:

$$p = 0.15217 \qquad q = 0.68989 \tag{28}$$

In Fig. 6 we compare the analytical relation Eq. (26) with the constants given in Eq. (28) with numerical results for m^*. The numerical data follow from the numerical optimization of the mass distribution for different chain length and different dissipative constants, including all data given in Figs. 2-5. As one can see from Fig.6, the results of the analytical theory and of the numerical optimization agree well, except for large dissipation values. We would like to stress that no fitting parameters have been used.

3 Conclusion

We investigated analytically and numerically the transmission of kinetic energy through one-dimensional chains of inelastically colliding spheres. For constant restitution coefficient, ϵ = const., the distribution of the masses which leads to optimal energy transfer, is an exponentially decreasing function which is independent on ϵ, i.e., it is the same as for elastic particles with $\epsilon = 1$.

For viscoelastic particles where ϵ depends on the impact velocity, the optimal mass distribution is not necessarily a monotonous function, but depending on the chain length n and on the material parameters of the spheres it may reveal a pronounced maximum.

We develop a theory which describes the total energy losses along the chain, so that the optimal mass distribution, minimizing the losses, may be obtained as a solution of a variational equation. We derived an expression relating the heaviest mass in the chain to the chain length and the dissipation constant. Having no fitting parameters, it is in good agreement with the numerical data.

It has been demonstrated before that for the case of "thermodynamically-large" granular systems the impact-velocity dependence of the restitution coefficient leads to qualitatively different behavior as compared to systems with $\epsilon = \text{const.}$ (e.g. [1–5, 10–14]). Our system demonstrates that the velocity dependence of the restitution coefficient leads to qualitative modifications in small and simple systems too. Therefore, in general, we believe that the assumption of a constant coefficient of restitution is an approximation which justification cannot be assumed *á priori* but has to be checked for each particular application.

References

1. N. V. Brilliantov and T. Pöschel, *Granular Gases with Impact-velocity Dependent Restitution Coefficient*, (in this volume, page 100).
2. N. V. Brilliantov and T. Pöschel, Phys. Rev. E **61**, 1716 (2000).
3. N. V. Brilliantov and T. Pöschel, Phys. Rev. E, **61**, 5573 (2000).
4. T. Schwager and T. Pöschel, Phys. Rev. E, **57**, 650 (1998).
5. N. V. Brilliantov and T. Pöschel, preprint
6. N. V. Brilliantov, F. Spahn, J.-M. Hertzsch, and T. Pöschel, Phys. Rev. E, **53**, 5382 (1996).
7. R. Ramírez, T. Pöschel, N. V. Brilliantov, and T. Schwager, Phys. Rev. E. **60**, 4465 (1999).
8. N. V. Brilliantov and T. Pöschel, Phys. Rev. E (Nov. 2000, in press), cond-mat/9906138.
9. I. S. Gradshtein and I. M. Ryzhik, *Table of Integrals, Series and Products*, 5-th ed., edited by A. Jeffrey (Academic Press, London, 1965).
10. S. Luding, E. Clément, A. Blumen, J. Rajchenbach, J. Duran, Phys. Rev. E **50**, 4113 (1994).
11. S. Luding, E. Clément, J. Rajchenbach, and J. Duran, Europhys. Lett. **36**, 247 (1996).
12. H. Salo, J. Lukkari, and J. Hanninen, Earth, Moon, and Planets **43**, 33 (1988).
13. J. O. Petzschmann, U. Schwarz, F. Spahn, C. Grebogi, and J. Kurths, Phys. Rev. Lett. **82**, 4819 (1999).
14. F. Spahn, U. Schwarz, and J. Kurths, Phys. Rev. Lett. **78**, 1596 (1997).

III

Vibrated Granular Media

Experimental Studies of Vibro-fluidised Granular Beds

Ricky D. Wildman[1], Jonathan M. Huntley[1], and Jean-Pierre Hansen[2]

[1] Department of Mechanical Engineering, Loughborough University, Loughborough LE11 3TU, United Kingdom. e-mail: J.M.Huntley@lboro.ac.uk
[2] Department of Chemistry, University of Cambridge, Cambridge CB2 1EW, United Kingdom

Abstract. A series of experiments has been performed on two-dimensional and three-dimensional granular gases. The concept of granular temperature has been used to characterize the kinetic energy of the grains. Further tests of the analogy with gas kinetic theory have resulted in investigations into the collective behaviour of the grains, and the diffusion coefficient. Positron Emission Particle Tracking has been used to follow the motion of a single grain within an experimental cell, and preliminary measurements of the granular temperature of a dilute three-dimensional granular gas have been made for the first time.

1 Introduction

Investigation into the behaviour of granular materials has uncovered a range of interesting phenomena, many of which have been studied extensively [1]. Theoretical understanding of granular flow began in the 1970's, based on analogies with the kinetic theory of gases [2, 3]. One barrier to the successful analysis of granular flow is that the material never reaches "thermal" equilibrium, rather, it forms a non-equilibrium steady state. Gas kinetic theory is generally concerned with the relaxation towards equilibrium; hence any analysis of granular flows must assume, and then justify, that the system is sufficiently close to equilibrium for such methods to be appropriate. In his review paper, Campbell [4] pointed out the need for experimental techniques capable of measuring microscopic properties such as granular temperature. In response to this, methods based on high-speed photography [5] and Diffusive Wave Spectroscopy [6] have since been developed to provide such data. These methods have helped to justify assumptions of near-equilibrium [5, 7].

The dissipative system closest to an atomic fluid is the homogeneous cooling state, where the granular gas is initially in a high energy state, but is allowed to cool freely through collisions. This has been the subject of much interest, with several theoretical and numerical studies [8–10], but is a difficult situation to study experimentally [11]. From an experimental viewpoint, a non-equilibrium steady state such as that seen in a vibro-fluidised granular bed is a more convenient system for study and a theoretical understanding of the microscopic processes is beginning to be formulated [12], complementing current experimental investigations [13].

2 Granular Temperature

In the following sections we use the definition that the granular temperature, T_0, is equivalent to the mean kinetic energy of a particle measured in the centre-of-mass frame,

$$T_0 = \frac{1}{2} m \overline{c^2} \quad (1)$$

where m is the mass of the particles and c is the speed. Both gravity and the non-equilibrium state break the rotational invariance of the system, so that the temperature measured in each direction, T_i, (where i can be equal to x or y in two dimensions and equal to x, y and z in three), is given by

$$T_i = m \overline{v_i^2} \, . \quad (2)$$

3 Experimental Details

Experiments on two dimensional fluidised beds were carried out using an electro-magnetically driven shaker. The moving platform can attain a maximum peak-to-peak displacement of 25.4 mm, and a maximum velocity and acceleration of 1.06 m s^{-1} and 70 g, respectively. A cell made up of two glass plates, 165 mm wide by 285 mm high, was mounted onto the platform. The plate separation was controlled by spacers. The plate spacing was adjusted to just over the particle diameter, d, typically being 5 mm, creating an idealized two dimensional granular powder [14]. In general, steel ball bearings were used as the granular media in the following two-dimensional experiments.

A Kodak Ektapro 1000 high-speed video camera system was used to record the two-dimensional motion of the particles, with the resulting images stored digitally on memory boards within the camera. Up to 1600 full-size images of 239 x 192 pixels can be recorded at a framing rate of up to 1000 frames per second, giving a total of 1.6 seconds of recording time. The framing rate could be increased up to a maximum of 12000 frames per second with a proportional decrease in the number of rows of pixels. Images are transferred by GPIB to a Sun Ultra workstation, where the in-plane coordinates of the particle centres are located using image processing software. There are three steps to the image analysis.

1. The edges of the particles are detected by means of a Sobel filter.
2. The positions of the centres are calculated by Hough transformation of the resulting image.
3. The positions of all the detected particles within the field of view are tracked from frame to frame, from which velocity distribution functions can be calculated.

These procedures are described in detail by Warr *et al.* [5].

4 Packing Fraction and Velocity Distributions

A natural consequence of the dissipation of the energy during collisions is that the granular temperature decays with increasing altitude. A side-effect of this is that the grains are no longer operating in an iso-thermal atmosphere. Moreover excluded volume correlations between grain positions are particularly important in the high density regime, i.e. at lower altitudes. Thus, the exponential decay of density in the ideal gas situation is no longer observed; the packing fraction rises initially, reaches a plateau and then decays exponentially (see Fig. 1).

This same dissipation effect also changes the form of the velocity distribution (see Fig. 2). The tails of the distribution tend to be overpopulated in comparison to a Maxwell velocity distribution [11], the y-component velocity distribution is wider than the x [14, 19], and furthermore the distribution is asymmetrical at positions close to the base, due primarily to the heat flux originating from the vibrating base [15].

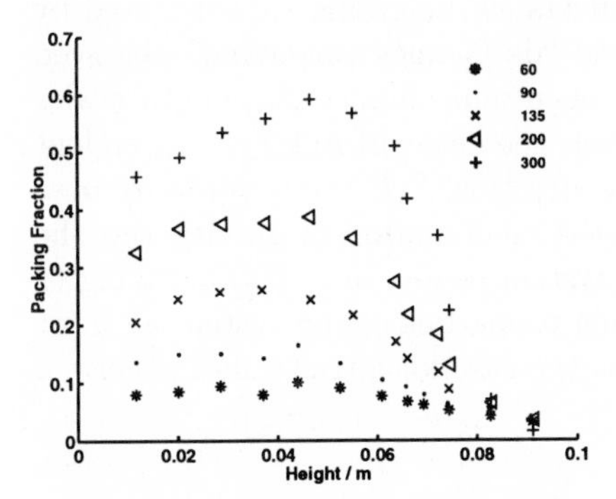

Fig. 1. 2-D Packing fraction profiles for a range of grain numbers. $N = 60$, 90, 135, 200 and 300.

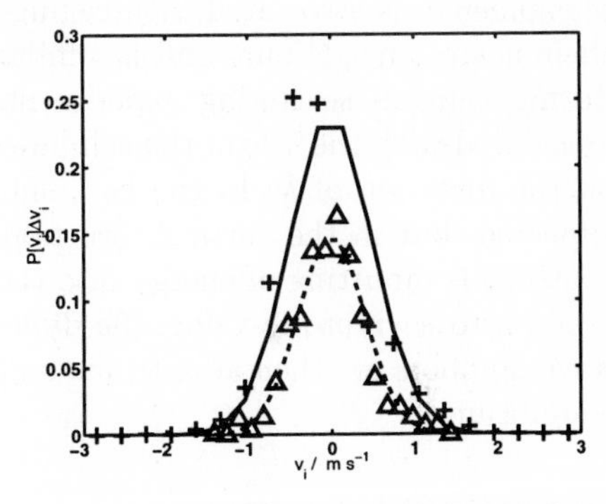

Fig. 2. Maxwell-Boltzmann distributions fitted to experimentally obtained velocity distributions. Triangles denote the x-component of the velocity, crosses the y-component. $N = 90$, altitude above the base, $y = 21$ mm.

5 Local Density Fluctuations

The collective motion of the grains is characterized by the density autocorrelation function $F(\boldsymbol{k}, t)$, or intermediate scattering function, defined as

$$F(\boldsymbol{k}, t) = \langle \rho_{\boldsymbol{k}}(t) \rho_{-\boldsymbol{k}}(0) \rangle \tag{3}$$

where $\rho_{\boldsymbol{k}}$ is the collective density variable. This variable is defined by

$$\rho_{\boldsymbol{k}}(t) = \frac{1}{\sqrt{N_n}} \sum_{i=1}^{N_n} e^{j\boldsymbol{k}\cdot\boldsymbol{r}_i(t)} \tag{4}$$

where $\boldsymbol{k}$ is the wave number and $\boldsymbol{r}$ is the position vector of the i-th grain at time t. N_n is the number of grains in the field of view of the n-th frame. The initial value of the intermediate scattering function leads directly to the static structure factor $S(\boldsymbol{k})=F(\boldsymbol{k},t=0)$. Figure 3, from a study performed by Warr and Hansen [16], shows a comparison of the static structure factor in eight different directions with the result of a theoretical analysis by Baus and Colot for elastic hard spheres in thermal equilibrium [17]. The comparison is rather good and is an indication that the static properties of a fluidised granular bed are similar to those of a fluid in thermal equilibrium. The intermediate scattering function characterizes the decay of local fluctuations in density on the scale of $\lambda \cong 2\pi/k$. On the shortest scale, $\lambda < d/2$, $F(\boldsymbol{k},t)$ is isotropic and decays exponentially. At intermediate wave numbers, where λ is on a scale of the mean grain separation, the decay of $F(\boldsymbol{k},t)$ slows down significantly, compared to wave numbers above and below the 'Bragg' peak (Fig. 4). This phenomenon is associated with caging effects as the grains are frustrated by their nearest neighbours and is similar to "de Gennes narrowing" observed during neutron scattering experiments on atomic fluids. On length scales associated with the size of the cell however, the decay of $F(\boldsymbol{k},t)$ is dependent on the direction of $\boldsymbol{k}$. In the horizontal direction $F(\boldsymbol{k},t)$ decays to zero as expected, but in the vertical direction, the combination of gravity and the continuous inputting of energy into the system results in $F(\boldsymbol{k},t)$ decaying to a non-zero asymptotic value; the dynamic properties of the system at short wave numbers are then anisotropic unlike the case for a true fluid in thermal equilibrium.

6 Mean Squared Displacement

The anisotropy in the velocity distributions, and by implication, the granular temperature, is expected to be reflected in the displacements of the grains. A simple measure of the spatial behaviour of the grains is provided by the mean squared displacement. The mean squared displacement can be described by a Brownian motion model of the diffusing grains:

$$\left\langle |\boldsymbol{r}(t) - \boldsymbol{r}(0)|^2 \right\rangle = \frac{2\overline{c^2}}{\beta^2}\left(\exp(-\beta t) + \beta t - 1\right) \tag{5}$$

where β is known as the friction coefficient [18]. At short times, i.e., $t \ll \tau_E$, where τ_E is the mean collision time, (5) approximates to

$$\left\langle |\boldsymbol{r}(t) - \boldsymbol{r}(0)|^2 \right\rangle = \overline{c^2} t^2 \,. \tag{6}$$

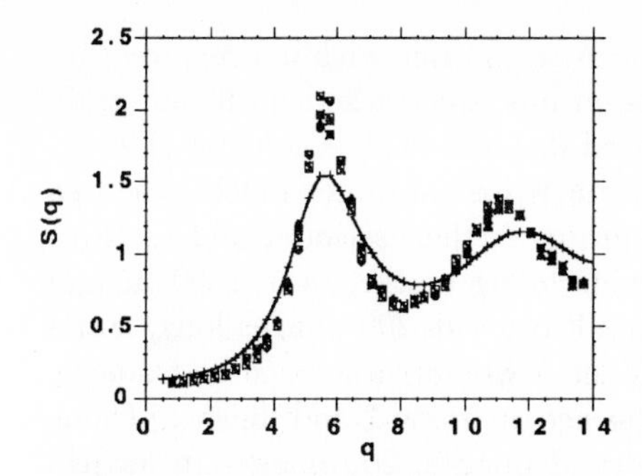

Fig. 3. Static structure factor, $S(q)$ as a function of the dimensionless parameter $q = kd$, where k is the modulus of the wave vector and d is the particle diameter. These data sets were obtained by taking angular wedge cuts through the static structure factor surface evaluated for each wave vector separately. Eight separate cuts, each with a wedge angle of 30 degrees, were taken regularly spaced around the plane. All data sets superimpose, indicating near perfect isotropy in the static structural properties. The solid line shows the theoretical parameterization due to Baus and Colot [17]

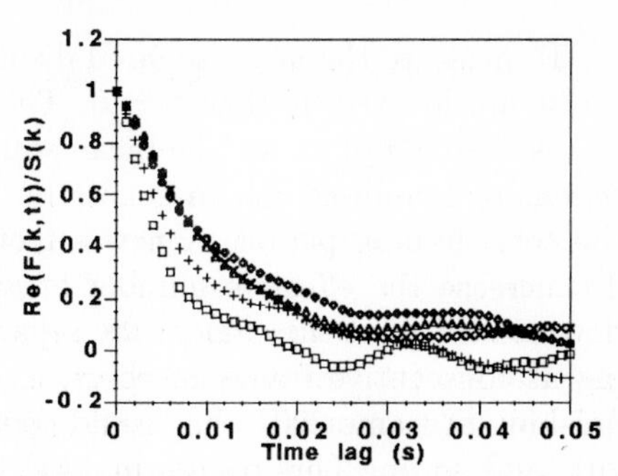

Fig. 4. Real part of the intermediate scattering function, $F(\boldsymbol{k}, t)$, normalized by the static structure factor, $S(\boldsymbol{k})$, as a function of the time lag. de Gennes narrowing (slowing down in the decay of density fluctuations) can be seen as one moves over the Bragg peak. Plusses, crosses, triangles and squares correspond to $\boldsymbol{k} = (-5, 4)$, $\boldsymbol{k} = (-7, 5)$, $\boldsymbol{k} = (-7, 6)$, $\boldsymbol{k} = (3, 10)$ and $\boldsymbol{k} = (-10, 10)$, respectively, where, with reference to Fig.3, q values are given by 3.246, 4.309, 4.709, 5.891 and 7.346. Here values of integers n_x and n_y that describe $\boldsymbol{k}$, where $\boldsymbol{k} = 2\pi\left(\frac{n_x}{239}\,\frac{n_y}{192}\right)$, are shown in the brackets.

At low to intermediate packing fractions, $\nu < 0.5$, the mean collision time is significantly greater than 1 millisecond, a time which corresponds to the framing interval of the high speed camera. Therefore the mean squared speed can be extracted from the mean squared displacement behaviour by means of a polynomial fit [19]. After long times, i.e. $t \gg \tau_E$, (5) reverts to the Einstein equation [20]:

$$\left\langle |\boldsymbol{r}(t) - \boldsymbol{r}(0)|^2 \right\rangle = 4Dt \tag{7}$$

where the diffusion coefficient D, is given by

$$D = \frac{\overline{c^2}}{2\beta} \, . \tag{8}$$

Granular temperature can be calculated from the diffusion coefficient using kinetic theory methods, assuming the validity of Enskog kinetic theory (section 8). Where it is possible to measure both granular temperature and diffusion coefficient, investigation of the mean squared displacement leads to a test of these kinetic theory methods.

To measure the mean squared displacement, grains within a region of interest are located at time $t = 0$. These grains are tracked until one grain leaves the field of view. The mean squared displacement is calculated at intervals of 1 millisecond. In general, a grain is lost at times much less than the total filming period. A new set of grains is then selected and tracked. To increase the effective number of grains being tracked, the 1600 images downloaded from the camera are separated into sets 200 images long, starting at consecutive frame numbers, e.g., set 1 would commence at frame 1, finishing at frame 200, set 2 would commence at frame 2 and finish at frame 201, and so on. This results in 1400 sets of images, 200 images in length. Each image is split into six horizontal segments, allowing the mean squared displacement to be measured at different heights, although the upper most and lowest segments are disregarded as the grains in these regions have a tendency to escape from the field of view at extremely short times.

Grains are selected for tracking at the start of each set by creating a box 100 by 10 pixels (approximately 26.4×2.7 mm^2) at the centre of each horizontal segment. Any grain located within this box at the start of each set of images is tracked until one of the grains is lost or 200 frames have been analyzed. In this way, if for example 5 grains, on average, were found within the box at each start frame, then this would result in 7000 grains being tracked in total. This technique allows the grains to move beyond their original segment when the tracking ceases. If the tracked grains move into regions of substantially different packing fraction or granular temperature from that observed initially, then the mean squared displacement will deviate strongly from the expected asymptotic behaviour predicted by Eq. (7). In this case a measurement of D is not taken. This approach results in a method that is "self-correcting", and allows measurements to be taken only when any effect of de-localization of the grains is small.

Experiments were carried out at a vibration amplitude of 2.12 mm and a frequency of 50 Hz. The total number of grains in the cell, N, took values ranging from 90 to 480. Figure 5 shows the mean squared displacement of the grains for $N = 300$ at a height of about 25 mm from the base. At low densities grains will leave the field of view before diffusive behaviour is fully established (i.e., a linear relationship between mean square displacement and time is not observed), requiring that D is measured by fitting Eq. (5) to the experimental data using non-linear regression. As expected, the mean squared displacement in the y-direction is greater than that in the x-direction, complementing the anisotropy observed in the velocity distributions [14].

7 Granular Temperature Profiles

The granular temperature profiles show a similar form for all numbers of particles: the particles are highly energetic close to the base of the cell, and due to dissipation during collisions the granular temperature is reduced with increasing altitude. An example of this behaviour for granular temperature measured in the x-direction ($N = 300$) is shown in Fig. 6. Three methods were used to measure the granular temperature, (A) by analysis of the short time mean squared displacement behaviour, based on Eq. (6), (B) by calculating the mean squared velocity explicitly from the measured velocities and (C) by fitting a Maxwell-Boltzmann distribution to the measured velocity distribution. Comparison of these methods shows that the methods are self-consistent (see e.g. Fig.6), and that the method of analysis of the ballistic regime is at least as robust with regard to statistical errors as the methods dependent on the calculation of the velocity [19]. As expected from the anisotropy observed in the velocity distributions and mean squared displacement behaviour, $T_Y > T_X$.

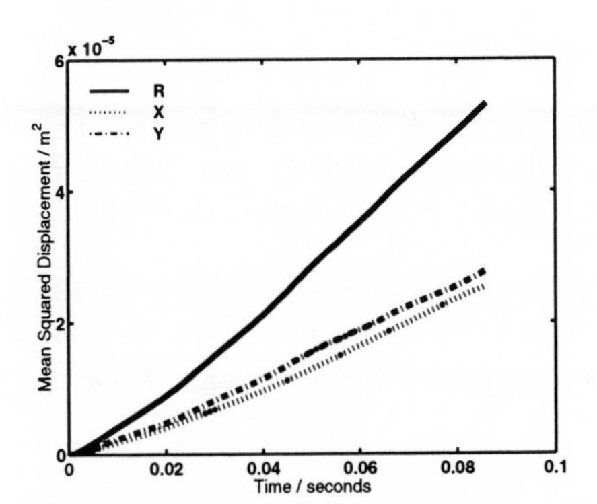

Fig. 5. Mean squared displacement of grains within a two dimensional cell. $N = 300$, $y = 25.0$ mm, $R = (x^2 + y^2)^{1/2}$.

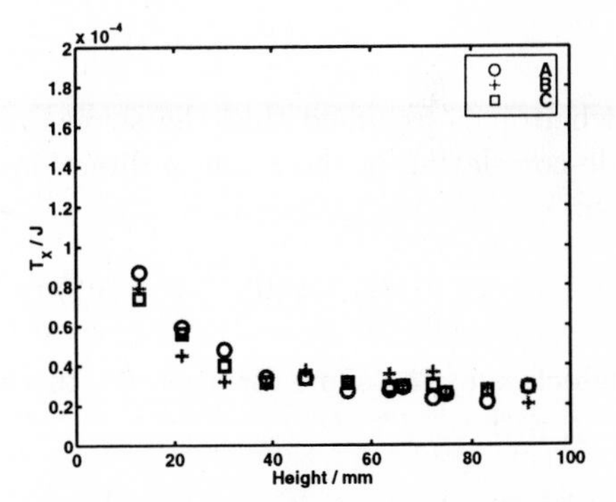

Fig. 6. Granular temperature profile (measured in the x-direction) as a function of height. Determined using three differing methods (see text). $N = 300$.

8 Self-Diffusion

According to Enskog kinetic theory, the temperature of a two-dimensional fluid in thermal equilibrium is related to the diffusion coefficient by [21]

$$D = \frac{d}{8\nu g(d)} \left(\frac{\pi T_0}{m} \right)^{\frac{1}{2}} \tag{9}$$

where $g(d)$, the radial distribution function at contact, can be estimated by [22]

$$g(d) = \frac{16 - 7\nu}{16(1-\nu)^2} \,. \tag{10}$$

It is of interest to examine the extent to which a granular gas, undergoing inelastic collisions and in a state far from equilibrium, mimics the behaviour of such a fluid in thermal equilibrium. Equipartition of energy no longer holds as $T_Y > T_X$ [23]. Therefore the mean square velocity in the y-direction will exceed that in the x-direction, and one may expect the vertical and horizontal diffusion coefficients to differ. A rough estimate of the difference may be gained by starting from the exact relation [20]:

$$\left\langle |x(t) - x(0)|^2 \right\rangle = 2 \int_0^t dt' \int_0^{t'} dt'' \left\langle v_x(t') \, v_x(t'') \right\rangle . \tag{11}$$

Within the Enskog theory of uncorrelated binary collisions [20],

$$\left\langle v_x(t') \, v_x(t'') \right\rangle = \left\langle v_x^2 \right\rangle \exp\left(- |t' - t''| / \tau_E\right) \tag{12}$$

where it is assumed that the same Enskog mean collision time governs the de-correlation in the x and y directions. Substitution of (12) into (11) leads to:

$$\left\langle |x(t) - x(0)|^2 \right\rangle = 2 \left\langle v_x^2 \right\rangle \tau_E \left[t + \tau_E \left(\exp\left(-t/\tau_E\right) - 1 \right) \right] \tag{13}$$

which is of the same form as (5). In the long time limit, (13) leads back to (7), with

$$D_x = \frac{1}{2} \left\langle v_x^2 \right\rangle \tau_E = \frac{T_x}{2m} \tau_E \tag{14}$$

and similarly

$$D_y = \frac{1}{2} \left\langle v_y^2 \right\rangle \tau_E = \frac{T_y}{2m} \tau_E \tag{15}$$

and

$$D = D_x + D_y = \frac{1}{2m} \left(T_x + T_y \right) \tau_E = \frac{T_0}{m} \tau_E = \frac{d}{8 \nu g(d)} \left(\frac{\pi T_0}{m} \right)^{\frac{1}{2}} \tag{16}$$

so that (14) and (15) may be rewritten as

$$D_x = D \frac{T_x}{T_x + T_y} \quad ; \quad D_y = D \frac{T_y}{T_x + T_y} . \tag{17}$$

The Enskog theory leading to (17) does not account for correlations between successive collisions and effects of collective motion. These are included in MD simulations, which provide estimates of the ratio of the actual diffusion constant to its Enskog prediction as a function of the packing fraction for the case of smooth, hard elastic discs [21].

A dimensionless diffusion coefficient D_0 is generated by normalising the diffusion coefficient by the ratio of the grain diameter squared to the mean collision time:

$$D_0 = \frac{D\tau_E}{d^2} = \frac{\pi\sqrt{2}}{64\nu^2 g(d)^2} . \tag{18}$$

The dimensionless diffusion coefficients in the x and y directions, D_{0x} and D_{0y} respectively, are calculated in the same manner and are also functions of packing fraction only. Experimentally, τ_E is calculated from the measured granular temperature.

The measured and predicted diffusion coefficients were determined for packing fractions up to $\nu \sim 0.8$. The predicted diffusion coefficient was calculated following measurements of the granular temperature [Method A]. The diffusion coefficient was also measured directly using two methods: (*i*) by fitting the Einstein equation (7) to the linear portion of the mean squared displacement versus time plots [Method B], and (*ii*) by fitting (5) to the same plots using a non-linear regression technique [Method C].

Figure 7 shows the experimentally determined dimensionless diffusion coefficients together with the dimensionless diffusion coefficients calculated, via Chapman-Enskog theory (9), from the measured granular temperature. Results from all methods of measuring D fall reasonably well onto a single line over the whole range of packing fractions. It should be noted that Method A results fall on to the theoretical line by definition. The dilute phase of a granular gas is expected to behave like a nearly ideal gas at large distances from the base. In this phase the diffusion coefficient is high, reflecting the low packing fraction. Cage effects become important at high densities, reducing the diffusion coefficient due to a combination of correlated collisions and enclosure. At high packing fractions the measured D deviates from the theoretical predictions as the collective behaviour becomes more significant, and the grains become caged. The diffusion coefficient is seen to drop off rapidly beyond a packing fraction of about 0.75 which is similar to observations reported in studies of discs in motion on an air-table [24]. The dimensionless diffusion coefficients for the x and y directions are observed to follow a broadly similar pattern (see e.g. Fig. 8 showing D_{0X} versus packing fraction) [13].

An alternative way of presenting the results is to normalize D by the predicted value of the diffusion coefficient and to plot this against a measure of packing fraction. Figure 9 shows the ratio D/D_E (where D_E denotes predicted diffusion coefficient) against the volume expansion of the bed above the minimum attainable volume. A value of $V/V_0 = 1$ indicates that the maximum packing fraction ($\nu = 0.91$) has been reached. Increases in V/V_0 indicate an expansion of the bed. A rolling average of the data points shows qualitative agreement with MD results [21], which were obtained for an elastic hard disc gas. The figure suggests that above a value of $V/V_0 \sim 1.3$ ($\nu < 0.7$), the measured and predicted self-diffusion coefficients are within 10-20% of each other (i.e. $D/D_E = 1 \pm 0.2$). At higher densities the figure suggests that

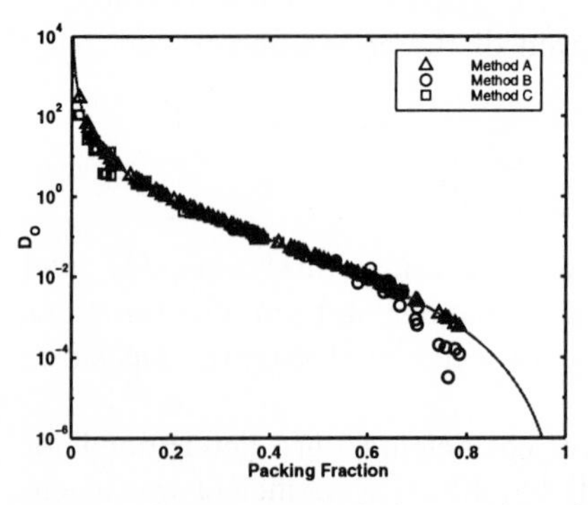

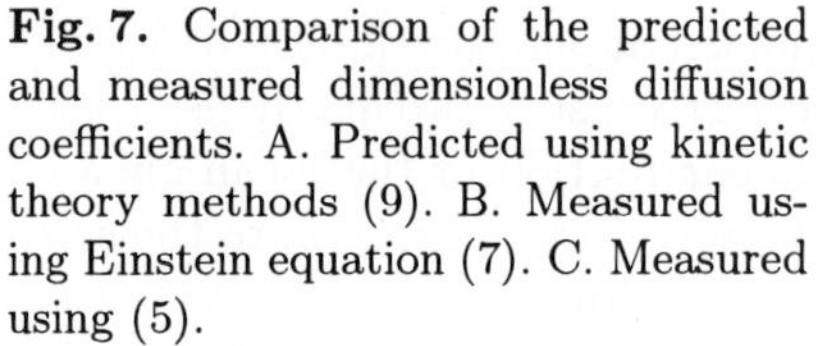

Fig. 7. Comparison of the predicted and measured dimensionless diffusion coefficients. A. Predicted using kinetic theory methods (9). B. Measured using Einstein equation (7). C. Measured using (5).

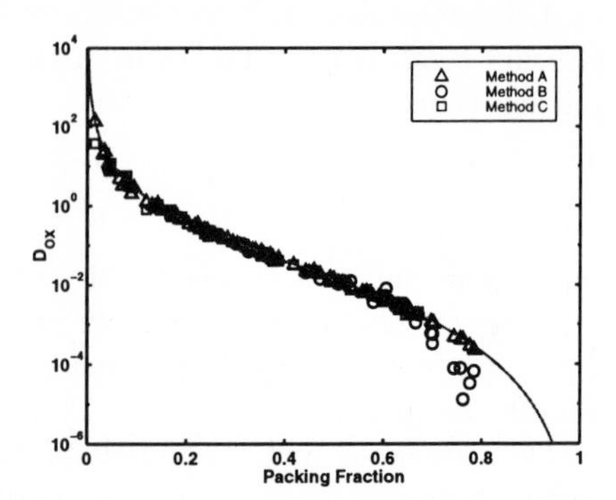

Fig. 8. Comparison of the predicted and measured x-component of dimensionless Diffusion coefficients.

diffusion drops to zero faster than is expected from MD studies [21], and that diffusion will cease to occur at a packing fraction $\nu \sim 0.83$. At intermediate densities, the range of the packing fraction for which D/D_E rises above one in MD studies differs from that observed experimentally. Collective effects, which tend to be significant when $D/D_E > 1$, are not appreciable until the packing fraction reaches values of the order of 60%, compared to the MD studies of hard elastic discs where collective behaviour is already important at $\nu \sim 0.40$. Despite these differences, the agreement between the MD and experimental results is in some ways surprisingly good. The MD results are used under the assumption that the system is analogous to a gas of smooth, hard elastic discs. There are, of course, a number of important differences between such a model and the system being studied here. The granular gas is not in equilibrium and there is a "heat" flux in the upward direction. Consequently the system may be in a state far from that modeled in MD studies [21].

A third way of presenting the results is to calculate granular temperature from measured D values using (9), and to compare against temperature values measured directly from the mean squared displacement in the ballistic regime. Figure 10 shows a granular temperature profile for $N = 300$ grains. Reasonable agreement of the three methods of measuring the granular temperature is observed for these intermediate packing fractions ($\nu \sim 0.5$). This suggests that indirect measurement of T from D may be valid, at least for intermediate packing fractions. This approach could be useful when interpreting results from experimental techniques which have insufficient time resolution to measure T directly.

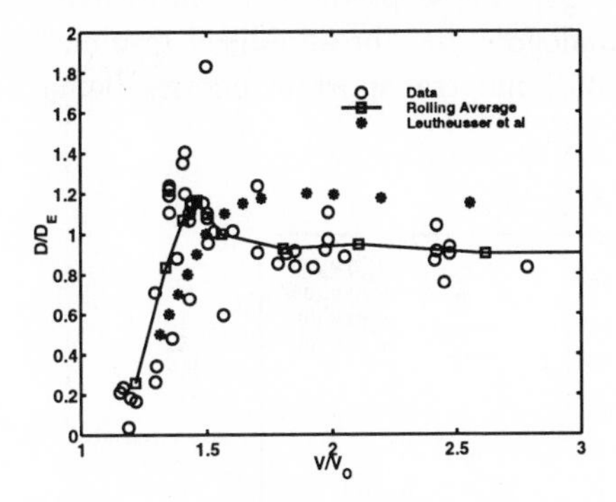

Fig. 9. Ratio of measured to predicted diffusion as a function of volume expansion of the bed above that predicted at maximum packing. Star data points are results from MD simulations [21].

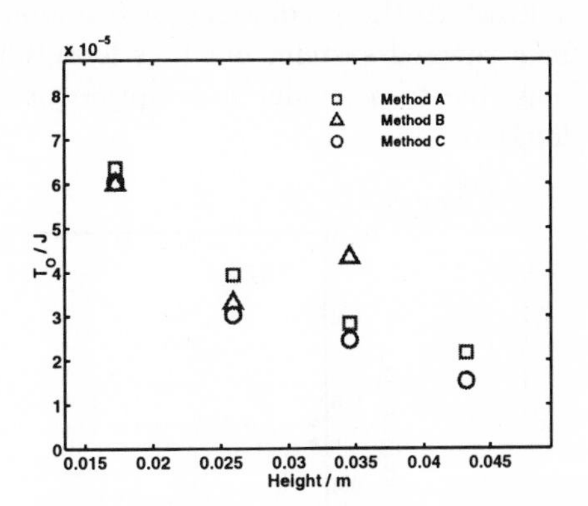

Fig. 10. Comparison of three methods (see text) of measuring the granular temperature.

9 Velocity Auto-correlation Function

The velocity of the grains was estimated using

$$\boldsymbol{\nu}_f = \frac{\boldsymbol{r}_{f+1} - \boldsymbol{r}_{f-1}}{2\Delta t} \tag{19}$$

where f is the current frame number and Δt is the time between two successive frames. The normalized velocity cross-correlation (NVCF) and the auto-correlation functions (NVAF) $Z(t)$ are given by

$$Z_{ij}(t) = \int \frac{\langle v_i(t) v_j(0) \rangle}{\langle v_i(0) v_j(0) \rangle} dt \tag{20}$$

where i and j indicate the direction ($i \neq j$ for cross-correlation, $i = j$ for auto-correlation). The correlation functions were calculated for a range of grain numbers ($N = 40$ to 480). The NVCF were found to be approximately zero for all times. An example of a NVAF for $N = 300$ and $y = 20$ mm is shown in Figure 11, with the short time behaviour of the same function shown in Figure 12. This indicates that the relaxation times in the x and y directions are similar, justifying assumptions made in (14) and (15). At times of the order of several mean collision times the motion imparted by the shaker dominates the y-velocity auto-correlation functions as evidenced by the sinusoidal variation at a period of 0.02 seconds. The x-velocity component is de-coupled from the shaker motion and no such sinusoidal signal is seen. In vibro-fluidised beds, the velocity auto-correlation function appears to decay monotonically (for times less than the period of vibration of the base), in contrast to the predictions of Campbell [25]. These predictions were made for a sheared system, but it is loosely analogous to vibro-fluidised granular beds due to a similar assumption of collisional transport of energy being dominant.

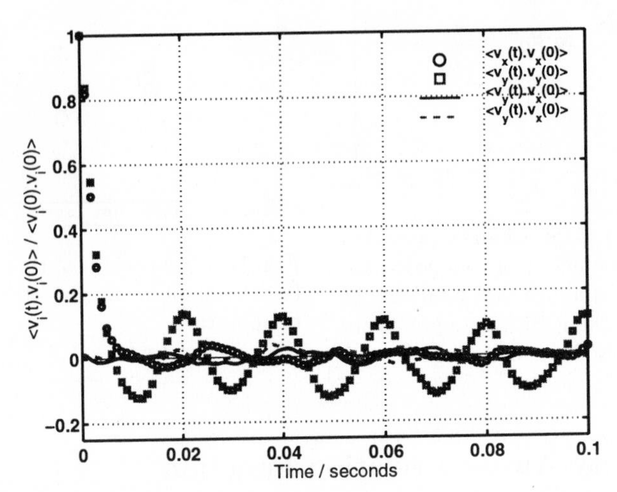

Fig. 11. Normalized velocity correlation functions up to $t = 0.1$ seconds. $N = 300$, $y = 20$ mm.

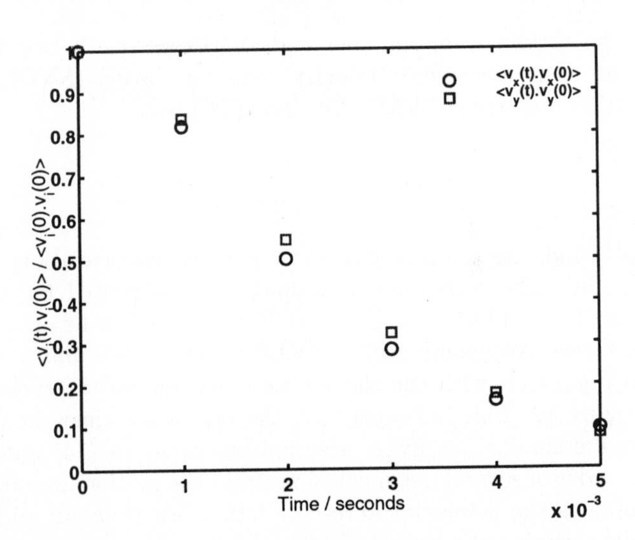

Fig. 12. Normalized velocity auto-correlation functions up to $t = 5$ milliseconds. $N = 300$, $y = 20$ mm.

10 Positron Emission Particle Tracking

In three dimensions, techniques such as high speed photography cannot probe the behaviour at positions beyond the surface of the granular bed. To analyse the motion of individual grains in this situation, a technique such as Positron Emission Particle Tracking (PEPT) must be used. The following experiments were carried out at the PEPT facility based at the University of Birmingham, U.K.. The basic approach is that one follows the trajectory of a single positron-emitting tracer particle. As the positrons annihilate with electrons, pairs of back-to-back 511 keV photons are produced which are then detected in coincidence by a pair of large position-sensitive detectors (see Fig.13). This coincidence defines a line along which the positron annihilated. In theory, just two measured coincidences define lines that cross at the tracer particle position. In practice, many of the detected events are corrupted (e.g. because one of the photons has scattered before detection) so one needs to detect a reasonable number (of order 100) in order to distinguish the valid events (which essentially cross at a point) in the presence of this background [26]. As a result, typical measurement parameters are a time resolution of 5 milliseconds and a measurement accuracy of 5 mm.

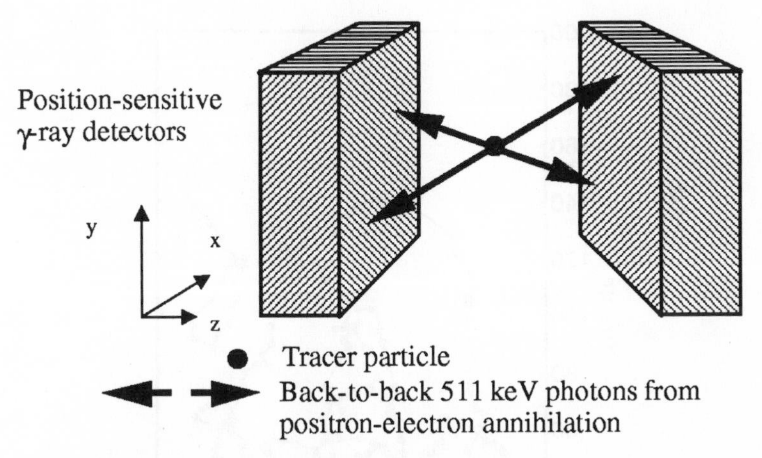

Fig. 13. Schematic of the PEPT facility showing a location event.

Whereas high speed photography is a wholefield technique, PEPT data are only obtainable from a single point (i.e., from the tracer particle). On the other hand, the data acquisition process is automated and the particle remains active for several hours. In that time it will normally visit even relatively small volume elements in the flow several times allowing a detailed composite map of the flow field to be built up.

A range of numbers of ballotini glass spheres ($N = 300$ to 2800, $d = 5$ mm) were placed into a cylinder 300 mm high, 140 mm in diameter. The experimental cell was shaken at a frequency of 50 Hz, and amplitude 1.91 mm, using the same shaker facility as described in section 3. During each experiment a single tracer particle (taken from the grains to be used in each experiment) was tracked for about one hour.

Figure 14 shows the first 500 location events projected on to the $X-Y$ plane for a relatively dilute system ($N = 300$, corresponding to ≈ 0.5 grain layers). This gives a qualitative indication of the motion of the tracer particle. The essentially random motion of the grain suggests that the system is ergodic, i.e., that a time average is equivalent to an ensemble average. Under these circumstances, a reasonable estimate of the packing fraction distribution can be made by assuming that the packing fraction within any small volume element is proportional to the time-averaged fractional residence time spent by the tracer particle within the element. In view of the axial symmetry of the experimental arrangement, the natural volume element is a horizontal slice. With this in mind, the fraction of time spent in a horizontal slice, 5 mm thick, was measured. Figure 15 shows that the packing fraction behaviour is somewhat similar to that seen in two dimensions: the packing fraction rises initially with increasing altitude, reaches a plateau value and then decays in an exponential manner.

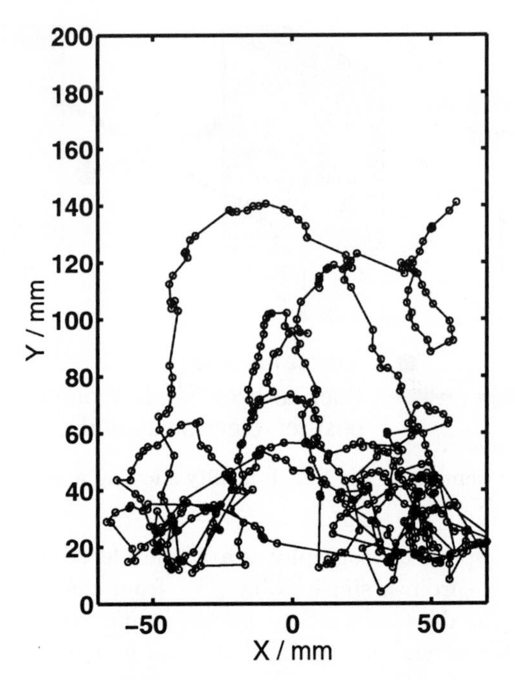

Fig. 14. Projection of tracer position on to the $X-Y$ plane for the first 500 location events. $N = 300$.

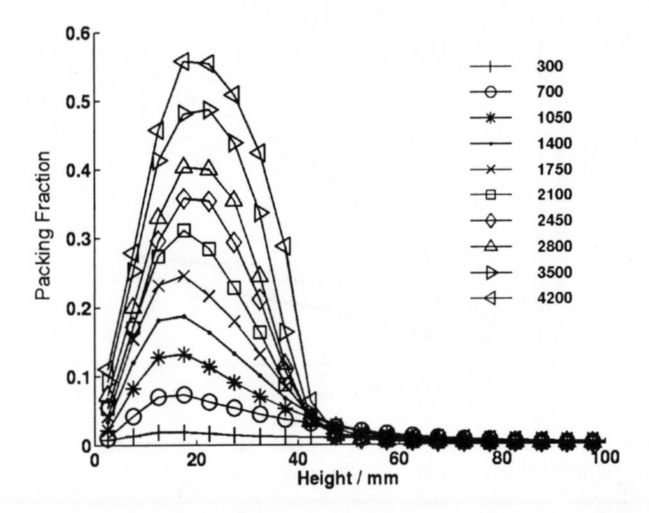

Fig. 15. Packing fraction profiles for a range of grain numbers.

With the current resolution it is not possible to measure the velocity of a grain between the successive collisions. However, by using the methods developed for two dimensional systems, the granular temperature can be measured by indirect means; the mean squared displacement at short times can be used to analyse the free motion of the grains. Unfortunately, with an average temporal resolution of around 5 milliseconds, the "ballistic" regime can only be probed at very low packing fractions ($\nu \sim 0.02, N = 300$). Figures 16 and 17 show the mean squared displacement for $N = 300$ and $N = 2800$ respectively. Both figures show the mean squared displacement of a grain initially within a horizontal slice 20 mm thick centred at $y = 30$ mm. As in the 2D case, the x and y behaviour is quite different. The z behaviour (i.e., in the horizontal direction normal to the PEPT detector plate) is not shown as the error involved in the location of a grain moving in this direction is often three times as great as the error associated with the other directions. At extremely high packing fractions ($\nu \sim 0.6$, Fig. 17) the mean square displacement is seen to saturate indicating that the diffusion is close to zero.

The granular temperature, calculated from the mean squared displacement in the ballistic regime, for a system containing 300 grains is shown in Fig. 18. The granular temperature profiles are reminiscent of those seen in two dimensions (see Fig. 10): the grains are most energetic close to the base and this energy is dissipated with increasing altitude.

11 Conclusions

A series of experiments has been conducted to investigate the dynamics of steady state flows in vertically vibrated granular beds. Analogies with gas ki-

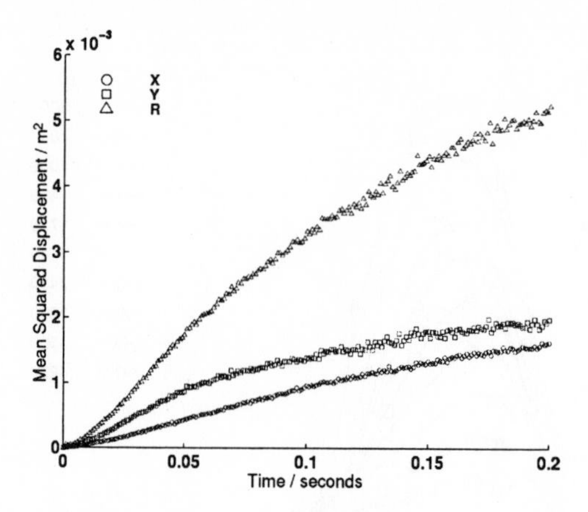

Fig. 16. Mean squared displacement of a grain tracked around a cylindrical cell using PEPT techniques. $N = 300$, $y = 30$ mm.

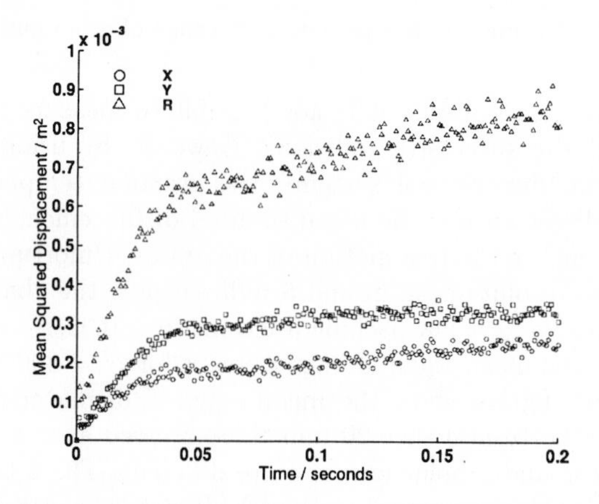

Fig. 17. Mean squared displacement of a grain tracked around a cylindrical cell using PEPT techniques. $N = 2800$, $y = 30$ mm.

netic theory have been shown to be useful, although dissipation causes some significant quantitative difference. Packing fraction profiles are not simply exponential decaying functions of height, and the granular temperature is neither isotropic nor isothermal. Velocity distribution functions, mean squared displacement, and self-diffusion properties of the grains reflect the anisotropy; in general, the y-component of the variable under investigation is larger than the corresponding x-component. The measurement of mean squared displacement allowed two novel methods of measuring granular temperature to be developed. In particular, analysis of the ballistic motion of the grains at short times produced an accurate measure of the granular temperature. Self-diffusion can also be used to determine granular temperature profiles as the

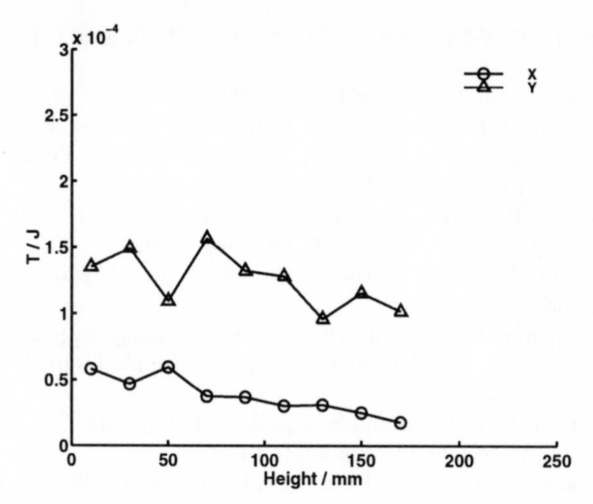

Fig. 18. Granular temperature profiles measured in both the x and the y directions for a three dimensional granular gas.

usual scaling between T and D predicted by kinetic theory methods has been found to hold to within 20% over a wide range of packing fractions (up to $\nu \sim 70\%$). PEPT has allowed the granular temperature to be measured in three dimensions for dilute granular gases for the first time, and packing fraction profiles have been determined for a range of grain densities.

Acknowledgements

The authors are grateful to D. Parker and D. Allen for providing access to the PEPT facility at the University of Birmingham. Funding from the Department of Trade and Industry, Engineering and Physical Sciences Research Council, Unilever Plc, ICI Plc, Zeneca Plc, Schlumberger Cambridge Research, and Shell International Oil Products is also gratefully acknowledged.

References

1. H. M. Jaeger, S. R. Nagel and R. P. Behringer, Phys. Today, **49**, 32 (1996).
2. J. T. Jenkins and S. B. Savage, J. Fluid Mech. **130**, 187 (1983).
3. S. B. Savage, J. Fluid Mech. **92**, 53 (1979).
4. C. S. Campbell, Annu. Rev. Fluid Mech. **22**, 57 (1990).
5. S. Warr, G. T. H. Jacques and J. M. Huntley, Powder Technol. **81**, 41 (1994).
6. N. Menon and D. J. Durian, Science **275**, 1920 (1997).
7. K. Helal, T. Biben and J. P. Hansen, Physica A **240**, 361 (1997).
8. S. E. Esipov and T. Pöschel, J. Stat. Phys. **86**, 1385 (1997).
9. T. P. C. van Noije and M. H. Ernst, Granular Matter **1**, 57 (1998).
10. N. V. Brilliantov and T. Pöschel, Phys. Rev. E **61**, 1716(2000); (see also in this volume, page 100)

11. W. Losert, D. G. W. Cooper, J. Delour, A. Kudrolli and J. P. Gollub, cond-mat/9901203
12. V. Kumaran, J. Fluid Mech. **364**, 163 (1998).
13. R. D. Wildman, J. M. Huntley and J. P. Hansen, Phys. Rev. E **60**, 7066 (2000).
14. S. Warr, J. M. Huntley and G. T. H. Jacques, Phys. Rev. E **52**, 5583 (1995).
15. V. Kumaran, Phys. Rev. E **57**, 5660 (1998).
16. S. Warr and J. P. Hansen, Europhys. Lett. **36**, 589 (1996).
17. M. Baus and J. L. Colot, J. Phys. C **19**, L643 (1986).
18. J. L. Doob, in "Noise and Stochastic Processes" (ed. N. Wax), Dover Publishing, New York, 1954.
19. R. D. Wildman and J. M. Huntley, Powder Technology, in press.
20. J. P. Hansen and I. M. McDonald, "Theory of Simple Liquids", 2nd Edition, Academic Press, London, 1986.
21. E. Leutheusser, D. P. Chou and S. Yip, J. Stat. Phys. **32**, 523 (1983).
22. D. Henderson, Molecular Physics **30**, 971 (1975).
23. S. McNamara and S. Luding, Phys. Rev. E **58**, 2247 (1998).
24. L. Oger, C. Annic, D. Bideau, R. Dai and S. B. Savage, J. Stat. Phys. **82**, 1047 (1996).
25. C. S. Campbell, J. Fluid Mech. **348**, 85 (1997).
26. D. J. Parker, A. E. Dijkstra, T. W. Martin and J. P. K. Seville, Chem. Eng. Sci. **52**, 2011 (1997).

Pattern Formation in a Vibrated Granular Layer

Eric Clément and Laurent Labous

Laboratoire des Milieux Désordonnés et Hétérogènes - UMR 7603, Université Pierre et Marie Curie - Boîte 86, 4, Place Jussieu, F-75252 Paris.
e-mail: Eric.Clement@ccr.jussieu.fr

Abstract. We present a numerical study of a surface instability occuring in a bidimensional vibrated granular layer. The driving mechanism for the formation of stationary waves is closely followed. Two regimes of wavelength selection are identified: a dispersive regime and a saturation regime. For the latter, a connection is established between the pattern formation and an intrinsic instability occuring spontaneously in dissipative gases.

Granular assemblies under vertical vibrations show a very broad and interesting phenomenology and their study was undertaken on various different viewpoints either experimentally, numerically or theoretically (for extensive reviews on the subject see [1, 2] and references therein). In a series of experiments Melo et al. [3, 4] reported a pattern forming instability taking place in a vibrated thin layer of grains. The apparent phenomenology of the patterns (squares, stripes, hexagons, localized structures) is strongly reminiscent of the outcome of a parametric instability occurring in vibrated fluid layers called the Faraday instability [5] (see Ref. [6] for a modern viewpoint and further references). Experiments showing surface patterns were also performed in a granular layer confined to two-dimensional (2D) cells [7] and the dispersion relation of the excited standing waves was related quantitatively to the dispersion relation observed in three dimensions (3D). Numerical simulations were performed reproducing qualitatively the phenomenon, by using event-driven algorithms in 2D [8] and in 3D [9] geometries, and also using a soft-particle algorithm in 2D [10]. Note that a recent simulation of a simplified toy-model for the horizontal momentum transfer also reproduces the pattern formation [11] and, in some limit, the dispersion relation. Theoretical models were proposed to render this pattern forming instability [12–15] but those models, though displaying the observed phenomenology, do not render properly the measured dispersion relation. Therefore, there is so far no clear vision of the basic mechanisms driving this instability. Here, we report some results on an extensive study we performed using an optimized version of an event-driven algorithm already used by Luding et al. [8]. The system we investigate consists of N beads in a container of size L with periodic boundary conditions, constrained to move in 2D. The bottom plate moves vertically with a trajectory $z(t) = A \sin \omega t$ (A is the amplitude and $f = 1/T = \omega/(2\pi)$ the frequency for period T.) The collision interactions stem from a collision matrix described in Ref. [8] which physical foundations can be found in Refs [16]. The collision parameters are a frontal restitution coefficient r, a tangential restitution coefficient β (with a maximal value β_o) and a friction coefficient ε. To avoid as much as possible the inelastic collapse [17], the frontal restitution coefficient is taken to decrease with velocity: $\varepsilon(u) = 1 - \varepsilon_0 u^{1/5}$ with the relative velocity in the normal direction u. A dissipation cut-off is introduced for small impact velocities (for $u < u_0 = 10^{-6}\mathrm{ms}^{-1}, \varepsilon = 1$). We use $\varepsilon_0 = 0.4$ for particle-particle interactions (for bead-plate collisions we set $\varepsilon = 1$). The other physical parameters are $\beta_0 = 0.0$ and $\mu = 0.2$. This choice is made to get as close as possible to the values for aluminum beads used in the experiment of Clément et al. [7]. Also, we checked that this choice of parameters is non-critical as long as there is (i) enough dissipation to avoid fluidization of the layer [18], (ii) a dissipation cut-off to avoid inelastic collapse [17], (iii) some friction between the beads and with the bottom plate, in order to stabilize the patterns. The N spheres of diameter d are initially packed in the cells with horizontal width L, the layer thickness is defined as $H = \sqrt{\frac{3}{2}} N_h d$ with $N_h = N/(L/d)$. The algorithm efficiency is improved by the implementation of a time delayed procedure applied to the search of the event sequence [19]. Such a procedure reduces the computing time to $\mathcal{O}(N \log N)$ instead of $\mathcal{O}(N^2)$ for a standard event-driven algorithm[18]. For a relative acceleration $\Gamma = A\omega^2/g$ situated in a moderate range beyond the threshold $\Gamma \simeq 2.5$ up to $\Gamma = 4$, an instability occurs and a stationary pattern is obtained with a wave length λ roughly constant (an increase of a few percent can sometimes be evidenced). The instability stops for values of the acceleration around $\Gamma = 4.2$, due to a well known problem of matching between the downwards velocity of the plate and the velocity of the falling layer. The instability is resumed for larger accelerations but with a free flight of the layer lasting more than one period (around $\Gamma = 6$). The impact frequency is then one half of the vibration frequency. The pattern is made of peaks such that minima and maxima exchange positions at each period of excitation (see Fig. 1). As already noticed in Refs. [3] and [7], two important phases for the layer response can be considered: the free flight phase (lasting about $T/2$) where the peak pattern is forming and the energy input phase, where the plate is in contact with the layer and provides energy into the system. In general, the peak zone collides at a phase slightly delayed with respect to the minimum zone. This is due to the general presence of an arch at the bottom of the layer as it was observed experimentally [7] and numerically [8].

In Fig. 2, we represent the pressure and the density fields superposed for two different frequencies. We observed that the arch amplitude as well as the peak amplitude decrease when the frequency is increased. The later impacts of the higher density regions are very energetic in the sense that the velocity fluctuations become large. The general reason for this high pressure impact is both due to the higher compaction state inside the peaks and to a larger number of particles in the vertical direction (peak amplitude), hence

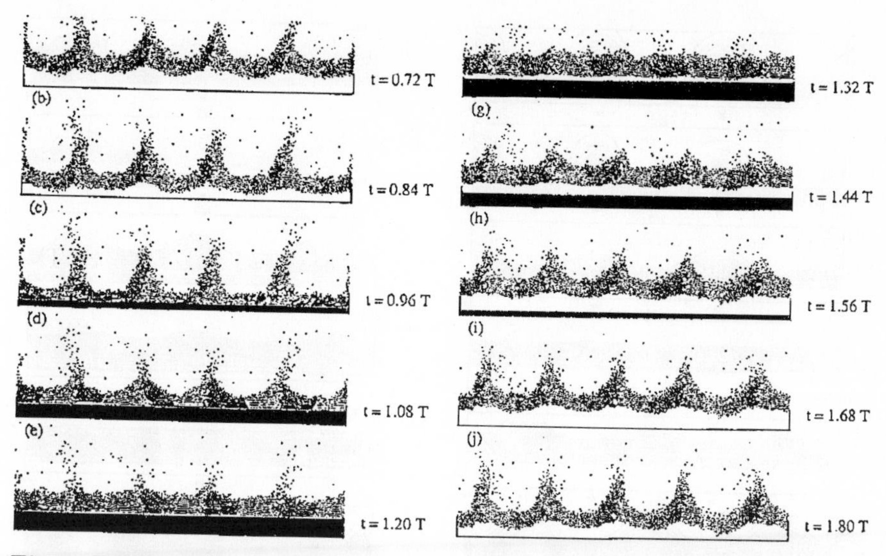

Fig. 1. Display of the vibrated layer during two vibration periods. Dark and light particles have a horizontal velocity to the right and to the left, respectively. The simulation parameters are $N_h = 12$, $f = 15\,\mathrm{Hz}$, $d = 1\,\mathrm{mm}$ and $\Gamma = 3.6$.

creating an important momentum flux on the plate. These regions of high pressure transfer large horizontal momentum to regions of low pressure and, therefore, two horizontal energetic masses flow from the former peak positions and collide head-on at the place where a dip was formerly present. Due to the presence of the bottom plate, this collision results in a net momentum flux in the upward direction. The spatial distribution of extra upward momentum will mark the place for a new peak when the layer leaves the plate again. We see that this non-uniform driving – in time and in space – is fundamentally different from the driving mechanism in fluids (namely a *uniform* acceleration modulation). We also noticed that just before the free flight regime, regions where the former peaks were present and where the pressure is high, are still in contact with the plate and subsequently will fall down with a small initial velocity, while the regions with a high upward momentum flux have at this instant almost no mechanical contact with the plate. This velocity difference contributes to the peak formation as well as the distortion of the layer thus creating the arches. Now, we investigated the wavelength selected for $\Gamma = 3.6$ where the patterns are fully developed. The wavelength of the pattern, $\lambda = 2\pi/k$, is monitored using the horizontal density auto-correlation function technique described in Ref. [8], which marks the presence of the peaks. In Fig. 3(a), we present the simulation results for $N_h = 6$ and 12. The quantity $\omega^2/4gk$ is plotted as a function of $Hk = 2\pi H/\lambda$. In analogy to fluids, we estimate this representation to be a rather natural one since for the Faraday instability, corresponding to the parametric excitation of grav-

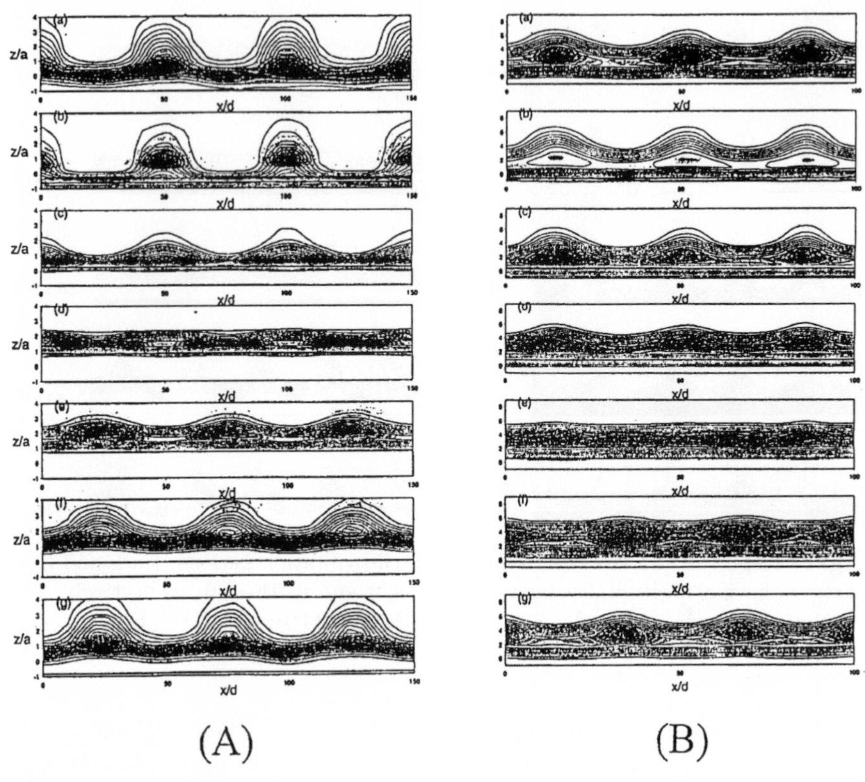

Fig. 2. Displayed are the pressure fields (iso-lines) and the density fields (gray scale) for a time series during one period. The simulation parameters are $d = 1\,\mathrm{mm}$ and $\Gamma = 3.6$. (A): $N_h = 6$, $f = 10\,\mathrm{Hz}$ such that $Hk = 0.68$; (B): $N_h = 12$, $f = 15\,\mathrm{Hz}$ such that $Hk = 1.86$.

ity waves, one would get the dispersion relation (at the edge of instability): $\omega^2/4gk = \tanh(Hk)$ [20]. This plot separates between a large channel regime where $Hk \gg 1$ and a shallow channel regime where $Hk \ll 1$. We observe, for the ganular layer, two different regimes: (i) a dispersive regime (we mean here that the wavelength depends on the impact frequency) at low frequencies and the wavelength selected is surprisingly close to the dispersion relation for gravity waves in a fluid and (ii) a saturation regime at larger frequencies with a cross-over depending on the layer height. Now, both regimes are examined independently and computations were made in a large range of bead sizes, frequencies, layer heights and accelerations. The dispersion regime is presented in Fig. 3(b) and the saturation regime in Fig. 3(c). For the dispersive regime (lower frequencies), we observe a collapse of the whole data set around a straight slope bounded by two extreme values: $0.3 < Hk < 2.8$ and $0.4 < \omega_c^2/4gk < 1.5$. The data collapse is interrupted at high frequencies by the wavelength saturation: $\lambda_{sat}\,(N_h, d)$ reported in Fig. 3(c). The empirical

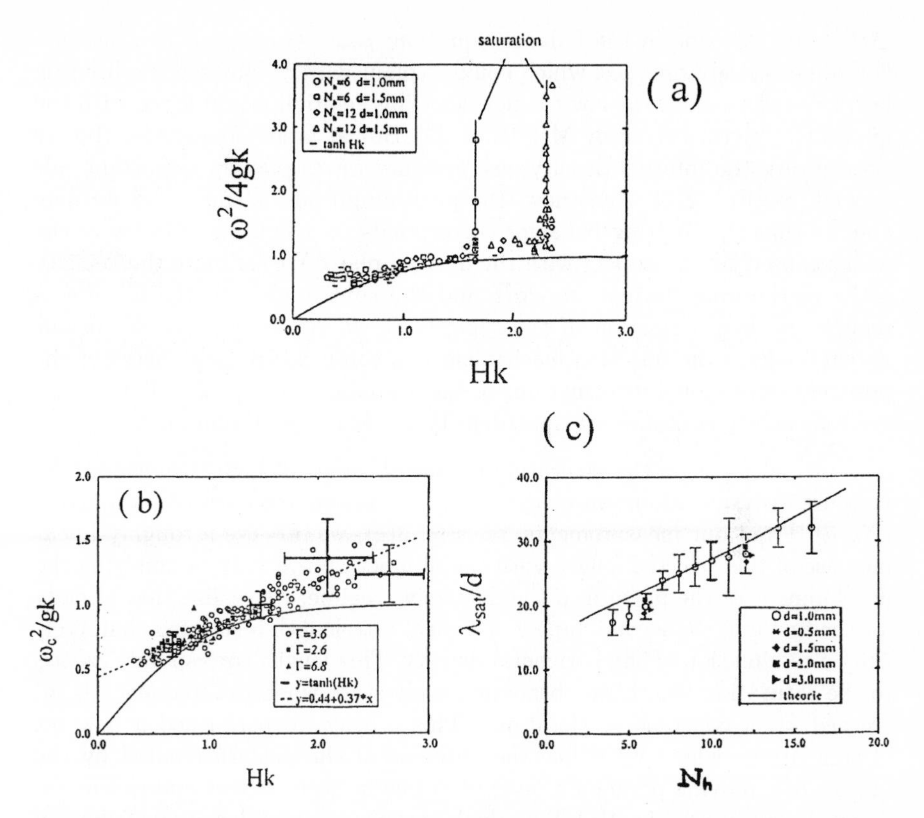

Fig. 3. Selected wavelength $\lambda = 2\pi/k$ displayed in the form $\omega_c^2/4gk$ as a function of Hk; $\omega_c/2\pi$ is the impact frequency. The continuous line in (a) and (b) are: tanh Hk. **(a)**: plot for: $N_h = 6$, $d = 1\,\text{mm}$ (○); $N_h = 6$, $d = 1.5\,\text{mm}$ (□); $N_h = 12$, $d = 1\,\text{mm}$ (◇); $N_h = 12$, $d = 1.5\,\text{mm}$ (△); **(b)**: dispersive part for $\Gamma = 3.6$(○); $\Gamma = 2.6$(■); $\Gamma = 6.8$(▲) (for many layer heights and bead diameters), the dotted line is the empirical linear best fit. **(c)**: saturation part λ_{sat}/d as a function of the number of layers N_h, for $d = 0.5\,\text{mm}(*)$, $d = 1\,\text{mm}$ (○), $d = 1.5\,\text{mm}$ (◆), $d = 2\,\text{mm}$ (○), $d = 3\,\text{mm}$ (▲); the straight line is the theoretical prediction.

best fit of the data is the straight line of Fig. 3(b): $\omega_c^2/4gk = A + BHk$, with $A \simeq 0.44$ and $B \simeq 0.37$. We verified that these results are in good agreement with the wavelength selection observed experimentally in 2D [7] and in 3D [3, 9] (within the data scatter of course). Note that the presence of a restoring mechanism due to gravity (peaks are collapsing on the plate) is consistent with a standard mechanical picture where the average momentum density or the mass fluxes transferred during the energy input phase ($\approx \rho V_{impact}/T$) are driven by a pressure difference on the scale of a wavelength ($\approx \Delta P/\lambda$). If we estimate that the pressure difference in this regime scales with the peak amplitude $p \simeq 4a$ (see Ref. [7] for an experimental determination), i.e.

$\Delta P \approx \rho g p$, we obtain the balance equation: $\rho a \omega^2 \approx \rho g a/\lambda$. This relation is a dimensional argument which could explain why we observe the limiting law: $\omega^2/4gk \to const.$ at low frequencies. This relation would agree with the qualitative picture given by Melo et al. [3]. But at larger frequencies (before saturation), the internal density and pressure waves play an important role as well, see Fig. 2(b), since now, the peak amplitude is small and we may assume that the limiting behavior corresponds to a limiting velocity of the waves caused by the impact with the bottom plate. We estimate the velocity to be of the magnitude: $c \approx \sqrt{gH}$ and the contribution to the dispersion relation is then a relation of the type: $\omega = ck$. As a consequence, in our understanding, the selection mechanism is a complex interplay between the possibilities of global deformations of the granular layer (peak collapse) and the internal dynamics of pressure/density waves due to the impact.

Now, we focus on the saturating regime obtained at high frequencies. We measure the saturation wavelength λ_{sat} for various couples of parameters (N_h, d). From our measurements, see Fig. 3(c), we observe a roughly linear increase of this selected length with the number of layers. By monitoring the development of the pattern in the dispersive regime, we realize that already for the first and the second impact, a typical wavelength is selected characterizing a modulation of the horizontal density. This length corresponds *exactly* to the saturation length we obtain at steady-state when the frequency is increased (at a constant acceleration). This is why we performed a new set of simulations where we follow the outcome of the pattern created by the impact of a moving plate on a layer of grains initially at rest and in the absence of gravity, see Fig. 4(a). We calculate the horizontal density distribution $\sigma(x) = \int \rho(x,z)\,dz$. The power spectrum of this distribution $S_k =< \widetilde{\sigma}_k \widetilde{\sigma}_k^* >$, is monitored as a function of time in Fig. 4(b). We observe a band of unstable modes with a fastest growing wavelength characterized by a wave number: $k_s = 2\pi/\lambda_s$. In Fig. 4(c), we display this selected wave length λ_s as a function of the saturation wavelength for various sets of experiments $\lambda_{sat}(N_h, d)$. We observe that $\lambda_s \simeq \lambda_{sat}(N_h, d)$. Now we relate this early pattern selection to the general issue of a granular stripe instability. The dynamics of an impacted layer of dissipative grains was studied in detail [21]. We identified two important phases as a result of the impact: (i) just after the impact, an upwards compression wave and a downwards dilation wave cross vertically the layer at very high speed. These waves do not cause global distortion of the layer but are extremely dissipative; (ii) after the passage of those waves an expansion of the layer follows, characterized by a vertical velocity gradient and subsequently, the layer looses contact with the plate. During the expansion phase, the layer is unstable and a wavelength characterizing a density modulation in the horizontal direction shows up.

In Fig. 5(a), we represent the value of this wavelength for different *constant* coefficients of restitution ε (now ε is independent of the collision velocities). We can separate between two limiting regimes such that :

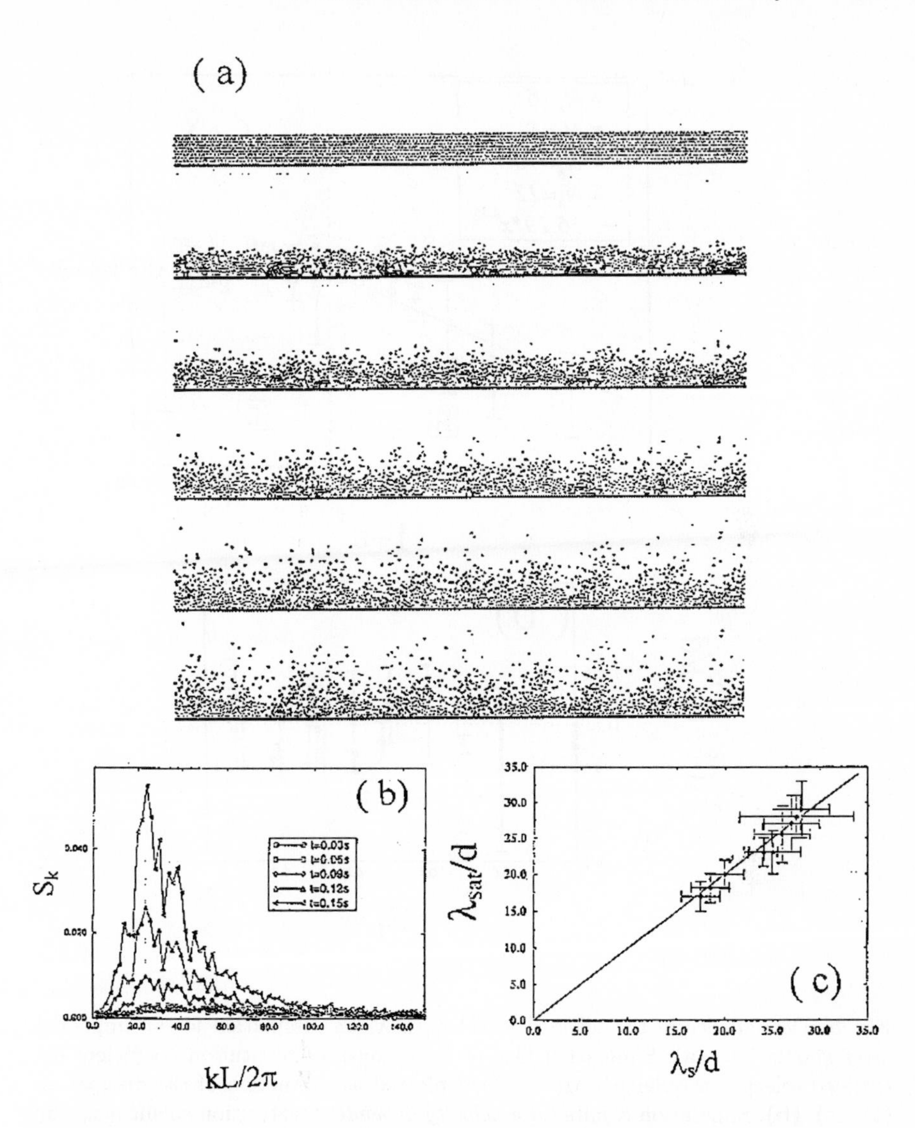

Fig. 4. Pattern formation in a granular layer impacted from the bottom (without gravity). **(a)**: Evolution of the grains after the impact. Dark particles have a horizontal velocity to the right and light ones move to the left. **(b)**: Time evolution of the structure factor $S(k)$ after the impact. **(c)**: Comparison between the selected wavelength λ_s and the saturation wave length λ_{sat} for different layer heights N_h. The straight line is $\lambda_{sat} = \lambda_s$.

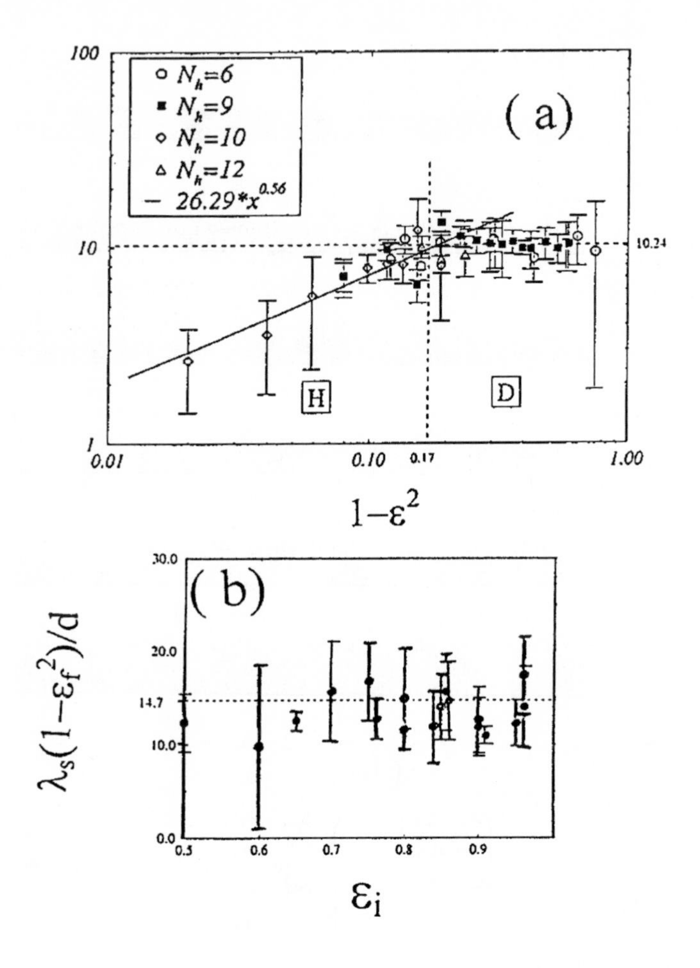

Fig. 5. Dependence of the selected wavelength λ_s on dissipation for an impacted layer simulation. **(a)**: Simulation results for a *constant* restitution coefficient ε ; rescaled selected wavelength $\lambda_s(1-\varepsilon^2)/d$ plotted as a function of the dissipation $(1-\varepsilon^2)$. **(b)**: Simulation results for a *velocity dependent* restitution coefficient; the rescaled wavelength $\lambda_s(1-\varepsilon_f^2)/d$ is plotted as a function of the initial restitution ε_i.

$$\lambda_s/d \approx 1/\left(1-\varepsilon^2\right)^\beta \tag{1}$$

where $\beta \simeq 1$ in the strong dissipation regime (i.e. $1-\varepsilon = \mathcal{O}(\infty)$) and $\beta \simeq 0.5$ in the weak dissipation limit (i.e. $1-\varepsilon \to 0$). Note that such scaling relations between a structural length scale and a restitution coefficient were already identified in the case of freely evolving granular gases. The scaling with $\beta = 1$ is naturally occuring in the formation of 1D structures and clusters with long lasting multiple contacts [17] and the scaling with $\beta \simeq 0.5$ is naturally occurring at the onset of a structural instability in weakly dissipative granular gases described by a dissipative hydrodynamics [22]. Importantly, these regimes show a selected wavelength value λ_s, which is *independent* of the number of layers N_h. Now, we do the same study but with a restitution coefficient depending on the collision velocity. We have in the early stage an impact with a layer moving at a constant velocity: U_0 and, therefore, a typical initial restitution coefficient : $\varepsilon_i = 1 - \varepsilon_0 \left(U_0\right)^{1/5}$. But after a large number of impacts, the average kinetic energy of the layer is decreased and we get a typical final restitution coefficient : $\varepsilon_f = 1 - \varepsilon_0 \left(U_f\right)^{1/5}$ characterized by a typical collision velocity U_f just before the expansion phase. A systematic study[21] of the scaling behavior of the initial temperature T before the expansion phase shows that we have

$$T^{1/2} \approx \frac{U_0}{N_h} \exp\left[-\xi\left(N_h-1\right)(1-\varepsilon)\right], \tag{2}$$

with $\xi \simeq 1.1$. Therefore, using the relation $T = U_f^2$, we estimate the final restitution coefficient with the relation

$$\frac{1-\varepsilon_f}{1-\varepsilon_i} = \left(U_f/U_0\right)^\alpha \tag{3}$$

For our simulations we use $\alpha = 1/5$. In Fig. 5(b) we plot the rescaled wavelength $\lambda_s\left(1-\varepsilon_f^2\right)^{1/2}/d$ as a function of the initial restitution coefficient ε_i. Then we estimate

$$\lambda_s/d = \frac{C}{\left(1-\varepsilon_f^2\right)^\beta} \tag{4}$$

with $\beta \simeq 0.5$ and $C \simeq 15$. The $\beta = 1/2$ exponent indicates that the selection mechanism for the wavelength rather corresponds to the weak dissipation limit we identified previously (though the dissipation was quite high initially). As a consequence, putting together Equations (2), (3), and (4) we get a mean-field theoretical prediction for the saturation wave length which is displayed in Fig. 3(c). In the limit of $\varepsilon_i \to 1$ we obtain an asymptotic formula for the selected wavelength for any α and N_h

$$\lambda_s/d \approx \frac{1}{(1-\varepsilon_i)^{1/2}} N_h^{\alpha/2} \exp\left[\frac{\alpha}{2}\xi\,(N_h-1)\,(1-\varepsilon_i)\right]. \tag{5}$$

We realize that the exponential growth of the selected length as a function the number of layers is damped by the weak value of the coefficient α characterizing the velocity dependence of the dissipation. Thus, both antagonistic effects give an approximately linear increase of the wavelength. As a consequence, the presence of an intrinsic instability due to the dissipative character of the granular layer prevents the selected wavelength to decrease (when the frequency is increased) at a value smaller than the intrinsic dissipative length. The vibrations are only here to sustain the motion due to this imposed wavelength and drive a dynamics such that larger densities in the layer create larger agitations and thus larger pressures. As a consequence, an horizontal flow develops towards lower pressure regions and the alternative horizontal motion is sustained at the pace of the vertical impacts with the plate. The dependence of the saturated wavelength with the number of layers is related to the dissipation properties of the grains which depend on the collision velocities between two grains.

In conclusion, we presented some results of a numerical study on a pattern forming instability occuring in a 2D vibrated layer of dissipative grains. We focus on mechanisms leading the formation of stationary oscillating surface peaks which are separated by a well defined wavelength. We identify two distinct regimes. The first regime (dispersive) corresponds to a periodic excitation of the layer where the gravity restoring force plays an important role in competition with internal density and pressure waves created by repeated impacts with the bottom plate. The dispersion relation is such that we have in general a relation of the type: $\omega_c^2/4gk = O(1)$ with a value smaller for thin layers and larger for thick layers (sizes beeing compared to the selected wavelength). At larger frequencies (acceleration being constant) this dispersion relation is interrupted by a saturation regime where the wavelength is now independent of the frequency. We show how this new wavelength selection mechanism relates to an instability occurring spontaneously in dissipative gases. We also stress on the influence of the detailed microscopic dissipation laws affecting the values of the selected wavelength.

Acknowledgements

We are grateful to S. Luding for many interesting discussions.

References

1. H. M. Jaeger, S. R. Nagel and R. P. Behringer, Rev. Mod. Phys. **68**, 1259 (1996).
2. E. Clément, "Granular Packing under Vibration" in *Physics of Dry Granular Media*, H. J. Herrmann, J.-P. Hovi and S. Luding, Kluwer Acad. Publisher, Dordrecht, Holland (1998) p.585.
3. F. Melo, P. Umbanhowar and H. Swinney, Phys. Rev. Lett. **72**, 172 (1994); ibid **75**, 3838 (1995).
4. P. Umbanhowar, F. Melo and H. Swinney, Nature **382**, 793 (1996).
5. M. Faraday, Philos. Trans. R. Soc. **121**, 299 (1831).
6. S. Fauve, in *Dynamics of non-linear and disordered systems*, G. Martinez-Mekler and T. H. Seligman, World Scientific, Singapore, (1995) p.67.
7. E. Clément, L.Vanel, J. Duran and J. Rajchenbach, Phys. Rev. E **53**, 2972 (1996).
8. S. Luding, E. Clément, J. Rajchenbach and J. Duran, Europhys. Lett, **36**, 247 (1996).
9. C. Bizon, M. D. Shattuck, J. B.Swift, W. D. McCormick and H. Swinney, Phys. Rev. Lett. **80**, 57 (1998).
10. K. M. Aoki and T. Akiyama, Phys. Rev. Lett. **77**, 4166 (1996).
11. T. Shinbrot, Nature (London) **389**, 574 (1997).
12. L. S. Tsimring and I. S. Aronson, Phys. Rev. Lett. **79**, 213 (1997).
13. D. H. Rothman, Phys. Rev. E. **57**, 1239 (1998).
14. E. Cerda, F. Melo and S. Rica, Phys. Rev. Lett.**79**, 4570 (1997).
15. J. Eggers and H. Riecke, Phys. Rev. E **59**, 4476 (1999).
16. W. Goldsmith, *Impact, the Theory and Physical Behavior of Colliding Solids* (Edward Arnorl, London, 1960); O. Walton et al., J. Rheol. **30**, 949 (1983). S. F. Foerster et al., Phys. Fluids **6**, 1108 (1994). L. Labous et al. Phys. Rev. E **56**, 5717 (1997).
17. B. Bernu and R. Mazighi, J. Phys. A **23**, 5745 (1990); S. McNamara, W. R. Young, Phys. Fluids A **5**, 34 (1993).
18. S. Luding, H. J. Herrmann and A. Blumen, Phys. Rev. E **50**, 3100 (1994).
19. B. D. Lubachevsky, J. Comp. Phys. **94**, 255 (1991).
20. L. D. Landau and E. Lifschitz, Fluid Mechanics (Pergamon Press, London, 1963).
21. L. Labous, Thèse de doctorat, Université de Paris VI (1998).
22. I. Goldhirsch and G. Zanetti, Phys. Rev. Lett. **70**, 1619 (1993); S. McNamara and W. R. Young, Phys. Rev.E **50**, R28 (1994).

Experimental Study of a Granular Gas Fluidized by Vibrations

Éric Falcon[1], Stéphan Fauve[1], and Claude Laroche[2]

[1] Laboratoire de Physique Statistique, École Normale Supérieure, 24, rue Lhomond, 75231 Paris Cedex 05, France. e-mail: Eric.Falcon@lps.ens.fr
[2] Laboratoire de Physique, École Normale Supérieure de Lyon, 46 allée d'Italie, 69364 Lyon Cedex 07, France

Abstract. We report experimental results on the behavior of an ensemble of inelastically colliding particles, excited by a vibrated piston in a vertical cylinder. When the particle number is increased, we observe a transition from a regime where the particles have erratic motions (granular "gas") to a collective behavior where all the particles bounce like a nearly solid body. In the gaslike regime, we measure the pressure at constant volume, and the bed expansion at constant external pressure, as a function of the number N of particles. We also measure the density of particles as a function of the altitude, and find that the "atmosphere" is exponential far enough from the piston. From these three independent measurements, we determine a "state equation" between pressure, volume, particle number and the vibration amplitude and frequency.

1 Introduction

Once fluidized a vibrated granular medium looks like a gas of particles that can be described using kinetic theory of usual gases. However, granular gases basically differ from ordinary ones mostly due to the inelasticity of collisions, i.e., nonconservation of energy. While over the years many attempts based on kinetic theory [1] have been made to describe such dissipative granular gases, no agreement has been found so far both with experiments [2, 3] and numerical simulations [3–6], for the dependence of the "granular temperature", i.e., the mean kinetic energy per particle, on the parameters of vibration [7–10]. The aim of this study is to guess possible gas-like state equations for such a dissipative granular gas and to observe new kinetic behaviors which trace back to the inelasticity of collisions, e.g., the tendency of such media to form clusters. Although this feature has probably been known since the early observation of planetary rings [11] and although various cluster types in granular flows have been observed numerically [12], there exist only a few recent laboratory experiments. One experiment, with a horizontally shaken 2–D layer of particles, displayed a cluster formation, but the coherent friction force acting on all the particles was far from being negligible [13]. We performed a similar experiment by exciting a 3–D granular medium with a vertically vibrated piston. We did observe clustering, but we could not rule out a lock-in mechanism involving the time scale connected with gravity and the period of vibration [14]. We thus repeated this experiment in a low-gravity environment, where inelastic collisions were the only interaction mechanism, and observed a motionless dense cluster that confirms that the inelasticity of collisions alone can generate clustering [15].

This paper is devoted to the study of the low-density situation, where clustering does not occur. It is organized as follows. Section 2 is devoted to the presentation of our experimental setup. The experimental results are presented in Secs. 3–5. We report in Sec. 3 (resp. Sec. 4) the measurements of pressure (resp. volume) of a gas of spherical particles excited by a vibrating piston and undergoing inelastic collisions. At constant external driving, we show that the pressure passes through a maximum for a critical number of particles before decreasing for large N. The density of particles as a function of the altitude is studied in Sec. 5, where we observe an exponential atmosphere far enough from the piston. From measurements of Secs. 3–5, we show in Sec. 6 that the dependence of the "granular temperature", T, on the piston velocity, V, is of the form $T \propto V^{\theta}$, where θ is a decreasing function of N. Finally, we discuss our results in the light of previous works [2–10] and give our conclusions in Sec. 7.

2 Experimental Setup

The experiment consists of a transparent cylindrical tube, with an inner diameter of 60 mm, filled from 20 up to 2640 stainless steel spheres, 2 mm in diameter, roughly corresponding to 0 up to about 5 particle layers at rest. An electrical motor, with eccentric transformer from rotational to translational motion, drives the particles sinusoidally with a 25 or 40 mm amplitude, A, in the frequency range from 6 to 20 Hz. A lid in the upper part of the cylinder, is either fixed at a given height, h (constant–volume experiment) or is mobile and stabilized at a given height h_m due to the bead collisions (constant–pressure experiment). The heights h and h_m are defined from the lower piston at full stroke.

When the vibration is strong enough and the number of particles is low enough, the particles display ballistic motion between successive collisions like molecules in a gas, see Fig. 1(a). When the density of the medium is increased, the gaslike state is no longer stable but displays the formation of a dense cluster bouncing like a nearly solid body, see Fig. 1(b). This paper is devoted to the study of the gaslike state. Its aim is to determine experimentally a "state equation" between pressure, volume, particle number and the vibration parameters (amplitude and frequency).

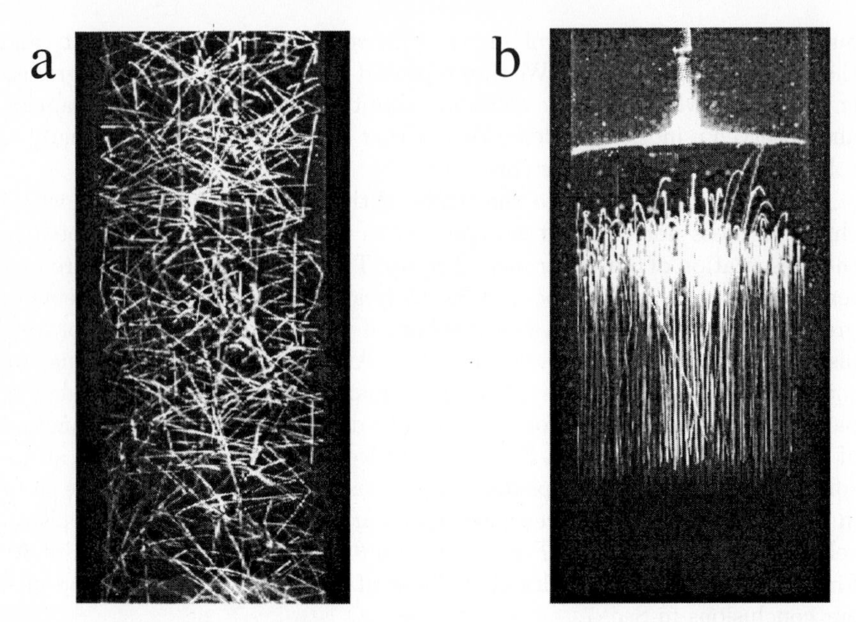

Fig. 1. Transition from a dissipative granular gas to a dense cluster: (a) $N = 480$; (b) $N = 1920$, respectively corresponding to roughly 1 and 4 particle layers at rest. The parameters of vibration are $f = 20$ Hz and $A = 40$ mm. The driving piston is at the bottom (not visible), the inner diameter of the tube being 52 mm

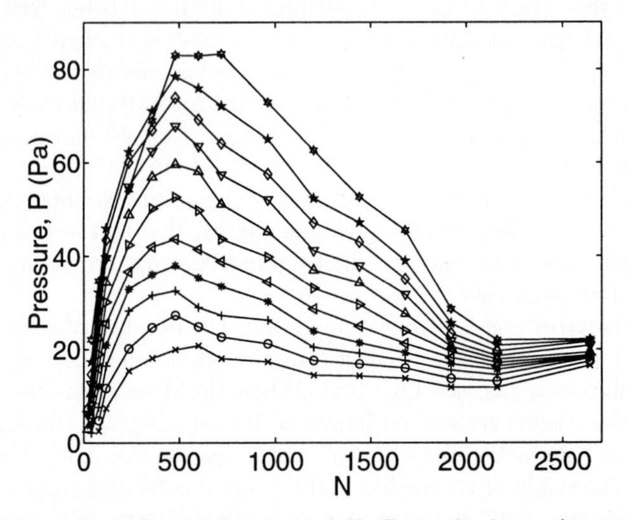

Fig. 2. Mean pressure P as a function of N. From the lower (×–marks) to the upper (hexagrams) curve, vibration frequency f varies from 10 to 20 Hz with a 1 Hz step. For all these experiments, $h - h_0 = 5$ mm and $A = 25$ mm, h_0 being the bed height at rest. One single layer of particles at rest corresponds to $N = 600$. Lines join the data points

3 Pressure Measurements

Time averaged pressure measurements have been performed as follows. Initially, a counterwheight of mass 46 g balances the lid mass. The piston drives stainless steel spheres in erratic motions in all directions, see Fig. 1(a). Particles are hitting the lid all the time, so that to keep it at a given height h we have to hold the lid down by a given force, Mg, where M is the mass of a weight we place on the lid and g the acceleration of gravity. At a fixed h, *i.e.* at a constant–volume, Fig. 2 shows the time averaged pressure $P = Mg/S$ exerted on the lid as a function of the number N of beads in the container, for different frequencies of vibration, S being the area of the tube cross-section. At constant external driving, *i.e.* at fixed f and A, the pressure passes through a maximum for a critical value of N roughly corresponding to 0.8 particle layers at rest. This critical number is independent of the vibration frequency. A further increase of the number of particles leads to a decrease in the mean pressure since more and more energy is dissipated by inelastic collisions. Note that gravity has a small effect in these measurements that are performed for $V^2 \gg gh$, where $V = 2\pi fA$ is the maximum velocity of the piston. For N such that one has less than one particle layer at rest, most particles perform vertical ballistic motion between the piston and the lid. Thus, the mean pressure increases roughly proportionally to N. When N is increased such that one has more than one particle layer at rest, interparticle collisions become more frequent. The energy dissipation is increased and thus the pressure decreases.

4 Volume Measurements

We now consider the bed expansion under the influence of collisions on a circular wire mesh lid placed on top of the beads leaving a clearance of about 0.5 mm between the edge of the lid and the tube. Due to the bead collisions, the lid is stabilized at a given height h_m from the piston at full stroke. Although the lid mass is roughly 50 times smaller than the total mass of the beads, the lid proves to be quite stable and remains horizontal. The expansion, $h_m - h_0$, of the bed is displayed in Fig. 3 as a function of N for different vibration frequencies. h_0 is the bed height at rest. At fixed f, the expansion passes through a maximum for a critical value of N roughly corresponding to 0.6 particle layers at rest. This critical number is independent of the vibration frequency. When N is further increased, the expansion decreases showing, as for pressure measurements, an increase in dissipated energy by inelastic collisions. Note that the height h_m of the granular gas is much larger than for the pressure measurements of Fig. 2. Consequently, gravity is important here.

As already found experimentally in 1–D [3] and 3–D [16, 17] and numerically [3, 18], we observe that, at N fixed, the granular medium exhibits, see

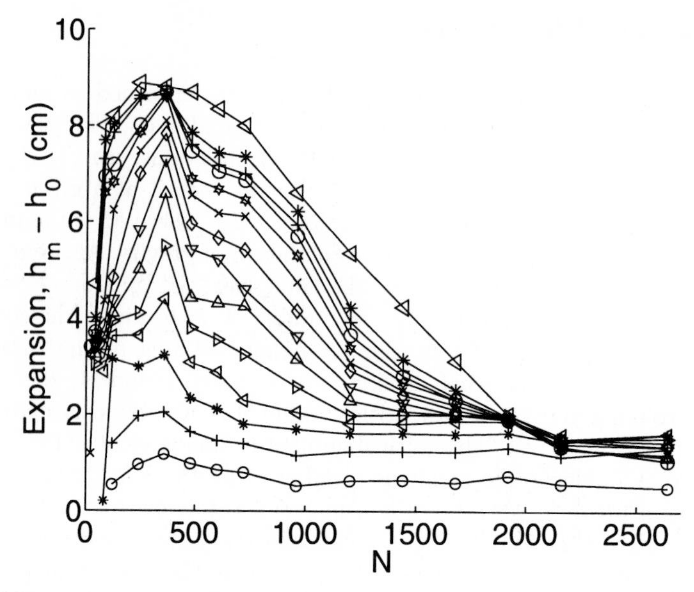

Fig. 3. Maximal bed expansion, $h_m - h_0$, as a function of N, for various frequencies f of vibration. From the lower (o–marks) to the upper (◁–marks) curve, f varies from 7 to 20 Hz with a 1 Hz step and $A = 25$ mm. One single layer of particles at rest corresponds to $N = 600$. Lines join the data points

Fig. 4(a), a sudden expansion at a critical frequency corresponding to a bifurcation similar to that exhibited by a single ball bouncing on a vibrating plate [3, 16]. Moreover, Fig. 4(a) shows that this critical frequency depends on the number of layers, n. When n increases above 0.4, a transition from the 1–D–like behavior to a 3–D one is observed: the expansion at the critical frequency becomes less abrupt and tends to increase regularly with f, see Figs. 4(a)–(b).

5 Density Measurements

Time averaged density measurements at a given height z are performed by means of two closely coupled coils, $\delta z = 5$ mm in height, and 64 mm in inner diameter, the cylindrical tube now being 52 (resp. 62) mm in inner (resp. outer) diameter. An 1.5 kHz a.c. voltage is applied to the primary coil, the turns ratio of the transformer being roughly equal to 2. Steel spheres moving accross the permanent magnetic field of the primary coil, generate an inductive voltage variation accross the secondary one. This root mean square a.c. voltage ΔU is a function of the mean inductance and mutual variations which are proportional to the mean number of particles in the volume delimited by the sensor at altitude z from the piston surface at full stroke. We have calibrated the sensor with steel spheres at rest and we have

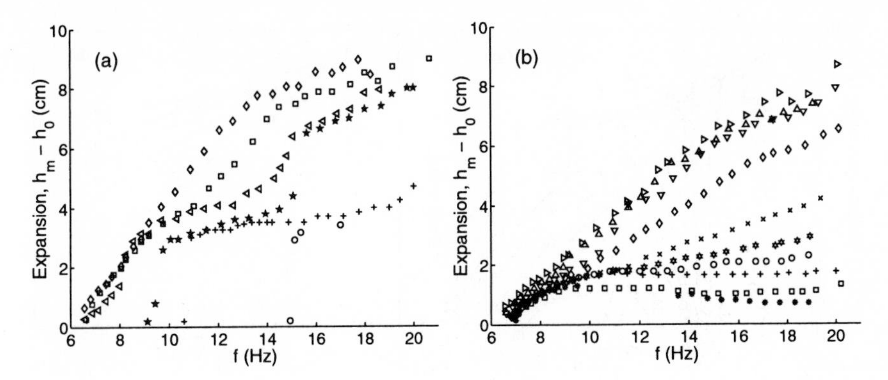

Fig. 4. Maximal bed expansion, $h_m - h_0$, as a function of f, for various numbers of particles N. (a) From the *lower* (○–marks) to the *upper* (◇–marks) curve, $N = 20$, 40, 80, 120, 240 and 360. (b) From the *upper* (▷–marks) to the *lower* (∗–marks) curve, N = 480, 600, 720, 960, 1200, 1440, 1680, 1920, 2160 and 2640. For all these experiments $A = 25$ mm. One single layer of particles at rest corresponds to $N = 600$

checked that $\Delta U \propto N$. We have also found that the effect of spheres outside of the sensor volume decays exponentially with the distance to the sensor, with a 10±1 mm decay length independent of N, for our range of N. Particles density as a function of altitude is shown in Figs. 5(a)–(c) for 3 different total numbers of particles and for various frequencies. For each N, the profile density in log–linear axes displays a decay (at low f), a plateau (at intermediate f) or a dip (at high f) near the piston and an exponential decay in the tail at high altitude, whatever f.

As for an isothermal gas, the atmosphere is found to be exponential far enough from the piston, but on very different length scales, i.e., few cm (resp. km) for our experiment (resp. for air). Such a dense upper region supported on a fluidized low-density region near the piston has been also reported numerically [18] and predicted theoretically [19]. Although the dip in the density profiles at the bottom was already observed in a 2–D granular gas experiment, non-negligible coherent friction force acting on all the particles did not allow determination of the granular temperature dependence on the piston velocity V from exponentional Boltzmann distributions fitted to tails of density profiles [2]. We can fit an exponential curve to the tail of the profile density. From the decay rate, ξ, in the fitted exponential and using kinetic theory [2], we can extract the dependence of the granular temperature T on the piston velocity. In Fig. 6, we plot $-1/\xi$ (which is proportional to T) against V with log-log-axis together with power law fits of the form $T \propto V^{\theta}$ where θ is n–dependent. Note that this power law, being observed only on a small range of velocities, one cannot rule out another functional behavior. In particular, the faster increase of T at high velocity is not significant because of imprecision on the exponential decay of the density (see Fig. 5).

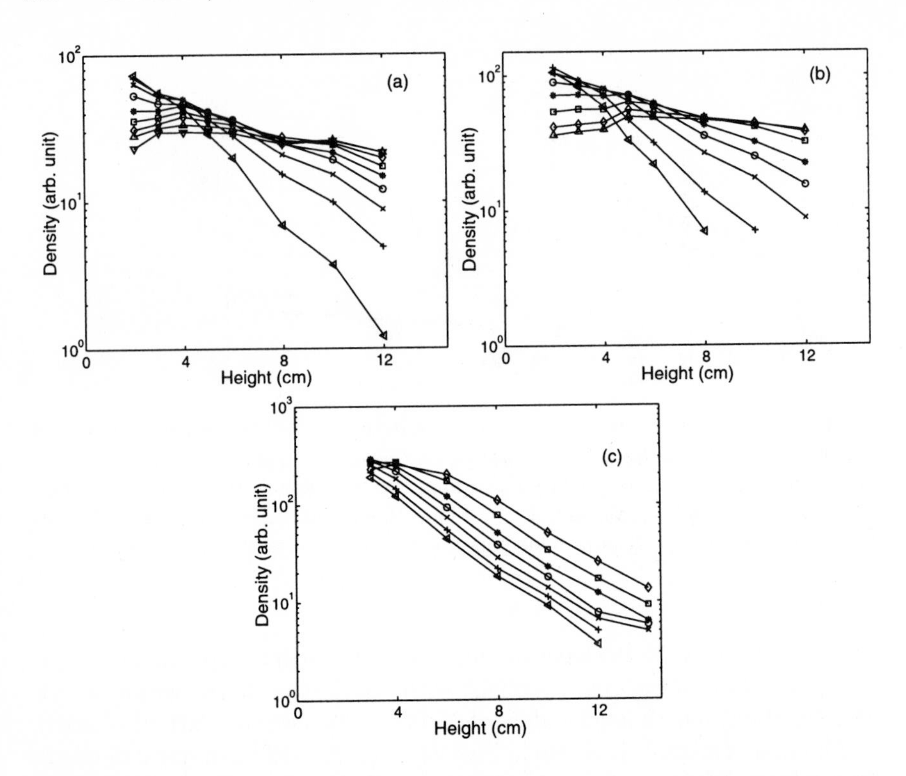

Fig. 5. Mean density as a function of the height, for various frequencies f of vibration and 3 numbers of particles (a) $N = 480$: $f = (\triangleleft)$ 5, (+) 6, (×) 7, (∘) 8, (∗) 9, (□) 10, (◇) 11, (△) 12 and (▽) 13 Hz; (b) $N = 720$: $f = (\triangleleft)$ 5, (+) 5.6, (×) 7.1, (∘) 8.6, (∗) 10.1, (□) 12.2, (◇) 14 and (△) 15.2 Hz; (c) $N = 1440$: $f = (\triangleleft)$ 7, (+) 8, (×) 10, (∘) 11.4, (∗) 12.8, (□) 15 and (◇) 17 Hz. For all these experiments, $A = 40$ mm. One single layer of particles at rest corresponds to $N = 480$. Lines join the data points

6 Towards a State Equation

In order to use the above measurements to determine a state equation, we have to find the appropriate dependence $T = T(V, n)$ of the granular temperature as a function of the vibrating velocity V and the number of particle layers n. It is known that for a fixed number n of granular layers at rest, one has in the low-density limit $P\Omega \propto T$ [20], where P is the mean pressure and Ω the volume. Taking into account this law for small densities and our observation $T \propto V^{\theta(n)}$ from density measurements in an exponential atmosphere, we have plotted $\log(P)$ and $\log(h_m - h_0)$ as functions of $\log(V)$. On the reported frequency range, these curves are straight lines, the slopes of which give $\theta(n)$. The behavior of $\theta(n)$ for the experiments at constant volume in Sec. 3 (resp. constant pressure in Sec. 4) is displayed in Fig. 7 with

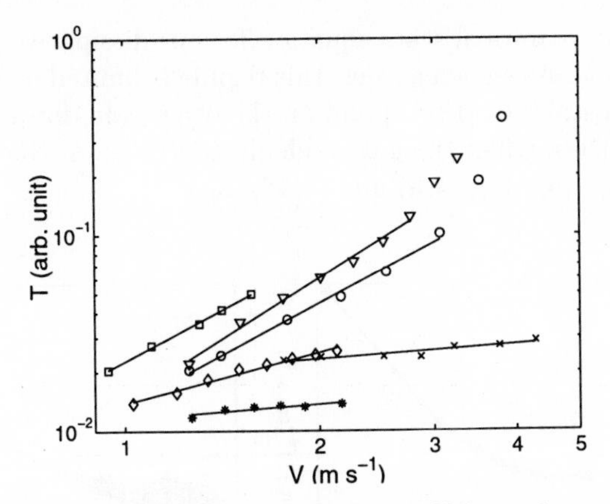

Fig. 6. Log-log plot of granular temperature versus V for various numbers of layers: ($\bigtriangledown$) 0.8, ($\square$) 1, ($\circ$) 1.2, ($\diamond$) 2, ($\times$) 2.4 and ($*$) 3. Experiments ($\square$), ($\diamond$) and ($*$) (resp. ($\bigtriangledown$), ($\circ$) and ($\times$)) are performed for $A = 25$ mm (resp. $A = 40$ mm). Power law fits of the form V^θ are dispayed in solid lines

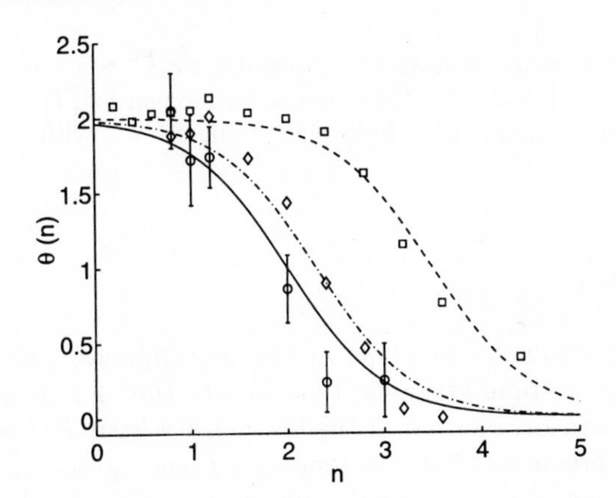

Fig. 7. Evolution of the exponent θ as a function of the number of layers, n, from pressure ($\square$), volume ($\diamond$) and density ($\circ$) measurements. Fits are $\theta(n) = 1 - \tanh(n - n_c)$ with $n_c = 2$ (*straight line*), 2.3 (*dot-dashed line*) and 3.5 (*dashed line*)

$\square$–mark (resp. $\diamond$–mark) together with the one in $\circ$–mark extracted from exponential density profiles of Sec. 5. The three curves, obtained with different experimental conditions and independent measurements have the same shape which could be simply fitted by $\theta = 1 - \tanh(n - n_c)$ where $n_c = 3.5$ (resp. 2.3) and 2.

We can now use the observed law $T \propto V^{\theta(n)}$ to scale the pressure and bed expansion measurements of Figs. 2 and 3. The results are displayed in Figs. 8(a)–(b) and show a rather good collapse of all the data on a single curve. We have thus shown that the law, $P\Omega \propto T$, together with $T \propto V^{\theta(n)}$, provide a correct empirical state equation for our dissipative granular gas in the kinetic regime. As shown earlier, this regime is limited at high density by the clustering instability [13–15] and on the other side, for a fixed too small number of particles, when the gas suddenly contracts on the piston below a critical frequency (see Fig. 4(a) and [3, 16, 18]).

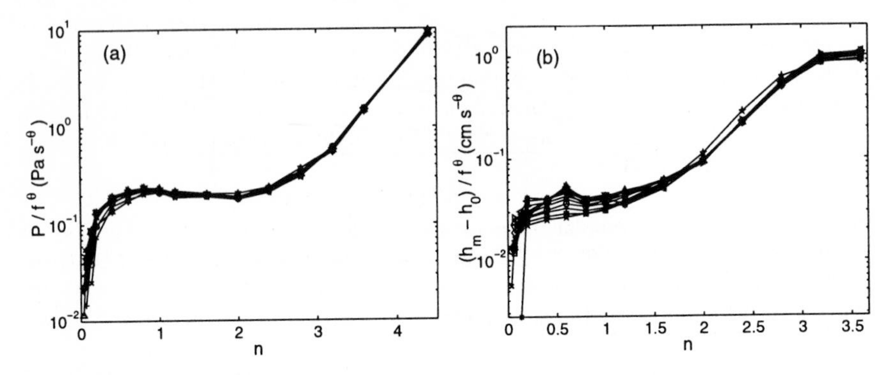

Fig. 8. (a) Mean pressure P from Fig. 2 rescaled by f^{θ} as a function of n. $\theta(n) = 1 - \tanh(n - n_c)$ with $n_c = 3.5$. (b) Maximal bed expansion, $h_m - h_0$, from Fig. 3 rescaled by f^{θ} as a function of n. $\theta(n) = 1 - \tanh(n - n_c)$ with $n_c = 2.3$

7 Conclusion

The aim of this study was an experimental determination of the state equation of dissipative granular gases. It is known that for a fixed number of granular layers at rest, one has in the low-density limit $P\Omega \propto T$ [20], where P is the mean pressure, Ω the volume and T the "granular temperature". However, the dependence of T on the vibration amplitude, A, and frequency, f, of the piston and on the number of particles, n, is still a matter of debate [7–10]. Kinetic theory [2, 9] or hydrodynamic models [7] show $T \propto V^2 n^{-1}$, whereas numerical simulations [3–6] or experiments [2, 3] give $T \propto V^{\alpha} n^{-\beta}$, with $1.3 \leq \alpha \leq 2$ and $0.3 \leq \beta \leq 1$, where $V = 2\pi f A$ is the maximum velocity of the piston. These previous conflicting results may be explained by the V–dependence of T that we have found from our measurements: $T \propto V^{\theta(n)}$, with θ continuously varying from $\theta = 2$ when $n \to 0$, as expected from kinetic theory, to $\theta \simeq 0$ for large n. We emphasize that we have only considered the dependence of the granular temperature on the vibration velocity V. For the dependence on the number of particles, another term of the form $n^{-\beta}$ should exist, that we cannot determine from these measurements.

References

1. J. T. Jenkins, S. B. Savage, A theory for the rapid flow of identical, smooth, nearly elastic, spherical particles. Fluid Mech. **130**, 187–202 (1983); C. S. Campbell, Rapid granular flows. Ann. Rev. Fluid Mech. **22**, 57–92 (1990).
2. S. Warr, J. M. Huntley, G. T. H. Jacques, Fluidization of a 2–D granular system: Experimental study and scaling behavior. Phys. Rev. E **52**, 5583–5595 (1995).
3. S. Luding, E. Clément, A. Blumen, J. Rajchenbach, J. Duran, Studies of columns of beads under external vibrations. Phys. Rev. E **49**, 1634–1646 (1994).
4. S. Luding, H. J. Herrmann, A. Blumen, Simulations of 2–D arrays of beads under external vibrations: Scaling behavior. Phys. Rev. E **50**, 3100–3108 (1994).
5. S. Luding, Granular materials under vibration: Simulations of rotating spheres. Phys. Rev. E **52**, 4442–4457 (1995).
6. H. J. Herrmann, S. Luding, Modeling granular media on the computer. Continuum Mech. Thermodyn. **10**, 189–231 (1998).
7. J. Lee, Scaling behavior of granular particles in a vibrating box. Physica A **219**, 305–326 (1995).
8. S. McNamara, S. Luding, Energy flows in vibrated granular media. Phys. Rev. E **58**, 813–822 (1998).
9. V. Kumaran, Temperature of a granular material "fluidized"by external vibrations. Phys. Rev. E **57**, 5660–5664 (1998).
10. J. M. Huntley, Scaling laws for a 2–D vibro-fluidized granular material. Phys. Rev. E **58**, 5168–5170 (1998).
11. P. Goldreich, S. Tremaine, The dynamics of planetary rings. Ann. Rev. Astron. Astrophys. **20**, 249–283 (1982).
12. M. A. Hopkins, M. Y. Louge, Inelastic microstructure in rapid granular flows of smooth disks. Phys. Fluids A **3**, 47–57 (1991); S. McNamara, W. R. Young, Inelastic collapse and clumping in a 1–D granular medium. Phys. Fluids A **4**, 496–504 (1992); I. Goldhirsch, G. Zanetti, Clustering instability in dissipative gases. Phys. Rev. Lett. **70**, 1619–1622 (1993).
13. A. Kudrolli, M. Wolpert, J. P. Gollub, Cluster formation due to collisions in granular material. Phys. Rev. Lett. **78**, 1383–1386 (1997).
14. É. Falcon, S. Fauve, C. Laroche, Cluster formation, pressure and density measurements in a granular medium fluidized by vibrations. Eur. Phys. J. B **9**, 183–186 (1999).
15. É. Falcon, R. Wunenburger, P. Évesque, S. Fauve, C. Chabot, Y. Garrabos, D. Beysens, Cluster formation in a granular medium fluidized by vibrations in low gravity. Phys. Rev. Lett. **83**, 440–443 (1999).
16. C. E. Brennen, S. Ghosh, C. R. Wassgren, Vertical oscillation of a bed of granular material. J. Appl. Mech. **63**, 156–161 (1996).
17. M. L. Hunt, S. Hsiau, K. T. Hong, J. Fluid Eng. **116**, 785–791 (1994).
18. Y. Lan, A. D. Rosato, Macroscopic behavior of vibrating beds of smooth inelastic spheres. Phys. Fluids **7**, 1818–1831 (1995).
19. D. A. Kurtze, D. C. Hong, Effect of dissipation on density profile of 1–D gas. Physica A **256**, 57–64 (1998).
20. S. McNamara (1998) Private Communication.

Computer Simulation of Vertically Vibrated Granular Layers in a Box with Sawtooth-Shaped Base

Zénó Farkas, András Vukics, and Tamás Vicsek

Department of Biological Physics, Eötvös University, Budapest, Pázmány P. Stny 1A, 1117 Hungary. e-mail: zeno@biol-phys.elte.hu

Abstract. Motivated by recent advances in the investigation of fluctuation-driven ratchets and flows in excited granular media, we have carried out event driven Molecular Dynamics simulations to explore the horizontal transport of granular particles in a vertically vibrated system whose base has a sawtooth-shaped profile. The resulting transport of the granular material exhibits novel collective behavior, both as a function of the number of layers of particles and the driving frequency; in particular, under certain conditions, increasing the layer thickness or the driving frequency leads to a *reversal of the current*. The direction of the transport may also depend on the coefficient of restitution. We give a simple explanation for the mechanism responsible for determining the direction of the granular flow.

1 Introduction

The best known and most common transport mechanisms involve gradients of external fields or chemical potentials that extend over the distance traveled by the moving objects. However, recent theoretical studies have shown that there are processes in far from equilibrium systems possessing vectorial symmetry, that can bias thermal noise type fluctuations and induce macroscopic motion on the basis of purely local effects. This mechanism is expected to be essential for the operation of molecular combustion motors responsible for many kinds of biological motion; it has also been demonstrated experimentally in simple physical systems [1, 2], indicating that it could lead to new technological developments such as nanoscale devices or novel types of particle separators. Motivated by both of these possibilities, as well as by interesting new results for flows in excited granular materials [3–8], we have carried out a series of simulational studies that explore the manner in which granular particles are *horizontally* transported by means of *vertical* vibration.

In the corresponding theoretical models – known as "thermal ratchets" – fluctuation-driven transport phenomena can be interpreted in terms of overdamped Brownian particles moving through a periodic but asymmetric, one-dimensional potential in the presence of nonequilibrium fluctuations [9–12]. Typically, a sawtooth-shaped potential is considered, and the non-linear fluctuations are represented either by additional random forces or by switching between two different potentials. Collective effects occurring during the fluctuation-driven motion have also been considered [13–15], leading to a number of unusual effects that include current reversal as a function of particle density. Here we investigate an analogous transport mechanism for granular materials. By carrying out extensive event driven simulations on granular materials vibrated vertically by a base with a sawtooth profile, it is possible to achieve a fascinating combination of two topics of considerable current interest – ratchets and granular flows. A number of recent papers have focused on vibration-driven granular flow, and the details of the resulting convection patterns have been examined, both by direct observation [4, 5, 7, 8] and by magnetic resonance imaging [6, 16]. Granular convection has also been simulated numerically by several groups; the study most closely related to the present work deals with the horizontal transport that occurs when the base is forced to vibrate in an asymmetric manner [17].

We have carried out a series of real experiments also; the experimental setup and results can be found in [18].

2 Event Driven Simulations

Simulations have been useful in the studies of granular systems [19, 20]. Here we perform event driven simulations of inelastic particles with an additional shear friction in a two-dimensional system whose base has a sawtooth-shaped profile. In event driven simulations the difficult part of the algorithm is determining the next event (e.g., the next collision), and the motion of the particles is calculated analytically between two events [23–25]. The event driven simulations represent an alternative to the more common Molecular Dynamics calculations assuming differentiable interaction potentials between the particles [26].

The particles are modeled as disks with a given radius R_i. These disks can rotate, their moment of inertia is $m_i R_i^2/2$ (m_i is the mass of the ith particle). The motion of the beads is simulated in a two-dimensional cell (see Fig. 1). To realize the topology of the experiment, periodic boundary conditions are applied in the horizontal direction. The height of the cell is chosen in such a way that the number of particles hitting the upper wall is negligible. The base is oscillating sinusoidally with frequency f and amplitude A. The geometry of the base can be seen in Fig. 2 in detail. The shape of a tooth can be described by three parameters: its height h, width w and asymmetry parameter a. This asymmetry parameter is $a = l/w$, where l is the distance of the projection of the sawtooth's top point from the left side point of the unit. The top point of a tooth is rounded by an arc which smoothly joins the two sides.

We consider collisions during which the particles can stick to or slide along each other's surface. For calculating the velocities of two particles after a collision, a simple model is used which corresponds to recent collision models [27–29] with $\beta_0 = 0$, i.e., without tangential restitution. The algorithm de-

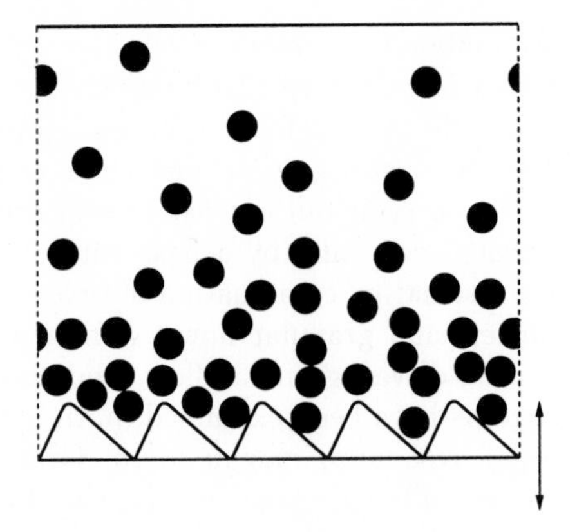

Fig. 1. Visualization of the arrangement used in the simulations. The upper wall is fixed, the sawtooth shaped base is oscillating sinusoidally. Periodic boundary condition is applied in horizontal direction.

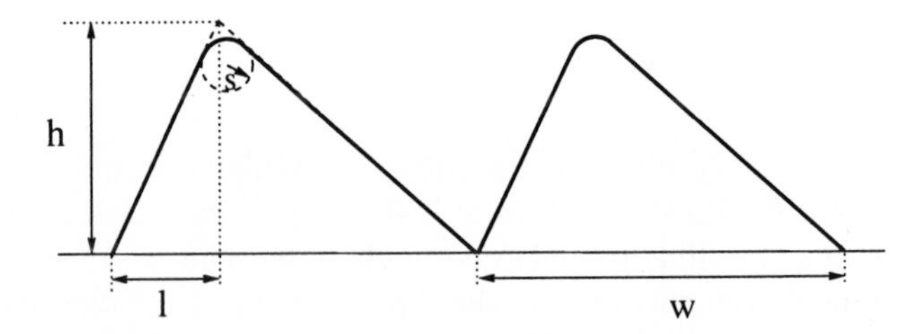

Fig. 2. The geometry of the sawtooth base. The shape of a tooth can be described by three parameters: its height h, width w, and asymmetry parameter $a = l/w$, where l is the horizontal projection of the left part of the tooth. The top of the teeth are rounded by an arc with radius s smoothly joining to both sides.

scribed in [25] has been implemented in our simulation with some extensions, e.g. gravitational field, rotating particles, sinusoidally moving objects. For details, the reader is referred to [18].

A typical problem arising in event driven granular simulations is called inelastic collapse [30]. One possibility to avoid the numerical difficulties associated with it is to make the coefficient of normal restitution (e_n) velocity dependent [30, 31]. We use a very simple rule in our simulation: e_n is constant if $v'_n > v_n^{min}$ (v'_n is the relative normal velocity of the colliding particles after the collision), however, if v'_n would be less than v_n^{min}, it is set to this minimal normal velocity: $v'_n = v_n^{min}$. Therefore, the relative normal velocity cannot be less than v_n^{min} between two objects after their collision. Using $v_n^{min} = 5$ mm/s is enough to avoid inelastic collapse, and still this extra rule does not influence the dynamics of the system, as far as the transport is concerned. We checked this by determining the proportion of collisions when this rule had to be used and checking the change in the average velocity as the threshold value was lowered. We found that in most situations the number of collisions when this rule has to be applied is negligible, and even in dense systems we did not detect a change (beyond error bars) in the transport velocity when v_n^{min} was decreased. The simulation program was written in C programming language, and it ran on personal computers with Linux operating system.

3 Results

The system we studied is a complex one, with many parameters. Since it is impossible to explore the dependence of its behavior on each of the parameters, some of them were fixed in all of the simulations. (Still there remained quite a few parameters to vary, the results shown in this section are selected from simulations with more than 5000 different parameter sets.) We fixed two parameters: the width of a sawtooth was $w = 6$ mm and the amplitude of the oscillation was $A = 2$ mm. According to previous results [32], the *direction* of the horizontal transport does not depend very much on the friction coefficient μ. It may, however, depend on the coefficient of restitution e_n, but due to the great number of other parameters, e_n was kept at fixed value of 0.8 in all parameter sets except for one series when it was varied from 0.4 to 1.0. For the sake of simplicity, both e_n and μ were chosen to be the same in particle-particle and particle-wall collisions, although the simulation allows setting different values for different types of collisions. The radii of the particles varied from 1.05 mm to 1.155 mm uniformly, to avoid the formation of a hexagonal structure which often appears in two dimensions when the system freezes. The masses were equal, hence it did not matter what their actual value was.

3.1 Dependence on the Sawtooth Shape

First we investigated how $\langle v_x \rangle$ depends on the shape of the sawtooth (other parameters were kept fixed), where the average has been taken over both time and the individual particles. We intended to find sawtooth shapes resulting in negative transport. According to the experiments, negative transport occurs only for a few particle layers ($n = 0 \ldots 3$) and, as the number of particles is increased, the direction of the transport becomes positive. We chose $n = 1$, and the frequency was $f = 25$ Hz ($\Gamma = 5.0$, where $\Gamma = (2\pi f)^2 A/g$ is the dimensionless acceleration). The width of the cell was $24w = 144$ mm, therefore one layer contained about 70 particles. These simulations lasted for 50 internal seconds, the following results represent averages of two runs for each parameter sets (with different initial conditions).

In Fig. 3 $\langle v_x \rangle$ is plotted as a function of a (the height was fixed at four different values). The first thing worth to note is that $\langle v_x \rangle$ tends to zero as the shape becomes symmetrical (i. e. a tends to 0.5). The shape of the plots depends on the height h of the sawtooth. At $h = 3$ mm the velocity is positive and decreases monotonically, while at $h = 5$ mm it is negative and has a minimum at $a \simeq 0.25$, and when the height is 9 mm, it is still negative but monotonically increasing. These curves are helpful in understanding the mechanism of negative transport.

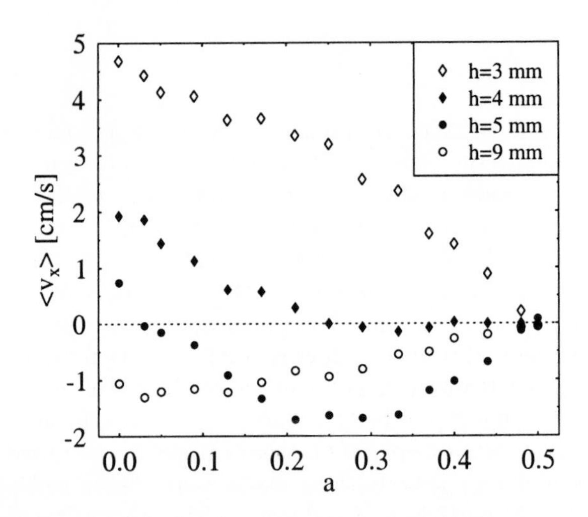

Fig. 3. The horizontal transport as a function of the asymmetry of the sawtooth. The shape of the curves depends on the height h of the sawtooth.

In Fig. 4 the dependence of the horizontal transport on the height of the sawtooth can be seen. The shapes of the graphs are similar in case of $a = 0.05$ and $a = 0.333$. $\langle v_x \rangle$ is positive if h is small, and decreases and turns to negative with increasing h. Finally, it becomes zero for high h values. More detailed investigations showed that it is true for nearly all possible values of a (except for $a \simeq 0.5$, since then the transport vanishes). In conclusion, the horizontal transport can be negative if the height of the sawtooth is between 4 and 16 mm and if the asymmetry parameter a is less then about 0.4. In other cases it is positive or zero (if the sawtooth is symmetric). It is important to note that these results are valid only for a small number of particles. As it will be shown, increasing the number of particles the negative transport vanishes and turns into positive in all cases.

3.2 Dependence on the Layer Width

The results in the previous section show that if there are only few particles, the transport can be negative or positive, depending on the shape of the sawtooth. Now we show what happens if more and more particles are added into the system.

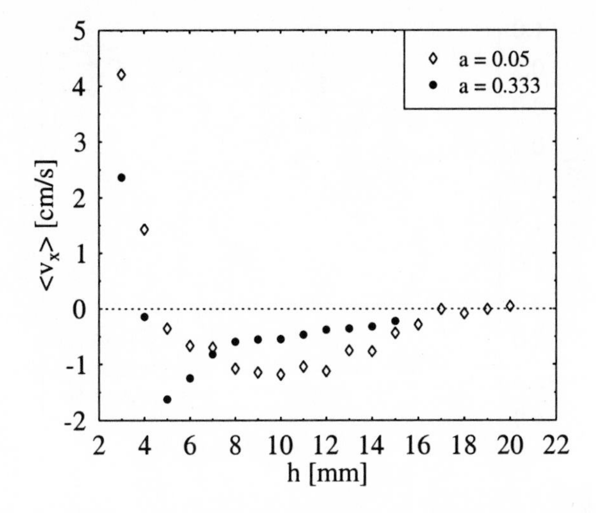

Fig. 4. Horizontal transport as a funcion of the height of the sawtooth. The shape does not depend very much on the asymmetry of the sawtooth.

In Fig. 5 one can see that negative $\langle v_x \rangle$ becomes zero and turns to positive as the layer width is increased. (Unfortunately the event-driven simulation becomes unfeasible as the density increases with increasing particle number, therefore $n \simeq 5$ is the maxium layer width we can have in our simulation.) If the sawtooth shape is such that even for a few particles the transport is positive, then with increasing particle number the transport decreases (see Fig. 6).

3.3 Frequency Dependence

Another interesting question is what happens if the frequency is varied. A sawtooth shape producing negative transport for few particles was chosen. Then with different layer widths the horizontal transport shows interesting behavior as a function of frequency. The results are shown in Fig. 7.

Generally, the negative transport becomes stronger with increasing frequency. In three cases ($n = 1.05$, $n = 1.75$, and $n = 2.8$) the transport is *reversed*, from positive to negative. This phenomenon is again very illuminating when we try to explain the mechanism of negative transport.

3.4 Dependence on the Coefficient of Restitiution

We have demonstrated that the transport properties of the granular layer in our system depend on the various parameters in a non-trivial way. In particular, increasing the layer width, through its effect on the enhanced clustering of the beads, resulted in a change of the direction of the current.

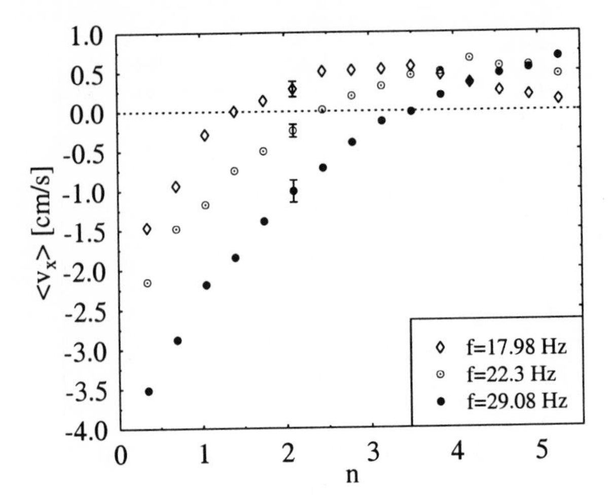

Fig. 5. The horizontal transport as a function of layer width at three different frequencies, when the shape of the sawtooth is: $h = 6$ mm and $a = 0.25$. The system width was $10w = 60$ mm, one run lasted for 30 internal seconds. The data are averaged over 7 runs.

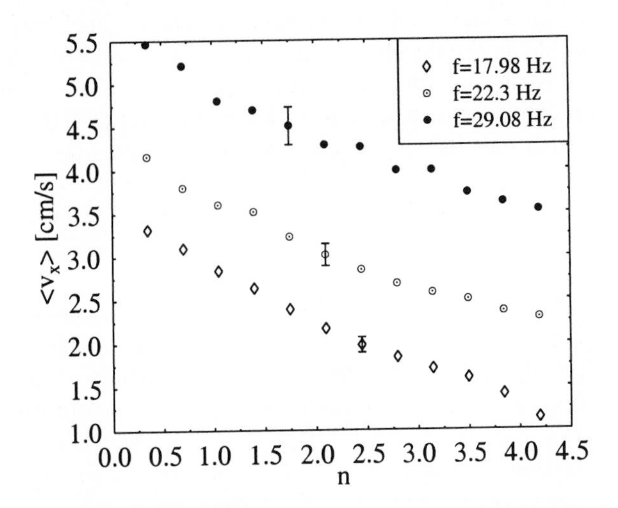

Fig. 6. The horizontal transport as a function of layer width at three different frequencies. The shape of the sawtooth: $h = 3$ mm and $a = 0.1$. The data shown here are averaged over 7 runs, one run lasted for 30 internal seconds.

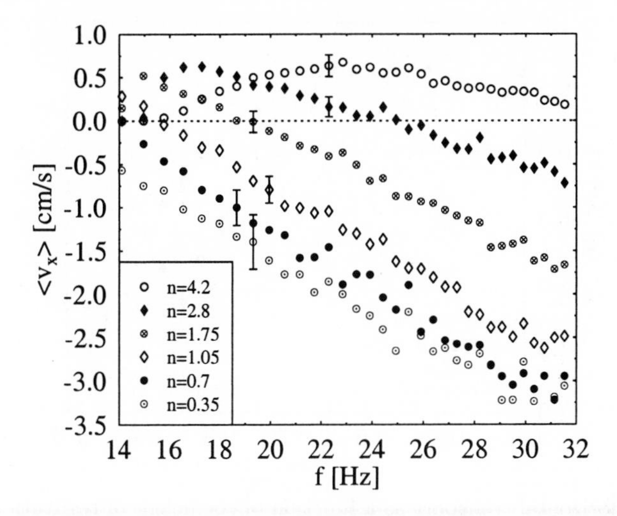

Fig. 7. Horizontal transport as a function of the driving frequency, with six different layer widths, where the relation between the frequency and the dimensionless acceleration is $f = (\Gamma g)^{1/2} A^{-1/2} (2\pi)^{-1}$, with $A = 2$ mm. The shape of the sawtooth: $h = 6$ mm, $a = 0.25$. The system width is $10w = 60$ mm, one run lasted for 30 internal seconds. The plots for $n = 0.35$, $n = 0.7$, $n = 1.05$, $n = 1.75$, $n = 2.8$, and $n = 4.2$ are averages over 30, 18, 12, 7, 7, 7 runs, respectively.

Thus, we are motivated to investigate the effect of changing the coefficient of restitution as well. Except for these series, e_n is fixed at 0.8, but now it varies from 0.4 to 1.0. It can be seen in Fig. 8 that as the dissipation is decreased the direction of transport reverses from positive to negative. The magnitude of the velocity reaches a local maximum near $e_n = 0.95$, and decreases until 1.0.

4 Discussion

According to our studies of a simplified geometrical model the following qualitative argument can be used to explain the observed current reversal as a function of the particle number: There is an intermediate size and asymmetry of the teeth for which a single ball falling from a range of near-vertical angles bounces back to the left (negative direction) in most of the cases. This effect is enhanced by rotation, due to friction between the ball and the tooth. However, if there are many particles present, this mechanism is destroyed, and on average, the direction of the motion of particles will become positive (the "natural" direction for this geometry); this corresponds to the usual ratcheting mechanism characterized by larger distances traveled by the particles along the smaller slope with occasional jumps over to the next valley between the teeth [9–12, 22]. There is no net current for symmetric teeth in our case,

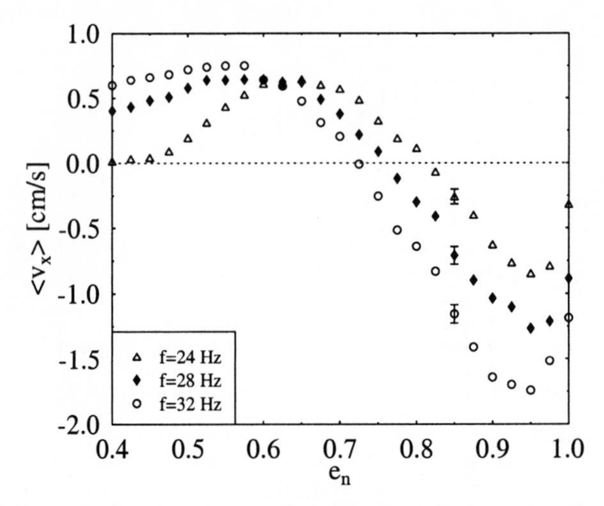

Fig. 8. The horizontal transport as a function of coefficient of restitution for three different driving freqencies. The layer width is $n = 2.8$, the shape of the sawtooth and the system width are the same as in Fig. 7. The data shown here are an averaged over 20 runs.

however the motion of a single particle is very interesting in that situation as well [21].

The reversal of the current as a function of the frequency can be interpreted in a similar manner. For smaller frequencies there are many particles close to the base and the current is positive. For larger frequencies the granular state is highly fluidized, the density considerably decays and the mechanism leading to negative transport for a small number of particles (discussed above) comes into play.

To obtain a deeper insight into the process of horizontal flow, the momentum density field and mass distribution is plotted in case of *positive* (Fig. 9) and *negative* (Fig. 10) transport. In Fig. 9 it can be seen that the particles slide down the side with the smaller slope, and when the base is in rising phase, they are given a positive net horizontal impulse. It is worth noting that most particles do not collide with the opposite side. On the other hand, if the sawtooth is higher and more symmetric, the off-bouncing particles collide with the opposite side of the sawtooth, therefore their horizontal impulse is reveresed, and the net current becomes negative (see Fig. 10).

The remarkable result that emerges from both experiment and simulation is that the flow direction can change as the layer thickness varies. This is entirely unexpected and requires further investigation; the only related behavior of which we are aware is the alternating current direction in a model of collectively moving interacting Brownian particles in a "flashing" ratchet potential [13].

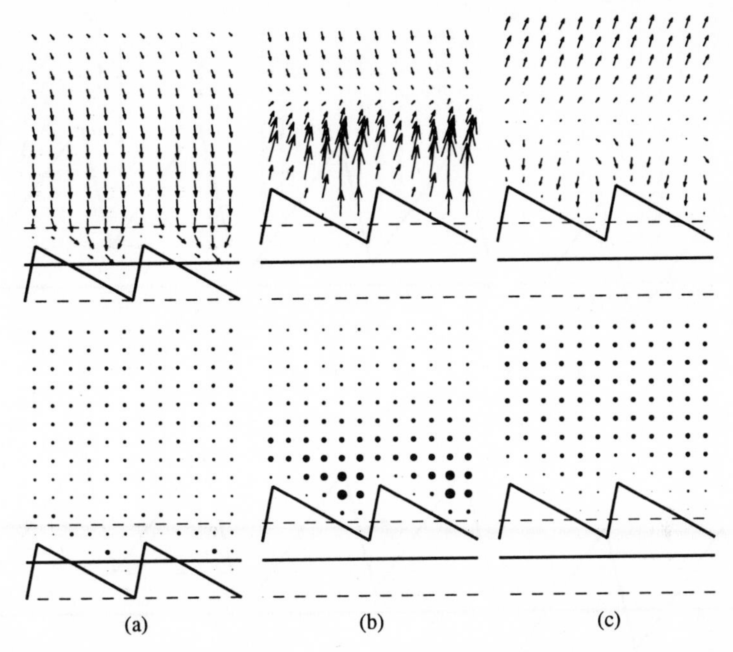

Fig. 9. Momentum density field (upper row) and mass density field (lower row) in case of *positive* horizontal transport ($\langle v_x \rangle = 3.9 \pm 0.3$ cm/s, $h = 3$ mm, $a = 0.1$, $f = 25$ Hz, $n = 0.7$). The arrows represent the average momentum vector in a box (with width $w/6$) starting from the center of the box, the radii of the disks are proportional to the average mass. (The momentum and mass carried by a particle whose center is in the box are taken into account when calculating the average in the box.) The momentum and the mass are averaged for these spatial boxes and phase frames of the oscillation (with length $2\pi/6$). In the jth frame, the phase $\phi = 2\pi f t$ of the oscillation $A \sin(2\pi f t)$ is $j 2\pi/6 \leq \phi < (j+1)2\pi/6$. (a) $j = 4$, (b) $j = 0$, (c) $j = 2$. The actual height of the simulation cell is three times larger than shown here.

In conclusion, we have investigated granular transport in a system inspired by models of molecular motors by means of numerical simulation, and have observed that the behavior depends in a complex manner on the parameters characterizing the system. These results ought to stimulate further research into this fascinating class of problems.

Acknowledgements

Useful discussions with I. Derényi are acknowledged. One of the authors (T. V.) has had extensive interactions with D. Rapaport about the standard (not event driven) simulations of the present system (to be published). This work was supported in part by the Hungarian Research Foundation Grant No. T019299 and MKM FKFP Grant No. 0203/1997.

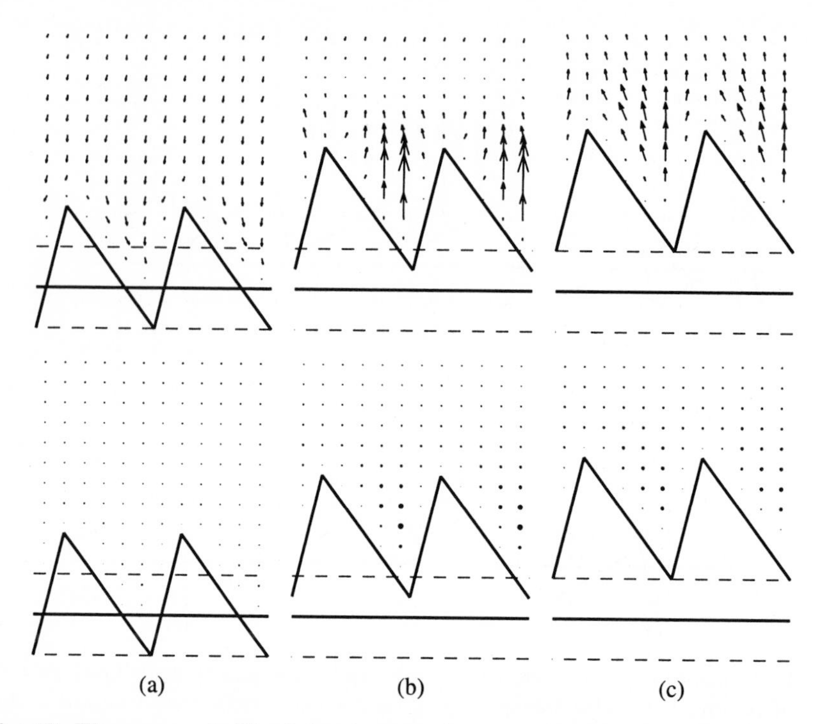

Fig. 10. The same as in Fig. 9, but in case of *negative* horizontal transport ($\langle v_x \rangle = -3.2 \pm 0.3$ cm/s, $h = 6$ mm, $a = 0.25$, $f = 30$ Hz, $n = 0.35$). (a) $j = 4$, (b) $j = 0$, (c) $j = 1$. The ratio of the absolute value of the average momentum in a box and the length of the arrow is the same as in Fig. 9.

References

1. J. Rousselet, L. Salome, A. Ajdari, and J. Prost, Nature **370**, 446 (1994).
2. L. P. Faucheux, L. S. Bourdieu, P. D. Kaplan, and A. J. Libchaber, Phys. Rev. Lett. **74**, 1504 (1995).
3. H. M. Jaeger, S. R. Nagel, and R. P. Behringer, Rev. Mod. Phys., **68**, 1259 (1996).
4. S. Douady, S. Fauve, and C. Laroche, Europhys. Lett. **8**, 621 (1989).
5. E. Clément, J. Duran, and J. Rachjenbach, Phys. Rev. Lett. **69**, 1189 (1992).
6. E. E. Ehrichs, H. M. Jaeger, G. S. Karczmar, J. B. Knight, V. Yu. Kuperman, and S. R. Nagel, Science **267**, 1632 (1995).
7. H. K. Pak and R. P. Behringer, Phys. Rev. Lett. **71**, 1832 (1993).
8. J. B. Knight, H. M. Jaeger, and S. R. Nagel, Phys. Rev. Lett. **70**, 3728 (1993).
9. A. Ajdari and J. Prost, C. R. Acad. Sci. Paris **315**, 1635 (1992).
10. M. O. Magnasco, Phys. Rev. Lett. **71**, 1477 (1993).
11. R. D. Astumian and M. Bier, Phys. Rev. Lett. **72**, 1766 (1994).
12. C. R. Doering, W. Horsthemke, and J. Riordan, Phys. Rev. Lett. **72**, 2984 (1994).
13. I. Derényi and A. Ajdari, Phys. Rev. E **54**, R5 (1996).

14. I. Derényi and T. Vicsek, Phys. Rev. Lett. **75**, 374 (1995).
15. F. Jülicher and J. Prost, Phys. Rev. Lett. **75**, 2618 (1995).
16. M. Nakagawa, S. A. Altobelli, A. Caprhan, E. Fukushima, and E-K. Jeong, Exp. Fluids **16**, 54 (1993).
17. J. A. C. Gallas, H. J. Herrmann, and S. Sokolowski, J. Phys. (France) II **2**, 1389 (1992).
18. Z. Farkas, P. Tegzes, A. Vukics, and T. Vicsek, cond-mat/9905094.
19. G. C. Barker, in *Granular Matter: An Interdisciplinary Approach*, edited by A. Mehta (Springer, Heidelberg, 1994), p. 35.
20. H. J. Herrmann, in *3rd Granada Lectures in Computational Physics*, edited by P. L. Garrido and J. Marro (Springer, Heidelberg, 1995), p. 67.
21. J. Duran, Europhys. Lett., **17**, 679 (1992).
22. A. L. R. Bug and B. J. Berne, Phys. Rev. Lett. **59**, 948 (1987).
23. M. P. Allen and D. J. Tildesley, *Computer Simulation of Liquids* (Clarendon Press, Oxford, 1987).
24. D. E. Wolf, in *Computational Physics: Selected Methods – Simple Exercises – Serious Applications*, edited by K. H. Hoffmann and M. Schreiber (Springer, Heidelberg, 1996), p. 64.
25. B. D. Lubachevsky, J. Comput. Phys. **94**, 255 (1991).
26. D. C. Rapaport, *The Art of Molecular Dynamics Simulation* (Cambridge University Press, Cambridge, 1995).
27. O. R. Walton and R. L. Braun, J. Rheol. **30**, 949 (1986).
28. S. F. Foerster, M. Y. Louge, H. Chang, and K. Allia, Phys. Fluids **6**, 1108 (1994).
29. S. Luding, Phys. Rev. E **52**, 4442 (1995).
30. S. McNamara and W. R. Young, Phys. Fluids A **4**, 496 (1992).
31. S. Luding, E. Clément, J. Rajchenbach, and J. Duran, Europhys. Lett. **36**, 247 (1996).
32. Z. Farkas, Diploma thesis, Eötvös University, 1998.
33. W. H. Press, S. A. Teukolsky, W. T. Vetterling, and B. P. Flannery, *Numerical Recepies in C* (Cambridge University Press, Cambridge, 1992).

Resonance Oscillations in Granular Gases

Alexander Goldshtein, Alexander Alexeev, and Michael Shapiro

Faculty of Mechanical Engineering, Technion, Haifa 32000, ISRAEL.
e-mail: mergcga@aluf.technion.ac.il

Abstract. The paper is devoted to investigation of a new phenomenon: resonance oscillations in granular gases produced by a piston vibrating at one end of a closed tube. The main feature of this phenomenon, predicted by a hydrodynamic model and confirmed by computer simulations, are: (*i*) a column of a granular gas oscillates periodically with the frequency f/n, where f is the vibrational frequency, n is a positive integer number, (*ii*) the oscillation patterns are governed by the shock waves propagating across the column; (*iii*) the averaged kinetic energy per particle is proportional to f^2, and it strongly depends on the vibrational amplitude; (*iv*) the maximal value of this kinetic energy pumped by these external vibrations is of order f^2L^2, where L is the tube length

1 Introduction

Hydrodynamic models of inelastically colliding granules (granular gases) has become a subject of growing research activity, motivated by applications as well as fundamental interests in granular materials. Experiments show that under the action of external forces, e.g., vibrations [1], granular materials may flow like liquids; hence it is natural to use hydrodynamic models for the description of such flows. Because of kinetic energy dissipation due to granular inelastic collisions, an external source of energy needs to sustain the fluidized (gas-like) hydrodynamic state. Without such an energy source the kinetic energy decays and, eventually, inelastic collapse can occur [2, 3].

This paper is devoted to fluidization of layers of granular materials by external vibrations. From the hydrodynamic point of view kinetic energy can be transferred from a vibrating plate to a granular material by "heat flux" and wave propagation mechanisms. The heat flux mechanism [4] prevails when the mean free path, λ, of the moving granules exceeds the vibrational amplitude, A. The applicability of this mechanism for vibrofluidization was verified for one-dimensional [5] and two-dimensional granular systems [6–8]. Computer simulations [5–7] and the experimental data [8] showed that the hydrodynamic model of Haff [4] based on this mechanism breaks down, unless the system is very dilute and consists of almost elastic granules [6, 7]. Here we discuss the wave mechanism, prevailing for $A \gg \lambda$ and show that it allows to fluidize more dissipative granular systems than does the heat flux mechanism.

Previous studies on the wavy motion of granular materials were concerned with semi-infinite granular layers [9, 10], or systems affected by the gravity force [11, 12]. The results of the latter studies cannot be compared with those obtained for finite layers and without gravity [5–8]. We compare the two fluidizing mechanisms acting in a granular gas agitated by a vibrating piston in the absence of gravitation. We start our consideration from discussion of the resonance problem for a granular gas.

2 Hydrodynamic Model

The resonance oscillations were described for the molecular gases [13–15]. An analytical solution was obtained for the following conditions: (*i*) perfect gas (viscosity and thermal conductivity are negligibly small) moves within a tube closed at one end by a rigid plug and at the other end a piston oscillating sinusoidally; (*ii*) vibrational amplitude A is much smaller than the tube length L; (*iii*) vibrational frequency f is equal to f_r/n, where n is a positive number, f_r is the resonance frequency related to the speed of sound, a_0 by

$$f_r = a_0/(2L) \tag{1}$$

Under the above conditions, the gas oscillations are amplified and shock waves develop. A space-time diagrams for the first ($n = 1$) and second ($n = 2$) resonance are presented in Fig. 1(a) and (b), respectively. One can see that the number of shock waves prevailing at any moment within a resonance tube is equal to the number n. Each of these waves travels with a constant speed a_0 in either direction and strikes the plug and piston once per n vibrational periods. It means that the value $1/n$ describes the number of wave-piston interactions per period. We use this value for classification of the patterns depicted in Figs. 3(a)-(c).

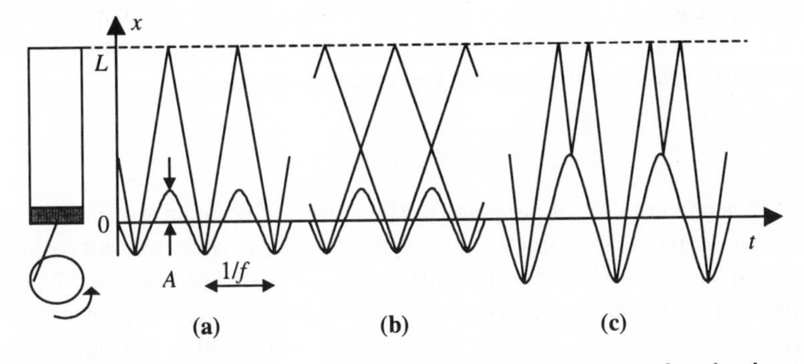

Fig. 1. Schematic of patterns of gas resonance oscillations in a closed tube: wavy line – piston path, zigzag line – shock wave path. (a) first resonance ($f = f_r$, $A \ll L$), (b) second resonance ($f = 2f_r$, $A \ll L$), (c) half resonance ($f = f_r$, $A \sim L$).

The first and second resonance oscillations were experimentally registered and theoretically investigated [13–15] under condition (*ii*). In this case the distance between the plug and piston changes insignificantly during one period, any periodic shock wave travels the same way to and fro, and can interact with the piston not more than once per period. For large enough vibrational amplitudes the distance between the plug and piston can be small and additional interactions between the piston and the shock wave may take place, see Fig. 1(c). By the analogy with the first and second resonance, when any shock wave interacts with a piston, respectively, once per one or two periods, we will call the pattern depicted in Fig. 1(c), where a shock wave hits a piston twice per period, the half resonance pattern. Below, we show, see Figs. 4, that the same patterns prevail also for 1-D discrete model of granular columns.

The resonance oscillation of a molecular gas [13–15] is a special case of resonance phenomena known in physics. In a more general sense, resonance is a phenomenon exhibited by a physical system acted upon by an external periodic driving force, in which the resulting amplitude of oscillation of the system becomes large when the frequency of the driving force approaches a natural free oscillation frequency. Resonance can be understood as any phenomenon which is greatly enhanced when a physical parameter describing the system is equal or very close to a given characteristic value. For example in quantum mechanics resonance absorption of neutrons is characterized in terms of energy rather than frequency. Below we show that resonance oscillations of granular gases (in contrast to molecular gases) may be achieved when the driving amplitude is close to a (characteristic) resonance amplitude.

The problem of resonance oscillation of a conservative gas has been solved using the Euler inviscid hydrodynamic equations by series expansion in terms of small parameter $\delta = \sqrt{A/L}$ [13–15]. This solution may be generalized for the case of dilute granular gas consisting of smooth inelastic hard spheres [16]. Equating the energy dissipation rate of the tube via the volumetric sink term (see [17–19]) to the time-averaged power generated by the piston (see [13]), we get the necessary condition for resonance oscillations of a nonconservative granular gas:

$$\delta^3 = (A/L)^{3/2} = 0.284\left(1 - e^2\right) N\,, \tag{2}$$

where N is the number of particles in the volume $L\sigma^2$, e, σ are the restitution coefficient, a particle diameter, respectively. For e close to unity the right hand side of (2) is close to the parameter $N(1-e)/2$ used in [5, 20] as a characteristic of the dissipative properties of granular systems. According to (2) the total energy generation is of the order δ^3, which is consistent with the weak shock wave approximation implicitly employed in the solution.

Condition (2) implies that for a fixed energy dissipation the resonance occurs only for large enough amplitudes δ predicted by (2). We will call the resonance amplitude the driving amplitude satisfying condition (2). Since $\delta < 1$ condition (2) predicts also the maximal energy dissipation allowing the resonance oscillations. It is remarkable that in contrast to the conservative gases, the resonance condition for the granular gas is independent of the vibrational frequency, f and the speed of sound a_0. It means that f and L define the resonance speed of sound, a_r through relation (1) as $2Lf$. Therefore, any vibrational frequency, will be the resonance one, since any initial state with an arbitrary kinetic energy, E_i will eventually evolve to the resonance state, where the rate of dissipated and generated energy are balanced. In this state the kinetic energy per particle in the system is given by

$$E_r = 3.6(Lf)^2 . \tag{3}$$

3 Computer Simulation

Here we present the results of computer simulations of the motion of N identical sizeless inelastic particles ($\lambda = L/N$) constrained to move along a line between a harmonically oscillating piston and a resting plug. The boundaries collide with the first and N-th particle in an elastic manner, interparticle collisions are governed by models of inelastic hard spheres (IHS) [4] or inelastic soft spheres (ISS) [21]. The results of the simulations presented here were obtained by a driven-event method [5–7, 22] for the IHS, by the fourth degree Runge-Kutta numerical integration scheme for the ISS model.

In the initial state treated, the particles were uniformly distributed with velocities randomly chosen between $-\sqrt{2E_i}$ and $\sqrt{2E_i}$, where $E_i \gg E_r$ (we discuss effect of E_i below). The average kinetic energy, $E(t)$ and average relative particle energy, $E_{rel}(t)$ per particle per second (the latter is a more appropriate characteristic of vibrofluidization since it distinguishes between the relative particles' motion and their motion as a whole [23]) defined as

$$E(t) = \frac{1}{2tN} \int_0^t \sum_{i=1}^{N} \dot{x}_i^2 \, dt \,, \quad E_{rel}(t) = \frac{1}{2t(N-1)} \int_0^t \sum_{i=1}^{N} (\dot{x}_i - \dot{x}_{i-1})^2 \, dt \,, \tag{4}$$

were used to characterize the system's state. We found that after large enough time of the system evolution these energies converge to constant amounts (say) E, E_{rel}, which slightly depend on initial velocities of particles. E, E_{rel} were found to be proportional to f^2 (cf. (3)). We scaled all system's parameters with appropriate combinations of the tube length and the vibrational frequency, and found that any IHS system's state is uniquely characterized in terms of A and the disippative parameter $D = N(1 - e)$ (cf. (2)). For the ISS columns we modified the parameter D using the effective restitution coefficients e_{eff}, accounting for the energy losses of two "soft" particles having the relative kinetic energy before the collision equals to E_r (defined by (3)).

Figure 2 presents E versus the dimensionless vibrational amplitude for several $D \leq 1$. Each curve has several global and many local maxima. Our simulations show that the local maxima in Fig. 2 are caused by the effect of initial conditions, whereas the global maxima are practically independent of these conditions. To illustrate this we presents in Fig. 3 dependence of

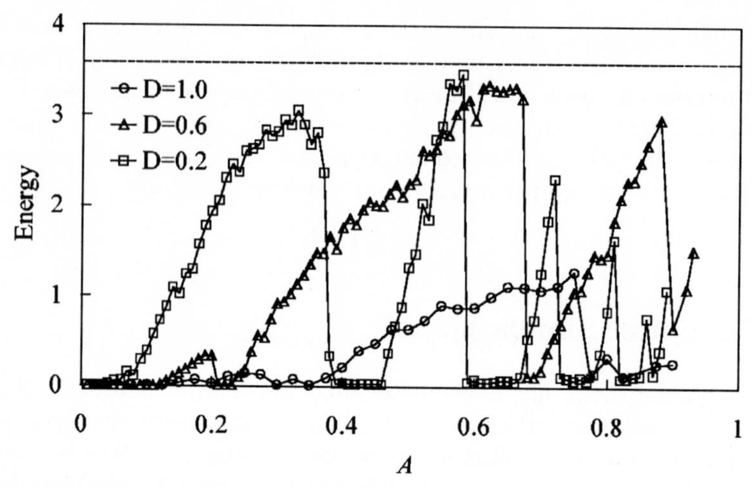

Fig. 2. Kinetic energy versus vibrational amplitude for $N = 100$. Dashed line corresponds to the hydrodynamic prediction E_r (see (3)).

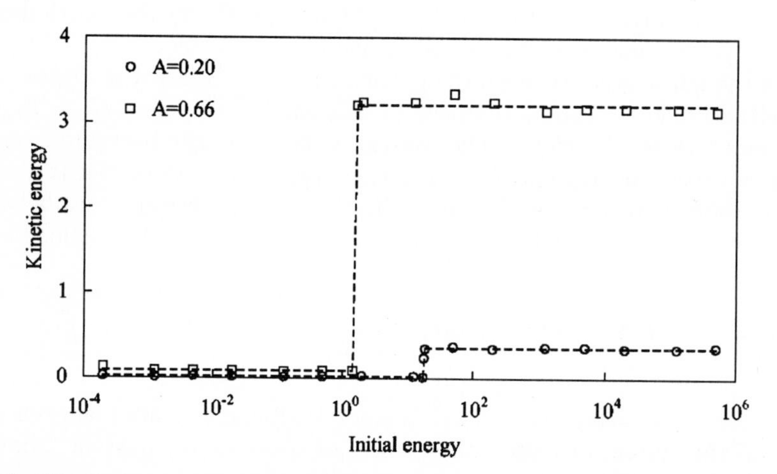

Fig. 3. Kinetic energy, E versus initial kinetic energy, E_i for $D = 0.6$, $N = 100$. Dashed lines give analytic approximations by the Heaviside function $\Theta(E_i)$ defined as follows $\Theta(E_i) = E_{min}$ for $E_i < E_{tr}$, $\Theta(E_i) = E_{max}$ for $E_i > E_{tr}$. $E_{min} = 0.016, 0.08$, $E_{max} = 0.35, 3.3$, $E_{tr} = 18, 1.6$ for $A = 0.2, 0.66$, respectively.

the kinetic energy, E versus initial kinetic energy, E_i calculated for $D = 0.6$, $N = 100$ and for several vibrational amplitudes. One can see that E is almost independent of E_i for intervals $E < E_{tr}$ and $E > E_{tr}$ (see Fig. 3), where kinetic energy may be approximated by constants E_{min} and E_{max}, respectively. However, in a small vicinity of E_{tr} E rapidly changes (practically by jump). The lower kinetic energy (E_{min}) corresponds to irregular motion of small number of particles (i.e., the cluster patterns registered in [5]), whereas

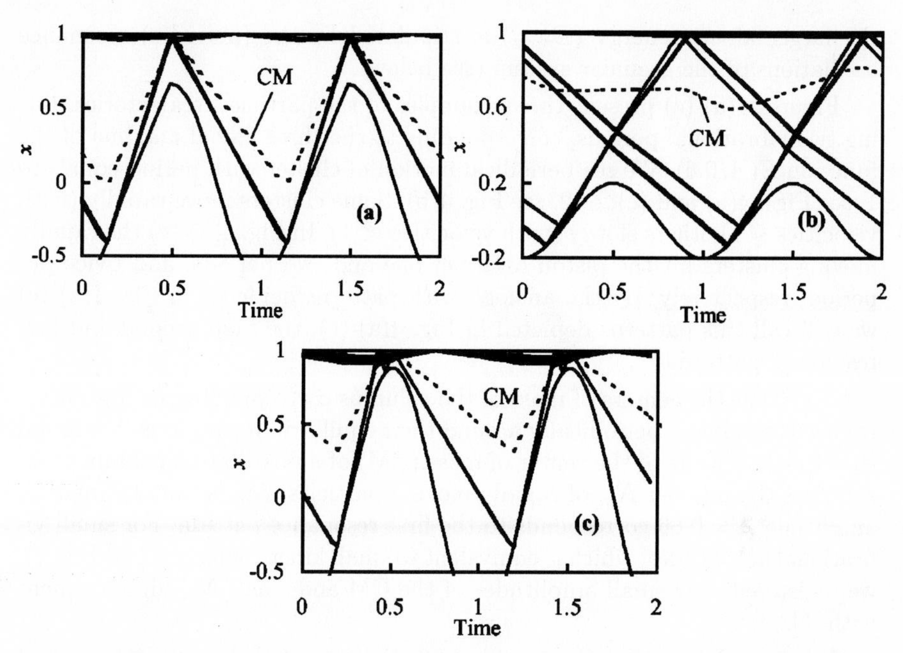

Fig. 4. Trajectories of particles' resonance oscillations ($D = 0.6$, $N = 100$). Piston moves sinusoidally - $A \sin 2\pi t$: (a) first resonance pattern ($A = 0.66$), (b) second resonance pattern ($A = 0.2$), (c) half resonance pattern ($A = 0.88$); CM - position of center of mass. The number of rapidly moving particles $N_m = 55, 75, 51$ for (a), (b), (c), respectively.

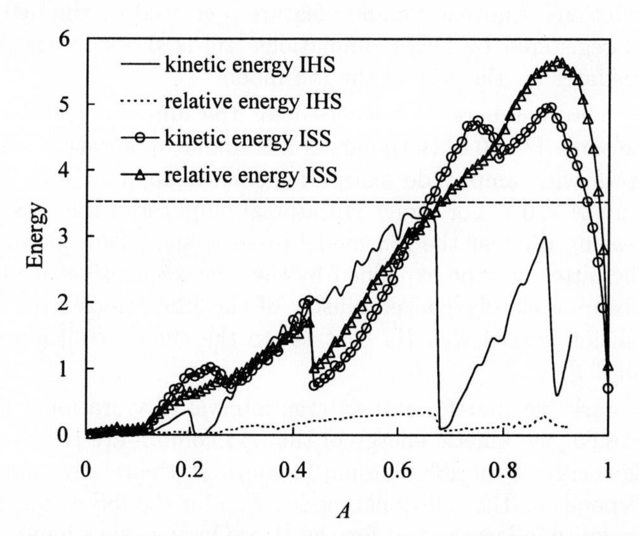

Fig. 5. Kinetic and relative energy versus vibrational amplitude for a column of IHS and ISS ($D = 0.6$, $N = 100$). Dashed line corresponds to equation (3).

the larger kinetic energy (E_{max}) is stipulated by the (periodic) resonance oscillations of the granular system (see below).

Figures 4(a)-(c) present three examples of the particles' trajectories during two vibrational periods, corresponding to the three global maxima of the function $E(A, 0.6)$. We see periodical motion of cluster with period equals to 1, see Figs. 4(a) and (c), or 2, see Fig. 4(b). Some clusters move rapidly (with velocities ~ 1) others slowly (with velocities $\ll 1$). In Fig. 4(a)-(c) the rapidly moving clusters hit the piston once per one and two periods, and twice per period, respectively. By the analogy with patterns depicted in Fig. 1(a)-(c) we will call this patterns depicted in Fig. 4(a)-(c), the first, second and half resonance patterns.

Note that the centers of mass of the columns corresponding to Figs. 4(a)-(c) oscillate almost periodically around their equilibrium positions. We found that the amplitude of the center of mass (CM) of an oscillation column characterizes the number N_m of rapidly moving particles. N_m is the maximal for amplitude $A = 0.66$ corresponds to the first resonance pattern. For small vibrational amplitudes, which is equivalent to small kinetic energies (see Fig. 2), we registered very small amplitudes of the CM and small N_m, in agreement with [5].

It is noteworthy that the maximal kinetic energy, E that can be pumped into the system by the vibrational excitation, only slightly depends on the energy dissipation for small and moderate D (see Fig. 2) In spite of the difference between the simulated 1-D granular system and the 3-D granular gas model, the hydrodynamic prediction E_r given by (3) agrees with the DEM simulations. Another common feature predicted by the hydrodynamic model and registered by DEM simulations [16] is shock waves. We discuss these waves later for the case of the ISS model.

In Fig. 5 we compare E calculated for the different collisional models but for the same $D = 0.6$. Both curves has similar qualitative behavior (the energy grows with amplitude except for several points where it decreases abruptly) for $A < 0.7$. For larger vibrational amplitudes the IHS model predicts decreasing whereas the ISS model predicts significant increasing of the energy. The latter may be explained by the "detachment" effect [23, 24], i.e. particles forming closely packed cluster of the ISS detach from each other after a collision with a wall (in contrast to the cluster collision of the IHS with a wall [22]).

It is remarkable that for rather large interval of vibrational amplitudes, $0.3 < A < 0.65$, the kinetic energy of the IHS column are relatively close to the kinetic energy of the ISS column. In contrast, the relative kinetic energy strongly depends on the collisional model. E_{rel} for the ISS column about one order of magnitude larger than for the IHS. On the other hand, both energetic characteristics (E, E_{rel}) are very close for the ISS column. It means that for the ISS system major part of the particle energy is associated with their relative motion whereas for the IHS system most of the energy is associ-

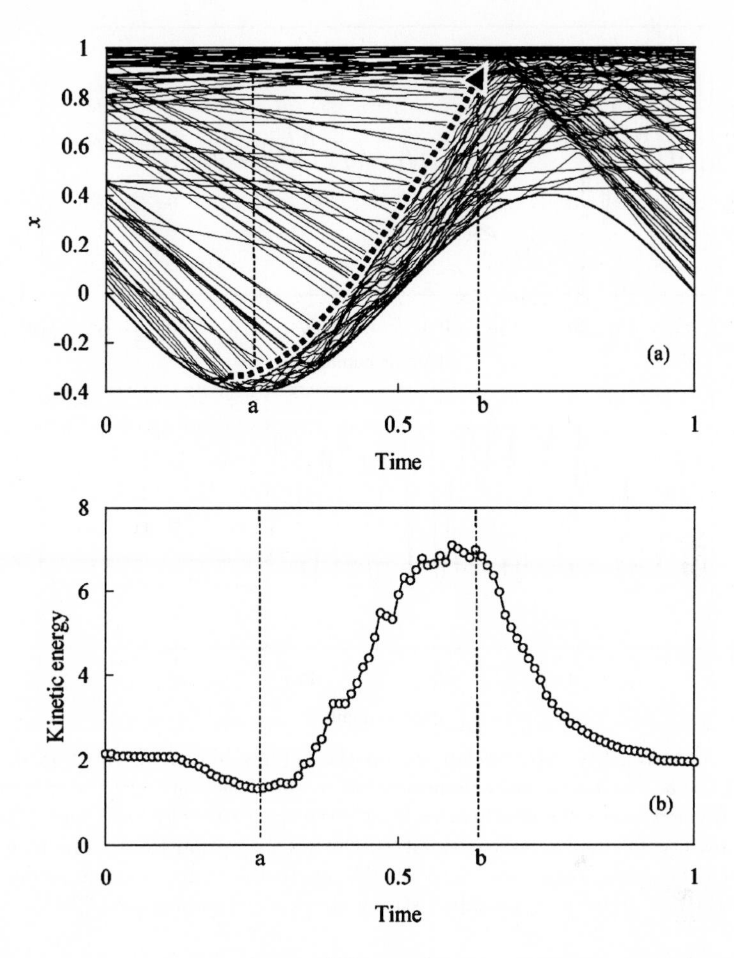

Fig. 6. Kinetic energy evolution, ISS trajectories and piston path during one vibrational period. (A = 0.4, D = 0.6, N = 100): (a) particles trajectories and piston path, wavy line – piston path, thin solid lines – particles trajectories, dashed line with arrow – shock wave front. The particles situated below the dashed line are involved in intensive relative motion (their trajectories are zigzag lines) by the shock wave generated by the piston. The rest of the particles are less agitated (their trajectories are straight lines). (b) kinetic energy of the column vs. time.

ated with their motion as a whole (cluster-like motion). The latter difference in the system behavior, apparently, results from the difference between the collisional models. Any collision of the IHS dissipates the same portion of their relative energy before the collision, i.e., $1 - e^2$, since the restitution coefficient e is constant. For the ISS energy losses decrease with decreasing relative energy, since the restitution coefficient is a decaying function of the initial energy [21].

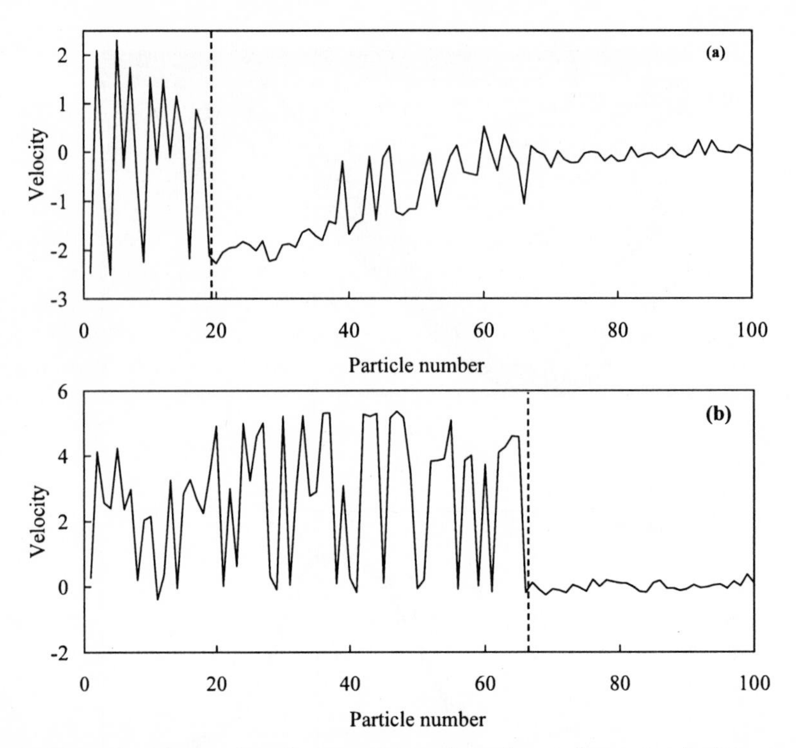

Fig. 7. Particle velocity distribution within the column shown in the Fig. 6: (a) $t = a$, (b) $t = b$. The particles are enumerated in the direction from the piston to the plug. Dashed line is the shock wave front separating particles with high ("hot" gas) and low ("cold" gas) relative velocities. During the period from moment a till b the front propagates from "hot" into "cold" gas. In the hydrodynamic terms the process described above may be called strong shock wave propagation [25].

For the ISS model each of global maxima of the kinetic energy is related to the first or half resonance patterns [16]. An example of the first resonance pattern for the ISS together with the kinetic energy evolution is presented in Fig. 6(a) and (b). In contrast to the IHS model, see Fig. 4(a), now trajectories of each particle are well distinguished and fill practically all the phase volume in the plane $x - t$, see Fig. 6(a). At the moment a the kinetic energy reaches its minimum, see Fig. 6(b) and the piston approaches its outstroke position, see Fig. 6(a). At the moment b the column detaches from the piston, and the energy reaches its maximum. Velocity distributions within the column for the moments a and b are depicted in Figs. 7(a) and (b), respectively. One can distinguish in these figures shock wave patterns. Between the moments a and b, the piston pushes the shock wave towards the plug. This shock wave compresses the column of particles, increases the number of agitated particles from 20 to 65 (see Fig. 7) and their energy from 1.5 to 7, see Fig. 6(b). The rest of the vibrational period the particles lose their energy, the column expands and the process repeats again.

The applicability of the hydrodynamic description to a particulate system depends on the kinetic energy distribution between the particles, rather than on the total kinetic energy. In Fig. 8 the kinetic energy distribution for the ISS and the IHS columns is presented for $D = 0.6$. All curves are normalized in such a way that the area below the curve equals to one. The amplitudes $A = 0.4$ and $A = 0.66$ correspond to the maxima of energy (see Fig. 5) related to the first resonance pattern of the ISS and the IHS column, see Fig. 4(a), respectively. For the IHS column one can easily distinguish two groups of particles with significantly different energies. All particles in each group possess (practically) the same energy. For the resonance amplitude, about 75 particles move rapidly ($N_m = 75$) leading to large oscillations of the center of mass, see Fig. 4(a) and the discussion after it) others form a slowly moving cluster. For any non-resonance amplitude N_m is smaller. For example, increasing of the vibrational amplitude from $A = 0.66$ to $A = 0.8$ *decreases* the number of rapidly moving inelastic hard spheres from $N_m = 75$ to $N_m = 22$ (see Fig. 8). For the ISS particles energy variation within the column is much stronger than for the IHS, i.e., the ISS system avoid cluster formation.

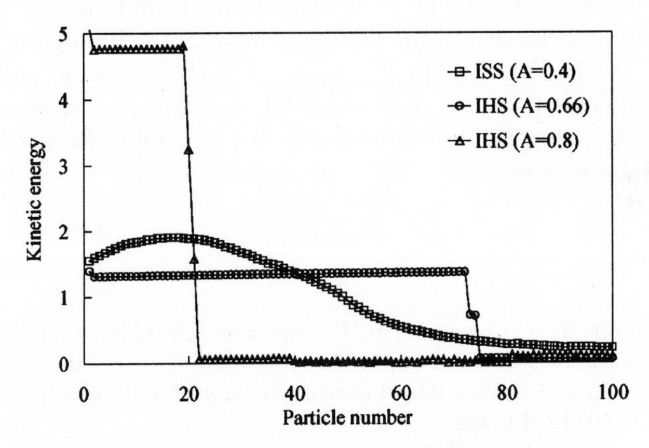

Fig. 8. Kinetic energy distribution for D=0.6, N=100.

4 Conclusions

The results of our computer simulations of vibrated inelastic granular columns, performed for the IHS and the ISS collisional models, have many common features and, thus, are independent of the collisional models:

(i) For large enough piston vibrational amplitudes the columns oscillate periodically with the period equals to the vibrational period;

(ii) The maximal value of the kinetic energy of the columns driven by the external vibrations is proportional to the square of the vibrational frequency, f and strongly depends on the vibrational amplitude in a non-monotonic manner. The maximal values of the kinetic energy, achievable by means of external vibrations, are of the order of f^2L^2;

(iii) The periodic oscillations of the granular columns corresponding to the maximal values of the kinetic energy are governed by the shock waves.

It was demonstrated that the resonance phenomenon provides the maximal vibrofluidization of inelastic granular column (i.e., the maximal kinetic energy, E pumped by the piston and the number of rapidly moving granules, N_m).

All the features (i)-(iii) are also predicted on the basis of a simple hydrodynamic model of resonance oscillations of a three-dimensional granular gas. It remains to assess the applicability of this (and more sophisticated hydrodynamic models) to 2D and 3D vibrated granular gases and other collisional models.

Acknowledgements

This research was supported by the Israel Science Foundation, by the Center of Absorption in Science and the Gilliady Program for Immigrant Scientists Absorption, by the Fund for the Promotion of Research at the Technion The authors are indebted to Prof. T. Pöschel for the assistance in preparation of the paper. A. G. gratefully acknowledges Dr. S. Luding for his hospitality and fruitful discussions.

References

1. H. M. Jaeger, S. R. Nagel, and R. P. Behringer. Granular solids, liquids, and gases. Rev. Modern Physics **68**, 1259 (1996) and references therein.
2. I. Goldhirsch, and G. Zanetti. Clustering instability in dissipative gases. Phys. Rev. Lett. **70**, 1619 (1993).
3. S. McNamara, and W. R. Young. Inelastic collapse and clumping in a one-dimensional granular medium. Phys. Fluids A **4**, 496 (1992)
4. P. K. Haff. Grain flow as a fluid-mechanical phenomenon. J. Fluid Mech. **134**, 401 (1983)
5. Y. Du, H. Li, and L. P. Kadanoff, Breakdown of hydrodynamics in a one-dimensional system of inelastic particles Phys. Rev. Lett. **74**, 1268 (1995).
6. E. L. Grossman, T. Zhou, and E. Ben-Naim. Towards granular hydrodynamics in two dimensions. Phys. Rev. E **55**, 4200 (1996).
7. S. McNamara, and J. L. Barrat. Energy flux into a fluidized granular medium at a vibrating wall. Phys. Rev. E **55**, 7767 (1997).
8. A. Kudrolli, M. Wolpert, and J. P. Gollub. Cluster formation due to collision in granular materials. Phys. Rev. Lett. **78**, 1383 (1996).

9. A. Goldshtein, M. Shapiro, and C. Gutfinger. Mechanics of collisional motion of granular materials. Part III. Self-similar shock-wave motion. J. Fluid Mech. **316**, 29 (1996).
10. A. Goldshtein, M. Shapiro, and C. Gutfinger. Mechanics of collisional motion of granular materials. Part IV. Expansion wave. J. Fluid Mech. **327**, 117 (1996).
11. A. Goldshtein, M. Shapiro, L. Moldavsky, and M. Fichman. Mechanics of collisional motion of granular materials. Part II. Wave propagation through vibrofluidized granular layers. J. Fluid Mech., **287**, 349 (1995).
12. A. V. Potapov, and C. Campbell. Propagation of inelastic waves in deep vertically shaken particle beds. Phys. Rev. Lett. **77**, 4760 (1996).
13. B. Betchov. Non-linear oscillations of the column of a gas. Phys. Fluids **1**, 205 (1958).
14. W. Chester. Resonant oscillations in closed tubes. J. Fluid Mech. **18**, 44 (1964).
15. A. Goldshtein, P. Vainshtein, M. Fichman, and C. Gutfinger. Resonance gas oscillations in closed tubes J. Fluid Mech. **322**, 147 (1996).
16. A. Goldshtein, A. Alexeev, and M. Shapiro. Hydrodynamics of resonance oscillations of columns of inelastic particles Phys. Rev. E **59**, 6967 (1999).
17. C. K. K. Lun, S. B. Savage, D. J. Jeffery, and N. Chepurniy. Kinetic theories for granular flow: inelastic particles in Couette flow and slightly inelastic particles in a general flow field. J. Fluid Mech. **140**, 223 (1984).
18. J. T. Jenkins, and M. W. Richman. Grad's 13-moment system for a dense gas of inelastic spheres. Arch. Rat. Mech. Anal. **87**, 355 (1985).
19. A. Goldshtein, and M. Shapiro. Mechanics of collisional motion of granular materials. Part I. General hydrodynamic equations. J. Fluid Mech. **282**, 75 (1995).
20. B. Bernu, and R. Mazighi. One dimensional bounce of inelastically colliding marbles on a wall. J. Phys. A: Math. Gen. **23**, 5745 (1990).
21. N. V. Brilliantov, F. Spahn, J.-M. Hertzsch, and T. Pöschel. Model for collisions in granular gases. Phys. Rev. E **53**, 5382 (1996) and references therein.
22. S. Luding, E. Clement, A. Blumen, J. Rajchenbach, and J. Duran. Studies of columns of beads under external vibrations. Phys. Rev. E **49**, 1634 (1994).
23. A. Alexeev, A. Goldshtein, and M. Shapiro. Vibrofluidization of highly dissipative granular layers: one-dimensional computer simulations. Powder Tech., submitted, (1999)
24. S. Luding, E. Clement, A. Blumen, J. Rajchenbach, and J. Duran. Anomalous energy dissipation in molecular-dynamics simulations of grains: The "detachment" effect. Phys. Rev. E. **50**, 4113 (1994).
25. Ya. B. Zel'dovich, and Yu. P. Raizer. Physics of shock waves and high-temperature hydrodynamic phenomena. Academic Press, New York and London (1967).

IV

Granular Astrophysics

Dynamical Evolution of Viscous Discs. Astrophysical Applications to the Formation of Planetary Systems and to the Confinement of Planetary Rings and Arcs

André Brahic

Université Paris VII Denis Diderot, Equipe universitaire Gamma-Gravitation, C. E. de Saclay, DSM/DAPNIA/SAp, Bâtiment 709, L'Orme des Merisiers, 91191 Gif sur Yvette cedex, France. e-mail: brahic@discovery.saclay.cea.fr

Abstract. Planetary rings belong to the most prominent examples of granular gases in astrophysical systems. They are subject of permanent scientific investigation since centuries. So far, however, neither the detailled mechanisms of formation of planetary rings nor their complex spatial and dynamic behavior, originating from inelastic particle collisions, resonances, gravitational perturbations and others, are completely understood. We give a review on the present state of knowledge about planetary ring dynamics and particularly focus on the discussion of unsolved theoretical problems and unexplained observations.

1 Introduction

Over the last three centuries, astronomers have studied the motion of dimensionless particles following Newton's laws. Using mainly perturbation methods and various expansion techniques, they have developed celestial mechanics to such a level that most of the problems relative to the motion of planets and satellites are now solved. In the middle of the XVIIth century, René Descartes suggested that collisions may have played a role at the time of formation of planets and satellites, but he did not make any quantitative study. At the beginning of the XXth century, Henri Poincaré was the first to notice the importance of inelastic collisions among celestial bodies and to claim that such collisions should have important dynamical consequences. One century later, astronomers have observed collisions everywhere in the Universe at each scale from meteoritic bombardment of the planets to encounters between galaxies. The solar system itself is the result of collisions between kilometre-size planetesimals. Particles of planetary rings are suffering endless mutual collisions. Collisions between galaxies play a major role in the evolution of clusters of galaxies. Interstellar cloud collisions increase the number of regions of stellar formation.

The evolution of a N-body system of particles of finite size cannot be fully studied by analytical methods. It can be described by a Boltzmann equation with a collisional term. Analytical solutions can be found providing severe assumptions that turn the problems into academic ones.

Up to now, three-dimensional numerical simulations of inelastically colliding particles are practically the only way to explore the behaviour of colliding systems. Following quite different approaches, Trulsen and his collaborators on the one hand and Hénon, Brahic and their collaborators on the other hand were the first to perform in 1970 such simulations. The particle velocity dispersion and the global evolution of systems of colliding particles have been studied in the 70s. In the 80s with the spatial exploration of Saturn's rings and the discovery of ring systems around Jupiter, Uranus and Neptune, it became clear that these systems of colliding particles are strongly perturbed and confined by small satellites. Simulations of perturbations of flat discs by external and embedded satellites have been extensively investigated. In the 90s, applications to real systems such as planetary arcs and rings, the proto solar system nebula and galactic discs have been examined.

At the present time, the evolution of a real physical system of colliding particles is far from being understood. Inelastic collisions, resonances and gravitational perturbations all play an important role, but cannot be modeled together. It is not possible to include in the simulations all the assumptions making the model close to a real system. As a consequence, several authors have developed a number of quite different approaches. It is out of question in a few pages to give an exhaustive review of the remarkable results, which have been obtained in only three decades. This presentation can be considered just as an introduction to the subject. Only some results obtained by our group, a number of unsolved theoretical problems and few unexplained observations are described here. Technical details are not developed here. Of prime importance are the confinement of planetary arcs, the precession of narrow rings and the accretion of planetesimals.

Granular gases and molecular dynamics are two fields of research where cross-fertilization can be found with the study of viscous discs in the astrophysical context. A number of efficient algorithms have been developed in molecular dynamics, but cannot be directly used in astrophysical problems. The main reason is the nature of the interaction between particles. Contrary to the case of electro-magnetic forces, there is no screen effect for gravitational forces. A large number of particles far from a studied location cannot be neglected in the gravitational problem. Moreover, collisions between celestial bodies can be catastrophic leading to the destruction of the bodies involved. Molecules are not destroyed in a molecular gas!

Granular flows, which occur in laboratory experiments or in the geophysical context such as snow and rock avalanches, correspond to systems with a mean density much larger than the values found in the astrophysical context. Even, if collisions are much less frequent between bodies in the solar system celestial bodies than on the Earth, two kinds of quite different studies can be distinguished depending of the optical depth τ of the system, i.e. of the mean free path of a given particle. If τ is small, i.e. if the mean free path of a particle is large compared to the size of the system, a particle suffers in mean one collision only after several revolutions. If τ is large, a particle suffers about one or more collisions per revolution. The first case corresponds to systems such as the proto-solar nebula or the galactic disc. The second one corresponds to many places in the planetary rings. The behaviour of the system is not the same for different values of τ and some numerical methods are more appropriate than others for large or small values of τ. For example, it is clear that a hydrodynamics approach is more appropriate for a system of particles suffering many collisions per revolution. Inversely, a N-body approach is more suitable for systems in which the mean free path is large. A statistical approach can be well adapted to a system having a large number of particles. A deterministic approach may be more accurate for the study of complex dynamical phenomena such as resonances or confinement mechanisms.

2 A Simple Model for an Unconfined Disc of Colliding Particles

Three-dimensional (3d) gravitating systems of colliding masses are found in many astrophysical contexts, from planetary rings to galaxies and from accretion discs around compact stars to the proto-solar nebula. At least ten different kinds of models have been developed in Europe, United States and Japan. In fact, real systems are too complex to be precisely simulated. It should be noticed that the algorithms do not simulate in fact real systems, they can be considered just as numerical experiments similar to laboratory experiments. Scaling factors must be used to extrapolate the results to real celestial bodies. Simulations take into account a smaller number of particles than the number involved in real systems. Computing time increases with the number of interactions. Thus, it increases at least as the square of the number of particles.

A number of authors use a Monte-Carlo approach. They take at random two colliding particles following a given statistical law. Corresponding calculation times are several orders of magnitude faster as compared with a deterministic approach. These statistical methods are efficient to study local effects, but the statistical law is either arbitrary or difficult to compute taking into account the large number of parameters. They are less efficient to study collective effects or the role of resonances and most of the models are uni- or bi-dimensional.

Particle-in-the-box approaches can be easily 3d and have the advantage to follow each particle individually. The idea is to divide the 3d space into a number of boxes in which particles interact without considering neighbouring boxes. A large number of particles can be included and some important results have been obtained with this method, but the interaction law is rather naïve and some effects due to particular interaction laws can be completely ignored.

In the purely deterministic approach, the evolution of each particle is individually followed over the whole 3d space and all interactions are calculated.

Hidden statistical effects, which can be found with the previous approaches, or small but systematic effects due to oversimplyfied collision laws can be avoided, but the price to pay is an enormous calculation time. Only a small number of particles can be considered and statistical fluctuations are much bigger.

Since 1970, we have used a deterministic model, which has been regularly improved in order to solve new problems. The initial idea was to study the simplest but non-trivial model and to separately investigate different mechanisms. Coupled mechanisms are studied only after the understanding of each one separately. A more complex model is not necessarily more realistic. After the study of the collective behaviour of a disc of one hundred colliding particles in the 70s, perturbations by an external satellite have been studied in the 80s. In the 90s, we have studied the evolution of few thousand particles perturbed by externals. We have mainly followed the evolution of radially and azimuthally confined arcs, as they are observed in planetary rings, and accretion mechanisms perturbed by an initial large body, which is supposed to have existed at the very beginning of the solar system history.

2.1 The Standard Model

Here, we give some information on the method and on the behaviour of a disc of colliding particles by means of a simple illustrative model, which we have called the "standard model", and which has the following characteristics: attraction between particles of finite dimension has been neglected, and so particle orbits are Keplerian around a central mass point. Each collision is assumed to be instantaneous; consequently simultaneous collisions of several particles can be entirely neglected. All particles are spheres having the same mass and radius. The total mass of the particles has been neglected with respect to the central mass. After a collision, the perpendicular component of the relative velocity of two colliding particles is multiplied by a coefficient k which lies between 0 and -1, whilst the grazing component is multiplied by a coefficient k' which lies between 0 and +1. The couple (k, k') gives the amount of energy lost during a collision. $(k, k') = 3D(-1, +1)$ corresponds to a perfectly elastic collision and $(k, k') = 3D(0, 0)$ corresponds to a completely inelastic collision.

Real collision laws between celestial bodies are rather poorly known and a more complex law should not be necessarily more realistic. The method, however, can be easily extended to more complex situations, for example, a different collision model or different potential fields or even self-gravitating systems.

The principal difficulty is to establish whether or not two given particles will in fact collide. In the field of molecular dynamics, Alder and Wainwright (1959), for example, first set up a list of possible two-particle collisions. They then find which collision is the earliest, and the new velocity vectors of these particles are calculated. The list is then modified: all entries involving either of these two particles are deleted, and the collisions, which now become possible, are added. The algorithm then continues as above. This method applies particularly well to molecular dynamics, because straight-line segments connect successive binary collisions. The procedure used by Trulsen (1972) in order to study the early solar system is somewhat analogous to that of Alder and Wainwright: he also sets up a list of all collisions possible at a given moment. Considering the elliptical tubes, which are swept by each particle in space, does this. The minimum distance between points on two elliptical orbits is calculated. If this distance is smaller than the sum of the particle radii, a collision is said to be possible, and the time at which this collision would occur is calculated. This algorithm has been used in the case of Keplerian orbits, but cannot be easily generalized to more complex gravitational fields.

In the gravitational approach, there is no explicit analytic relation giving the distance between two particles as a function of time. In particular, as a consequence of the gravitational field, the particle trajectories are for example elliptical around a central body such as the Sun or Saturn, or more complicated still, as is the case for self-gravitating systems. So, we have been led to use an approximate method, which is different from the approach used in molecular dynamics or by Trulsen. The distance between two particles can be expanded as a polynomial in terms of time. This calculation is done for each pair of particles and an iterative method is used to check if two particles collide during a given interval of time. The calculation is considered to have converged when the difference between two successive iterations is sufficiently small. It is carefully checked if the result obtained does represent a real root and not simply a minimum distance and if the collision time found is included in the chosen interval. We determine which particle pair would collide first during this particular interval. New orbital elements are then calculated for the two colliding particles, and we continue the algorithm from the current collision time. If during some time interval no collision occurs, the algorithm continues the search in the next time interval, and so on. If we consider sufficiently small time intervals, the method is quite accurate. Only five or six iterations are usually sufficient to find the collision time of two particles, nevertheless, the collision time takes about 70% of the overall computing time.

The evolution of all dynamical quantities of interest is surveyed by taking snapshots of the system from time to time. A check of the accuracy of the computation is that the total energy per unit mass should be conserved. As a consequence of computer rounding errors, the orbital parameters calculated after each collision are slightly in error. These errors accumulate for successive collisions and the total energy changes slowly. After more than 10,000 collisions, the relative error in the total energy is less than 10^{-5}. Any small initial uncertainty changes the future evolution of every particle. We have checked that, although individual orbits become completely different, the macroscopic properties of the system are not affected. The results obtained with different computers, with different rounding error procedures, are difficult to compare in details because the individual behaviour of each particle is completely changed from one to another, but give the same overall statistical properties for an initially given system.

A number of simple tests can significantly limit the number of collision candidates. For example, couples of particles which have well separated orbits have not to be considered as well as couples of particles which are azimuthally far from each other during the given time interval. More sophisticated tests are used; nevertheless a large amount of computing time is used in this approach. It is the price we have to pay to study small effects such as perturbations by small satellites or resonances effects. The computing time increases at least as the number of potential collision pairs. It is impossible to follow the movement of millions of particles, even using the largest computers currently available. Fortunately, the results for a few hundreds or a few thousands particles can be scaled in such a way as to simulate the evolution of a much more realistic system. As far as the system may be described in terms of a simple Boltzmann equation, it can be shown (Brahic, 1976, 1976a) that a change of the number N of particles or their size r affects only the speed of evolution, which is proportional to $N(r/R)^2$, where R is some characteristic dimension of the system. Thus, a system consisting of many small particles can be realistically simulated by a system containing fewer, but larger particles; if $N(r/R)^2$ is the same for both systems, the time scale of evolution is the same. A given physical system can be in principle simulated by a model with arbitrary values of N and r. There are however some limitations in practice. The procedure does become unsatisfactory if r is too large, because the gravitational force acting on each of two particles, which have just collided, will not be the same, introducing a distortion, which increases towards the centre of the system. Too small a number of particles is also unsatisfactory, because statistical fluctuations, proportional to $1/N^{1/2}$, are then too large. A minimum of few hundreds particles have to be used.

The collisional dynamics of a differentially rotating disc of particles have been studied by Lynden-Bell and Pringle (1974), Brahic (1975, 1976a), Goldreich and Tremaine (1978) and Lin and Papaloizou (1979). The main results can be found in their papers and can be summarized in the following way: after a very fast flattening of the order of few tens of collisions per particle, the system reaches a quasi-equilibrium state in which collisions still continue and in which the thickness of the newly formed disc is finite. By this, we mean that the centres of the particles do not lie in the same plane, Fig. 1. Under the combined effect of differential rotation and of inelastic collisions, the disc spreads very slowly; particles move both inwards and outwards carrying out some angular momentum. In the absence of external confining forces, the spreading time is of the order of the time it takes particles to random walk a distance equal to the ring width. The larger are the particles, the larger is the rate of spreading. The particles oscillate through the central plane of the disc at a frequency of the order of Ω and the mean number of collisions suffered by a particle passing through the disc is of the order of the normal optical depth. The viscous stress converts orbital energy into random kinetic energy. Since the collisions are imperfectly elastic, the energy in random motions is dissipated as heat. The rate of damping depends on the details of the collision process and the one-dimensional velocity dispersion σ reaches an equilibrium related to the vertical thickness h by $h \approx \sigma/\Omega$.

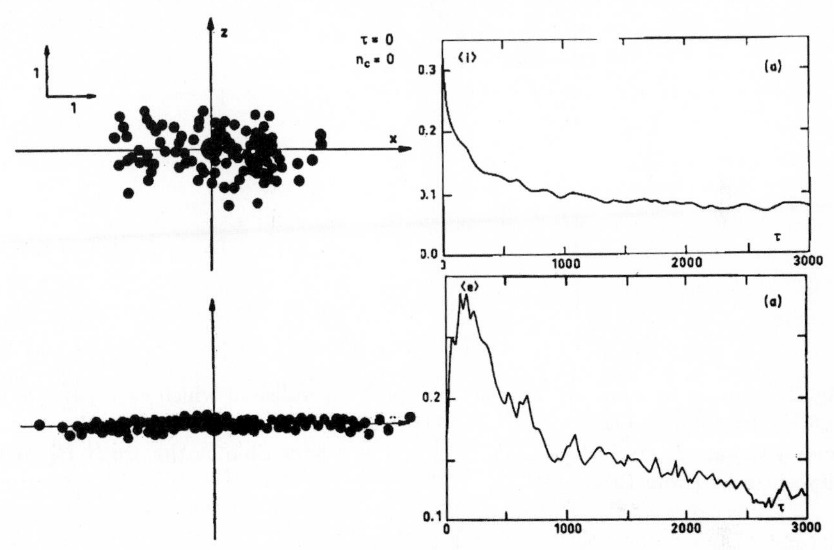

Fig. 1. Evolution of a three-dimensional system of colliding particles. The pictures on the left show the projection onto a plane containing the initial angular momentum vector of one hundred particles initially and after 2 500 mutual collisions. The graphs on the right show the evolution of the mean inclination $\langle i \rangle$ and of the mean eccentricity $\langle e \rangle$ as a function of time (from Brahic, 1977).

In an isolated disc made of a gravitating system of colliding particles, the total angular momentum is conserved but the total energy decreases. Inelastic collisions spread out the particles and extend the disc inwards and outwards. The system broadens under the combined effect of differential rotation and of collisions (equivalent to "friction" for a gas). This phenomenon (Brahic, 1975, 1977) is analogous to that described by various authors for an accretion disc around a compact object (Prendergast and Burbidge, 1968; Lynden-Bell and Pringle, 1974). The energy which is continually lost as a consequence of inelastic collisions is obtained at the expense of bodies moving inwards and outwards. Taking into account the conservation of angular momentum, the energy lost by the particles moving inwards is larger than the energy gained by the particles moving outwards: in spreading, the disc loses a small amount of its total energy. The time scale for this process is very long: it can be longer than the age of the Universe for rings around Saturn without nearby satellites (Brahic, 1977). This image of a slowly spreading disc is in good agreement with a homogeneous ring system with smooth edges, but observed rings present considerable radial structure. The discovery of narrow rings and sharp edges indicate either that rings are young or that confinement mechanisms are at work.

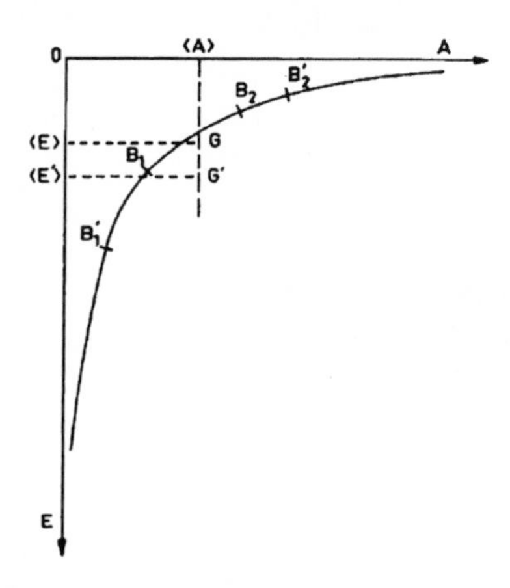

Fig. 2. Angular momentum A versus energy E for a disc in which each particle is on a Keplerian orbit $\left(E \approx -1/A^2\right)$. The spreading disc is represented at the initial time by the arc B_1B_2 with barycentre G and at a later time by the arc $B'_1B'_2$ with barycentre G' (from Brahic, 1977).

In average, radial spacing between the orbits of the particles is small as compared with the size of the particles, Keplerian motion introduces a differential rotation and collisions take place even if the orbits are circular, since an inner particle revolves faster and will overtake an outer one. During a collision, part of the relative velocity is transformed into vertical and radial motion. After some time, equilibrium is established between the radial and the vertical velocities. The thickness of the disc depends on the coefficient of restitution. For very inelastic collisions, the system is completely flattened. For intermediate collisions, the thickness of the disc is of the order of a few times the size of the particles for a system of equal sized particles. For almost elastic collisions, the thickness increases and it becomes three-dimensional.

For Keplerian motion, the circular velocity v of a particle at distance R from the centre is equal to $v = \sqrt{GM/R} = \omega R$ where G is the gravitational constant, M is the central mass and ω is the rotation frequency at a distance R from the centre. The radial spacing between two colliding particles is of the order of magnitude of the radius r of the particles and their mean relative velocity $\langle v_r \rangle$, due to differential rotation, is of the order of $r\omega$. The energy transferred into random motion is thus of the order of $\omega^2 r^2$.

After a collision, the energy loss is equal to $\left(1-k^2\right) v_r^2$. Therefore, in equilibrium:

$$\left(1-k^2\right)\left\langle v_r^2\right\rangle \approx \left\langle (r\omega)^2 \right\rangle , \tag{1}$$

and in first approximation:

$$\langle v_r \rangle \approx r\omega \Big/ \sqrt{1-k^2} \ . \tag{2}$$

Since we have assumed that the random velocities are uniformly distributed over all directions, the vertical and horizontal random velocities are also of the order of $r\omega / \left(1-k^2\right)^{1/2}$. As a consequence, the mean inclination $\langle i \rangle$ and the mean eccentricity $\langle e \rangle$ are of the order of $\langle v_r \rangle / v$ and, in first approximation:

$$\langle i \rangle \approx \frac{\alpha}{\sqrt{1-k^2}} \frac{r}{R} , \qquad \langle e \rangle \approx \frac{\beta}{\sqrt{1-k^2}} \frac{r}{R} , \tag{3}$$

where α and β are dimensionless constants of the order of unity. For reasonable values of $k (0 > k > -0.5)$, $\langle i \rangle$ and $\langle e \rangle$ are of the order of r/R. The energy, which is continually lost as a consequence of inelastic collisions, is obtained at the expense of the bodies moving inwards and outwards. Indeed, as a consequence of the conservation of angular momentum, the energy lost by particles moving inwards is larger than the energy gained by the particles moving outwards: in spreading, the disc loses a small amount of its total energy. For Keplerian motion, the angular momentum A and the energy E per unit mass of one particle are respectively proportional to $R^{1/2}$ and $-1/R$. In consequence, the curve $E = f(A)$ is concave. Assuming that the orbits are circular, the population of particles is represented at a given time by a set of points on the arc $B_1 B_2$ on Figure 2. The total energy and the angular momentum per unit of mass of the disc are respectively $\langle E \rangle$ and $\langle A \rangle$. The point of coordinates $\langle A \rangle$ and $\langle E \rangle$ is the barycentre G of the arc $B_1 B_2$ taking into account the mass distribution. After a while, the disc has broadened into an arc $B_1' B_2'$. Since the angular momentum A is constant and the total energy decreases, G moves down on a vertical line, that is away from the curve $E = f(A)$.

2.2 Further Models

After the study of this simple model, several additional parameters have been included in order to systematically explore a number of astrophysical problems and the results have been compared to the behaviour of the standard model. For example, in a real system, most collisions are between particles of very unequal size. Realistic models should include a distribution of masses. No real equipartition of energy is observed, but rather a kind of segregation between massive and less massive particles. The mean inclination and the mean eccentricity of the large particles is smaller than the mean eccentricity of the small particles. The more elastic the collisions, the bigger the separation.

The internal rotation of particles have been taken into account and the numerical studies show that exchanges between rotational and orbital energies lead to an equilibrium state of rotational properties after few collisions per particle and to a state in which the rotation axes are randomly oriented.

Models in which the rebound coefficients k and k' depend on the particle relative velocity have been considered. The final velocity distribution in the disc is very sensitive to the detailed collision law. Accretion of particles having small relative velocities and fragmentation of particles with large relative velocities are important effects to be included in the simulation. It is relatively easy to include accretion. It is much more difficult to take into account fragmentation phenomena, not only because the physics of fragmentation is badly understood, but also because the number of particles may increase so rapidly that computing times can reach enormous values. The first results show that fragmentation has probably played a major role in the dynamics of disc such as the proto-solar nebula.

When the oblateness of the central body is taken into account, a disc of colliding particles, which is not initially in the equatorial plane of the central body, is quickly brought back to this plane.

More realistic gravitational fields do not drastically change the results. On the contrary, encounters between particles, i.e. when two particles have a close fly-by without physical contact, play a major role in the dynamical evolution of the system, especially for large optical depth systems. Encounters are in some sense similar to elastic collisions, increasing the relative velocity of particles. Close encounters make the system 3d whereas inelastic collisions make the system quite flat.

All these results can be found in more than one hundred articles (see for example Borderies, Brahic, Brophy, Burns, Charnoz, Esposito, Goldreich, Greenberg, Hänninen, Lin, Lissauer, Lukkari, Marouf, Petit, Porco, Shu, Sicardy, Salo, Stewart, Thébault, Tremaine, Weidenschilling, Wetherill, ...), some of them are quoted in the reference list.

Among all the effects which can be added to the simple standard model, it turns out that the most important is the influence of a large body revolving around the system, for instance, a satellite in the case of Saturn's rings or a large proto-Jupiter in the case of the early solar system. We make some comments on this subject in the following paragraphs.

3 Disc – Satellite Interactions

One of the applications of the dynamical evolution of discs of colliding particles is the study of planetary rings, even if individual particles of the simulation have nothing to do with the real particles of planetary rings. Compared to the size of the system, ring-particles are several orders of magnitude smaller and their number is many orders of magnitude larger. But, using appropriate scaling factors, both systems can be fruitfully compared.

It was striking to note that the standard model and real rings seem to evolve quite differently. The former corresponds to a homogeneous disc with smooth edges like Saturn's rings seen by ground-based telescopes. The space exploration of the latter revealed heterogeneous discs with sharp edges and even confined rings and arcs. The discovery of narrow rings around Uranus (Elliot et al., 1977, 1981, 1984; Nicholson et al., 1978), the thousand of structures observed around Saturn in 1980 and 1981 by Voyager spacecrafts and the discovery of Neptune's arcs (Hubbard et al., 1984) were big surprises for astronomers (see part 4). These rings are far from being understood, but it seems that interactions between rings of colliding particles and external or embedded satellites play a major role.

Through gravity, a satellite can alter a disc particle's orbit making it elliptical. The overall effect of this on a population of particles in a ring is to increase the density of particles in some places and to decrease it in other places and finally to give rise to spiral density waves, similar to those in galaxies. If the perturbing satellite is exterior to the ring, this wave moves outwards, carrying negative energy and angular momentum. An isolated ring particle would return to its original position, but the rings are dense enough to make collisions between particles frequent, and particles involved in such collisions finally move inward towards the planet. The outside satellite tends to take angular momentum from the ring and put in its orbit. Conversely, an inner satellite, moving faster than the particles of the ring adds energy to nearby particles and, thus, transfers angular momentum to the ring from its orbit. This exchange of angular momentum has been studied by Lin and Papaloizou (1979), Goldreich and Tremaine (1979, 1980, 1981, 1982, 1983), Salo and Lukkari (1984), Salo (1985), Salo et al. (1988, see also this volume, page 330), Hänninen and Salo (1992), Sicardy and Brahic (1990), Thébault and Brahic (1999), Charnoz et al. (2000) and others. A satellite near a planetary ring can push the ring away and simultaneously the satellite is repulsed.

It is not intuitively obvious how a force of attraction can act to repel another body. This repulsion mechanism between a disc and a satellite depends on the presence of many ring particles that can often collide with each other. Changing a particle's angular momentum and thus its orbit depends on its having collisions or gravitational interactions (close encounters) with other particles. The physics involving many colliding particles sets the study of ring dynamics apart from classic celestial mechanics, in which few or no collisions are assumed, and from fluid dynamics, in which much more frequent collisions are known to occur than in a planetary ring. The mean free path of an individual particle is indeed much shorter in a fluid.

Satellites exert torques on the ring material at the locations of their low order resonances. A back of an envelope approach may illustrate this phenomenon in the case of a circular ring and a circular satellite orbit. This very crude calculation is developed below and cannot be considered as a demonstration. It is just a way to point out the physics of the problem. A satellite S is responsible for the build up of the eccentricity e and the decrease of the semi-major axis a of a test particle P, see Fig. 3. In a rotating frame of reference linked to the satellite, Ω, E and J represent respectively the angular velocity, the energy and the angular momentum. The indices p and S are relative to the test particle and to the satellite respectively. The test particle and the satellite are on very close orbits, such that $a_S - a_p \ll a_S$ where a_S and a_p are the semi-major axis of the satellite and of the particle respectively. The mass m_p of the particle is negligible with respect to the mass m_S of the satellite: $m_p \ll m_S$. The impulsion approximation is used to simulate the interaction between the test particle and the satellite, this means that the gravitational attraction of the satellite on the particle is considered only at closest approach when the central mass, the satellite and the test particle are aligned.

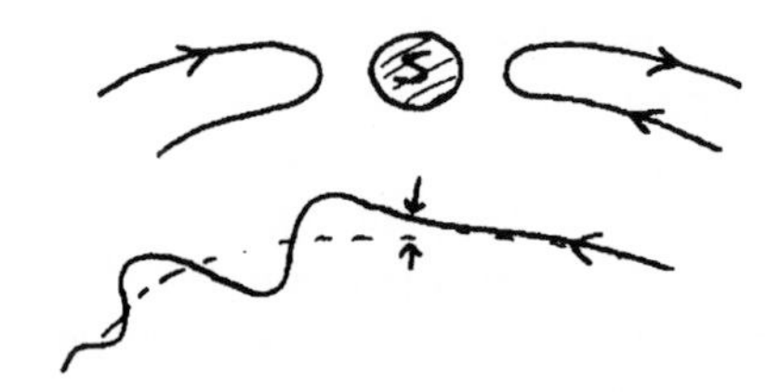

Fig. 3. Test particle path in a frame of reference rotating with the perturbing satellite S.

For the satellite moving on a circular orbit, the Jacobi constant $E_p - \Omega_S J_p$ is conserved. This can be easily shown by the computation of the derivative $dE_p/dt\,[dE_p/dt = \Omega_S/\left(dJ_p/dt\right)]$. For a Keplerian orbit one has

$$E = \frac{GM}{2a} \quad \text{and} \quad J = \sqrt{GMa\left(1 - e^2\right)}\,. \tag{4}$$

Using the impulsion approximation, the relative variations of energy and angular momentum can be easily calculated:

$$\frac{\Delta E}{E} = \frac{-\Delta a}{a} \tag{5}$$

$$\frac{\Delta J}{J} = \frac{1}{2}\left[\frac{\Delta a}{a} - \frac{\Delta e^2}{1 - e^2}\right] = \frac{1}{2}\left[\frac{\Delta a}{a} - \left(\Delta e^2\right)\right], \tag{6}$$

because $e = 0$.

Thus

$$\frac{\Delta J}{J} = -\frac{1}{2}\left[\frac{\Delta E}{E} + \Delta e^2\right]. \tag{7}$$

For a circular orbit, $\Omega_p = -2E/J$. Using the Jacobi integral, $\Delta E = \Omega_S \Delta J$, it is easy to derive:

$$\frac{\Delta J}{J}\left[1-\frac{\Omega_S}{\Omega_p}\right] = -\frac{1}{2}\Delta e^2\,, \tag{8}$$

so, J being equal to $ma^2\Omega_p$ one has:

$$\Delta J = -\frac{\Omega_p^2 a^2}{2\left(\Omega_p - \Omega_S\right)}\Delta e^2 \tag{9}$$

per unit mass.

During the encounter which lasts a time Δt equal to $\Delta t = X/\left(\Omega_p X/2\right) = 2/\Omega_p$, the particle receives an impulsion ΔP perpendicular to its motion, such as:

$$\Delta P = F\Delta t = \frac{GM_S\, m}{X^2}\frac{2}{\Omega_p}\,. \tag{10}$$

The trajectory is deflected by an angle $\alpha = \Delta P/P$, P being equal to $P = ma\Omega_p$. Thus:

$$\alpha = \frac{F\left(\Delta t\right)}{P} = \frac{\Delta P}{P} = \left(\frac{GM_S m}{X^2}\right)\left(\frac{2X}{\Omega_p X}\right)\left(\frac{1}{m\Omega_p a}\right) = \frac{2GM_S}{X^2\Omega_p^2 a}. \tag{11}$$

The orbit, which was initially a circle, becomes an ellipse of eccentricity $\Delta e \approx \alpha$ after the encounter with $\Delta e^2 \approx \left(\Delta e\right)^2 \approx \alpha^2$. Thus:

$$\Delta J = \frac{\Omega_p^2 a^2}{2\left(\Omega_S - \Omega_p\right)}\left(\frac{2GM_S}{X^2\Omega_p^2 a}\right)^2 = \frac{\Omega_p^2 a^2 2a}{3\Omega_p X}\frac{4G^2M_S^2}{X^4\Omega_p^4 a^2}\,. \tag{12}$$

ΔJ has the same sign as X, i.e.

$$\Delta J = \frac{8a}{3}\frac{\left(GM_S\right)^2}{\Omega_p^3 X^5}\mathrm{sign}\left(\Omega_S - \Omega_p\right). \tag{13}$$

A particle initially located on a circular orbit loses angular momentum after an encounter with an external satellite and, conversely, gains angular momentum after an encounter with an internal satellite. The torque per unit mass of the ring due to the satellite is equal to:

$$T = \frac{\Delta J}{\Delta t}\,, \quad \text{with} \quad \Delta t = \frac{4\pi}{3}\frac{a}{\Omega_p\left|X\right|}\,. \tag{14}$$

Making the sum over all particles and assuming that the satellite has close encounters with particles, which are always on circular orbits, the torque T is thus equal to

$$T \approx \frac{\left(GM_S\right)^2}{\Omega_p^2 X^4}\mathrm{sign}(X)\,. \tag{15}$$

Contrary to encounters, collisions tend to circularize the orbits. This is a justification of the crude assumption that the satellite has always encounters with ring particles located on circular orbits. The torque is maximum on the resonances. Summarizing over all resonances and assuming that particles lose memory between two collisions, the torque on a ring of mass $M = \sigma r \Delta r$, where σ is the surface mass density, due to a satellite of mass M_S is thus equal to:

$$T \approx \left(\frac{GM_S}{\Omega_p X^2}\right)^2 M_r \operatorname{sign}(X) . \tag{16}$$

Satellites play a dominant role in determining the morphology of rings of particles. A more sophisticated calculation indicates that they exert torques on the ring material at the locations of their low order resonances. For the simplest case of a circular ring and of a circular satellite orbit, the strongest torques occur where the ratio of the satellite orbit period to the ring particle period equals $q/(q \pm 1)$ where q is an integer. The torque is of the order of:

$$T_q \approx \pm q^2 \frac{G^2 M_S^2 \sigma}{\Omega^2 r^2} , \tag{17}$$

where Ω, r and σ are the orbital frequency, the radius and the surface mass density evaluated at the resonance location. The sign $\pm$ indicates that angular momentum is always transferred outwards in these interactions since the sum of the total mechanical and gravitational energy decreases.

The spacing between neighbouring resonances from a nearby satellite is very small. The widths of individual resonances are greater than the frequency separation so that they overlap. Thus, it is useful to sum the discrete resonance torques and to define the total torque on a narrow ringlet of width Δr:

$$T \approx \pm \frac{G^2 M_S^2 \sigma r \Delta r}{\Omega^2 X^4} , \tag{18}$$

where X is the separation between the satellite and the ringlet, such that $r \gg X \gg \Delta r$.

The presence of dissipation is necessary. Although the expression for torque does not reveal explicitly any dependence on dissipation, T_q would vanish without dissipation. The exact nature of the dissipation does not affect the expression for the torque. Particle collisions are evidently the main source of dissipation.

This resonance torque does not act on isolated test particles. For example, there are not enough collisions in the asteroid belt and such a mechanism is not responsible for the formation of Kirkwood gaps (Kirkwood, 1872) since Jupiter is too far! The rate of transfer of angular momentum can also be calculated by a perturbative approach without reference to individual resonances. The gravitational force on a ring particle due to a satellite is only effective at close encounters. Since, the tangential component of the relative velocity of the particle with respect to the satellite is reduced during encounter, angular momentum is exchanged with the net result that the ring experiences a torque. During encounter, a particle initially moving on a circular orbit acquires a radial velocity and thereafter moves on a Keplerian ellipse. In a frame of reference corotating with the perturbing satellite, all particles initially moving in circular orbits must follow similar pathes after encounters. Thus, each perturbing satellite generates a standing wave. In the inertial frame, each particle moves on an independent Keplerian ellipse, but the pericentres of these elliptical orbits and the phase of the particles on the orbits are such that the locus of the particles is a sinusoidal wave which moves through the ring with the angular velocity of the perturbing satellite. The damping of these waves, by collisions, results in a net exchange of angular momentum between the satellite and the ring particles. Torques not only transfer angular momentum but also energy, they can also change the eccentricity of the rings. Elliptic rings are probably associated with eccentric satellites.

The standard model developed by our team has been improved to take into account the perturbations of an external satellite. The Keplerian motion of the particles around the central mass is then perturbed. The use of a classical integrator taking into account the potential of the planet and the comparatively small potential of the satellite is not adapted to this problem. Indeed, cumulative errors due to the integrator could mask the weak effects of the perturbing satellite. Rather, we have used the orbital elements of the particles and we estimate their perturbations through Gauß equations. Using this model in the frame of planetary rings, Sicardy (1991) has shown that a combination of corotation and Lindblad resonances can confine and maintain arcs around a planet (Fig. 4). Thébault (1997) has written an enhanced version of this algorithm and has applied the results to the evolution of a disc of planetesimals perturbed by an early proto-Jupiter (Thébault and Brahic, 1999), Fig. 5. It now consists of a 4th order Runge-Kutta applied to Gauß perturbing equations. This method makes use of the fact that the orbits are always close to Keplerian ones, by treating mostly the perturbing part of the potential. For the case considered here, i.e. a perturbed and optically thin collisional disc, it achieves very satisfying performances compared to other, intrinsically more sophisticated, integrators (Burlisch-Stoer, Hal-Levison), considering in particular the precision and time steps constraints induced by the deterministic collision search.

4 Planetary Rings and Arcs

At present, the main astrophysical application of the simulation of viscous discs is the study of planetary rings. Between 1977 and 1989, space exploration and stellar occultations observed from ground-based observatories have revealed thousands of unexpected structures inside rings. As recently as 1977

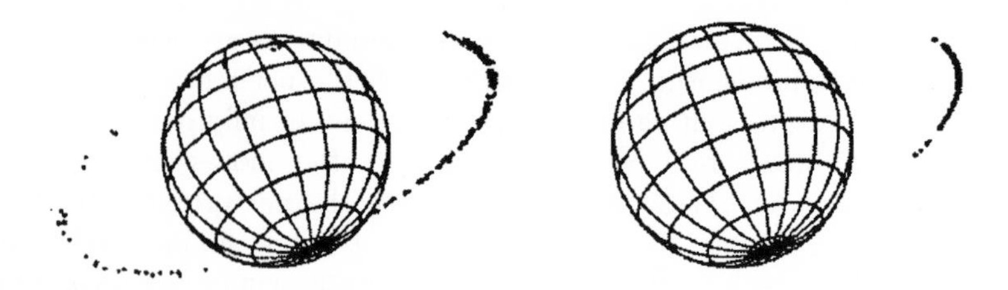

Fig. 4. Application to ring arcs. The left figure shows that a free arc of colliding particles spreads very rapidly due to the Keplerian differential motion. On the right, a shepherding satellite may confine the arc (from Sicardy, 1991).

we knew only one ringed planet in the solar system and scientists were struggling with the question: "Why does only Saturn have rings?" That question is clearly obsolete now. Instead, we turned to ask: "Why are the ring systems so different from one planet to the other? Do they share common features? What are the most efficient dynamical processes shaping the rings? Are planetary rings young or old?" For the 400th anniversary of Saturn's rings discovery, the Cassini mission will just be achieved with probably more new problems than solved questions. Rings are now considered as natural laboratories closer and much easier to observe than circum-stellar discs, accretion discs around neutron stars or black holes, galactic discs or the proto-solar nebula. Most of the information on rings can be found in two books (Brahic, 1984; Greenberg and Brahic, 1984) published after the 1982 I.A.U. colloquium on planetary rings. Information that is more recent can also be found in journals such as Icarus and Planetary and Space Science. There is not enough room here to make a complete review on rings. We give only a number of general remarks and a list of unsolved problems in the last part.

For a long time, planetary rings have fascinated men's imagination. They give a kind of three-dimensional perception, which lacks to most of astronomical objects, and they are beautiful! For anyone who has a pair of binocular or a small telescope, watching Saturn's rings is a pleasure! With lunar crescents and spiral galaxies, rings are among the objects of the sky the most widely represented by artists. They are a symbol of mysterious beauties of the heavens and of the future of science-fiction novels. They are even used by advertising companies to sell cars or electronic material.

But it is not for their beauty that astronomers are interested by rings. Even if they represent only a very small fraction of the solar system mass, they are a very peculiar and a very interesting class of objects. Their interest is not proportional to their mass, they are like perfume where a small amount can reveal much. Many physical phenomena widely spread in the Universe are at work inside the planetary rings. For astronomers, rings are a close natural laboratory ideal to study several important mechanisms which explain the behaviour of celestial objects far away from us, like spiral galaxies, or far in the past, like the primitive solar system.

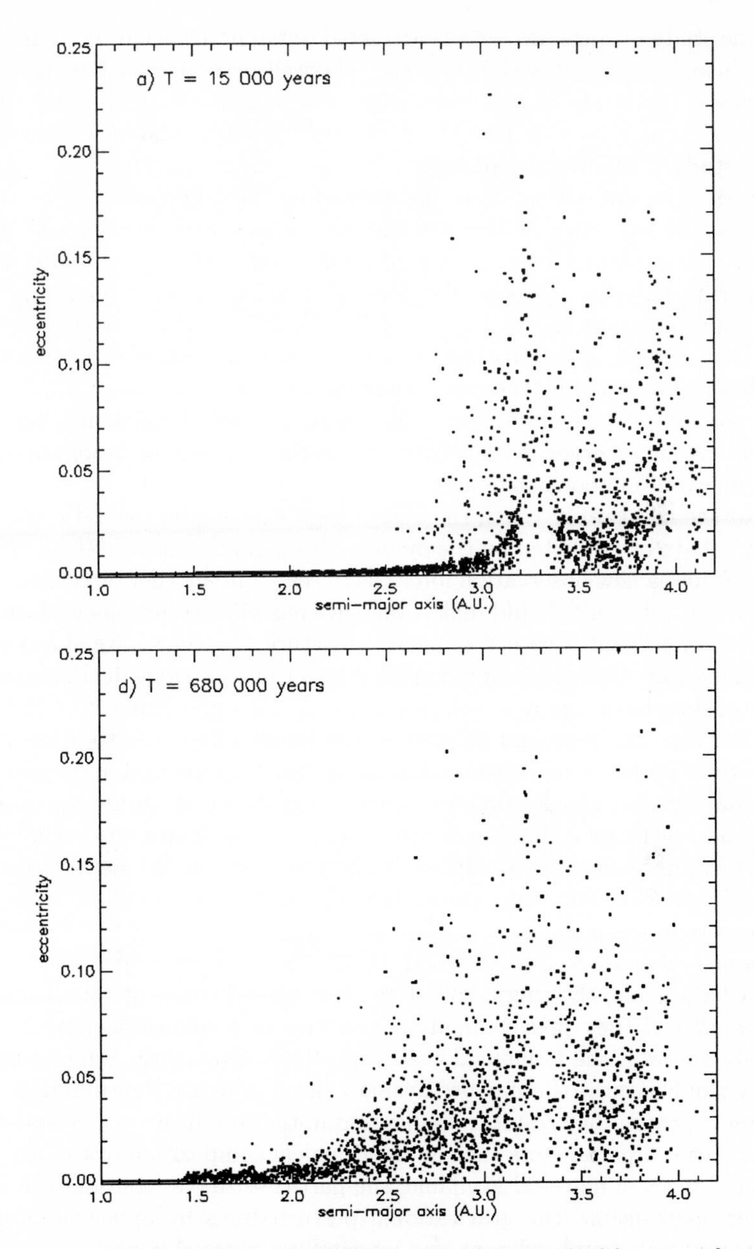

Fig. 5. Application to a disc of 2 500 planetesimals perturbed by an early proto-Jupiter. In a (e, a) diagram, the top figure shows that particles located near the 2:1 resonance are quickly excited by the perturber. Due to collisions, this increase of eccentricities, and thus of the relative velocities of the particles, propagates inwards (bottom figure), inhibiting accretion. This collisional diffusion mechanism is described in part 5 (from Thébault and Brahic, 1999).

The study of rings has always attracted outstanding scientific minds such as Galileo, Huygens, Cassini, Laplace, Maxwell or Poincaré. For more than 350 years, the study of planetary rings is a dynamic and evolving subject bringing regularly to us new theories, new disputes, and new discoveries. Their study is full of anecdotes.

Even if we are still far from understanding the origin and the behaviour of rings, the existence of ring systems around a planet is now well understood: it is a natural consequence of tidal forces and mutual collisions between ring particles. Despite the flood of new information on morphology and optical properties, we have very little evidence about what rings are, how they formed, and how they behave. Answers to such questions can only be obtained by building theoretical models and comparing their implications with past and future observations. A general picture of collisional flattening and spreading emerges, with structure governed in part by resonances with known or still unknown satellites.

For the last twenty years, rings have been challenging theoreticians: they never behave as predicted by the models. Every time a theory seems to bring the solution, a new observation forces us to search in another direction. Gravitation and collisions should make homogeneous planetary rings with smooth edges: they present a wealth of radial structure and they have sharp edges! Collisions and resonances should make circular narrow rings in the equatorial plane of the planet: many are elliptical or inclined and, furthermore they precess as a rigid body around the planet! Keplerian differential rotation should make rings perfectly homogeneous in the azimuthal direction and rapidly destroy any longitudinal asymmetry: broken rings, arcs and clumps are observed inside narrow rings! A number of first order problems is still unresolved. From time to time, papers are published with titles such as "Planetary rings explained", or "The rings of Uranus: theory", or "A simple model of Saturn's rings", or "An explanation for Neptune's ring arcs" giving the feeling that planetary rings are well understood. Reality is not so simple!

At large scale, the rings around the four giant planets are strikingly dissimilar even if they present many similarities at a small scale. Up to now, nobody knows if the tenuous Jovian ring, the spectacularly bright ring system around Saturn, the narrow Uranian rings, and the Neptunian arcs are different because they result from different initial conditions or because of different physical parameters such as the chemical composition of the particles, the total mass of material available, the particle-size distribution, the initial angular momentum, the gravitational perturbations by external satellites, the meteoritical bombardment or dust satellite material supply.

There are thousands of unexplained features in planetary rings, but the most mysterious seem to be the survival of azimuthal brightness asymmetries such as arcs, clumps within arcs, broken rings and the slow precession of elliptical rings around the planet.

4.1 Ring Radial Structure, Narrow Rings and Confinement Mechanisms

In a period of only a few years, between 1979 and 1981, our best resolution on Saturn's rings improved by a factor 10^4 with the fly-by encounters of Pioneer 11, Voyager 1 and 2. Contrary to what was expected, rings present considerable radial structure. Rather than being the smooth, continuous structures apparent in Earth-based images of Saturn's rings, planetary rings are more commonly characterized on the one hand by sets of narrow ringlets with sharp edges, sometimes slightly inclined or elliptical, sometimes kinky or broken, and on the other hand by dense rings with density waves or bending waves running through them.

A number of ring features are presented in the following Figures 6-15.

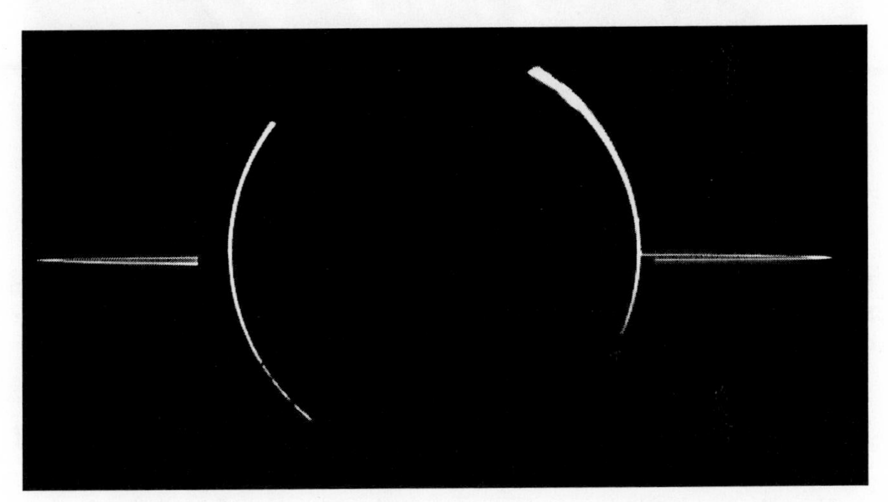

Fig. 6. Photograph of Jupiter main ring (NASA/JPL document)

Many narrow rings are approximately 10 kilometres large around Saturn, Uranus and Neptune. Some are extremely narrow. For example the width of the Uranus' γ ring is of the order of 800 meters for a circumference which is larger than 500,000 kilometres. The morphology of ring edges varies significantly. Some edges, such as the A ring outer edge, the Maxwell ringlet edges and Encke division edges, are extremely sharp. In several cases, the optical depth goes from zero to several tenths on a scale smaller than one kilometre. Stellar occultations by the Uranian rings show that, at some places, the transition from opaque to transparent is made in a radial distance of less than forty meters. Several narrow ringlets and gaps can easily be associated with nearby satellites. But many of them are not apparently linked with observed satellites. There are probably some unseen moonlets embedded in the ring systems. The outer A ring of Saturn and the main Jovian ring have both an

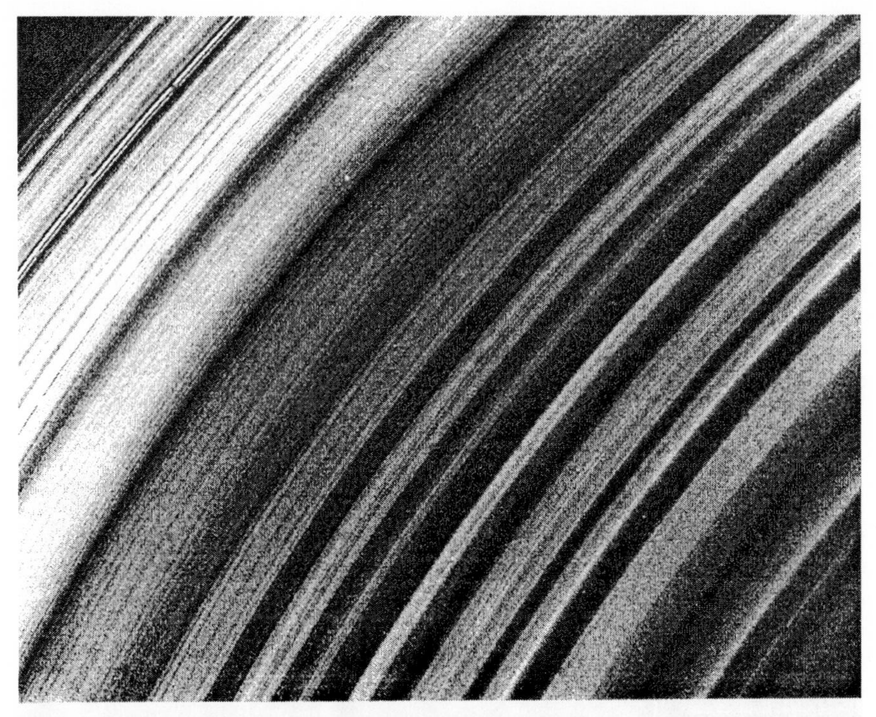

Fig. 7. Saturn's rings structures. In this 6,000 kilometre-wide photograph, the narrowest features seen here are about ten kilometres across (Voyager 2 - NASA/JPL document).

abrupt outer boundary with a small satellite orbiting just at the outside of the ring, like a guardian satellite.

Narrow rings and sharp edges require the presence of a confinement or a repulsion mechanism in order to halt the radial spreading. Goldreich and Tremaine (1979) have proposed that the torques exerted by the inner and outer satellites would be sources and sinks of the angular momentum transport outward through the ring. Viscous stress arises from inter-particle collisions and differential rotation. Depending on the mass and the distance of the perturbing satellites, there are several variants of the shepherding mechanism. Nearby and massive satellites produce wakes, which are perturbations which damp between successive close encounters of the ring particles with the satellite. Smaller and more distant satellites are rather responsible for perturbations, which can be described in terms of discrete resonances. These perturbations, extending around the entire circumference of the ring, are concentrated at resonant semi-major axes and are well separated by unperturbed regions from neighbouring resonances. A shepherd satellite may have just one resonance near the ring's edge or several resonances within the ring. Planetary rings can support leading and trailing spiral density waves which

Fig. 8. Composite photograph of an eccentric ringlet in Saturn's C ring (Voyager 2 - NASA/JPL document).

are controlled by a combination of the Coriolis force and of the ring's self gravity. This phenomenon is similar to density waves in Messier 51. Close to a resonance, the long spiral waves have wavelengths several orders of magnitude larger than the inter-particle spacing. These waves can exist only on the satellite side of the resonance and propagate toward and away from the resonance. The satellite excites the long trailing wave at the resonance and this wave carries away all of the angular momentum (positive or negative) which the resonance torque gives to the disc. The wave damps due to non-linear and viscous effects close to the resonance. The particles on the satellite side of the resonance move towards the resonance. If the resonance torque is sufficiently large, a gap opens on the satellite side of the resonance.

Goldreich and Tremaine (1980) have calculated the rate at which angular momentum and energy are transferred between a disc of colliding particles and a satellite, which orbit the same central mass in order to understand their mutual evolution. They only use the linear approximation and they assume that the satellite has a small eccentricity. A satellite on a circular orbit exerts a torque on the disc in the immediate vicinity of its Lindblad resonances and angular momentum is transferred outwards from the disc to

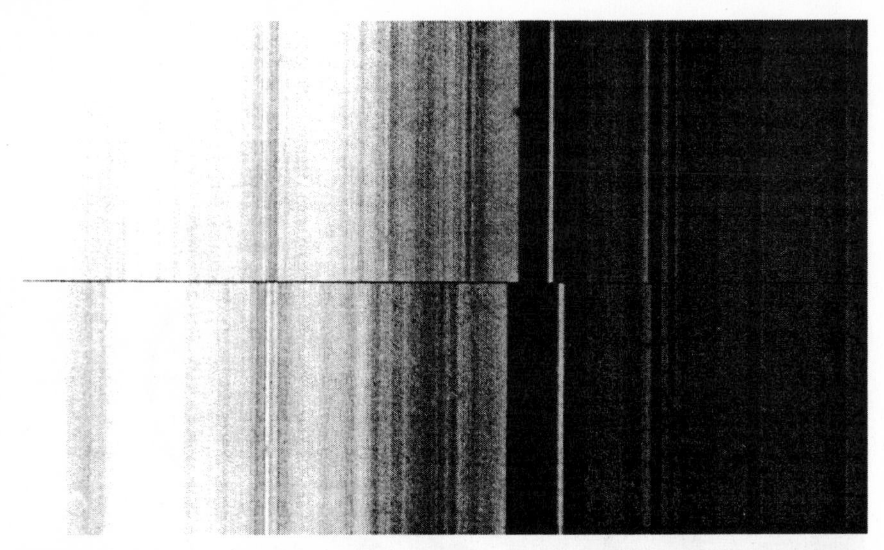

Fig. 9. Composite photograph of the outer edge of the Saturn's B ring and the inner part of Cassini division on opposite sides of the planet. Although the major features line up in the two images, there are differences in the location of some fine structures. The edge of the B ring differs by about 50 kilometres between the two images. This is probably due to the gravitational influence of Mimas (Voyager 2 - NASA/JPL document).

Fig. 10. Photograph of two kinky and discontinuous ringlets inside the Encke division taken on opposite sides of the planet (Voyager 2 - NASA/JPL document).

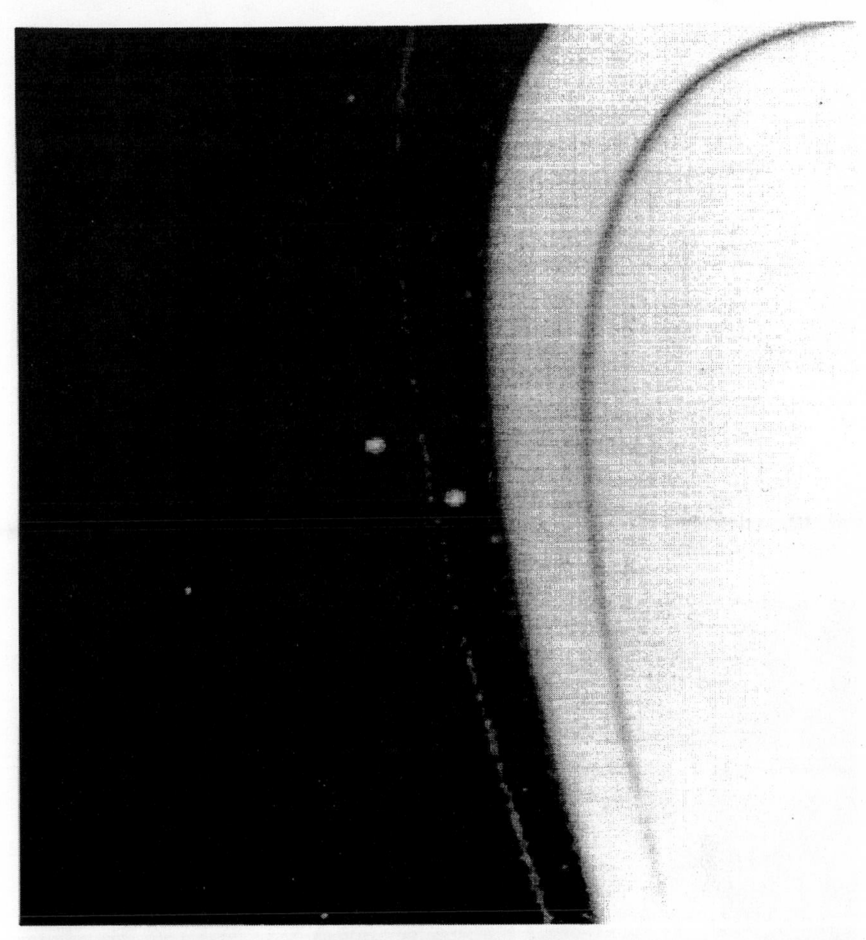

Fig. 11. Saturn's F ring and its two shepherding satellites (Voyager 2 - NASA/JPL document).

an external satellite or from an internal satellite to the disc. A satellite on an eccentric orbit exerts a torque on the disc both at Lindblad resonances and corotation resonances. In general, torques from Lindblad resonances increase the satellite's eccentricity, while those from corotation resonances damp it.

These results can provide an explanation for the formation of the Cassini division and the confinement of narrow rings by small satellites, which orbit within the ring system. The Encke gap is carved out of the Saturn's A ring by Pan, a small satellite which transfers angular momentum from its inner to its outer edge (Cuzzi and Scargle, 1985). Wakes are seen in the ring material adjacent to these edges. The outer edge of Saturn's A ring is located at the 7:6 resonance of the co-orbital satellites Janus and Epimetheus while the outer edge of the B ring is located at the 2:1 resonance of Mimas. The discovery

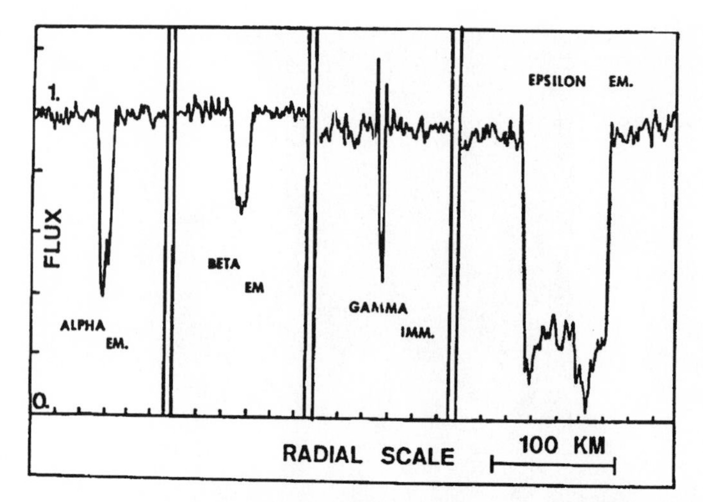

Fig. 12. The α, β, γ and ϵ rings of Uranus (from left to right) observed during stellar occultations show quite different structures (from Brahic, 1982).

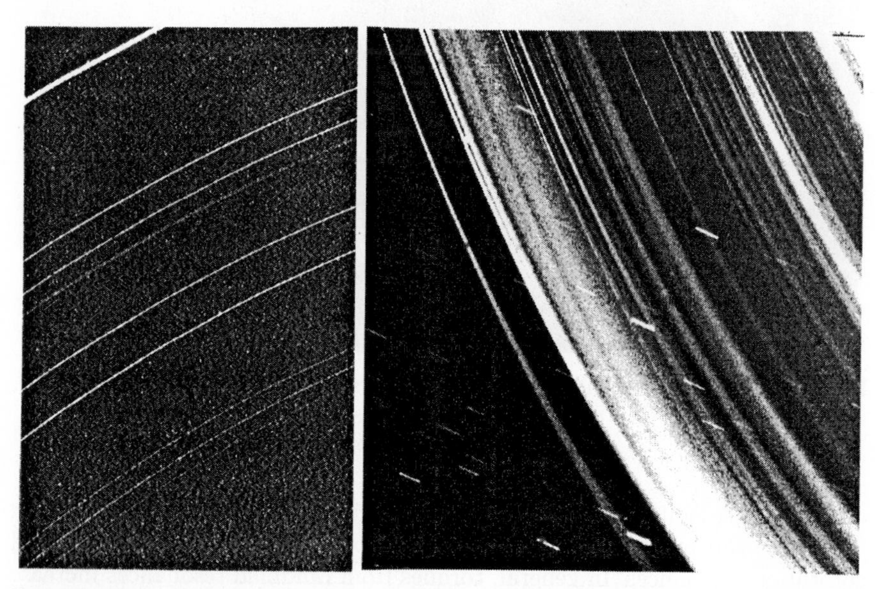

Fig. 13. Uranian rings in backscattered light (left) and in forward-scattered light (right). The ten classical rings are seen on the left image. The right image highlights micron-sized dust particles (Voyager 2 - NASA/JPL document).

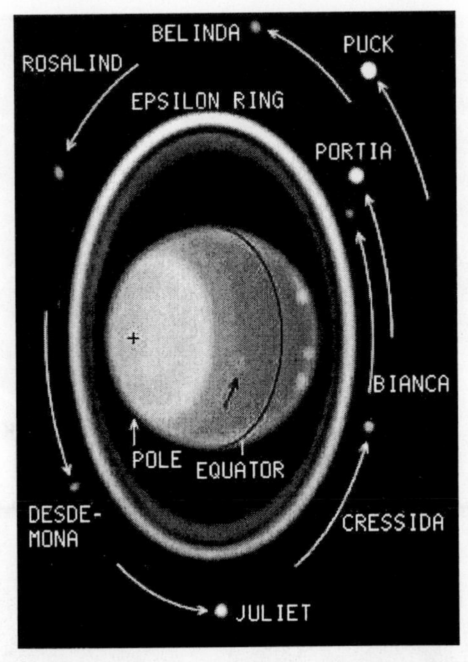

Fig. 14. Uranus rings and close satellites (NASA/ST document).

of the Saturn's F ring shepherds (Smith et al., 1981) and of Uranus' ϵ ring shepherds (Smith et al., 1986, 1987) were the most spectacular success for the shepherding theory.

Numerous wave-like features can be seen in the Voyager 1 radio occultation profiles of Saturn's rings (Marouf et al., 1986). Many are the signature of spiral density waves and bending waves excited by gravitational resonances with Saturn's satellites. Twenty nine clearly discernable wave-like features can be seen in the Saturn's ring radio occultation profile, many of which are coincident with known Lindblad and vertical resonances. For example, a spiral bending wave excited by the Titan 1:0 nodal inner vertical resonance is located in the C ring between 77 500 and 77 600 kilometres from the planet's center. This wave propagates outwards and the wave pattern rotates in a retrograde direction, opposite to the orbital motion of the ring particles (Rosen and Lissauer, 1988). In the C ring, a Mimas 4:1 density wave can be recognized. In the B ring, a Janus-Epimetheus 2:1 density wave can be observed. In the A ring, a Janus-Epimetheus 4:3 density wave and a Janus-Epimetheus 5:4 density wave propagates in the inner part of this ring. The source of several wave-like features, in particular within the Saturn's C ring, remains in doubt.

The discovery of narrow rings around the giant planets does not close the list of surprises. Planetary rings should be circular, because differential precession across an eccentric ring would soon lead to particle collisions that

Fig. 15. The rings of Neptune (Voyager 2 - NASA/JPL document).

would circularize the rings. Nevertheless, some rings are elliptical and have variable widths. Since many rings in both Uranus and Saturn systems are manifestly eccentric, there must be some mechanism to prevent this rapid circularization. Other rings are normally circular, but they are inclined. The boundaries of the ϵ ring can be fit by aligned Keplerian ellipses. Several Uranian rings are not circular.

The remarkable thing is that the elliptic rings precess slowly around the planet, just as they should due to the planet's oblateness. The best example is given by the Uranus' ϵ ring which precesses at a rate of 1.364^o per day as a solid body or by the Maxwell gap ringlet which precesses at a rate of 14.68^o per day at a rate in very good agreement with the rate determined by planetary oblateness alone. Whatever is the origin of the ring eccentricities, differential precession due to the quadrupole moment of the central planet should destroy the apse alignment. Nevertheless, several Uranus rings and Saturn's narrow ringlets undergo locked precession.

Goldreich and Tremaine (1979) have explored four different possible mechanisms for maintaining uniform precession of elliptic rings such as the ϵ ring. Apse alignment can be maintained by the self-gravity of the ring if the satellite is massive enough. A nearly guardian satellite could force uniform precession in the ring. Smooth pressure gradients, due to inter-particle collisions, cannot produce uniform precession without an unreasonably large ring thickness. In a system of colliding particles, the possibility of a discontinuity analog to a shock which cannot be modeled by smooth pressure gradients. Shock-like phenomena could force uniform precession, but have not been analyzed.

4.2 Planetary Arcs

The discovery of Neptune's arcs by Brahic and Hubbard (Hubbard et al., 1984, 1985, 1986, Brahic and Hubbard, 1989, Brahic et al., 1986) has led to a revision of the sequence of observations of the Voyager 2 spacecraft during its Neptune fly-by. Then, the rings have been observed in excellent conditions during 15 days in August 1989. The results have stimulated the search for azimuthal asymmetries in other planetary rings. Arcs, with sharp edges both in the azimuthal and in the radial direction, have been discovered around Neptune, Uranus, and Saturn (Ferrari and Brahic, 1993). Using the Voyager images, the quantitative study of the inhomogeneities in ten typical narrow rings of the Saturn's, Uranus', and Neptune's ring system reveals embedded arcs in many of them. They extend from 3,000 km to 10,000 km in azimuth and have abrupt edges in the longitudinal direction. The arcs are 3 to 7 times brighter than the rest of the ringlet, they are themselves made of clumps and substructures, 300 to 500 km long, distant for about 500 to 2,000 km one to each other. The photometric analysis shows that arcs are rich in dust and contain more large particles than the ring itself. Neptune's arcs have been observed at least during 6 years using ground-based observations (Sicardy, Roques, and Brahic, 1991), but there is no data on the stability of the arcs' substructures.

Keplerian differential rotation should quickly erase any clump or local inhomogeneity inside the rings. Surprisingly, apparently stable azimuthal brightness asymmetries are visible at different scales in the planetary rings. The more striking example is given by the arcs of Neptune: "Liberté", "Egalité" and "Fraternité". A detailed analysis of Voyager images reveals that many azimuthal brightness variations are visible at different scales: the arcs themselves appear to be made of small clumps of particles embedded in a more continuous and fainter ring.

Large-scale brightness asymmetries are visible in Jupiter's and Saturn's rings. Azimuthal asymmetries have been found by Camichel (1958) in Saturn's ring *A* and later observed from many ground-based observatories as well as from Pioneer 11 and Voyager 1 and 2 spacecrafts. In Saturn's ring *A*, the first and the third quadrants (as measured counter-clockwise from superior geocentric conjunction) are 10% to 15% dimmer than the second and fourth quadrants. This azimuthal asymmetry is far from uniform in the *A* ring and it preserves the same orientation to the observer, no matter where he is located with respect to the Sun and Saturn. This phenomenon has not been entirely explained, but is widely attributed to mutual shadowing of ring particles in the density "wakes" of larger ring members (Colombo, Goldreich and Harris, 1976).

Azimuthal brightness asymmetries have also been found in both, the main ring and in the nearby surrounding halo of Jupiter. On close-up views of Jupiter main ring, the far arm appears to be about 10% brighter than the near arm. Because a complete high resolution view of the whole Jupiter's ring is not available, it is impossible to know if it is a quadrant asymmetry similar to that observed for Saturn's *A* ring. Anyway, the mutual shadowing of ring particles cannot be an explanation in this case of Jupiter's diaphanous ring system: the particle density is too low! So, the cause of this asymmetry remains mysterious (Showalter et al., 1987).

At a much smaller scale, brightness asymmetries have been found in narrow ringlets. Some, like some Uranian rings, are the result of the ringlet's variable width. Others, like the Adams ring, are due to arcs or clumps of particles embedded in a continuous ring. Brightness variations as a function of longitude have been measured for Uranus' rings by Svitek and Danielson (1987). The observed substantial variations can be explained by the varying width of the rings. Neptune arcs are the more striking example of azimuthal brightness asymmetries in narrow rings. Whereas the Le Verrier ring seems rather constant at the resolution of Voyager 2 images, many azimuthal brightness variations are visible in the Adams ring at different scales: the arcs appear as small clumps of particles embedded in a continuous and faint ring.

The Voyager 1 and 2 spacecrafts found that the *F* ring of Saturn contains a remarkable diversity of features like kinks, clumps, and braids with typical longitudinal spacings estimated to range from 5,000 to 130,000 kilometres (Smith et al., 1981, 1982). The *F* ring's peculiar structure has been widely attributed to the gravitational perturbations from its two adjacent shepherding satellites Pandora and Prometheus. Cuzzi and Burns (1988) have re-analyzed the five abrupt depletions in the flux of trapped magnetospheric electrons observed by Pioneer 11 within a 2,000 kilometres wide band surrounding Saturn's *F* ring. Two of the five observed decreases are probably due to absorption by the *F* ring. However, none of the other three are likely to be caused by Pandora, Prometheus, or the *F* ring. They infer that the observed depletions of charged particles are caused by clumps of material with low optical depth of the order of 10^{-4} to 10^{-3}. This hypothetical belt of colliding objects could partly explain the observed structure of the *F* ring. This hypothesis raises the possibility that the *F* ring is not necessarily "shepherded" by Pandora and Prometheus over millions years, but is merely one of a series of transient events occurring at the very edge of the Saturn's Roche zone. Such moonlet belts could play an important role in order to explain ring arcs seen around Saturn, Uranus, and Neptune.

The Encke gap does contain two narrow, discontinuous ringlets which appear kinked and clumpy in the Voyager images, with morphology and particle size reminiscent of Saturn's *F* ring. The similarity of the Encke ringlets led

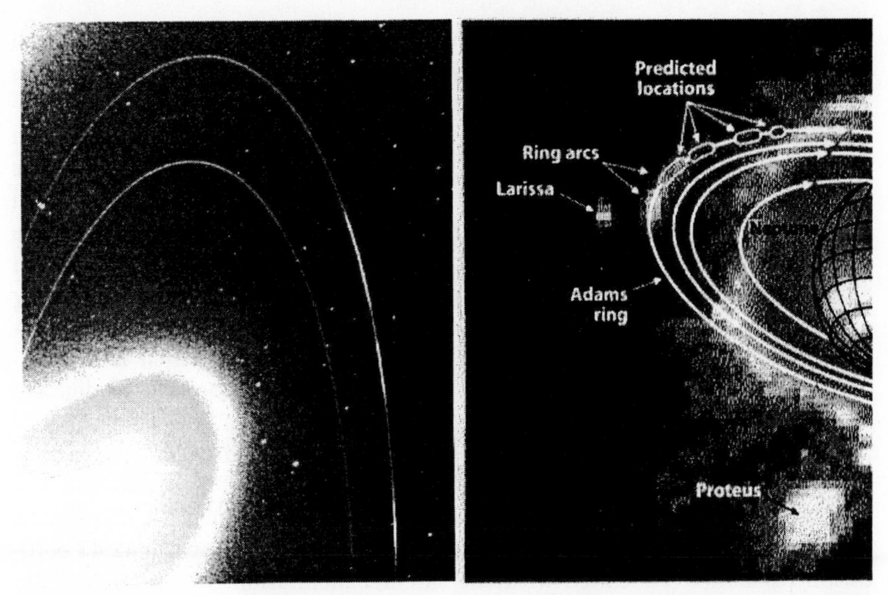

Fig. 16. At left, photograph of the Neptunian arcs taken in 1989 by the Voyager 2 spacecraft (NASA/JPL document). At right, photograph of the arcs taken in 1998 by the space telescope (NASA/ST document). The arcs are not at the position predicted by theoretical models.

to a suspicion that a moonlet (or more), embedded within the gap, could produce the observed ringlet structure and also keep the gap clear by gravitational torques (Cuzzi and Scargle, 1985; Hénon, 1981, 1984; Showalter et al., 1986).

A systematic quantitative analysis of the azimuthal brightness variations has been made by Ferrari (1992) for a number of narrow ringlets. Liberté, Egalité and Fraternité arcs have been observed over one week by the Voyager 2 spacecraft. They are respectively 4,400, 5,500, and 11,000 kilometers long. They have sharp edges in the longitudinal direction and their structure does not depend on the phase angle of observation. There are made of clumps which are 4 to 7 times more opaque than the ring in which they are embedded. A fourth arc lies just before Liberté arc and many clumps are also visible between the arcs.

Differential rotation due to Keplerian motion should quickly destroy the arcs and clumps: the time required to spread over 360° an azimuthal structure, that is 20 kilometers wide in the radial direction, is only about 5 years. Stable, non transient ring arcs would obviously require a longitudinal confinement mechanism. Three models have been proposed respectively by Lissauer (1985), Goldreich, Tremaine and Borderies (1986) and Lin, Papaloizou and Ruden (1987). In each case, azimuthal confinement is associated with a corotation resonance and a Lindblad resonance. A corotation resonance occurs at an orbital radius where the angular velocity matches the pattern speed of a perturbation potential. Test particles have equilibrium positions at azimuths where the perturbation potential is maximal. Inelastic collisions between particles cause energy loss and finally escape of the particles. In order to maintain ring particles in libration nearby a corotation resonance, the energy input is given by Lindblad resonances. A Lindblad resonance occurs at an orbital radius where a perturbation potential forces an orbiting test particle at its epicyclic frequency. In the Lissauer model, corotation resonances are located at the Lagrange points of a satellite. Lindblad resonances are provided by one or more additional satellites. Goldreich and Tremaine use only one satellite (providing both corotation and Lindblad resonances) which is inclined or eccentric with respect to the ring plane. Lin, Papaloizou and Ruden use also a single satellite, but corotation resonance arises from non-axisymmetric perturbations of the planet's gravitational potential. Porco (1991) has done a careful kinematical study of the rings and satellites of Neptune and has proposed that the Neptunian arcs are azimuthally confined by a resonant interaction with the nearby satellite Galatea. She found that the 42:43 corotation resonance associated with the satellite inclination, which falls within the Adams ring, can explain the confinement of the Liberté, Egalité, and Fraternité arcs. This model is not compatible with the July 1998 observations of Neptunian arcs with the Space Telescope (see Figure 16).

In fact, a complete dynamical study has to be done before the understanding of the arc's confinement. In particular, the strength of this resonance should be calculated in order to understand if it can explain the particularly sharp edges of the arcs. If this Galatea 42:43 resonance is responsible for the arc stability, it is not understood why some corotation sites should be populated while most of them are empty, why the arc's abrupt edges do not correspond to the corotation site boundaries, and how substructures are formed. It is still not known why some resonances should lead to strongly confined arcs while some other resonances which have apparently a strength of the same order of magnitude do not correspond to observed arcs. Even if corotation resonances can maintain already formed asymmetries, an explanation of the origin of such structures has still to be provided. Transient arcs should require the continuous creation, destruction, and replenishment of local concentrations of ring material.

4.3 Rings Origin

For a long time, it was believed that the planetary rings, which we see today are basically products of the same formation processes that gave rise to the regular satellite system of the giant planets or to the planets of the solar system. In fact, nobody knows if planetary rings are primordial or much more recent objects. Taking into account the number of efficient destruction mechanisms seen at work today in planetary rings, a number of astronomers believe that the rings are young. The discovery of narrow rings and sharp edges indicates either that rings are young or that confinement mechanisms are at work. No crucial observation can tell us if rings represent a failure of the innermost portion of a circum-planetary disc to accumulate into satellites inside the Roche limit or whether they are the result of the disruption of pre-existing satellites. In the first case, we can hope to discover some hints of initial conditions in the early solar system. In the second case, even if rings are young objects, the physical mechanisms they exhibit have a primordial importance for a better understanding of cosmogonical problems. Even, if planetary rings are not remnants of primordial accretion discs around planets, they have motivated critical thinking on dynamical processes relevant for the planetary origin: the tendency for an orbiting, collisionally interacting system to flatten into a disc, to spread by viscous shear, and probably to be truncated by large embedded bodies.

5 Other Discs and the Formation of Planets

Rings and discs, such as planetary rings, satellite systems, the solar system itself, the Milky Way, spiral galaxies, or accretion discs represent a fundamental class of celestial structures, but, despite the flood of new information on morphology, and optical and spectral properties, we have very little direct evidence about what discs are, how they form, or how they behave.

Can we learn something on planet and star formation from planetary ring studies? Astronomers do not agree on the answer and there is a lot of controversies on this subject. For many, planetary rings are probably examples of arrested growth, they are basically products of the same formation process that gave rise to the regular satellite systems surrounding each of the ringed planets, and they afford a good opportunity of studying some of the accretion mechanisms which operated at the time of the Sun and the planets formation. For some other astronomers, physical and chemical conditions are so different within planetary rings and the proto-planetary nebula that ring observations cannot be just naïvely extrapolated in order to describe the primitive nebula For them, it is not clear if colliding particles encircling a giant planet have anything to do with the primitive solar system. But, even if the scale of many phenomena is quite different in planetary rings and in the proto-planetary nebula, we can reasonably hope that both systems share several physical mechanisms in common.

It is interesting to note that, soon after their discovery, Saturn's rings became important in cosmological and cosmogonical speculations. Rings consisting of rotating matter were first introduced by Descartes (1644) with his theory of invisible vortices and an actual material ring was first postulated by Huygens twelve years later when he solved the problem of Saturn's telescopic appearance by imagining that this planet is surrounded by a ring. In 1750, Wright, in order to explain the Milky Way, mentioned the notion of a flat disc of stars revolving around a common center like the rings of Saturn.

Pierre Simon de Laplace began dynamical studies of Saturn's rings in 1787 and introduced the "nebular hypothesis" on the origin of the solar system. He proposed that the solar system resulted from a rotating nebula, which, as it contracted, had thrown off successive rings of matter. Each ring had aggregated and condensed into planets. In a similar way, proto-planets had thrown off rings that give rise to satellites. In this approach, Saturn's ring system was considered to be one or more unfinished satellites, just as the asteroid belt can be considered to be an unfinished planet. By the end of the nineteenth century, rings were phenomena intimately related in the minds of scientists to the origin of the solar system. At the end of the twentieth century, the origin of planetary rings is still unknown, but many scientists are convinced that, even if they were formed only recently, ring systems suffer a number of dynamical processes, which are relevant to planetary origin.

The dynamical processes of coagulation and fragmentation, density wave generation and propagation, gravitational instability, confinement mechanisms, ... all have received much study because they are important within planetary rings. It is clear that a number of these processes has played an important role during planetary formation. In this sense planetary ring studies have contributed to a better understanding of planetary formation processes.

The torque between a disc and a nearby satellite can play a role at the early stages of the solar system formation. The proto-planetary nebula forms a disc in orbit around the proto-Sun and interacts with the proto-Jupiter. Angular momentum is efficiently transferred between Jupiter and the proto-planetary disc and Jupiter's eccentricity is damped from the disc. The time scale of transfer is of the order of a few thousand years. This is so rapid that substantial changes in both the semi-major axis of Jupiter and the structure of the proto-planetary disc must have taken place at the beginning of the solar system formation.

In turn, the principles and techniques of several fields of astronomy have been applied for the modeling of the structure, the behaviour, and the evolution of planetary rings. For example, the explosion of new data on rings from planetary probes and sophisticated ground-based observing techniques has attracted astronomers whose past work has generally not focused on planets, but rather on the dynamics of galaxies and stellar accretion discs. Several concepts, from density waves to turbulence and stability, already studied in the frame of galactic dynamics and solar system formation, have recently been applied to planetary rings. They promise to reflect significantly on our understanding of the early evolution of the primitive nebula.

Superficially, apart from a difference in scale of a factor of about 10^5, planetary rings and the primitive nebula would seem to have many similarities. They are both made of innumerable discrete objects whose random motions are small compared to their circular speed. They are both spatially thin structures supported primarily by centrifugal equilibrium. They both have considerable internal structure. They are both subject to viscous shear. In both systems, large-scale electromagnetic interactions probably play a small role. Collective gravitational effects explain much of the structure of both objects.

There are, however, differences in the scale of the collective processes, which operate in planetary rings and in the primitive nebula. There is not only a difference in scale, but planetary rings have also a smaller particle mean free path, a higher surface density and a shorter particle orbit period. Time scales of evolution of dynamical processes are very much shorter for planetary rings. Planetary rings are relatively thinner than the primitive proto-planetary nebula. The thickness over radius ratio is of the order of 10^{-7} for planetary rings and of 10^{-2} for a proto-planetary disc. Ring particles have suffered 10^2 revolutions around the central body. This is much larger than the 10^4 to 10^8 revolutions of "particles" in the proto-planetary disc. Accretion is a rapid process around the proto-Sun, but is forbidden inside the Roche limit of the planets. The natural scale of self-gravitational disturbances in flattened distributions of matter with surface mass density σ and angular rotation speed Ω (Toomre, 1964) is much smaller for planetary rings than for a proto-planetary nebula. There are obvious differences in the physical nature of the constituent bodies.

Thousands of papers have been written on the origin of the solar system since four centuries. It is out of question to give a report on this subject here. Information can be found in several books (Brahic, 1982; Gehrels, 1978; Black and Matthews, 1985; Levy and Lunine, 1993). As an example, we just quote here an application of low optical depth viscous disc models on the early solar system, which is under study by Charnoz, Thébault and Brahic (2000). Some hints on the dynamical effect of an early proto-Jupiter on a disc of planetesimals in the inner solar system may be found with the three-dimensional simulation of a perturbed disc of test particles having mutual dissipative collisions.

Several analytical and numerical studies (see for example Greenberg et al. 1978, Wetherill and Stewart 1989, Barge and Pellat 1991, 1993, Greenzweig and Lissauer, 1990) have shown that the accretion process in a planetesimals swarm is not homogeneous, as initially supposed by Safronov (1969,1972). A few large objects always tend to emerge from the initial distribution. The number of these bodies being too small to have a significant global dynamical influence on the system, the relative velocities in the disc remains small, a few ms^{-1}, during the major part of the accretion process. As a consequence, the accretion rate of the growing bodies increases more and more because of a gravitational focusing factor, giving rise to the so-called runaway growth. These results were recently confirmed and further developed by Namouni et al. (1996) and Weidenschilling et al. (1997).

Nevertheless, in most of the studies, these accretion scenarios were obtained for ideally isolated discs, whose evolution is driven only by internal interactions. Such a simplification neglects all eventual external perturbations, in particular those due to growing proto-giant planets. The small size of Mars and the absence of planets between Mars and Jupiter have lead a number of authors to believe that accretion has been inhibited in this region at the beginning of the history of the solar system. Some of them have proposed that an early more or less massive Jupiter may have increased the velocity dispersion of the planetesimals in such a way that the accretion process has been aborted. Kortenkamp and Wetherill (1997) and Marzari et al. (1997) have numerically studied the coupled effect of the proto-Jupiter perturbations on a 2d swarm of planetesimals having no mutual interactions but submitted to gas drag. They have shown that the presence of a giant proto-planet may reduce the number density of bodies and may increase the encounter velocities in the resonant regions, mainly the 2:1, slowing down the accretion process in these parts of the disc.

Thébault and Brahic (1999) and Charnoz, Thébault and Brahic (2000) have considered the effect of proto-Jupiter perturbations coupled with collisions. Resonances between particles and an early proto-Jupiter are responsible for a collisional diffusion mechanism which drives the particle velocity distribution and the efficiency of the accretion mechanism. The larger is the mass of the proto-Jupiter, the larger is the effect. Mutual collisions may spread the strong resonant perturbations far outside the resonant regions. This mechanism is very efficient and can propagate the resonant perturbations far towards the inner part of the disc, especially if the proto-Jupiter is massive, for example of about $10^{-3} M_{\aleph}$, the final perturbed region extending until the orbit of Mars. In this perturbed area, relative velocities among the bodies exceed by far the escape velocities of kilometre-sized objects, in other words, they are much larger than the values of relative velocities in an unperturbed system where runaway accretion takes place. As a consequence, this diffusion mechanism can have a significant influence on the accretion process. The time scale of this mechanism is of the order of a few 10^5 years, a value comparable with that of runaway accretion of planetary embryos in unperturbed discs.

In addition to this diffusion mechanism, a number of bodies, which encounter the proto-Jupiter, are suffering significant trajectory deviations and are crossing the asteroid region with large relative velocity. This effect, studied by Charnoz, is very efficient to increase relative velocities in a disc of planetesimals. The coupled effect of diffusion mechanism and of trajectory deviation of Jupiter neighbours may inhibit planetary formation with a proto-Jupiter of only ten terrestrial masses. A new program, specially designed for the study of low optical depth discs, has been recently written by Charnoz. The first results indicate that the early proto-Jupiter should be massive enough to inhibit planet formation in the asteroid zone, but not too massive, otherwise even the Earth could not have been formed! The efficiency of this mechanism strongly depends on several parameters, which are until now badly known: perturber's mass and eccentricity, gas density, optical depth of the system, and loss of energy during the impacts.

Collisional diffusion, always occurs in a perturbed disc of interacting bodies. It would thus be useful to consider its possible occurrence and its possible role in other astrophysical systems where mutually colliding bodies may be gravitationally perturbed. One of the most interesting systems may be the newly discovered extra-solar planetary systems (e.g. Mayor and Quelloz 1995, Marcy and Buttler 1996). In all these systems, massive planets have been detected orbiting very close to their parent star. The detection of smaller terrestrial planets is still out of our current technical possibilities, but, as a first step, it should be interesting to study if terrestrial planets may be formed in systems which are so different from our solar system.

With a giant planet located close to the central star, collisional diffusion should be a very efficient mechanism depending on the initial location of the observed giant planets, which may have migrated inwards from outer regions, either as a consequence of planetesimals scattering (Fernandez and Ip 1984) or of tidal interaction with the disc (Ward 1997, Trilling et al. 1998). A numerical simulation, taking into account the time scale of the migration process, should be performed. A careful exploration of the numerous free parameters, i.e. speed and amplitude of the migration, formation time scale of the perturbing planets, initial planetesimals mass and optical depth, gas density, etc., should lead to reliable constraints on the possible formation, by accretion, of large planetary embryos in these planetary systems. But even if planetary embryos may be formed in such systems, which is far from being obvious, and the question of their survival on larger time scales is not yet solved.

New observational techniques have led to the discovery of discs around young stars. The most famous of them, around the star β Pictoris, have been extensively observed and studied. The existence of planets and comets is suspected around this star. Among all these discoveries, the observation of confined discs around stars (see Figure 17) is of special interest. Like a confined ring around Saturn, which is the result of the interactions with satellites, this disc is perhaps confined by an unseen planet. Again, disc-satellite interactions coupled with resonances and collisions are particularly important to explain these new observations.

6 Open Problems

In order to fully understand the rings' physics, there are many problems, which would have to be solved before a significant progress can be made. The particle size distribution and the ring thickness should be well known in order to create reliable dynamical models. The stability of tenuous and ethereal rings should be understood in order to check if they are ephemeral structures or if they are permanently linked with the main ring systems. Small particle effects and electro-magnetic interactions should be studied in order to clearly separate dynamical effects which affect all particles from radiation and magnetic effects which affect only small particles. We do not

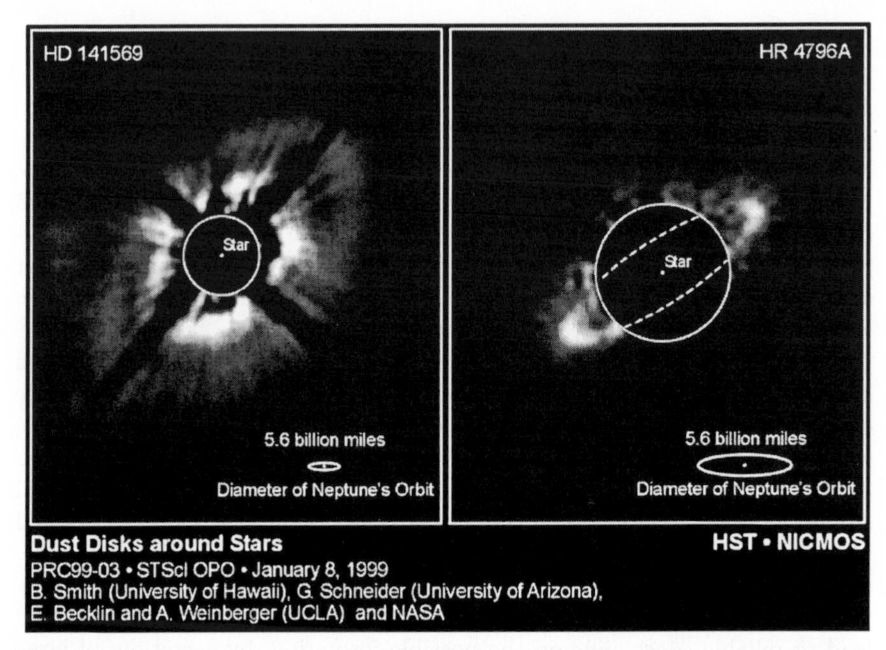

Fig. 17. This is not a picture of Saturn's rings, but the Space Telescope photograph, in the near infra-red, of discs of material around stars (NASA/ST document).

know how many one kilometre- to ten kilometre-sized objects are within or around the rings. A belt of large moonlets would have a very small optical depth and would be hard to detect, but should play an important dynamical role. Nearby satellites seem to play such an important dynamical role that a good knowledge of their number, mass and position should avoid the abusive use of "as yet unseen bodies" to explain the unexplained. It should be particularly important to know which features are time-variable within ring systems. It should be particularly interesting to detect any temporal change. That can be done for Neptune's arcs by a continuous set of ground-based occultation observations before a come back to Neptune. The Cassini mission to Saturn should provide a good opportunity to compare the Saturn's ring observations with Voyager data and to detect any change in the distribution of ring material. Variable satellite perturbations, pulsations, diffuse instability, etc. can produce changes in the morphology of those rings. The nature of viscosity is not known and the effects of meteoritical bombardment are poorly understood.

The full understanding of ring precession is probably a major progress which should be accomplished in the next future. Goldreich and Tremaine (1979) have studied several possible mechanisms for maintaining uniform precession in narrow rings such as the ϵ ring. But, Voyager radio occultation data challenge the hypothesis that self-gravity is responsible for maintaining apse alignment. But the understanding of the ring's precession will not be the end of the story. Indeed, a detailed analysis (French et al., 1986) shows that none of the Uranian rings is adequately modeled as having elliptical inner and outer edges in a fixed relative orientation: all of the rings have perturbed widths.

We now understand that a moon exerts coherent periodic perturbations on ring particles, which are in orbital resonance with it. These perturbations result in sharp outer edges of the A and B rings, as well as spiral density waves, which are observed propagating outwards from dozens of resonance locations within the rings. Nevertheless, there are still many open problems:

- For many sharp edges and many narrow rings, no shepherd satellite has been observed yet. There are more features still unexplained than features explained by existing satellites! It is always possible to assume the existence of small as yet undetected satellites which are large enough to confine the rings and small enough to escape detection.
- The strongest resonances in the Saturnian ring system often do not correspond to the largest gaps.
- The shepherding theory needs further refinements like the introduction of non-linear effects or the study of the detailed behaviour of ring particles near a sharp edge.
- Sharp edges are often too sharp! Many edges (for example Saturn's outer A and B rings, Encke division's edges, ...) are sharp on length scales which are about three orders of magnitude smaller than those over which the shepherding satellite torques act. Borderies, Goldreich and Tremaine (1982, 1983, 1985 and 1989) have introduced a major improvement to the shepherding mechanism: they have discovered that satellite perturbations can produce local angular momentum flux reversals in the direction of the viscous flux of angular momentum. This flux reversal can not only explain the presence of sharp edges, but is an essential feature of the shepherding mechanism.
- Saturn's F ring, Pandora, and Prometheus seem the best examples of the success of the shepherding theory, but a detailed analysis of the F ring structure shows that at least five components sometimes seem intertwined. The main component of the mass of this ring has not been detected. Thus the F ring does not provide yet a precise test of the theory.
- Cordelia and Ophelia have resonances coinciding with inner and outer edges of the ϵ ring. Goldreich and Porco (1987) have found that Cordelia has a 24:25 outer resonance with the inner edge of the ϵ ring, a 122:123 outer resonance with the λ ring and a 23:22 inner resonance with the δ ring while Ophelia has a 14:13 inner resonance with the outer edge of the ϵ ring and a 6:5 inner resonance with the γ ring. They thus proposed that Cordelia and Ophelia are shepherding the ϵ ring, that Cordelia is the outer shepherd for the δ ring and the λ ring and that Ophelia is the

outer shepherd for the γ ring. Apart from the resonances involving these two satellites and the ϵ, δ and λ rings, the fifteen known satellites cannot account alone for the structure of the rings of Uranus. The confinement of the 6, 5, 4, α, β and η rings has still to be explained and shepherding satellites have still to be discovered. It is possible that Uranus' rings could be maintained by a large number of small kilometre-sized satellites. Murray and Thompson (1990) have proposed that two nine-kilometre-sized, just below the ten kilometres threshold of detection in Voyager 2 images, satellites act as shepherds to three or more rings.

- There are many structures inside resolved narrow rings, which are not yet explained.
- Resonant angular momentum transfer to satellites seems to explain a number of observations, but why do some resonances explain abrupt outer edges (for example A and B rings) and nearby resonances of only slightly smaller strength do produce mere waves in the disc, with no hint of a gap?
- Some studies of ring-satellite resonances are just "numerology". Specially, when there are many close resonances, the coincidence is not really convincing as long as a dynamical study lacks.
- The short time variability of narrow rings has to be studied. We lack data over a long enough period of time. Additional stellar occultations by Uranian and Neptunian rings as well as the Cassini mission should bring some information on this topic.
- The torque due to gas drag associated with the distended Hydrogen atmosphere of Uranus has to be considered at least for the inner Uranian rings such as the 6, 5, 4, α and β rings. There is a problem in keeping α and β rings from collapsing into the atmosphere. They need huge shepherds, which have not been observed. The maximum shepherd torques on the ϵ ring from Ophelia and Cordelia is quite close to the minimum viscous torque and to the atmospheric torque. Perhaps, the flux reversal or a more accurate determination of the masses of the shepherds or an accurate calculation of the non linear torque at a ring edge may improve the situation.
- In the process of exciting density waves and maintaining sharp edges, moons exert a torque on ring particles, removing some of their angular momentum. Ring particles thus drift inwards, while moons move outwards. The associated time scales can be calculated (Goldreich and Tremaine, 1982). These timescales are uncomfortably short. There is a major problem: satellites just outside the Saturn's, Uranus', and Neptune's rings should have drift outwards until they fall into a resonance with external satellites like Rhea, Enceladus, Dione, Miranda, Ariel, Triton, etc. Such a situation is not observed. Are the rings young structures? Or is there any mechanism, which prevents the outward drift? For example, the decay of the A ring into Cassini division should take about 5,108 years. Janus and Prometheus move outwards at a velocity of about 3

centimeters per year and 70 centimeters per year, respectively. At the current speed, Janus was at the outer edge of Saturn's A ring $4 \cdot 10^8$ years ago and Prometheus $4 \cdot 10^6$ years ago.

- The local vertical thickness of a ring is an important dynamical parameter which characterizes the particle velocity dispersion. The controversy over the vertical structure of the rings, whether it is a single layer in thickness, a few layers or many layers thick, is not merely an academic exercise: it has implications for the dynamics and evolution of the rings. The "apparent" Saturn's rings thickness has been measured by Brahic and Sicardy (1981) from ground-based observations made during the transit of the Earth through the ring plane in March 1980. The observed brightness includes the warping of the disc, the contribution of large chunks and condensations, and the contribution of the E ring. Locally, the rings are extremely thin - perhaps as little as ten meters - (Lane et al., 1982, 1986; Marouf and Tyler, 1982; Porco et al., 1984; Zebker and Tyler, 1984). Dynamical arguments (Brahic, 1977; Brahic and Hénon, 1977; Brahic and Sicardy, 1981; Cuzzi et al., 1979, 1979a) imply that the characteristic thickness of a ring is a few times the diameter of the largest particles: dynamically, rings can be regarded as monolayers. But, for a true monolayer, there is no mutual shadowing by neighbouring particles unless the ring is observed edge-on. It seems that Saturn's ring is both a dynamical and an optical monolayer for the sub-system of the largest particles. But, if most of the mass resides in large particles which form an effective monolayer, most of the area, which plays the major role for optical observations, reside in a many-particles-thick layer of smaller particles (Brahic, 1977; Weidenschilling et al., 1984). Several estimates of the local thickness have been done. For example, modeling the observed Titan 1:0 nodal bending wave in Saturn's C ring with an existing bending wave theory gives a local ring thickness of less than about five meters (Rosen and Lissauer, 1988).
- The particle size distribution is also one of the main parameters which drives the ring's dynamics. Several estimates have been done using Voyager data as well as stellar occultation data and radar measurements of the reflection cross section of the rings. Observations of microwave opacity and near forwards scatter from Saturn's rings at wavelengths of 3.6 and 13 centimeters from the Voyager 1 ring occultation experiment give constraints regarding particle size distributions of the range of about one centimeter to a few meters (Marouf et al. 1986, Zebker and Marouf, 1985). A power law-type model with an index of the order of 3.3 to 3.4 is consistent with the data assuming an uniformly mixed set of particles in a many-particle-thick vertical profile. If the rings were assumed to be nearby a monolayer or if the vertical distribution of the particles were size dependent, the fractional opacity and the mass density estimates have to be lowered. Data from the Voyager radio occultation experiment seem to indicate that the main Saturnian rings are made of a larger num-

ber of smaller particles compared to the main Uranian rings which are populated by a smaller number of larger particles. It should be particularly important to know what is the distribution of ten meter- to one kilometre-sized objects in the rings.

- Tenuous and ethereal rings are also badly known. Rings are often embedded in halos or "atmospheres" and additional extremely faint rings extend well outside the Roche limit. For example, Jupiter is surrounded by an additional ring, which is far fainter than either of the other already faint components, and which is extending to a radius of 210,000 kilometers well farther than Amalthea. But, most of the ring material lies inside the orbit of Amalthea, though some material extends perhaps to the neighborhood of Thebe. This extended faint ring has a limited vertical extend which is smaller than 4,000 kilometers (Showalter et al., 1985). Plasma drag and the dominant destruction process, sputtering, should eliminate the tiny grains of this "gossamer" ring in a few thousand years. In order to remain visible today, ring material must be continually replenished from some source, either ejecta from Amalthea and Thebe or unseen bodies. Saturn's E ring extends from about 200,000 kilometers to far beyond 450,000 kilometers. There is a density peak near the orbit of Enceladus. Continued replenishment of the E ring by volcanic eruptions on Enceladus seems plausible. The full vertical thickness is of the order of 10,000 kilometers. The normal optical depth is of the order of $1.2 \cdot 10^{-6}$. The ten known Uranian rings are embedded in a highly structured dust disc. These faint rings share many common properties: they have low optical depths, they are immersed in a magnetospheric plasma, and they contain a significant fraction of micron-sized particles. Since dynamical evolution times and survival lifetimes for micron- to millimeter-sized grains are quite short, some replenishment mechanisms have to be found to explain the ethereal rings. Single-particle dynamics, rather than collective effects, most likely govern their form, and the majority of their particles have quite limited lifetime.
- Unseen material is probably present around giant planets. We do not know how many one kilometre- to ten kilometre-sized objects are within or around the rings. We do not know either how much material lies within the orbit of Mimas (or Amalthea, or Miranda, or Triton) and how much material shares the orbits of the main satellites. A belt of large moonlets should have a very small optical depth and should be hard to detect, but should play an important dynamical role. Nearby satellites seem to play such an important dynamical role that a good knowledge of their number, mass and position should avoid the abusive use of "as yet unseen bodies" to explain the unexplained.
- Temporal changes should be detected. It should be particularly important to know which features are time-variable within ring systems. It should be particularly interesting to detect any temporal change. That can be done for Neptune's arcs by a continuous set of ground-based occulta-

tion observations before a come back to Neptune. The Cassini mission to Saturn should provide a good opportunity to compare the Saturn's ring observations with Voyager data and to detect any change in the distribution of ring material. Variable satellite perturbations, pulsations, diffuse instability, ... can produce changes in the morphology of those rings.

- Electro-magnetic effects may be important in some cases. Small particle effects and magnetospheric interactions provide a wealth of phenomena which have only been appreciated after close looks of the ring systems from space missions. From observations of the relative strength of forward- and back-scattered light, it has become clear that micron- and submicron-sized grains dominate the populations of certain regions of planetary rings, including the Jupiter's halo and "gossamer" ring, the Saturn's E ring and the faint components of Neptune's rings. Also, the near radial spokes, that are seen to rotate across the dense Saturn's B ring and which seem to be elevated above the ring plane, are apparently composed of such small particles. Several physical phenomena are involved in magnetosphere-ring interactions and should be studied in detail. We give a few examples below. Plasma and energetic particles are absorbed by the particulates. Neutral atoms, molecules, ions and electrons are emitted upon the impact of energetic ions and simultaneously dust grains are eroded. The configuration of the magnetic field may cause a withdrawal of ring material into the central planet's atmosphere by impact ionization and sputtering. Dust particles are charged by electron and ion impact from the plasma as well as by the photoelectric effect from the solar ultra-violet radiation. Electrostatic disruption of individual grains as well as mutual repulsion and levitation may occur. Momentum exchange with the ambient plasma exerts a drag on the grains in the tenuous rings which causes radial drift towards or away from the planet. Stochastic fluctuations of the charge state of grains causes a diffusion of small dust particles throughout the magnetosphere.
- Improved observations of the rings and their environment should bring a better understanding of these phenomena. In particular, it would be important to reply to the following questions:
  - What is the nature of viscosity?
  - What other ices are in the rings besides H_2O?
  - What are the "non-icy" constituents? Are they homogeneously mixed or not?
  - Do "moonlet" compositions agree with typical "ring" compositions? Do they agree with each other?
  - What is the abundance, composition, and orbital distribution of the bombarding flux?
  - What is the structure and evolution of Saturn's dust cloud?
  - How efficient are the mechanisms of formation, transport, and destruction of dust?
  - What is the particle size distribution of the dust?

- What is the composition of the dust?
- What are the charged dust dynamics mechanisms?
- What are the effects of meteoritical bombardment? The principal erosion mechanism is bombardment by interplanetary dust. The dust ejects about 1,000 times its own mass, but most of the ejecta is re-accreted by the ring system. So, the net mass change of the ring system may be positive or negative. The composition of ring particles may be altered. The flux of micrometeoroids outside five astronomical units is poorly constrained; however, major changes in ring morphology over the age of the solar system are possible.

An even larger number of questions can be asked about gravitational perturbations by large bodies, initially located in the disc of the proto-solar nebula, or by planets around other stars. This field of research should be considerably developed during the XXIth century.

Planetary rings are both more common and more complex than suspected only twenty years ago. Facing such a complex behaviour, it is tempting to mix a few, well accepted, theoretical ideas and play with a large number of free parameters. It is striking to note that most of the recent models have very few characteristics in common and can barely be compared with one another. It would probably be more fruitful to study with a simple model unexplored fundamental physical mechanisms rather than to "play with parameters". A more complex model is not necessarily more realistic! We have also to be careful about the results of observations and their interpretation and we have not to be misled by irrelevant or non-significant observations. It is sometimes difficult to know if a selected sample of observations is really characteristic of the whole system. If a theoretical explanation is attractive enough, it is accepted even if the observational support is weak or non-existent. On the contrary, irrelevant observations detour researchers into dead-end theoretical considerations.

These planetary ring systems, interesting in their own right, also serve as prototypes for more massive disc systems such as the proto-planetary nebula, accretion discs around compact stars and spiral galaxies occurring elsewhere in astronomy. It is why their study is so important! But, for now, we have many more questions than answers. The confinement of arcs and the "rigid" precession of elliptical ringlets are probably among the main problems to solve today. Our best hope is the development of models studying simple fundamental mechanisms and continuous observations using each stellar occultation opportunity and space exploration like the Cassini mission and future missions to Neptune and Uranus. A real progress in our knowledge on rings will be made when we receive in the ground-based laboratories some individual ring particle samples in order to analyze them. Only then, we will be able to tell: "Give me one particle and I will explain the world!" Now, we are convinced that the importance of planetary rings is not proportional to their mass. They are like a perfume! A small amount of material gives a lot of information!

References

1. Alder, B.J. and Wainwright, T.E. 1959. Studies in molecular dynamics. I: General method. *J. Chem. Phys.* **31**, 459-466.
2. Barge, P. and Pellat, R. 1991. Mass spectrum and velocity dispersion during planetesimal accumulation: I. Accretion. *Icarus* **93**, 270-287.
3. Barge, P. and Pellat, R. 1993. Mass spectrum and velocity dispersion during planetesimal accumulation: II. Fragmentation. *Icarus* **104**, 79-96.
4. Boss, A. 1997. Giant planet formation by gravitational instability. *Science* **276**, 1836-1839.
5. Black, D.C. and Matthews, M.S. eds 1985. *Protostars and Planets II*, Tucson: University of Arizona Press.
6. Borderies, N., Goldreich, P., and Tremaine, S. 1982. Sharp edges of planetary rings. *Nature* **299**, 209-211.
7. Borderies, N., Goldreich, P., and Tremaine, S. 1983. Perturbed particle disks. *Icarus* **55**,124-132.
8. Borderies, N., Goldreich, P., and Tremaine, S. 1985. A granular flow model for dense planetary rings. *Icarus* **63**, 406-420.
9. Borderies, N., Goldreich, P., and Tremaine, S. 1989. The formation of sharp edges in planetary rings by nearby satellites. *Icarus* **80**,344-360.
10. Brahic, A. 1975. A numerical study of a gravitating system of colliding particles: Applications to the dynamics of Saturn's rings and to the formation of the solar system. *Icarus* **25**, 452-457.
11. Brahic, A. 1976. Thèse d'Etat, Université Paris VII.
12. Brahic, A. 1976a. Numerical simulation of a system of colliding bodies in a gravitational field. *J. Comp. Phys.* **22**, 171-188.
13. Brahic, A. 1977. Systems of Colliding Bodies in a Gravitational Field: I - Numerical Simulation of the Standard Model. *Astron. Astrophys.* **54**, 895-907.
14. Brahic, A. 1982. The rings of Uranus. In: *Uranus and the outer planets; Proceedings of the Sixtieth Colloquium, Bath, England, April 14-16, 1981.* Cambridge University Press, p. 211-236.
15. Brahic, A., ed. 1984. *Anneaux des planètes - Planetary rings*, Cepadues, Toulouse.
16. Brahic, A. and Hénon, M. 1977. Systems of colliding bodies in a gravitational field: II - Effect of transversal viscosity. *Astron. Astrophys.* **59**, 1-7.
17. Brahic, A. and Hubbard, W.H. 1989. The baffling ring arcs of Neptune. *Sky and Telescope* **77**, 606-609.
18. Brahic, A. and Sicardy, B. 1981. Apparent thickness of Saturn's rings. *Nature* **289**, 447-450.
19. Brahic, A., Sicardy, B., Roques, F., Mc Laren, R., and Hubbard, W.B. 1986. Neptune's arcs: Where and how many? *Bull. Amer. Astron. Soc.* **18**, 778.
20. Camichel, H. 1958. Mesures photométriques de Saturne et de son anneau. *Ann. d'Astrophys.* **21**, 231.
21. Charnoz, S., Thébault, P., and Brahic, A. to be published.

22. Colombo G., Goldreich, P., and Harris, A. 1976. Spiral structure as an explanation for the asymmetric brightness of Saturn's A ring. *Nature* **264**, 344-345.
23. Cuzzi, J.N., Burns, J.A., Durisen, R.H., and Hamill, P.M. 1979. The vertical structure and thickness of Saturn's rings. *Nature* **281**, 202-204.
24. Cuzzi, J.N., Durisen, R.H., Burns, J.A., Hamill, P. 1979a. The vertical structure and thickness of Saturn's rings. *Icarus* **38**, 54-68.
25. Cuzzi, J.N. and Scargle, J.D. 1985. Wavy edges suggest moonlet in Encke's gap. *Astrophys. J.* **292**, 276-290.
26. Cuzzi, J.N. and Burns, J. 1988. Charged particle depletion surrounding Saturn's *F* ring: Evidence for a moonlet belt? *Icarus* **74**, 284-324.
27. Descartes, R. 1644. Principia Philosophia. In *Oeuvres de Descartes*, eds C. Adams and P. Tannery, vol. VIII (Paris, 1905), p. 1-348.
28. Elliot, J.L., Dunham, E., and Mink, D. 1977. The rings of Uranus. *Nature* **267**, 328-330.
29. Elliot, J.L., French, R.G., Frogel, J.A., Elias, J.H., Mink, D.J., and Liller, W. 1981. Orbits of nine Uranian rings. *Astron. J.* **86**, 444-455.
30. Elliot, J.L. and Nicholson, P.D. 1984. The Rings of Uranus. in *Planetary Rings*, eds. R. Greenberg and A. Brahic (Tucson, Univ. of Arizona Press), 25-72.
31. Fernandez, J. A.; Ip, W.-H. 1984. Some dynamical aspects of the accretion of Uranus and Neptune – The exchange of orbital angular momentum with planetesimals. *Icarus* **58**, 109-120.
32. Ferrari, C. 1992. Thèse de l'Université Paris XI.
33. Ferrari, C. and Brahic, A. 1993. Azimuthal brightness asymmetries in planetary rings. I: Neptune's arcs and narrow rings. *Icarus* **111**, 193-210.
34. Ferrari, C. and Brahic, A. 1993. Azimuthal brightness variations in planetary rings: II - Arcs around Saturn and Uranus. *Icarus*, in press.
35. French, R.G., Elliot, J.L., and Levine, S.E. 1986. Structure of the Uranian rings. II. Ring orbits and widths. *Icarus* **67**, 134-163.
36. Gehrels, T., ed. 1978. *Protostars and Planets*, University of Arizona Press, Tucson.
37. Goldreich, P. and Porco, C.C. 1987. Shepherding of the Uranian rings. II Dynamics. *Astron. J.* **93**, 730.
38. Goldreich, P. and Tremaine, S. 1978. The velocity dispersion in Saturn's rings. *Icarus* **34**, 227-239.
39. Goldreich, P. and Tremaine, S. 1979. The excitation of density waves at the Lindblad and corotation resonances by an external potential. *Astrophys. J.* **233**, 857-871.
40. Goldreich, P. and Tremaine, S. 1980. Disk-satellite interactions. *Astrophys. J.* **241**, 425-441.
41. Goldreich, P. and Tremaine, S. 1981. The origin of the eccentricities of the rings of Uranus. *Astrophys. J.* **243**, 1062-1075.
42. Goldreich, P. and Tremaine, S. 1982. The Dynamics of Planetary Rings. *Ann. Rev. Astron. Astrophys.* **20**, 249-283.
43. Goldreich, P.; Tremaine, S.; Borderies, N. 1986. Towards a theory for Neptune's arc rings. *Astronomical Journal*, **92**, 490-494.
44. Goldreich, P. and Ward, W.R. 1973. The formation of planetesimals. *Astrophys. J.* **183**, 1051-1062.
45. Greenberg , R., Hartmann, W.K., Chapman, C.R., and Wacker, J.F. 1978. Planetesimals to planets: Numerical simulation of collisional evolution. *Icarus* **35**, 1-26.

46. Greenberg, R. and Brahic, A., eds 1984. *Planetary Rings.* Tucson: University of Arizona Press.
47. Greenzweig, Y. and Lissauer, J.J. 1990. Accretion Rates of Protoplanets. *Icarus* **87**, 40-77.
48. Hänninen, J. and Salo, H. 1992. Collisional simulations of satellite Lindblad resonances. *Icarus* **97**, 228-247.
49. Hayashi, C. 1981. Structure of the solar nebula, growth and decay of magnetic fields, and effects of magnetic and turbulent viscosities on the nebula. *Prog. Theor. Phys. Suppl.* **70**, 35-53.
50. Hénon, M. 1981. A simple model of Saturn's rings. *Nature* **293**, 33-35.
51. Hénon, M. 1984. A simple model of Saturn's rings - revisited. in *Planetary Rings*, A. Brahic ed., C.N.E.S., Cepadues, Toulouse, 363.
52. Heppenheimer, T. 1980. Secular resonances and the origin of the eccentricities of Mars and the asteroids. *Icarus* **41**, 76-88.
53. Hubbard, W. B., Brahic, A., Bouchet, P., Elicer, L.-R., Haefner, R., Manfroid, J., Roques, F., Sicardy, B., and Vilas, F. 1985. Occultation Detection of a Neptune Ring Segment. *Lunar and Planetary Science* **XVI**, 368-369.
54. Hubbard, W. B., Brahic, A., Sicardy, B., Elicer, L.R., Roques, F., and Vilas, F. 1986. Occultation detection of a Neptunian ring-like arc. *Nature* **319**, 636-640.
55. Hubbard, W. B., Vilas, F., Elicer, L.-R., Gehrels, T., Gehrels, J.-A., and Waterworth, M. 1984. Probable Ring of Neptune. *IAU Circ.* **1**, 4022.
56. Ip, W. H. 1987. Gravitational stirring of the asteroid belt by Jupiter zone bodies. *Beiträge zur Geophysik* **96**, 44-51.
57. Jewitt, D.C. and Danielson, G.E. 1981. The Jovian Ring. *J. Geophys. Res.* **86**, 8691-8697.
58. Kirkwood, D. 1872. On the formation and primitive structure of the solar system. *Proc. Amer. Phil. Soc.* **12**, 163.
59. Kortenkamp, S. and Wetherill, G. 1997. Gas drag effects on planetesimals evolving under the influence of Jupiter and Saturn. *Bull. of Am. Astr. Soc.* **29**, 28.06.
60. Lambert, J.H. 1761. *Cosmologische Briefe über die Einrichtung des Weltbaues.* Augsburg.
61. Lane, A.L., Hord, C.W., West, R.A.; Esposito, L.W., Coffeen, D.L., Sato, M., Simmons, K.E., Pomphrey, R.B., and Morris, R.B. 1982. Photopolarimetry from Voyager 2 – Preliminary results on Saturn, Titan, and the rings. *Science* **215**, 537-543.
62. Lane, A.L., Hord, C.W., West, R.A., Esposito, L.W., Simmons, K.E., Nelson, R.M., Wallis, B.D., Buratti, B.J., Horn, L.J., Graps, A.L., and Pryor, W.R. 1986. Photometry from Voyager 2 - Initial results from the Uranian atmosphere, satellites, and rings. *Science* **233**, 65-70.
63. Kant, I. 1755. *Allgemeine Naturgeschichte und Theorie des Himmels.* Königberg and Leipzig.
64. Laplace, P.S. de 1787. Mémoire sur la Théorie de l'Anneau de Saturne. *Mémoires de l'Académie Royale des Sciences de Paris.* 249.
65. Lecar, M. and Aarseth, S. 1986. A numerical simulation of the formation of the terrestrial planets. *Astrophys. J.* **305**, 564-579.
66. Lecar, M. and Franklin, F. 1997. The Solar Nebula, Secular Resonances, Gas Drag, and the Asteroid Belt. *Icarus* **129**, 134-146.
67. Lecavelier des Etangs, A. 1998. Planetary migrations and sources of dust in the β Pictoris disk. *Astron. Astrophys.* **337**, 501-511.

68. Levy, E.H. and Lunine, J.I. 1993 Protostars and Planets III, University of Arizona Press, Tucson.
69. Lin, D.N.C., Papaloizou, J.C.B., and Ruden, S.P. 1987. On the confinement of planetary arcs. *Mon. Not. R. Astron. Soc.* **227**, 75-95.
70. Lin, D.N.C. and Papaloizou, J.C.B. 1979. Tidal torques on accretion discs in binary systems with extreme mass ratio. *Mon. Not. R. Astron. Soc.* **186**, 799-812.
71. Lissauer, J. 1985. Shepherding model for Neptune's arc ring. *Nature* **318**, 544-545.
72. Lukkari, J. 1981; Collisional amplification of density fluctuations in Saturn's rings. *Nature* **292**, 433-435.
73. Lynden-Bell, D. and Pringle, J.E. 1974. The evolution of viscous discs and the origin of the nebular variables. *Monthly Not. Roy. Astron. Soc.* **168**, 603-637.
74. Marouf, E.A. and Tyler, G.L. 1982. Microwave edge diffraction by features in Saturn's rings – Observations with Voyager 1. *Science* **217**, 243-245.
75. Marouf, E.A., Tyler, G.L., Zebker, H.A., Simpson, R.A., and Eshleman, V.R. 1983. Particle size distributions in Saturn's rings from Voyager 1 radio occultation. *Icarus* **54**, 189-211.
76. Marouf, E.A., Tyler, G.L., and Rosen, P.A. 1986. Profiling Saturn's rings by radio occultation. *Icarus* **68**, 120-166.
77. Marzari, F., Scholl, H., Tomascella, L., and Vanzani, V. (1997) Gas drag effects on planetesimals in the 2:1 resonance with proto-Jupiter. *Planet. Space Sci.* **45**, 337-344.
78. Mayor, M. and Quelloz, D. 1995. A Jupiter-mass companion to a Solar-type star. *Nature* **378**, 355-359.
79. Marcy, G. and R. Buttler 1996. A planetary companion to 70 Virginis, *Astrophys. J.* **464**, L.147-L.151.
80. Namouni, F., Luciani, J-F, and Pellat, R. 1996. The Formation of Planetary Cores: A Numerical Approach. *Astron. & Astrophys.* **307**, 972-980.
81. Murray, C.D. and Thompson, R.P. 1990. Orbits of shepherd satellites deduced from the structure of the rings of Uranus. *Nature* **348**, 499-502.
82. Nicholson, P.D., Persson, S.E., Matthews, K., Goldreich, P., and Neugebauer, G. 1978. The Rings of Uranus: Result of the 10 April 1978 Occultation. *Astron. J.* **83**, 1240-1248.
83. Petit, J.M. and Hénon, M. 1987. A numerical simulation of planetary rings - I - Binary encounters. *Astron. Astrophys.* **173**, 389-404.
84. Petit, J.M. and Hénon, M. 1987. A numerical simulation of planetary rings - II - Monte Carlo model. *Astron. Astrophys.* **188**, 198-205.
85. Petit, J.M. and Hénon, M. 1988. A numerical simulation of planetary rings - III - Mass segregation, ring confinement, and gap formation. *Astron. Astrophys.* **199**, 343-356.
86. Pollack, J., Hubickyj, O., Bodenheimer, P., Lissauer, J., Podolack, M., and Greenzweig, Y. 1996. Formation of the giant planets by concurrent accretion of solids and gas. *Icarus* **124**, 62-85.
87. Porco, C.C., Nicholson, P.D., Borderies, N., Danielson, G.E., Goldreich, P., Holberg, J.B., and Lane, A.L., 1984. The eccentric Saturnian ringlets at 1.29 R_S and 1.45 R_S. *Icarus* **60**, 1-16.
88. Porco, C.C. 1991. An Explanation for Neptune's Ring Arcs. *Science* **253**, 995-1001.

89. Prendergast, K.H. and Burbidge, G.R. 1968. On the Nature of Some Galactic X-Ray Sources. *Astrophys. J. Lett.* **151**, L 83.
90. Rosen, P.A. and Lissauer, J.J. 1988. The Titan-1:0 nodal bending wave in Saturn's ring C. *Science* **241**, 690-694.
91. Safronov, V.S. 1969. *Evolution of the protoplanetary cloud and formation of the Earth and the planets.* Moscow, Nauka Press.
92. Safronov, V. S. 1972. Ejection of bodies from the solar system in the course of the accumulation of the giant planets and the formation of the cometary cloud. *The Motion, Evolution of Orbits, and Origin of Comets; Proceedings from IAU Symposium no. 45, held in Leningrad, U.S.S.R., August 4-11, 1970.* In: G.A. Chebotarev, E.I. Kazimirchak-Polonskaia, and B.G. Marsden (Eds.) International Astronomical Union. Symposium no. 45, Dordrecht, Reidel, p.329.
93. Salo, H. 1985. Numerical simulations of collisions and gravitational encounters in systems of non-identical particles. *Earth, Moon, and Planets* **33**, 189-200.
94. Salo, H. and Lukkari, J. 1984. Numerical simulations of collisions and gravitational encounters in systems of non-identical particles. *Earth, Moon, and Planets* **30**, 229-243.
95. Salo, H., Lukkari, J., and Hanninen, J. 1988. Velocity dependent coefficient of restitution and the evolution of collisional systems. *Earth, Moon, and Planets* **43**, 33-43.
96. Showalter, M.R, Burns, J.A., Cuzzi, J.N., and Pollack, J.B. 1985. Discovery of Jupiter's "gossamer" ring. *Nature* **316**, 526-528.
97. Showalter, M.R., Cuzzi, J.N., Marouf, E.A., Esposito, L.W. 1986. Satellite "wakes" and the orbit of the Encke Gap moonlet. *Icarus* **66**, 297-323.
98. Showalter, M.R., Burns, J.A., Cuzzi, J.N., and Pollack, J.B. 1987. Jupiter's ring system : New results on strcuture and particle properties. *Icarus* **69**, 458-498.
99. Sicardy, B. 1988. *Etude observationnelle, analytique et numérique des environnements planétaires.* Thèse d'Etat.
100. Sicardy, B. 1991. Numerical Exploration of Planetary Arc Dynamics. *Icarus* **89**, 197-219.
101. Sicardy, B and Brahic, A. 1990. The new rings – Contributions of recent ground-based and space observations to our knowledge of planetary rings. *Advances in Space Research* **10**, 211-219.
102. Sicardy, B., Roques, F., and Brahic, A. 1991. Neptune's rings, 1983-1989: Ground-based stellar occultation observations. I - Ring-like arc detections. *Icarus* **89**, 220-243.
103. Smith, B.A., Soderblom, L.A., Beebe, R. F., Boyce, J.M., Briggs, G., Bunker, A., Collins, S.A., Hansen, C., Johnson, T.V., Mitchell, J.L., Terrile, R.J., Carr, M.H., Cook, A.F., Cuzzi, J.N., Pollack, J.B., Danielson, G.E., Ingersoll, A.P.; Davies, M.E., Hunt, G.E., Masursky, H., Shoemaker, E.M., Morrison, D., Owen, T., Sagan, C., Veverka, J., Strom, R., and Suomi, V.E. 1981. Encounter with Saturn - Voyager 1 imaging science results. *Science* **212**, 163-191.
104. Smith, B.A., Soderblom, L.A., Batson, R., Bridges, P., Inge, J., Masursky, H, Shoemaker, E., Beebe, R., Boyce, J., Briggs, G., Bunker, A., Collins, S.A., Hansen, C.J., Johnson, T.V., Mitchell, J.L., Terrile, R.J., Cook, A.F.; Cuzzi, J., Pollack, J.B., Danielson, G.E., Ingersoll, A., Davies, M.E., Hunt, G.E., Morrison, D., Owen, T., Sagan, C., Veverka, J., Strom, R., and Suomi, V.E. 1982. A new look at the Saturn system: The Voyager 2 images. *Science* **215**, 505-537.

105. Smith, B.A., Soderblom, L.A.,B.A. Smith, L.A. Soderblom, R. Beebe, D. Bliss, J.M. Boyce, A. Brahic, G.A. Briggs, R.H. Brown, S.A. Collins, A.F. Cook II, S.K. Croft, J.N. Cuzzi, G.E. Danielson, M.E. Davies, T.E. Dowling, D. Godfrey, C.J. Hansen, C. Harris, G.E. Hunt, A.P. Ingersoll, T.V. Johnson, R.J. Krauss, H. Masursky, D. Morrison, T. Owen, J.B. Plescia, J.B. Pollack, C.C. Porco, K. Rages, C. Sagan, E.M. Shoemaker, L.A. Sromovsky, C. Stoker, R.G. Strom, V.E. Suomi, S.P. Synnott, R.J. Terrile, P. Thomas, W.R. Thomson, and Veverka, J. 1986. Voyager 2 in the Uranian system:imaging science results. *Science* **233**, 43-64.

106. Smith, B.A., Soderblom, L.A., D. Banfield, C. Barnet, A.T. Basilevsky, R. Beebe, K. Bollinger, J.M. Boyce, A. Brahic, G.A. Briggs, R.H. Brown, C. Chyba, S.A. Collins, T. Colvin, A.F. Cook II, D. Crisp, S.K. Croft, D. Cruikshank, J.N. Cuzzi, G.E. Danielson, M.E. Davies, E. De Jong, L. Dones, D. Godfrey, J. Goguen, I. Grenier, C.J. Hansen, C.P. Helfenstein, C. Howell, G.E. Hunt, A.P. Ingersoll, T.V. Johnson, J. Kargel, R. Kirk, D.I. Kuehn, S. Limaye, H. Masursky, A. Mac Ewen, D. Morrison, T. Owen, W. Owen, J.B. Pollack, C.C. Porco, K. Rages, P. Rogers, D. Rudy, C. Sagan, J. Schwartz, E.M. Shoemaker, M. Showalter, B. Sicardy, D. Simonelli, J. Spencer, L.A. Sromovsky, C. Stoker, R.G. Strom, V.E. Suomi, S.P. Synnott, R.J. Terrile, P. Thomas, W.R. Thomson, A. Verbiscer, and J. Veverka 1989. Voyager 2 at Neptune: Imaging Science results. *Science* **246**, 1422-1449.

107. Spitzer, L. 1939. The dissipation of planetary filaments. *Astrophys. J.* **90**, 675.

108. Svitek, T. and Danielson, G.E. 1987. Azimuthal brightness variation and albedo measurements of the Uranian rings. *J. Geophys. Res.* **92**, 14979-14986.

109. Thébault, P. 1997. *Les processus d'accrétion dans un disque de planétésimaux perturbé par un proto-Jupiter.* Thèse de doctorat.

110. Thébault, P. and Brahic, A., 1999. Dynamical influence of a Proto-Jupiter on a disc of colliding planetesimals. *Planet. Space Sci.* **47**, 233-243.

111. Toomre, A. 1964. On the gravitational stability of a disk of stars. *Astrophys. J.* **139**, 1217-1238.

112. Trilling, D.E., Benz, W., Guillot, T., Lunine, J.I.,Hubbard, W.B., and Burrows, A. 1998. Orbital Evolution and Migration of Giant Planets: Modeling Extrasolar Planets. *Astrophys. J.* **500**, 428-439.

113. Trulsen, J. 1972. On the rings of Saturn. *Astrophys. Space Sci.* **17**, 330.

114. Ward, W.R. 1981. Solar nebula dispersal and the stability of the planetary system I. Scanning secular resonance theory. *Icarus* **47**, 234-264.

115. Ward, W.R. 1997. Survival of Planetary Systems. *Astrophys. J.* **482**, L211-214.

116. Weidenschilling, S. 1975. Mass loss from the region of Mars and the asteroid belt. *Icarus* **26**, 361-3666.

117. Weidenschilling, S.J., Chapman, C.R., Davis, D.R., and Greenberg, R. 1984. Ring particles – Collisional interactions and physical nature. in *Planetary Rings*, eds. R. Greenberg and A. Brahic (Tucson, Univ. of Arizona Press), 367-415.

118. Weidenschilling, S. and Davis, R. 1985. Orbital resonances in the solar nebula: implications for planetary accretion. *Icarus* **62**, 16-29.

119. Weidenschilling, S., Spaute, D., Davis, R., Marzari, F., and Ohtsuki, K. 1997. Accretional evolution of a planetesimal swarm: 2. the terrestrial zone. *Icarus* **128**, 429-455.

120. Wetherill, G.W. 1980. Formation of the terrestrial planets. *Ann. Rev. Astron. Astrophys.* **18**, 77-113.
121. Wetherill, G. 1989. Origin of the asteroid belt. In *Asteroids II*, eds R. P. Binzel, T. Gehrels and M. S. Matthews, p. 661-680. University of Arizona Press, Tucson.
122. Wetherill, G. W. 1992. An alternative model for the formation of the asteroids. *Icarus* **100**, 307-325.
123. Wetherill, G. and Stewart, G. 1989. Accumulation of a swarm of planetesimals. *Icarus* **77**, 330-357.
124. Wetherill, G. and Stewart, G. 1993. Formation of planetary embryos: Effects of fragmentation, low relative velocity, and independent variation of eccentricity and inclination. *Icarus* **106**, 190-209.
125. Zebker, H.A., Marouf, E.A., and Tyler, G.L. 1985. Saturn's rings: Particle size distributions for thin layer model. *Icarus* **64**, 531-548.
126. Zebker, H.A. and Tyler, G.L. 1984. Thickness of Saturn's rings inferred from Voyager 1 observations of microwave scatter. *Science* **223**, 396-398.

Numerical Simulations of the Collisional Dynamics of Planetary Rings

Heikki Salo

Division of Astronomy, University of Oulu, FIN-90410 Oulu, Finland.
e-mail: heikki@sun4.oulu.fi

Abstract. Numerical simulations of planetary ring dynamics are reviewed, with main emphasis on local 3-dimensional simulations, which utilize a co-moving calculation cell with periodic boundary conditions. Various factors affecting the local balance between collisional dissipation and viscous gain of energy from the systematic velocity field are considered, including gravitational encounters and collective gravitational forces besides physical impacts. Simulation examples of the effects of particle size distribution, particles' spin motion, and different forms of the coefficient of restitution are given. Viscous stability properties are also briefly discussed: examples of both instabilities and overstabilities are given. In this context 2D-simulations are useful, eventhough physically unrealistic even for extremely flattened planetary ring systems.

1 Introduction

Planetary rings consist of numerous small (cm to meter-sized) icy particles orbiting the central planet on nearly circular, almost co-planar orbits. In dense rings the impacts between particles are very frequent, each particle colliding several times per orbital revolution. Even if the impact velocities are small, below 1 cm/s, a significant fraction of random kinetic energy is lost in each collision. An important ingredient in planetary ring dynamics, as compared to many other examples of granular matter, is the presence of a volume force: mean orbital speed depends on the planetocentric distance. This systematic velocity shear provides an energy source which balances the collisional dissipation, leading to a local steady-state in short time scales, in few tens of orbital periods. The steady-state properties, e.g. ring thickness, depend on the amount of dissipation. On the other hand, time scales of radial evolution are much longer, due to low viscosity related to strongly flattened rings.

In what follows we focus on dense rings and study their intrinsic evolution with N-body simulations, with parameter values corresponding to Saturn's rings. Technical aspects of the simulations are also summarized. Section 2 provides a brief sketch of basic collisional dynamics relevant to the numerical experiments; for excellent theoretical reviews on ring dynamics, see [9, 30], the latter dealing also with the influence of external satellites. Local simulations (Section 3) utilizing various laboratory measurements of the elasticity of ice

[3, 10]) yield a wide choice of possible models, ranging from a near monolayer ring to many particle thick multilayers. All these models, corresponding to different surface properties of particles, are in principle allowed by the upper limit for the local ring thickness (about 150 meters), derived from Voyager observations. However, they imply very different viscous stability properties: simulation examples of both viscous instability and overstability are provided in Section 5. The role of self-gravity (Section 4) depends also crucially on the elastic properties, via ring thickness: near monolayer models are susceptible to local gravitational instabilities, which manifest as small scale particle chains. As mentioned in the Summary, there is indirect observational evidence for such local inhomogeneities: this may offer a tool for constraining the actual elastic properties of ring particles.

2 Collisional Dynamics of Dense Rings

The fundamental quantity for collisional rings is the dynamical optical thickness τ, defined as the ratio of the total surface area of particles to the area of the ring they reside in. In dense rings τ is of the order of unity or larger. The importance of optical thickness stems from the fact that for a given size distribution and physical properties of particles, all local dynamical quantities of interest (impact frequency, kinetic temperature, vertical thickness, viscosity etc.) are functions of τ only.

2.1 Impact Frequency

The impact frequency between identical particles with radius r can be estimated as $\omega_c \propto n_s \sqrt{T} \sigma$, where n_s stands for the space number density of particles, $\sqrt{T}$ corresponds to 1-dimensional velocity dispersion, and $\sigma = 4\pi r^2$ is the collisional cross-section. The space density n_s can be approximated as n/H, where H is the geometric thickness of the ring, and $n = \tau / \pi r^2$ is the surface number density. Due to collisional partitioning of energy between horizontal and vertical motions, H is proportional to $\sqrt{T}/\Omega$, where Ω is the orbital angular velocity. The explicit $\sqrt{T}$ and n_s dependency thus cancels out, and ω_c depends only on τ. Analytical treatment [11], taking into account the anisotropic distribution of impact velocities and the Gaussian vertical profile, yields $w_c \approx 3\Omega\tau$, or about 18τ impacts/orbital period, in agreement with simulations. As a typical period in Saturn's rings is about 10 hours, each particle in dense rings collides several times/hour.

2.2 The Establishment of Local Energy Equilibrium

The dissipation in mutual impacts tends to decrease particle's orbital eccentricities and inclinations, and thus the kinetic energy stored in random motions. On the other hand, the colliding particles have typically slightly different semi major axis. Since the distribution of impact directions is more or less random, energy is transferred from the systematic velocity field to random motions.

The local kinetic temperature is thus determined by the balance between the viscous gain of energy from the orbital motion ($\dot{E}_{visc}$) and the loss of energy in dissipative impacts ($\dot{E}_{coll}$). The energy gain (per unit mass) is related to kinematic viscosity ν by $\dot{E}_{visc}=\nu(Rd\Omega/dR)^2 = \frac{9}{4}\nu\Omega^2$ in a Keplerian velocity field with $\Omega \propto R^{-1.5}$. The kinematic viscosity is usually divided into two parts [1], to the local viscosity ν_L due to the transport of angular momentum via particles' radial excursions, and to the non-local viscosity ν_{NL} arising from the angular momentum exchange in impacts between finite-sized particles at slightly different radial distance.

The basic expression for local viscosity is $\nu_L = \omega_c l^2$, where l is the radial mean free path. In high impact frequency regime l^2 is proportional to T/ω_c^2 while for $\omega_c/\Omega \to 0$, epicyclic motions set an upper limit $l^2 \approx T/\Omega^2$. A semi-empirical formula consistent with these limits yields (holds within 10% accuracy in simulations with $\omega_c/\pi\Omega < 1$)

$$\nu_L \approx \frac{1.35 T\omega_c}{(\pi\Omega)^2} \frac{1}{(\omega_c/\pi\Omega)^2+1} \tag{1}$$

The functional form of non-local viscosity is less clear. Analytical estimates [9] suggest that $\nu_{NL} \approx \omega_c r^2$, whereas simulations indicate that ν_{NL} also increases slightly with $\sqrt{T}/\Omega r$. In any case, ν_{NL} may become significant compared to ν_L if τ is moderately large and the system is not far from the monolayer state ($\sqrt{T}/r\Omega < 3$).

The energy loss in impacts can be approximated as

$$\dot{E}_{coll} = -1/2\ \omega_c T(1-{e_n}^2), \tag{2}$$

where e_n is the normal coefficient of restitution. An isotropic Gaussian distribution of impact velocities is assumed, and the loss due to friction is assumed to be negligible.

Two basic types of equilibrium behaviour can be expected, related to whether ν_L or ν_{NL} gives the dominant contribution to the energy gain. If $\nu_L >> \nu_{NL}$ (hot system with $\sqrt{T} >> r\Omega$) the equilibrium condition $\dot{E}_{visc} + \dot{E}_{coll} = 0$ implies

$$(1-{e_n}^2) = \frac{0.6}{(\omega_c/\pi\Omega)^2+1} \tag{3}$$

As ω_c depends only on τ, the equilibrium requires a specific value $e_n = e_{cr}$ for each τ. This equation is practically identical to the well-known Goldreich-Tremaine [8] formula $(1-{e_n}^2)(1+\tau^2) = 0.61$, derived from their semi-numerical solution of the collision integral. It also agrees with the treatment by Hämeen-Anttila [11], see also [30], where the collision integral is solved by an analytical approximation. On the other hand, in a cool system $\nu_{NL} \simeq \nu_L$, and the additional energy input due ν_{NL} requires larger dissipation and thus lowers the implied e_{cr}; in this case the value of steady-state e_n depends also on the ratio $\sqrt{T}/r\Omega$, and can be quite different from that of Eq. (3). The two different theoretical $e_{cr}(\tau)$ relations are shown in Fig. 1.

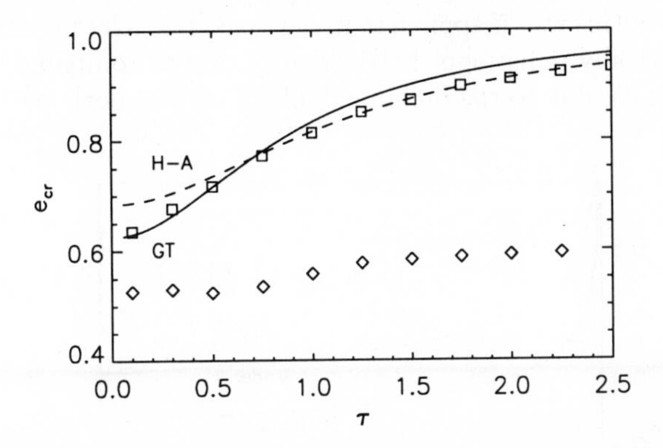

Fig. 1. Theoretical e_{cr} vs. τ dependency, according to [8] (GT), and [11] (H-A). Also shown are the effective values of e_n in two series of simulations: the upper points correspond to a 'hot' system (velocity dependent elasticity according to Model IV in Fig. 2), while the lower points correspond to a near monolayer (Model I with 1 meter particles). The effective e_n in simulations is measured by $\overline{e_n v_n^2}/\overline{v_n^2}$

Laboratory measurements of ice [3, 10] as well as theoretical work [4, 6, 29] indicate that e_n is a decreasing function of the perpendicular component (v_n) of impact velocity (Fig. 2). In this case T adjusts to a steady-state, where the effective value of e_n depends on τ via the energy balance equation. This steady-state is stable if $de_n/dv_n < 0$, as also implied by measurements: for example, if the instantaneous T exceeds the steady-state value, $|\dot{E}_{coll}| > E_{visc}$ and the system cools toward the steady state.

The average v_n and thus also T corresponding to $e_{cr}(\tau)$ depend on the functional form of $e_n(v_n)$. According to laboratory measurements, elasticity is very sensitive to the surface properties of particles; it may also depend on the particle size via the curvature of the impact point. As the actual properties of particles in planetary rings are not known, these measurements leave a wide margin for possible local steady-state properties. Especially, the most widely adopted elasticity model, that of Bridges et al. [3], with $e_n = 0.32{v_n}^{-0.24}$ (v_n expressed in cm/s) implies a near monolayer ring. Simulation examples with this (Model I) and the other models of Fig. 2 will be provided in Section 4.

What happens if e_n is constant, independent of impact velocity? In this case the critical value e_{cr} defined by Eq. (3) divides between stable and unstable temperature behaviour: if $e_n > e_{cr}(\tau)$, then $|\dot{E}_{coll}| < E_{visc}$ and T increases with time, leading to dispersal of the system. On the other hand, for $e_n < e_{cr}(\tau)$ the energy gain due to ν_L is less than required by dissipation, and T decreases. However, eventually the contribution by ν_{NL} balances the dissipation. The minimum value of $\sqrt{T}$ depends on the value of the constant e_n, but is typically few times $r\Omega$. Note that even in the case of $e_n(v_n)$ the behaviour can be quite similar to that of a constant $e_n < e_{cr}$, provided that $e_n(v_n)$ drops below e_{cr} already for $v_n \sim r\Omega$. Especially, Bridges et al. model behaves in this way if applied to meter-sized particles in Saturn's rings, yielding a practically indistinguishable steady-state as compared to $e_n \approx 0.5$ (see below, Fig. 4), due to the dominant role of ν_{NL} in both cases.

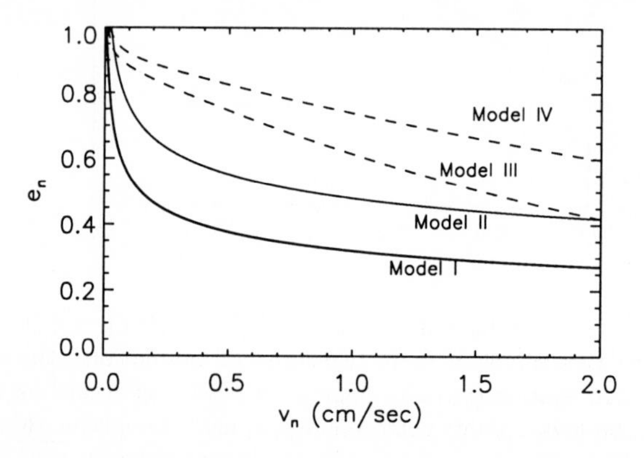

Fig. 2. Elasticity models. Solid lines denote fits to laboratory measurements with frosty ice particles, in two different temperatures [3], Model I; [10], Model II. Dashed lines correspond to measurements in [10], with particle sizes r=2.5 and 20 cm (Models III and IV, respectively), where particles have compacted frost on their surfaces

2.3 Radial Evolution

The conservation of total angular momentum, combined with the dissipative loss of total energy, implies that a ring system as a whole must spread with time (however, this does imply that *local* density variations will necessarily smooth out, see Section 5). The characteristic spreading time for a ring region with radial width W is $T_{spread} \simeq W^2/\nu$. Assuming $\sqrt{T} = kr\Omega$ and $\nu \approx T/\Omega$ (as for τ of the order of unity), $T_{spread} \approx 10/k^2(W/r)^2$ orbital periods. For example, for a region with $W \approx 10^3$ particle radii (about 1 km), $T_{spread} \approx 10^5$ orbital periods if the ring is near a monolayer state ($k \approx 3$). This is already very long compared to the time-scale of the establishment of local energy balance. For the whole of Saturn's rings, spreading times for the near monolayer models are of the order of the age of the Solar System [2].

3 Simulations of Collisional Rings

The local collisional dynamics of planetary rings is theoretically fairly well understood, and the steady-state properties can be estimated analytically from the kinetic theory [1, 8, 11, 21], even if size distribution and gravitational encounters are included [13]. Numerical simulations are still necessary for many reasons. Especially, they allow the accurate treatment of dense, flattened rings where finite-size effects become important. Also, the realistic inclusion of collective self-gravity (see Section 4), is still out of scope of analytical treatments.

3.1 Early Simulation Studies

The pioneering simulation studies of the collisional evolution of planetary rings were performed by Trulsen [34], Brahic [2], and Hämeen-Anttila & Lukkari [12]. All these simulations used the same approach: a complete ring of particles orbiting the central body in Keplerian orbits. Since only few hundred particles could be followed, these simulations were limited to low τ. Also, the particle sizes were large as compared to the width of the ring, which made it difficult to separate local evolution from the radial spreading.

Nevertheless, many basic characteristics of collisional systems were revealed, including the existence of the critical value for the coefficient of restitution [34], the minimum residual velocity dispersion of the order of $r\Omega$ [2], and the establishment of equilibrium with finite velocity dispersion for $e_n(v_n)$ [12]. These simulations also served as important checks for various analytical treatments, see Ref. [30].

3.2 Local Simulations

Local simulations offer a possibility to simulate dense rings, by restricting calculations to a small co-moving region orbiting inside the rings (Fig. 3). Due to systematic velocity shear individual particles tend to leave the calculation region, which is taken into account by periodic boundary conditions. The main advantage of the method is that radial spreading is prevented, thus facilitating the study of local steady-state properties as a function of fixed τ. This is justified, based on the large separation of local and radial time scales. Local method was first applied to planetary ring simulations by Wisdom and Tremaine [37] and to stellar disks by Toomre [33]. In the former study impacts between identical particles were taken into account, but not their mutual gravity, whereas the latter study treated only gravitational forces.

Local method. As the typical size of the calculation region is very small compared to the radial distance from the central body, linearized equations of motion can be used

$$\begin{aligned} \ddot{x} - 2\Omega\dot{y} + (\kappa^2 - 4\Omega^2)x &= F_x \,, \\ \ddot{y} + 2\Omega\dot{x} &= F_y \,, \\ \ddot{z} + {\Omega_z}^2 z &= F_z \,. \end{aligned} \tag{4}$$

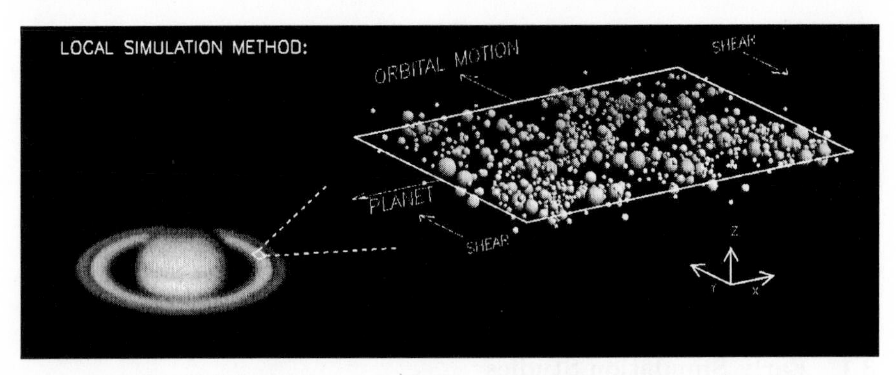

Fig. 3. Schematic representation of local simulation method

Here the x-axis points in the radial direction, the y-axis in the direction of orbital motion, and the z-axis is perpendicular to the equatorial plane. The reference point of the coordinate system moves with angular velocity of Ω in a circular orbit at a radial distance a. In the case of a central point mass, the epicyclic frequency κ and the frequency Ω_z of vertical oscillations are both identical to Ω. F_x, F_y, F_z denote additional forces, e.g. self-gravity.

The boundaries are treated by assuming that each particle with a position (x, y, z) has an infinite set of image particles at $(x + nL_x, y + mL_y - 3/2nL_x\Omega t, z)$, where m and n are integers, L_x and L_y denote the dimensions of the actual calculation region, and t the time reckoned from the beginning of the simulation. Eqs. (4) are invariant under this transformation. Each time a particle crosses the boundary, one of its images enters the box. If the crossing occurs across the inner or outer boundary, the velocity of the particle is modified by $\Delta\dot{y} = \pm 3/2\ \Omega L_x$, which corresponds to the difference of shear velocity across L_x. In this manner the evolution of the system is independent of the choice of the origin of the coordinate system. The results are also independent of the size of the calculation region, provided that it is large compared to the mean free path between impacts [23, 37].

Treatment of impacts. In most local simulations [20, 23–25, 37] the standard impact model with instantaneous velocity changes has been used. A tangential coefficient of restitution, e_t, can also be included, besides e_n, in which case the exchange of energy with particle's spin motion must also be taken into account. The velocity change in impact [21] is determined by

$$(\boldsymbol{v}_1)_{coll} = -e_n \boldsymbol{c}(\boldsymbol{c} \cdot \boldsymbol{v}_{coll}) + e_t \boldsymbol{c} \times (\boldsymbol{v}_{coll} \times \boldsymbol{c}), \tag{5}$$

where $(\boldsymbol{v}_1)_{coll}$ and $(\boldsymbol{v})_{coll}$ stand for the post- and pre-collisional velocity differences at the contact point and $\boldsymbol{c}$ for the unit vector joining the particle centers. In terms of $\boldsymbol{v} = \dot{\boldsymbol{R}}' - \dot{\boldsymbol{R}}$, the relative velocity of particle centers, $(\boldsymbol{v})_{coll} = \boldsymbol{v} - (r\boldsymbol{\omega} + r'\boldsymbol{\omega}') \times \boldsymbol{c} + (r + r')\Omega \boldsymbol{N} \times \boldsymbol{c}$, where r and $\boldsymbol{\omega}$ denote the particle radius and spin-vector (primed and unprimed symbols distinguish the two particles). The last term, $(r+r')\Omega \boldsymbol{N} \times \boldsymbol{c}$, where $\boldsymbol{N}$ is the unit vector in the direction perpendicular to the equatorial plane, arises due to the rotating coordinate system. The changes of the velocity and spin vector of an individual particle follow from the conservation of linear and angular momentum, yielding (for homogeneous spheres)

$$\dot{\boldsymbol{R}}_1 - \dot{\boldsymbol{R}} = \frac{m'}{m+m'}\{(1+e_n)\boldsymbol{cc}\cdot\boldsymbol{v}_{coll} + \frac{2(1-e_t)}{7}(\boldsymbol{v}_{coll} - \boldsymbol{cc}\cdot\boldsymbol{v}_{coll}\}, \quad (6)$$

$$(r\boldsymbol{\omega}_1 - r\boldsymbol{\omega}) = \frac{m'}{m+m'}\frac{5(1-e_t)}{7}\boldsymbol{c}\times\boldsymbol{v}_{coll}, \quad (7)$$

where m and m' are the masses of the impacting bodies. A simple model for slightly non-spherical shape has also been studied, by assuming that the tangent plane of impact is not exactly perpendicular to $\boldsymbol{c}$, but deviates from it by a small random amount in each impact [21].

Search of impact pairs. The speed of the collisional simulation depends crucially on the efficient search of impact pairs. For example, in Ref. [37] the fact is used that orbits between impacts are Keplerian epicycles, and iteratively one can solve for the intersection time of each pair of epicycles. The impact of the pair with the smallest impact time is executed and their post-impact elements are calculated, leading to updated intersection times with all the other particles. The system is thus moved from one impact to the next ("event-driven" method). This is fairly fast for small particle numbers ($\approx$ 50 in [37]), but as N increases, checking of all $N(N-1)/2$ pairs gets excessively slow.

For larger N it is advantageous to integrate the equations of motions, and during each time-step search for impacts with the help of a second-degree Taylor-polynomial [23]. This also allows inclusion of additional forces. Even then it is important to limit the number of pairs examined. This can be done quite efficiently by keeping track of the maximum pre-step separation actually leading to an impact, and by checking in each step only those pairs whose distance does not exceed this maximum multiplied by some threshold value. This threshold must be chosen in a manner which ensures that no impacts are lost, and it also must be dynamically adjusted as the temperature of the system evolves. In practice, the number of pairs examined is proportional to $\sqrt{T}/(r\Omega)N$. Note that the actual integration needs not to be performed by Taylor-series: for example in [24] a fourth order Runge-Kutta integration is utilized, and the impact locations initially estimated by the second-degree expansion are iteratively improved to correspond to the full accuracy of the integration. It must be stressed that it is important to take correctly into account impacts taking place over boundaries. For small calculation regions this fraction can be quite significant, and the omission of such impacts will modify the energy balance.

3.3 Results of Local Simulations

Simulations of identical particles. Fig. 4 displays typical examples of time evolution in collisional simulations. Regardless of the initial values, the system establishes a steady-state where T and other dynamical quantities stay on a constant level. The timescale to reach steady-state is typically of the order of 50-100 impacts/particle. The only exception is a case with constant $e_n > e_{cr}$ for which no steady-state exists. In the case of constant $e_n < e_{cr}$ the steady-state $\sqrt{T}$ is proportional to $r\Omega$. For $e_n(v_n)$ the steady-state depends drastically on the parameters of the elasticity model, as anticipated in Section 2. Once the steady-state is obtained, various dynamical quantities can be collected very accurately by averaging over long time spans. Especially, both local and non-local viscosity can be determined from the flow of angular momentum, averaged over all particle orbits and impacts, respectively [37].

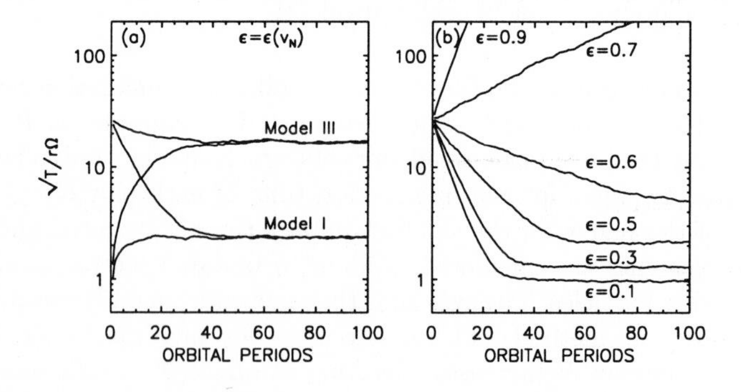

Fig. 4. Examples of time evolution in simulations. Frame (a) compares two $e_n(v_n)$ models of Fig. 2, starting from different initial values. Temperature is calculated by $T = (c_1^2 + c_2^2 + c_3^2)/3$, where c_1, c_2 and c_3 denote the principal axis components of velocity ellipsoid. In each simulation $r = 1$ m, Saturnocentric distance $a = 100\ 000$ km, and $\tau = 0.1$. In (b), evolution with different constant values of e_n is studied: for $\tau = 0.1$, $e_{cr} \approx 0.65$. $e_t = 1$ is assumed in these and all subsequent simulations, unless otherwise indicated

In Fig. 5 some interesting dynamical properties are plotted as a function of τ. The main effect of increased density is the reduction of $\sqrt{T}$, basically because E_{visc} becomes less effective as the mean free path between impacts is reduced. In Fig. 5 this is studied in terms of the geometric thickness H, which is roughly proportional to $\sqrt{T}$ (except in flat systems with large τ, where the piling of particles thickens the system although T is almost constant). The elastic models of Fig. 2 lead to even a factor of 10 difference in H. This difference also reflects in the local and, to a lesser degree, in the non-local contribution to ν. For a cool system the product $\tau\nu$ is monotonically increasing, because of the dominant role of ν_{NL}, whereas for a hot system it may have a decreasing portion at large τ regime, due to the reduction of

ν_L. As discussed in Section 5, negative $d(\tau\nu)/d\tau$ leads to radial instability, whereas strong enough positive slope may indicate overstability.

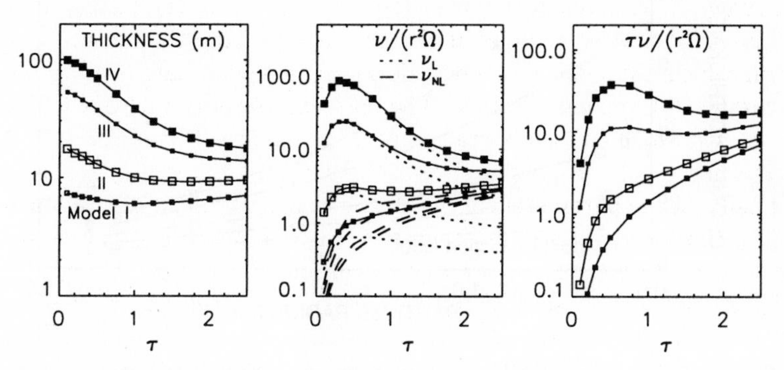

Fig. 5. Dependency of geometric thickness and viscosity on τ, for the various elastic models. Geometric thickness is defined by $H^2 = 12\overline{z^2}$. As in Fig. 4, $r = 1$ m and $a = 100\ 000$ km

Simulations with size distribution. According to Voyager measurements, the size distribution of particles in Saturn's ring can be approximated with a power-law [18]

$$dN/dr \propto r^{-q}, \quad 1 \text{ cm} < r < 5 \text{ m} \tag{8}$$

with $q \approx 3$. Size distribution leads to a steady-state where kinetic temperature of small particles exceeds that of the large ones (Fig. 6). However, the system is still far from equipartition of energy: depending on the elastic model, the kinetic energy of smallest particles is about 2-20% of that of the largest ones. In simulations, due to limited N the width of the distribution is smaller than in reality. Simulation tests [24] indicate that the effect of this truncation is small as long as $r_{max}/r_{min} > 10$, and correct r_{max} is used. In the case of constant $e_n < e_{cr}$, or with the Bridges-model, the thickness of the layer of the largest particles is proportional to r_{max}. The same scaling holds for the whole distribution as the behaviour of small particles is governed mainly by the dynamical state of the largest ones. The viscosities experienced by the small particles exceed those for the large ones, but otherwise the $\nu(\tau)$ behaviour is similar to that in the case of identical particles. Especially, simulations show no signs of selective viscous instability among small particles, which in principle could be possible if the system were closer to equipartition of energy (see [30]).

Fig. 6 illustrates also the effects of friction, with $e_t = 0$. This represents the maximal amount of extra dissipation due to friction, and is able to reduce H to about one half. However, measurements indicate that in reality

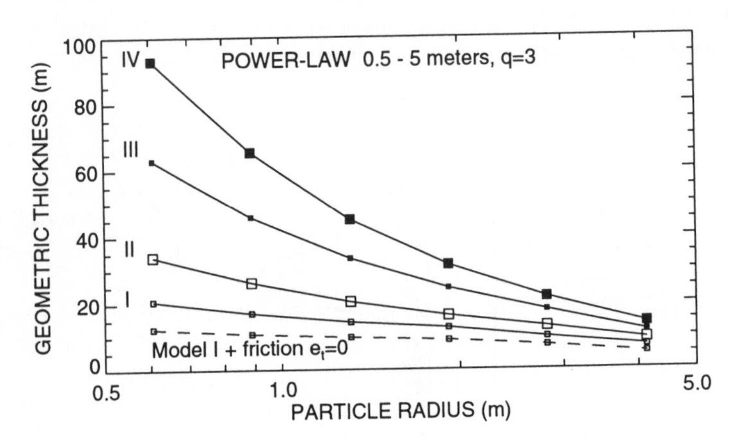

Fig. 6. Geometric thickness H in simulations with size distribution, for $\tau = 1.0$. Particles are divided to six logarithmic size-groups. Also indicated is the maximal effect of tangential friction, in combination with the Bridges-model (Model I).

e_t is close to unity [31]. Friction also leads to partitioning of energy between random motions and spin degrees of freedom: in general, for identical particles $E_{spin}/E_{random} \approx 2(1-e_t)/(9+5e_t)$ [21]. In the case of size distribution, small particles obtain relatively larger share of spin-energy, but the above ratio holds for the distribution as a whole. Spin-axis also acquire a residual mean orientation, $\overline{\omega_z} \approx 0.3\Omega$, indicating rotation in the same sense as the orbital motion. However, for smallest cm-sized particles this mean is insignificant as compared to the dispersion of ω_z (roughly $\propto 1/r$), so that their spin-axis are practically randomly oriented.

4 Self-Gravity

4.1 Expectations

The inclusion of particles' mutual gravitational forces affects the local dynamics of collisional systems on several, partially competing ways. For low τ the main effect comes from close binary encounters, which correspond to totally elastic impacts: the kinetic energy of the encountering pair is conserved, while the deflection of orbits transfers energy from the systematic to random motions. This extra heating will tend to increase $\sqrt{T}$, until it becomes comparable to the escape velocity from particle surfaces [5]. This corresponds to minimum $H \approx 10$ meters for the parameter values of Fig. 5 (assuming solid icy density, $\rho = 0.9$ g/cm^3). If collisions alone are able to maintain larger H, the effect of encounters is negligible.

For a system with larger τ, the mean vertical self-gravity can exceed the corresponding component of the central force. For identical particles this takes place if

$$H < 4\tau r(\rho/\rho_{plan})(a/r_{plan})^3, \tag{9}$$

where r_{plan} and ρ_{plan} are the radius and internal density of the planet (60 000 km and 0.7 g/cm^3 for Saturn). Inclusion of this extra force tends to reduce H quite markedly, both because of increased frequency of vertical oscillations and via increased impact frequency leading to enhanced dissipation. For the parameters studied in Fig. 5, the above condition is fulfilled for the coolest model when $\tau > 0.3$.

Intuitively, planar components of self-gravity may be expected to have less importance, due to partial cancellation of forces. However, as shown in Ref. [32], a differentially rotating disk is locally unstable against the growth of axisymmetric disturbances, if its radial velocity dispersion falls below the critical value $c_{cr} = 3.36G\Sigma/\kappa$. The first radial wavelength to become unstable corresponds to $\lambda_{cr} = 4\pi^2 G\Sigma/\kappa^2$. Even for Toomre parameter $Q_T = c_{rad}/c_{cr} \approx 2 - 3$, such system is susceptible to the growth of local, non-axisymmetric disturbances [14]. In Keplerian potential this gravitational instability manifests as trailing density enhancements, forming about 25° angle with respect to tangential direction. In the case of stellar systems, the heating accompanying these disturbances eventually suppresses them: in the case of particulate rings collisional dissipation can lead to a statistical steady-state, where new structures continuously emerge and dissolve. The condition $Q_T < 2-3$ is comparable to Eq. 9, indicating that whenever the vertical field is important, the system is also susceptible to collective horizontal instabilities.

4.2 Simulations of Self-Gravitating Systems

The first simulations of self-gravity, performed with azimuthally complete systems, included only gravitational encounters [17]. They illustrated the extra heating due encounters but had little significance for dense rings. On the other hand, in Ref. [37] the mean vertical self-gravity was included by applying a constant enhancement factor $\Omega_z/\Omega = 3.6$ in the equations of motion, approximating the expected situation in Saturn's B-ring. In these simulations the vertical self-gravity led to a reduced vertical thickness and strong enhancement in the impact frequency, as expected, and when compared to the non-gravitating case, a much larger non-local viscosity was obtained.

When self-gravity is correctly included, see Fig. 7, the collective gravitational instability overshadows the effects of enhanced vertical field and leads to dramatically different behaviour: near monolayer systems develop transient density enhancements, "wakes", each forming and dissolving in timescales of few orbital periods [25]. The scattering of particles by these wakes leads to enhanced velocity dispersion and thickening of the ring: the steady-state corresponds to $Q_T \approx 2-3$, as anticipated. Spatial autocorrelation analysis shows close correspondence to Julian-Toomre [14] stellar wakes.

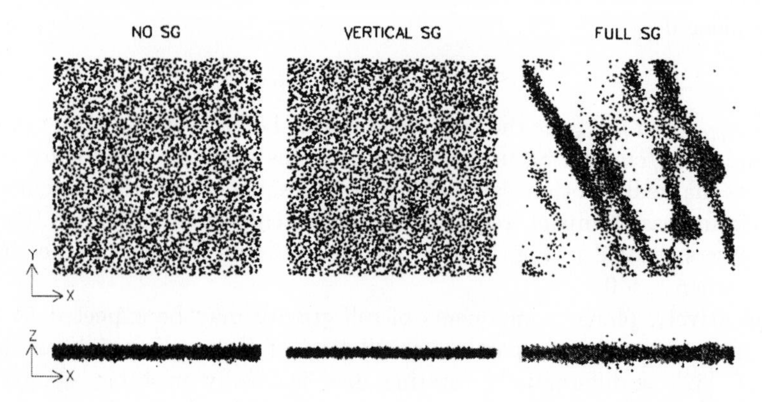

Fig. 7. Comparison of simulations either neglecting self-gravity (left), including just the vertical self-gravity (middle), or all components of self-gravity (right). In each case $a = 100\ 000$ km, $r = 1$ m, $\tau = 0.75$ and the Bridges et al. elasticity model is used. In the self-gravitating cases $\rho = 0.9$ g/cm^3. The size of the square-shaped calculation region is 125 m

Simulation of gravitating systems requires quite large calculation regions. The scale of the wakes is proportional to λ_{cr} (~ 100 meters), and in order to obtain wake amplitudes and directions which are independent of the periodic boundaries, L_x and L_y must exceed about $4\lambda_{cr}$. This implies

$$N \approx 2200\tau^3(\rho/0.9\ \mathrm{gcm}^{-3})^2(a/100\ 100\ \mathrm{km})^6(L_x/\lambda_{cr})(L_y/\lambda_{cr}) \qquad (10)$$

Simulations thus become computationally very demanding for large τ, especially when the outer edge of the rings ($a = 137\ 000$ km) is approached.

Whether or not collective wakes occur, depends crucially on the elasticity. If the collisions alone are able to keep $Q_T \geq 3$, no wakes can form. Applied to Saturn's rings this implies that local inhomogeneities may be expected only for rather dissipative cases, corresponding roughly to Bridges et al. elasticity model. On the other hand, smaller dissipation in impacts, as implied by any of the other three models studied, will suppress wakes.

At large distances gravitational accretion of particles is observed. The condition that the attraction between two contacting, radially aligned, synchronously rotating identical particles exceeds the tidal force due to planet is (see e.g. [35])

$$(a/r_{plan})^3 > 12(\rho_{plan}/\rho), \qquad (11)$$

whereas in the case of two very different sized particles the factor 12 is replaced by 3. For particles with $\rho = 0.9$ g/cm^3 this implies $a = 126\ 000$ and 80 000 km, respectively. However, in actual rings the non-zero velocity dispersion makes accretion more difficult [19, 25, 26], and the limiting distances for the formation of aggregates are shifted outward. Fig. 8 illustrates accretion: as a is increased the wakes begin to degrade into particle groups. For identical particles the group formation becomes efficient for $a \approx 140\ 000$ km, whereas in the case of size distribution this occurs already for $a > 125\ 000$ km.

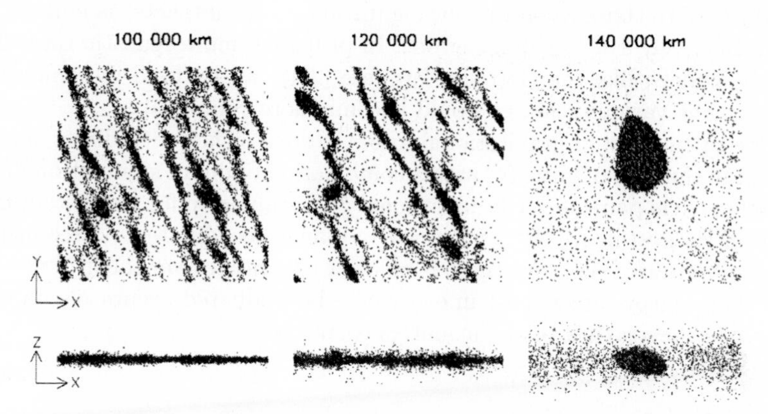

Fig. 8. Effects of self-gravity at different distances

4.3 Numerical Methods in Self-Gravitating Simulations

The simulation of self-gravity poses some technical problems, besides the large CPU-time consumption needed for the evaluation of mutual forces. Namely, as mentioned above, beyond a certain planetocentric distance the net force between a pair of particles in contact can become directed toward each other. This means that the method of treating collisions in terms of instantaneous velocity changes becomes insufficient: every now and then some particles will penetrate each other, leading to an artificial increase in the density of wakes and particle groups. If allowed, this would lead to unreliable results concerning particle accretion.

The solution is to include explicitly the pressure forces affecting in the impact. In Ref. [26] this is done in terms of the linear visco-elastic model, originally developed in Ref. [6] for the theoretical parameterization of measurements of velocity and size-dependent elasticity. The impact force is expressed as

$$F(\alpha) = k_1\alpha + k_2\dot{\alpha}\,, \tag{12}$$

where $\alpha = (r+r') - |\boldsymbol{R} - \boldsymbol{R}'| > 0$ is the penetration depth during the impact. Thus in the gravitational simulations, instead of searching for exact impact times, this extra force is included between colliding, slightly penetrating particles, and the motion is integrated through the impact with small time steps. The CPU-time consumption can be kept tolerable, provided that the impact duration, T_{dur} is not too long. The attractive feature of the above simple force model when applied to simulations is that the constants k_1 and k_2 can be easily tied to the desired T_{dur} and e_n [26]. Also $e_n(v_n)$ can be simulated, if the parameters are chosen according to the pre-impact velocity. The results are identical to those when calculating instantaneous impacts, as long as T_{dur} is less than, say 10^{-3} orbital periods. In principle, more realistic theoretical impact models, like the Hertzian law (see [29]) can be used in simulations, but then the various scalings would become more complicated.

The calculation of gravitational forces is done either with direct summation, or in the case of large N, with a combination of direct summation for the nearby particle pairs and a 3-dimensional grid evaluation for the distant gravity, with FFT utilizing the periodicity in planar coordinates. A tree-method has also been used [20], but there is no clear advantage over the grid method. Note that nearby forces must in every case be evaluated accurately, in order to include the gravitational encounters correctly.

5 Radial Evolution

The most puzzling features of Saturn's rings, revealed by Voyager fly-bys, are the radial density variations seen on all scales down to the resolution limit of few kilometers. Their origin has inspired a great deal of theoretical efforts, mainly on the role of perturbations due Saturn's inner satellites (see [9]). Indeed, the radial distances of the most regular density oscillations in the outer A-ring agree with locations of satellite resonances. However, resonances are too rare to explain the less regular variations in the densest B-ring. It appears inevitable that irregular density variations must have an intrinsic origin, related to some type of instability in dense collisional rings.

The originally proposed intrinsic mechanism was the viscous instability [11, 15, 16, 36], based on the assumption that there is a large collision-induced difference between the kinetic temperatures of rarefied (hot) and dense (cool) portions of the rings. In this case the dynamic viscosity, proportional to $\tau\nu$, may decrease with τ (as in the two 'hot' examples of Fig. 5), leading to collisional particle flux which is directed toward local density maxima. However, even if in principle possible for very elastic particles, the instability should lead to a bimodal density distribution, not consistent with observations.

A more promising possibility is the viscous overstability, which can take place if viscosity increases strongly enough with τ [27]. The linear stability analysis in [27] suggests that the B-ring should be overstable, with most overstable wavelengths of about 100 meters, having e-folding times of only few orbital periods. Superposition of such overstable waves, with amplitudes saturated by non-linear phenomena, might well yield similar structures as seen in the B-ring [28]. According to [27] the condition for overstability is that $\beta = (d\nu/d\tau)(\tau/\nu) > \beta_{cr}$ where β_{cr} is about 0.11. If self-gravity is included (however, assuming only strictly axially symmetric waves), β_{cr} is reduced to

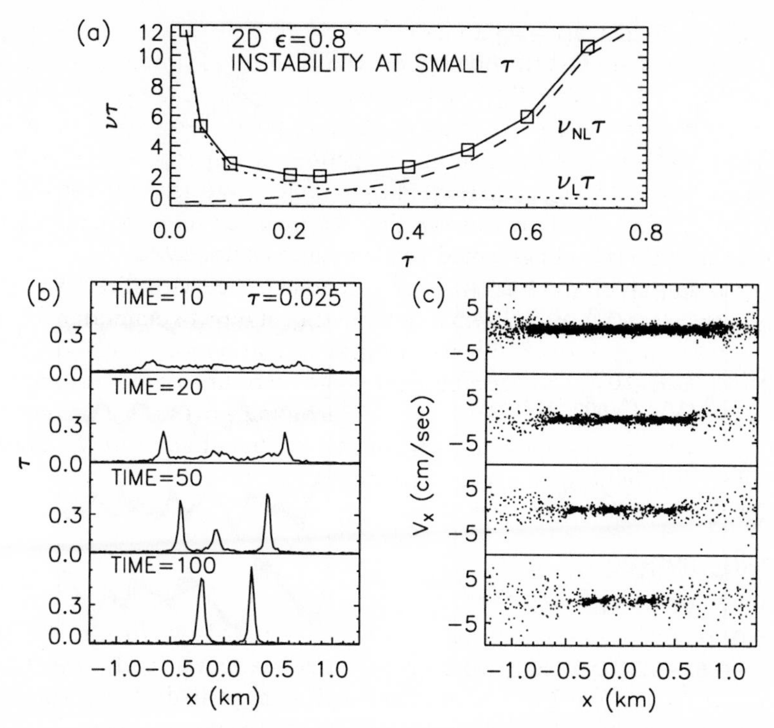

Fig. 9. Instability in 2D simulation with $e_n = 0.8$. Frame (a) shows the dependency of $\nu\tau$ vs. τ from a series of small-N simulations, while frames (b) and (c) display snapshots of radial density distribution and velocity profile in a simulation with radially extended calculation region ($L_x = 4000$ m, $L_y = 125$ m, $N = 4000$), for $\tau = 0.025$. Time is reckoned in orbital periods. Note that while illustrating the instability at small τ, this strictly 2D simulation has no relevance to real rings, even if they were in *near* monolayer state: if vertical motions were allowed, the system would rapidly disperse due to high e_n exceeding the 3D e_{cr}

even slightly negative values [27]. The condition for the viscous instability corresponds to $\beta < -1$.

With local simulations the radial evolution can be studied in two complementary ways. Firstly, it can be addressed by analytical means, e.g. with linear stability analyses similar to that in [27], utilizing viscosity vs. density dependency found in moderately small N simulations conducted for various $\tau's$. Secondly, the radial evolution, say in scales less than few kilometers, can also be studied directly, by a large N simulation with radially extended calculation region. This provides also a check for the results of stability analyses.

Two-dimensional simulations allow easy demonstration of both the above mentioned types of unstable behaviour. The basic difference as compared to 3-dimensional system is that in the 2D case $\omega_c \propto \sqrt{T}\tau$ (for small τ). This follows because the 2D system can not adjust its space density via vertical thickness.

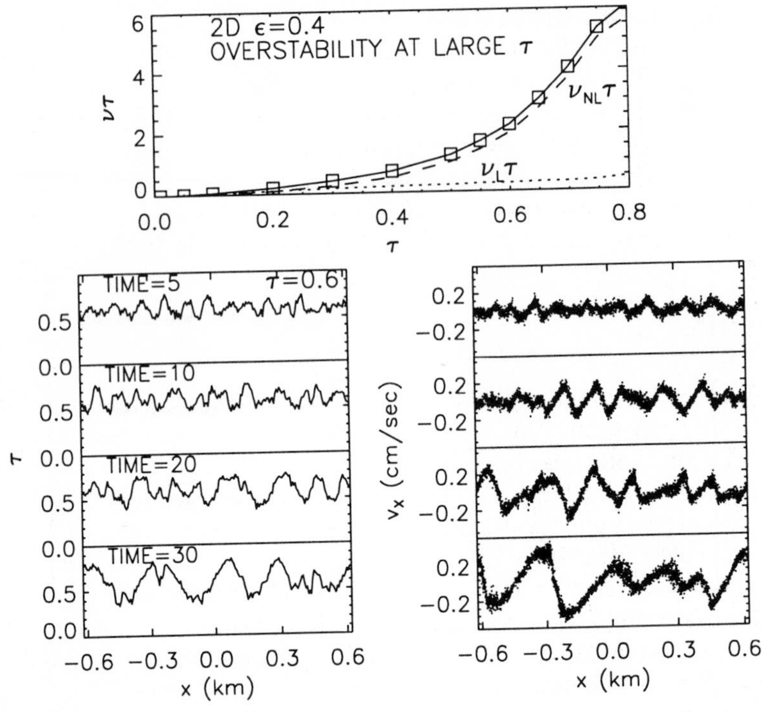

Fig. 10. Overstability in 2D simulation with $e_n = 0.4$. Two lower frames display evolution in a simulation with $\tau = 0.6$ ($L_x = 1240$ m, $L_y = 62$ m, $N = 15\,000$)

For the same reason, ω_c increases in a very non-linear fashion as τ approaches the 2D close packing limit $\tau_{max} = \pi/\sqrt{12} \approx 0.91$. Due to the temperature dependency of ω_c, even a constant e_n close to unity can be balanced for a fixed τ, by increasing ω_c via $\sqrt{T}$. For small τ and e_n close to unity, $\nu \propto \tau^{-2}$, and the system fulfills the condition of viscous instability. Direct verification of this low-τ 2D instability requires much smaller N than the large-τ 3D instability, while it still represents the same physical mechanism. On the other hand, for large τ the strong non-linear growth of ω_c implies similar growth in ν_{NL}, so that β can become very large, implying tendency for overstability. Both types of behaviour are confirmed by direct 2D-simulations (see Figs. 9 and 10.)

In the example of Fig. 10, $\beta \approx 2.5$ for $\tau = 0.6$, exceeding the theoretical critical $\beta_{cr} \approx 0.11$ by a wide margin. Two-dimensional simulations with smaller τ, corresponding to smaller β, indicate that overstability is not obtained once β falls below ≈ 1.2. This limit is considerably larger than the theoretical β_{cr} found in [27]. The discrepancy is not due to special characteristics of 2D simulations: non-gravitating 3D near monolayer simulations ($\beta \approx 1$) show no signs of overstability, whereas corresponding simulations with vertical gravity ($\beta \approx 1.3$) are unstable. These, still preliminary results indicate that the existing stability analyses overestimate the tendency for overstability. One possible indication might be that the bulk viscosity, ξ, of a collisional ring may in fact exceed the kinematic shear viscosity: according to formulas in [27], the theoretical β_{cr} would rise to the level implied by direct simulations if $\xi/\nu > 4$. However, even then the expected theoretical growth times would be shorter than found in simulations, suggesting that some other factor might be missing from the stability analysis. Studies in this direction are in progress, together with Dr. Frank Spahn's group in Potsdam.

6 Summary

The collisional and gravitational dynamics of dense planetary rings have been reviewed through simulation examples. In principle, numerical simulations, combined with theoretical considerations, yield a fairly complete picture of the dependency of the local steady-state properties on the optical thickness, elastic properties, and size distribution. The major uncertainty, concerning the actual local properties of rings is due to uncertainty in elastic properties. Laboratory measurements indicate that particles' elasticity is very sensitive to their surface properties, which leads to a large range of possible physical situations, characterized by different kinetic temperature, and consequently by different vertical thickness and filling factor.

A possible diagnostic tool for accessing local properties is provided by self-gravity, whose importance depends crucially on the velocity dispersion via Toomre's Q_T-parameter. For example, application of Bridges et al. elasticity measurements leads to the formation of trailing particle wakes in simulations. On the other hand, less dissipative models suppress wakes. There is indirect evidence for the presence of such local particle chains, provided by the observed azimuthal asymmetry of Saturn's A-ring. On low elevation images the brightness of the A-ring depends on a bisymmetric manner on ring longitude with respect to observing direction (see e.g. Ref. [7]). For example, there is a definite minimum at longitude of 246°, which can be accounted for by the presence of numerous small unresolved particle chains trailing on the average about 24° with respect to tangential direction. Detailed photometric modelling of simulation data (Salo et al., in preparation) and comparison with the observed brightness asymmetry can in principle constrain both the elasticity and the internal density of particles. Preliminary results seem to support near monolayer models with $\rho \sim 0.5$ g/cm^3.

The mechanisms related to irregular radial structures in dense rings are still not completely understood. The originally proposed explanation in terms of viscous instability seems now outruled. A more promising candidate is the viscous overstability [27, 28], taking place when viscosity increases with density. Simulation examples of both these types of unstable behavior were given, utilizing 2-dimensional simulations. However, comparison of theoretically predicted range of overstability with simulations indicate that current theoretical models still need significant revisions: according to simulations the condition for overstability is much more stringent than predicted. Nevertheless, overstability is still obtained in 3D gravitating simulations, supporting this explanation for the B-ring structure.

In comparison to other types of granular systems dense planetary rings have quite special characteristics. Basically this is due to the presence of volume-forces depending on the planetocentric distance. This establishes a linear shear profile, and provides a steady energy source balancing the collisional dissipation. Also, the proportionality of vertical thickness to velocity dispersion (due to collisional energy partitioning) leads to impact frequency which depends essentially only on the optical thickness.

References

1. S. Araki, S. Tremaine. The dynamics of dense particle disks. Icarus **65** 83–109 (1986).
2. A. Brahic. Systems of colliding bodies in a gravitational field. Astron. Astropys. **54**, 895–907 (1977).
3. F. G. Bridges, A. Hatzes, D. N. C. Lin. Structure, stability and evolution of Saturn's rings. Nature **309**, 333–335 (1984).
4. N. V. Brilliantov, et al. Model for collisions in granular gases. Phys. Rev. E **53**, 5382–5392 (1996).
5. J. N. Cuzzi, et al. The vertical structure and thickness of Saturn's rings. Icarus **38**, 54–68 (1979).
6. J. P. Dilley. Energy loss in collisions of icy spheres: Loss mechanism and size-mass dependence. Icarus **105**, 225–234 (1993).
7. L. Dones, J. N. Cuzzi, R. M. Showalter. Voyager photometry of Saturn's A ring. Icarus **105**, 184–215 (1993).
8. P. Goldreich, and S. Tremaine. The velocity dispersion in Saturn's rings. Icarus **34**, 227–239 (1978).
9. P. Goldreich, and S. Tremaine. The dynamics of planetary rings. Ann. Rev. Astron. Astrophys. **20**, 249–284 (1982).
10. A. P. Hatzes, F. G. Bridges, and D. N. C. Lin. Collisional properties of ice spheres at low impact velocities. Mon. Not. R. astr. Soc. **231**, 1091–1115 (1988).
11. K. A. Hämeen-Anttila. An improved and generalized theory for the collisional evolution of Keplerian systems. Earth, Moon, and Planets **31**, 271–299 (1978).
12. K. A. Hämeen-Anttila, and J. Lukkari. Numerical simulations of collisions in Keplerian systems. Astrophys. Space Sci. **71**, 475–497 (1980).
13. K. A. Hämeen-Anttila, and H. Salo. Generalized theory of impacts in particulate systems. Earth, Moon, and Planets **62**, 47–84 (1993).
14. W. H. Julian, and A. Toomre. Non-axisymmetric responses of differentially rotating disks of stars. Astrophys. J. **146**, 810–827 (1966).
15. D. N. C. Lin, and P. Bodenheimer. On the stability of Saturn's rings. Astrophys. J. **248**, L83–L86 (1981).
16. J. Lukkari. Collisional amplification of density fluctuations in planetary rings. Nature, **292**, 433–435 (1982).
17. J. Lukkari, and H. Salo. Numerical Simulations of Collisions in Self-Gravitating Systems. Earth, Moon, and Planets **31**, 1–13 (1984).

18. E. A. Marouf et al. Particle size distribution in Saturn's rings from Voyager I radio occultation Icarus **54**, 189–211 (1983).
19. K. Ohtsuki. Capture probability of colliding planetesimals: dynamical constraints on accretion of planets, satellites, and ring particles. Icarus **106**, 228–246 (1993).
20. D. C. Richardson. A new tree code method for simulation of planetesimal dynamics. Mon. Not. R. astr. Soc. **264**, 396–414 (1993).
21. H. Salo. Collisional Evolution of Rotating, Non-Identical Particles. Earth, Moon, and Planets **38**, 149–181 (1987).
22. H. Salo. Numerical Simulations of Collisions between Rotating Particles. Icarus **70**, 37–51 (1987).
23. H. Salo. Numerical simulations of dense collisional systems. Icarus **90**, 254–270. See also Erratum, Icarus **92**, 367–368 (1991).
24. H. Salo. Numerical simulations of dense collisional systems II: Extended distribution of particle sizes. Icarus **96**, 85–106 (1992).
25. H. Salo. Gravitational wakes in Saturn's rings. Nature **359**, 619–621 (1992).
26. H. Salo. The dynamics of dense planetary rings III: Self-gravitating identical particles. Icarus **117**, 287–312 (1995).
27. U. Schmit, and W. M. Tscharnuter. A fluid dynamical treatment of the common action of self-gravitation, collisions, and rotation in Saturn's B-ring. Icarus **115**, 304–319 (1995).
28. U. Schmit, W. M. Tscharnuter. On the formation of the fine-scale structure in Saturn's B-ring. Icarus **138**, 173–187 (1999).
29. F. Spahn, J.-M. Hertzsch, N. V. Brilliantov. The role of particle collisions for the dynamics in planetary rings. Chaos, Solitons and Fractals **5**, 1945–1946 (1995).
30. G. R. Stewart, D. N. C. Lin, and P. Bodenheimer. Collision induced transport properties in planetary rings. In Planetary Rings (Eds. R. Greenberg, A. Brahic) pp. 447–512. Univ. of Arizona Press, Tucson (1984).
31. K. D. Supulver, F. Bridges, D. N. C. Lin. The coefficient of restitution of ice particles in glancing collisions: experimental results for unfrosted surfaces. Icarus **113**, 188–199 (1995).
32. A. Toomre. On the gravitational stability of a disk of stars. Astrophys. J. **139**, 1217–1238 (1964).
33. A. Toomre. Gas-hungry Sc spirals. In Dynamics and interactions of Galaxies (Ed. R. Wielen), pp. 292–303. Springer, Berlin (1990).
34. J. Trulsen. Numerical simulation of jet streams I: The three-dimensional case. Astrophys. Space Sci. **17**, 241–262 (1972).
35. S. J. Weidenschilling et al. Ring particles: Collisional interactions and physical nature. In Planetary Rings, R. Greenberg, A. Brahic (Eds.) pp. 367–415. Univ. of Arizona Press, Tucson (1984).
36. W. Ward. On the radial structure of Saturn's rings. Geophys. Res. Lett. **8**, 641–643 (1981).
37. J. Wisdom, and S. Tremaine. Local simulations of planetary rings. Astron. J. **95**, 925–940 (1988).

Formation of Narrow Ringlets in Saturn's Rings

Jyrki Hänninen[1,2]

[1] Dept. of Physical Sciences, Division of Astronomy, University of Oulu, Finland
[2] Present address: Tuorla Observatory, Väisäläntie 20, FIN-21500 Piikkiö, Finland. e-mail: jyrki@astro.utu.fi

Abstract. The interaction between satellites and planetary rings takes mainly place in resonance locations, where the ratio of angular velocity of the ring particles and the satellite can be described with small integer numbers (e.g. 2:1, 4:3). In these locations the collisionally induced outward flow of the angular momentum is perturbed by the satellite. In some cases the perturbations may be strong enough to change the direction of the angular momentum flow. As a result the interparticle collisions will tend to increase density gradient, leading to formation of gaps, sharp edges, or narrow ringlets. Numerical 3-dimensional collisional simulations including satellite perturbations can capture the essence of this process.

1 Introduction

Saturn's rings exhibit abrupt density variations at all scales, but one of the most puzzling features of the Saturnian ring system are the narrow ringlets found in the isolated resonance locations. Although most of the radial ringlet structure consists of wavelike density variations, there exist also some truly isolated ringlets embedded in empty (optical depth $\tau < 0.01$) gaps. These ringlet-gap pairs are most characteristic to the C-ring and Cassini Division ($\tau \approx 0.1$). Many of these narrow ringlets (with typical widths of a few tens of kilometers) are found in the isolated resonance locations of different satellites. Their edges are extremely sharp: in several cases τ drops to zero within radial distance of 1 km. For example, the Prometheus 2:1 inner Lindblad resonance is located at the distance of 88,713.77 km. At the same radius a ringlet is found, embedded in the empty gap. There exists also resonances without ringlets and ringlets without any connection to the known satellite resonances [5].

In this paper I concentrate on the physics of satellite-ring interaction taking place in the isolated resonance locations. I will describe results from collisional 3D simulations in which gravitational perturbations of the satellite have been taken into account but not the ring self-gravity. The essential physics is not lost by the neglect of the self-gravity because of two reasons:

Firstly, in the low-density regions ($\tau < 0.5$ at the C-ring distance from the planet) the increase of the velocity dispersion due to self-gravity is of the same order of the magnitude or smaller than the velocity dispersion due to finite particle size [25].

Secondly, the ring self-gravity is not essential in the angular momentum exchange between the ring and the satellite [12, 20]: angular momentum exchange should occur if there is any process capable of creating a non-symmetric density response of the ring with respect to the rotating satellite potential. In planetary rings this process can be either the self-gravity of the ring particles or just interparticle collisions. These mechanisms also spread the influence of the interaction in the resonance zone into other parts of the ring [13].

In what follows I review shortly theoretical background of satellite-ring interaction. Then I discuss the necessary scaling of numerical simulations. Finally I deal with the formation of narrow ringlets in the simulations.

2 Collisional Dynamics of Unperturbed Rings

The collisional dynamics of unperturbed rings is reviewed in the article by Heikki Salo in this same book. So, for our purposes, it is enough to remind that the presence of viscosity and differential rotation induces a viscous stress that leads to an outward transfer of angular momentum. The efficiency of this process depends naturally on the magnitude of the viscosity ν. This angular momentum transport can also be described as a viscous torque [18]

$$T_\nu \simeq 3\pi\nu\Sigma\Omega r^2 \ , \tag{1}$$

where Σ stands for surface density and Ω for angular orbital velocity. This torque is exerted by the material inside radius r on the material outside this radius due to the viscous interaction. The torque T_ν is the rate at which angular momentum flows outward accross the radius r, and it is also referred to as viscous angular momentum luminosity L_H [2].

3 Satellite Resonances

The planetary ring and satellite interaction mainly occurs at isolated resonance locations. Resonances occur where the natural frequency of the ring particles equals the frequency of satellite's forcing. When satellite's circular frequency is denoted by Ω_s, vertical frequency by μ_s, and epicyclic frequency by κ_s, the disturbance frequency ω can be written as

$$\omega = m\Omega_s \pm n\mu_s \pm k\kappa_s \ , \tag{2}$$

where m, n, and k are non-negative integers. Vertical forcing can excite bending waves, but the horisontal forcing is observed to excite density waves and to open gaps. These horisontal resonances occur at the Lindblad resonances r_L where

$$\omega = m\Omega(r_L) \mp \kappa(r_L) \ , \tag{3}$$

and the vertical resonances occur at r_v where

$$\omega = m\Omega(r_v) \mp \mu(r_v) \ , \tag{4}$$

where the - sign denotes the inner resonance and the + sign denotes the outer resonance.

For the ring particles orbiting around Saturn, the various frequencies split so that $\mu > \Omega > \kappa$ due to Saturn's oblateness and the gravitational force of the ring itself. In practice this splitting is observed as dislocation of resonance radii of the various types of resonances. It also causes the orbital node to regress and the line of apsides to advance [27]. However, we concentrate only on the strongest ($k = 0$) horisontal resonances ($\mu_s = 0$), and because we treat the planet as a mass point ($\Omega = \kappa$), the resonance condition simplifies to

$$\frac{\Omega}{\Omega_s} = \frac{m}{m \mp 1} \ , \tag{5}$$

where - sign corresponds to resonance with external satellite (inner Lindbald resonance, ILR) and + sign corresponds to resonance with inner satellite (outer Lindblad resonance, OLR).

The perturbation of the satellite acts to increase random velocities of particles. The perturbation is strongest at discrete locations corresponding to above mentioned Lindblad resonances.

4 Angular Momentum Transfer

Besides affecting velocity dispersion, resonance perturbations also lead to the formation of azimuthal density variations with m-fold symmetry, stationary in the frame corotating with the satellite. Without interparticle collisions the density wave appears symmetric with respect to the satellite, and it consists of m equally spaced loops, following from the 180° phase shift in the alignment of particle orbits at various sides of the exact resonance radius.

For example, in the 2:1 ILR, particles inside the resonance radius are at periapse during the satellite conjuction while those outside pass the satellite at their apoapse. The particle-particle impacts remove the discontinuity in the alignment, leading to the formation of a trailing density wake. Besides the stationary density wake (in the coordinate system corotating with the satellite), the collision induced dissipation and the satellite perturbations force the particles outside the resonance radius to fall inward [14].

On the macroscopic level this can be intepreted as being due to the torque exerted on the satellite by the nonaxisymmetric density response of the ring, transferring angular momentum of the ring particles to the satellite (in the case of ILR). In terms of individual particle orbits, conservation of the Jacobi constant requires that the excitation of eccentrities is connected to the slight decrease in the semi-major axes a of the particles. If at least part of the eccentricity is damped by impacts, cumulative change of a is obtained [6].

When assuming that the gravitational potential of the perturbing satellite is small, the theoretical expression for the linear torque exerted by the satellite in a Lindblad resonance can be written as [12]

$$T_s^L = k_s \Omega^2 a^4 \Sigma_\circ \left(M_s/M_p\right)^2 \ , \tag{6}$$

where $\Sigma_\circ$ is the unperturbed surface density, and M_s and M_p denote the masses of the satellite and the planet, respectively. The constants $k_s = -14.8, -47.2, -96.5$ for ILR with $m = 2, 3, 4$, and $k_s = 1.2, 54.1, 106.4$ for OLR with $m = 1, 2, 3$. In the limit $m \gg 1$, $k_s \simeq \mp 8.5 m^2$ [2]. As a result the torque per square of the satellite mass and surface density of the ring should be constant.

It has been demonstrated with numerical N-body simulations that these theoretical estimates are very accurate, if the mass of a perturbing satellite is small enough not to create substantial surface density perturbation [14]. In the numerical simulations it has been found that the Eq. (6) is valid as long as

$$\tau \geq \tau_{crit} \simeq \frac{1}{3}\left(\frac{M_s}{M_p}\right)^{\frac{2}{3}} m^{\frac{4}{3}} \ , \tag{7}$$

while for smaller τ torque is reduced, proportional to impact frequency ω_c [14].

If the satellite perturbation is strong enough, the satellite generated surface density perturbation can increase up to $\Sigma/\Sigma_\circ \sim 2$ indicating that the torque is not linear. The nonlinear torque can be evaluated by assuming that the satellite generated surface density perturbation is of the order of unity and thus not proportional to the satellite mass [2]. As a result the torque is proportional to the first power of satellite mass instead of the second power. The linear theory is used to determine the critical satellite mass required to create the nonlinear density perturbation [10], and the resulting torque is obtained by multiplying the critical linear torque (torque caused by the critical satellite mass) by the ratio of the actual satellite mass to the critical mass, yielding

$$T_s^{NL} = k_s^{NL} \frac{\Sigma_\circ^2 \Omega^2 a^6 M_s}{M_p^2} \ , \tag{8}$$

where the value for the coefficient is $k_s^{NL} = -6.92$ in 2:1 ILR. In the collisional N-body simulations, the satellite exerted torque on the narrow ringlet was, indeed, observed to have same M_s/M_p-dependency [15]. However, the surface density dependency was observed to deviate from the theoretical estimate, yielding

$$T_s^{NL} \propto \frac{M_s}{M_p} \Sigma_\circ^q \ , \tag{9}$$

where $q \simeq 1.5$ [16].

5 Scaling of Numerical Simulations

The integration algorithm of our direct N-body simulation method is based on Aarseth's N-body integrator [1] which uses fourth-order force polynomial and individual time-step scheme in the integration of particle orbits. Calculation of particle-particle impacts can be done with very little extra CPU-time consumption, because mutual particle distances are easily calculated from the force polynomials [14]. As the present simulations concentrate on systems with low optical depth, self-gravity is neglected although it would be important in dense rings [24].

The impact model is the standard one: colliding bodies are assumed to be spherical and frictionless. Hence, the perpendicular component of the relative velocity is reversed and reduced by a factor of e in each impact:

$$\boldsymbol{b} \cdot \boldsymbol{v}' = -e\, \boldsymbol{b} \cdot \boldsymbol{v} \;, \tag{10}$$

where $\boldsymbol{v}'$ and $\boldsymbol{v}$ are the post- and precollisional relative velocities, and $\boldsymbol{b}$ is the unit vector in the direction joining the particle centers. For the coefficient of restitution, we use results from the laboratory experiments of the behaviour of ice at low temperatures and pressures [4]. The velocity-dependent e can be written in the form

$$e(v) = (v_n/v_c)^{-0.234} \;, \tag{11}$$

where $v_n = |\boldsymbol{b} \cdot \boldsymbol{v}|$ is the perpendicular component of the relative velocity and the constant v_c has value $v_c \simeq 0.01$ cm/s.

Collisional systems tend to establish balance between viscous gain of energy from the orbital motions and energy loss due to the interparticle collisions. As a result the effective value of the coefficient of restitution e will be adjusted to some fixed level depending on the optical depth. The random velocity dispersion of the collisional system scales differently on the two extreme limits: on the mass-point limit (elastic impacts, $R\Omega/v_c \ll 1$) the velocity dispersion is proportional to v_c, but on the limit of extremely flattened systems (soft impacts, $R\Omega/v_c \simeq 1$) the velocity dispersion results from the non-local viscous gain [26], and thus it scales according to $R\Omega$, where R is radius of a particle.

Both the local and non-local viscous gain are important in the realistic systems that have thickness of few particle radii. As the relative importance of the finite particle radius scales proportional to $R\Omega/v_c$, the numerical values obtained by some combination of R and v_c can always be scaled to other collisional systems, if the ratio is held fixed [24].

Same scaling can be expected to be applicable even if satellite perturbations are included. However, then we must also estimate the excess random velocity dispersion due to the satellite.

In order to estimate the effects of satellite perturbations let us consider an idealized case of close passage at circular orbit: a single encounter with the satellite creates a perturbed eccentricity ε_s [6, 17] for a ring particle

$$\varepsilon_s = 2.24 \left(\frac{a}{d}\right)^2 \frac{M_s}{M_p} , \tag{12}$$

where $d = a_s - a$ is the satellite-particle distance. The perturbed particle orbits are in same phase with each other just after the conjunction with the satellite. This collective motion is dispersed in time due to the radial eccentricity gradient and due to the differential rotation which causes streamlines to cross. These processes thus transform the collective motion of perturbed particles into the velocity dispersion. As a result the satellite generated residual velocity dispersion among ring particles is of the order of $c_s \simeq \varepsilon_s a\Omega$ [15].

The minimum velocity dispersion due to the finite size of ring particles is of the order of $c_f \simeq 3R\Omega$. Thus the relative importance of satellite perturbations to the residual velocity dispersion due to the finite particle size scales proportional to $c_s/c_f \sim M_s/R$. When this ratio is fixed, results that are obtained by some numerical values of M_s/M_p and R can be scaled to the other physical systems. In fact, a sufficient condition for this scaling is that $\varepsilon_s \sim M_s/M_p$, so that the above assumption of circular passage is not essential.

As we are interested in the satellite torques exerted on the simulation system, it is important to estimate how much the inaccuracy of orbital integrations might affect the angular momentum I. As I is not a conserved quantity in time dependent potential, we use the Jacobi constant $E_J = E - \Omega_s I$ which should be conserved for each individual particle between collisions (E denotes energy). In a collisionless simulation with typical satellite mass the observed error is about $\Delta E_J/E_J \simeq -3 \times 10^{-5}$ during 800 T_{orb} simulation. As the collisional calculations are performed with double precision their error is negligible. Accordingly, angular momentum changes significantly exceeding $\Delta I/I = -3 \times 10^{-5}$ during 800 T_{orb} simulation are considered reliable.

6 Ring Edges, Gaps and Narrow Ringlets

The angular momentum removal at ILR takes place in a narrow resonance zone, and will therefore try to clear a gap at the resonance location. On the other hand, viscous diffusion opposes this process and tends to smooth all density variations. A simple macroscopic criterion for the formation of the gap can be constructed by requiring that the time-scale t_{open} for angular momentum removal is shorter than the diffusion time-scale t_{close} [11]. From the condition for opening the gap $t_{open} < t_{close}$ we obtain

$$\frac{M_s}{M_p} > 0.1 \frac{\sqrt{\tau c^2}}{a\Omega} , \tag{13}$$

where it is assumed that $\tau \ll 1$ (c denotes velocity dispersion). The same type of condition follows from equating the viscous torque (1) with the satellite torque (6). Similar condition is also obtained if individual particle orbits

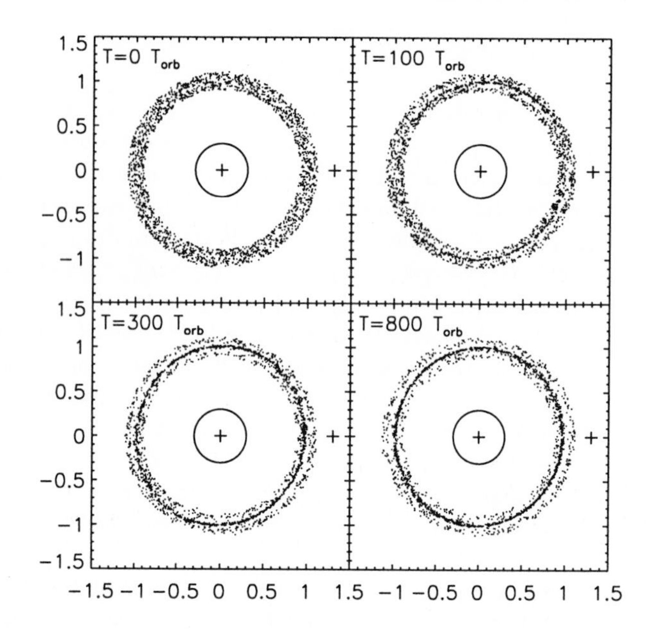

Fig. 1. Time evolution of collisional ring with 2000 spherical, frictionless, and massless particles with radius $R = 0.001$ and initial optical depth $\tau_\circ = 0.01$. The ring is perturbed with satellite of mass $\mu_s = M_s/M_p = 8.871 \times 10^{-5}$ on a circular orbit, 2:1 ILR falling on the unit distance. The satellite position is marked by the cross, but not in correct radial scale. The ring width is multiplied by a factor of two. In the numerical experiments the particle size is necessarily much larger than true size, if realistic values of optical depth are simulated. Therefore, to be able to compare the results to realistic systems, various parameters must be appropriately scaled. For 1 meter particles in the C-ring at Prometheus 2:1 ILR, $v_c = 4.2891 \times 10^{-4}$ has been adopted. Similarly to keep the ratio between satellite induced velocities and non-perturbed velocities constant, satellite mass has been scaled: the simulation mass $\mu_s = 8.871 \times 10^{-5}$ corresponds to the physical mass $\mu_{ph} = 1.0 \times 10^{-9}$

are inspected [8]. With these different approaches slightly different coefficient is acquired: it ranges from 0.1 to 0.5. This uncertainty is related to the difficulties in determining the exact value of viscosity [24, 29]. However, the Eq. (13) has been verified to be of the right order of magnitude with numerical N-body simulations [14, 15]. With large enough M_s the gap eventually appears as a well-defined outer boundary, a ring edge, as all the particles are removed from the outer parts of the system.

An example of the isolated narrow ringlet in the empty gap is the ringlet in the Prometheus 2:1 ILR located at the outer C-ring [23]. The narrow,

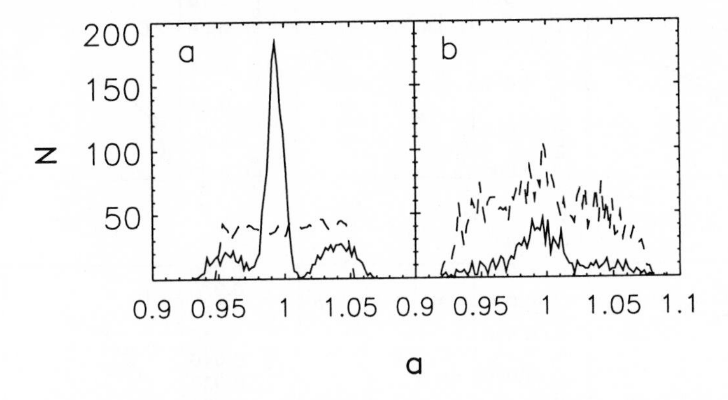

Fig. 2. (a) The radial density profile of the ring is shown at the beginning ($t = 0\,T_{orb}$, dashed line) and at the end ($t = 800\,T_{orb}$, solid line) of the collisional N-body simulation with 2000 massless particles. **(b)** Density profile at the end ($t = 800\,T_{orb}$) of a simulation with bimodal size distribution (actually $2N$ is plotted). The solid line shows the distribution of the population of 500 large particles ($R = 0.002$), and the dashed line shows the distribution of 2000 particles with $R = 0.001$. The initial optical depth of both the populations is $\tau_\circ = 0.01$

sharp-edged ringlet has mean optical depth $\tau \simeq 0.8$ and its width is ~ 16 km. The gap is ~ 30 km wide [22].

The time evolution of the ringlet formation in the collisional simulation is shown in Fig. 1. The simulation parameters have been scaled to the physical ring system at the distance of the Prometheus 2:1 ILR from Saturn.

The radial density profile of the formed ringlet is better seen in Fig. 2(a). The extreme sharpness of the narrow ringlet agrees with the observations: in several cases the optical depth goes from zero to several tenths in less than radial distance of 1 km [7, 21]. A typical feature observed also in the actual rings is the more dominant outer gap [19]. When reducing satellite mass in the simulations the inner gap will become weaker, and it will eventually disappear while the ringlet and outer gap still exist. In experiments with bimodal size distribution, see Fig.2(b), the large particles are more efficiently confined, in agreement with the observations where fraction of small particles is enhanced at the ringlet edges [28].

In the Fig. 3 it is studied how the ringlet steady-state properties depend on the mass of the perturbing satellite. The mass of the ringlet, see Fig. 3(a), is observed to depend on the radial width of the zone in which the satellite perturbations are capable of creating wakes that eventually transform into the ringlet [15]. The radial width of this zone is roughly proportional to $\sqrt{\mu_s}$. Because the observed ringlet width is independent of the satellite mass, see Fig. 3(d), the azimuthally averaged density has also to be proportional to $\sqrt{\mu_s}$, as demonstrated in Fig. 3(c).

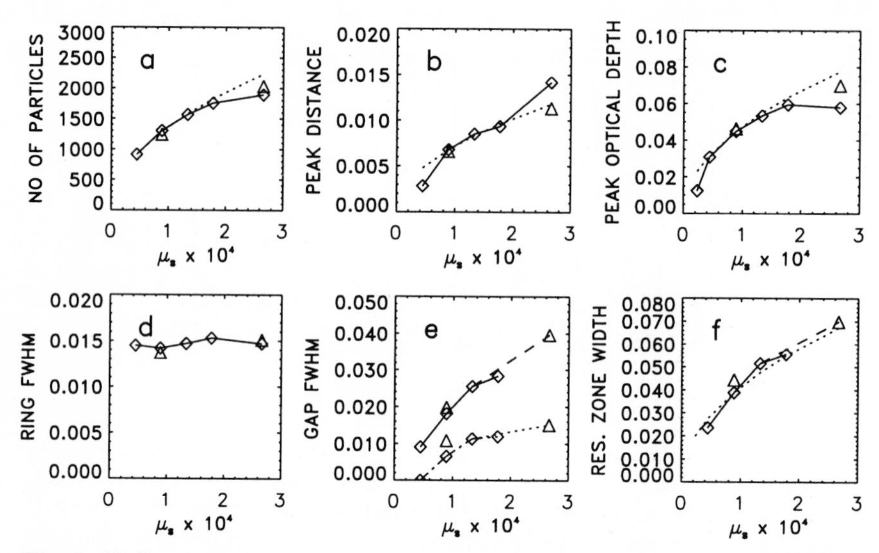

Fig. 3. Different ringlet and gap properties have been plotted as a function of the perturbing satellite mass μ_s. Note that in the x-axis are the simulated satellite masses. The corresponding scaled masses range from $\mu_{ph} = 1 \cdot 10^{-9}$ to $\mu_{ph} = 3 \cdot 10^{-9}$. **(a)** Number of particles in the ringlet at the end of the simulation run; **(b)** measured ringlet peak distance from the exact resonance radius; **(b)** peak optical depth τ; **(d)** full width of the half maximum (FWHM) density has been used as the criterion for the ringlet width; **(e)** widths of the outer and inner gaps; **(f)** width of the combined resonance zone. The dotted curves indicate indicate proportionality to $\sqrt{\mu_s}$, except in **(e)** in which the dotted line is related to the inner gap width. The diamonds show results from $N = 2000$ particles simulations, and the triangles are related to the simulations with $N = 3000$ particles

The dislocation of the ringlet, see Fig. 3(b), due to the strong energy dissipation in the wake stage is also proportional to $\sqrt{\mu_s}$. The smallest and largest satellite masses (in the simulations of $N = 2000$ particles) deviate from the $\sim \sqrt{\mu_s}$ curve, but in the former case there is no real isolated ringlet and in the latter case the ringlet suffers from the adequate number of particles in the simulation system. An additional simulation with $N = 3000$ particles follows the trend quite nicely. Because the dislocation results from the work done by the satellite on the ringlet, the ringlet having smaller mass has to drift farther inward in order to balance the energy consumed by the satellite.

The plot in Fig. 3(e) demonstrates how much larger the outer gap is compared to the inner gap. A good measure for the radial width of the resonance zone is the combined width of the ringlet and the outer and inner gaps. The size of this zone, see Fig. 3(f), is again proportional to $\sqrt{\mu_s}$, which is in agreement with the observed dependency of the final ringlet mass.

The ringlet evolution process can be understood in terms of work done by the satellite on the ring and energy dissipation due to interparticle collisions.

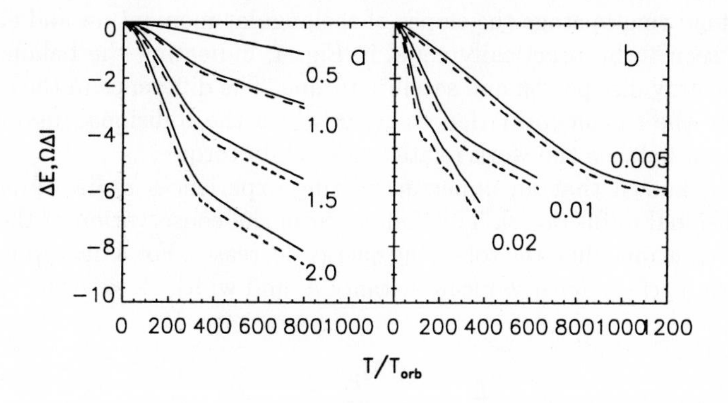

Fig. 4. The observed change in the ring angular momentum ΔI multiplied by Ω (dashed line), its slope being proportional to instantaneous satellite torque, displayed together with the cumulative change in the ring total energy ΔE (solid line). In the initial state $I = 2000$ and $E = -1000$. **(a)** The angular momentum change is compared between different satellite masses (the satellite masses are $\mu_{ph} = 0.5 \times 10^{-9}$, 1.0×10^{-9}, 1.5×10^{-9}, and 2.0×10^{-9}) with same initial $\tau_\circ = 0.01$. **(b)** Comparison between different initial optical depths ($\tau_\circ = 0.005$, 0.01, and 0.02) with the same satellite mass $\mu_{ph} = 2 \times 10^{-9}$

In Fig. 4 is shown the evolution of the ring total energy (solid line) which follows closely the $\Omega\Delta I$ curve (dashed line).

The work done by the satellite is related to the torque by $dE_s/dt = \Omega_s T_s$. Conservation of Jacobi constant for individual orbits between impacts requires that the satellite induced increase of random kinetic energy, dE_{kin}/dt, is related to the work by

$$\frac{dE_{kin}}{dt} = \left(1 - \frac{\Omega}{\Omega_s}\right) \frac{dE_s}{dt} . \tag{14}$$

In the simulations it was observed that during the nonlinear stage (after the ringlet has formed) the energy dissipation via collisions balances the velocity dispersion enhanced by the satellite perturbations,

$$\frac{dE_{coll}}{dt} + \frac{dE_{kin}}{dt} = 0 , \tag{15}$$

leading to dissipation rate

$$\frac{dE_{coll}}{dt} = (\Omega - \Omega_s) T_s \tag{16}$$

and to total energy decrease

$$\frac{dE}{dt} = \frac{dE_s}{dt} + \frac{dE_{coll}}{dt} = \Omega T_s . \tag{17}$$

In the final ringlet stage the slopes of the angular momentum and energy curves are seen to be practically equal in Fig. 4, indicating the balance between the energy dissipation and satellite torque. The difference in the curves is created in the ringlet formation stage, in which the collisional dissipation was unable to balance the work related to satellite torque.

It is well known that an unperturbed ring experiences radial spreading due to collisional diffusion [3]. This follows from the conservation of the ring angular momentum while the total ring energy decreases. For a narrow ringlet with uniform surface density, mean distance $\bar{a}$, and width $W \ll \bar{a}$ the ringlet energy is

$$E = -\frac{GM_p M_R}{2\bar{a}} \tag{18}$$

and the angular momentum is

$$I = \sqrt{GM_p \bar{a}} M_R \left(1 + \frac{1}{32}\left(\frac{W}{\bar{a}}\right)^2\right) , \tag{19}$$

where M_R stands for the ringlet mass and G denotes the gravitational constant. In the case of isolated and unperturbed ringlet $\Delta I = 0$ leading to the relation

$$\frac{dW}{dt} \simeq -\frac{8\bar{a}}{W}\frac{d\bar{a}}{dt} . \tag{20}$$

The ringlet thus expands rapidly, because $d\bar{a}/dt = 2\bar{a}^2/(GM_p M_R) dE/dt$ is negative.

However, if the ringlet is located at the resonance zone, torque is exerted on the ring, and the ΔI is not anymore zero. According to the simulations $dE/dt \simeq \Omega dI/dt$ leading to the relation

$$\frac{dW}{dt} \simeq \frac{3}{4}\frac{W}{\bar{a}}\frac{d\bar{a}}{dt} , \tag{21}$$

which indicates that the ringlet experiences slow contraction. Even though the maintenance of the narrow ringlet is explained by the above equation, the ringlet experiences a slow inward drift due to the dissipation that will eventually remove the ringlet from the resonance zone. However, the timescale for drifting accross the width of the resonance zone is of several orders of magnitude longer than the formation timescale [9, 16]. According to the simulation scaled to the values of Prometheus 2:1 ILR ringlet, the observed drift speed yields lifetime of about 8×10^6 years for the ringlet.

7 Summary

Even if there still remain unexplained structures in Saturn's rings, it is evident that the satellite perturbations are responsible for numerous observed features. The strongest ring-moon interaction takes place at the isolated resonance locations where different kinds of features are found. Beyond the scope of this paper are density and bending waves observed to propagate from several resonance locatations especially in Saturn's A-ring. In this process the ring self-gravity is essential. Also the outer edges of the A-ring and the B-ring coincide with strong isolated resonances.

There are also several examples of satellites orbiting very close to the rings. For example, the enigmatic narrow F-ring at the outer parts of the Saturnian ring system is shepherded by two moons, Prometheus and Pandora. The F-ring appears wavy, kinked, and braided due to the strong perturbations caused by close encounters of the shepherding satellites. Furthermore, inside the main rings there is orbiting a moonlet. This small embedded moonlet, Pan, is responsible for the maintenance of the Encke gap.

However, in this paper we have concentrated on the mechanism that probably explains many of the observed isolated ringlets in low density regions of Saturn's rings. It also works for OLR as well as for ILR. For example, the narrow F-ring was earlier thought to be sustained by the shepherding process. However, because the more massive moon Prometheus is closer to the ring, the satellite torques on the F-ring do not balance each other. One has thus to look for new explanations. The single-sided shepherding mechanism discussed in this article may also explain how the narrow core of the F-ring could be confined solely by Prometheus.

References

1. S. J. Aarseth. Direct Methods for N-Body Simulations. In *MultipleTime Scales* (J. E. Brackbill and B. I. Cohen, Eds.), 377–418. Academic Press, Orlando, Florida (1985).
2. N. Borderies, P. Goldreich, and S. Tremaine. Unsolved Problems in Planetary Ring Dynamics. In *Planetary Rings* (R. Greenberg and A. Brahic, Eds.), 713–734. Univ. of Arizona Press, Tucson (1984).
3. A. Brahic. Systems of Colliding Bodies in a Gravitational Field: I-Numerical Simulation of the Standard Model. Astron. Astrophys. **54**, 895–907 (1977).
4. F. G. Bridges, A. Hatzes, and D. N. C. Lin, Structure, stability and evolution of Saturn's rings. Nature **97**, 333–335 (1984).
5. J. N. Cuzzi, J. J. Lissauer, L. W. Esposito, J. B. Holberg, E. A. Marouf, G. L. Tyler, and A. Boischot. Saturn's Rings: properties and Processes. In *Planetary Rings* (R. J. Greenberg and A. Brahic, Eds.), 73–199. Univ. of Arizona Press, Tucson (1984).
6. S. F. Dermott. Dynamics of Narrow Rings. In *Planetary Rings* (R. Greenberg and A. Brahic, Eds.), 589–637. Univ. of Arizona Press, Tucson (1984).
7. L. Esposito, et al. Eccentric Ringlet in the Maxwell Gap at 1.45 Saturn Radii: Multi-Instrument Voyager Observations. Science, **222** 57–60 (1983).
8. F. Franklin, M. Lecar, and W. Wiesel. Ring Particle Dynamics in Resonances. In *Planetary Rings* (R. Greenberg and A. Brahic, Eds.), 562–588. Univ. of Arizona Press, Tucson (1984).

9. P. Goldreich, N. Rappaport, and B. Sicardy. Single-Sided Shepherding. Icarus **118**, 414–417 (1995).
10. P. Goldreich, and S. Tremaine. The Formation of the Cassini Division in Saturn's Rings. Icarus **34**, 240–253 (1978).
11. P. Goldreich, and S. Tremaine. Disk-Satellite Interactions. Astrophys. J. **241**, 425–441 (1980).
12. P. Goldreich, and S. Tremaine. The Dynamics of Planetary Rings. Ann. Rev. Astron. Astrophys. **20**, 249–283 (1982).
13. R. Greenberg. The Role of Dissipation in Shepherding of Ring Particles. Icarus **53**, 207–218 (1983).
14. J. Hänninen, and H. Salo. Collisional Simulations of Satellite Lindblad Resonances. Icarus **97**, 228–247 (1992).
15. J. Hänninen, and H. Salo. Collisional Simulations of Satellite Lindblad Resonances: II. Formation of Narrow Ringlets. Icarus, **108** 325–346 (1994).
16. J. Hänninen, and H. Salo. Formation of Isolated Narrow Ringlets by Single Satellite. Icarus **117**, 435–438 (1995).
17. W. H. Julian, and A. Toomre. Non-axisymmetric responsis of differentially rotating disks of stars. Astrophys. J. **146**, 810-830 (1966).
18. D. Lynden-Bell, and J. E. Pringle. The evolution of viscous discs and the origin of the nebular variables. Mon. Not. R. Astr. Soc. **168**, 603–637 (1974).
19. E. A. Marouf, G. L. Tyler, and P. A. Rosen. Profiling Saturn's Rings by Radio Occultation. Icarus **68**, 120–166 (1986).
20. N. Meyer-Vernet, and B. Sicardy. On the Physics of Resonant Disk-Satellite Interaction. Icarus **69**, 157–175 (1987).
21. C. Porco, et al. The Eccentric Saturnian Ringlets at $1.29R_s$ and $1.45R_s$. Icarus **60**, 1–16 (1984).
22. C. Porco, and P. Nicholson. Eccentric Features in Saturn's Outer C Ring. Icarus **72**, 437–467 (1987).
23. P. A. Rosen, G. L. Tyler, E. A. Marouf, and J. J. Lissauer. Resonance Structures in Saturn's Rings Probed by radio Occultation: II. Results and Interpretation. Icarus **93**, 25–44 (1991).
24. H. Salo. Numerical Simulations of Dense Collisional Systems. Icarus **90**, 254–270 (1991).
25. H. Salo. Simulations of Dense Planetary Rings: III. Self-Gravitating Identical Particles. Icarus **117**, 287–312 (1995).
26. H. Salo, J. Lukkari, and J. Hänninen. Velocity Dependent Coefficient of Restitution and the Evolution of Collisional Systems. Earth, Moon, and Planets **43**, 33–43 (1988).
27. F. H. Shu. Waves in Planetary Rings. In *Planetary Rings* (R. Greenberg and A. Brahic, Eds.), 513–561. Univ. of Arizona Press, Tucson (1984).
28. G. L. Tyler et al. The Microwave Opacity of Saturn's Rings at Wavelengths of 3.6 and 13 cm from Voyager 1 Radio Occultation. Icarus **54**, 160–188 (1983).
29. J. Wisdom, and S. Tremaine. Local Simulations of Planetary Rings. Astron. J. **95**, 925–940 (1988).

Granular Viscosity, Planetary Rings and Inelastic Particle Collisions

Frank Spahn[1], Olaf Petzschmann[1], Jürgen Schmidt[1], Miodrag Sremčević[2], and Jan-Martin Hertzsch[3]

[1] University of Potsdam, Am Neuen Palais 10, 14469 Potsdam, Germany. e-mail: frank@agnld.uni-potsdam.de
[2] University of Belgrade, Yugoslavia
[3] Brunel University, Uxbridge, UK

Abstract. The functional dependencies of the viscosity η on temperature and density, derived for granular gases under certain physical environments - force free, and in a central gravitational field - are compared and numerically checked. It is known that different physical conditions lead to different functional dependencies of the viscosity η on the granular temperature T and also the matter density. This is caused by gradients of volume forces which create curvatures in the particle trajectories (epicycles) which bound the free motion and limit the mean free path l to finite values even for vanishing particle density ρ, where in the force free case $l \propto \rho^{-1}$ diverges. This results in the known dependence $\eta \propto \sqrt{T}$ in the force-free case for nearly elastic collisions. In planetary rings the transport coefficients of momentum and energy are proportional to T. We check the validity of these expressions with numerical particle simulations. For planetary rings the dependence of the coefficient of restitution e on the impact velocity v_{imp} is crucial for their stability. Hence, we present models of the dynamics of the particle collisions, which account for the velocity-dependence $e(v_{imp})$ by a visco-elastic model for particle collisions, as well as for the sticking at very low impact velocities. The latter is a further improvement of previous models, and the results are in accordance with laboratory measurements.

1 Introduction

In the last decade the interest in the behaviour of granular matter has grown rapidly. Depending on the physical environment [19], granular assemblies behave like solids, e.g. in the inner part of a sandpile, like dense fluids, in avalanches [18] or also like gases occurring in shear cells or in space.

A crucial problem is the derivation of macroscopic equations which characterise the different states of granular matter. In this context even the state of stresses inside a resting sandpile is rather complex and causes many exciting and controversial discussions (see [4, 32]). The complexity and the nonconservative nature of the interactions between the granular particles lead to difficulties in the description of such non-equilibrium system. For instance, it seems to be impossible to derive relations for the transport coefficients of mass, momentum and energy which merely characterise the granular matter and which are applicable to every physical situation.

Another essential difference between granular and usual gases is the lack of a scale separation which is not found in the former as opposite to the latter. In granular assemblies it is possible that the size of the particles is of the order of the mean free path l, whereas in ordinary gases l is many orders of magnitude larger than the size of a molecule. Furthermore, in usual gases the state values change on scales orders of magnitudes larger than the mean free path. In contrast, in granular assemblies all the state variables may change on spatial scales which are comparable to the size of the granules, and thus, to the mean free path.

In order to obtain the macroscopic equations characterising the different states of granular matter, one has to derive the constitutive relations, e.g. for the stress in dependence of the strain in the case static granular assemblies like a sand pile, the pressure tensor as a function of the strain rate as well as the heat conduction in dependence on gradients of state variables in the case of excited granular configurations. These relations close the system of equations governing the configuration of the granules, i.e. continuum mechanics for static states or kinetic theory including its hydrodynamic approximation in order to describe granular gases.

In this review we will report on the differences of transport coefficients of granular matter in force free environments as well as in a gravitational central field like in planetary rings around the giant planets in the solar system.

In both contexts, the studies of transport phenomena are based on kinetic gas theory. Lun, Savage and colleagues [24], as well as Jenkins and Richman [20] have derived relations for the viscosity, heat conductivity, and cooling for granular assemblies in the force free case.

Kinetic theories for planetary rings have been pioneered by Goldreich and Tremaine [10] and extended by Araki [1, 2].

Although the properties of the granular grains have been assumed to be the same, the resulting viscosities have shown different functional dependencies, i.e. in force free case $\eta \propto \sqrt{T}$, and $\eta \propto T$ in planetary rings.

In the next Section we outline briefly the theoretical basis of kinetic theory leading to both different expressions of transport values.

In Section 3 the approaches of Jenkins and Richman [20] and Goldreich and Tremaine [10] are compared with numerical experiments and applied to the dynamics of planetary rings [34, 38]. The major goal of this investigation is to check which of these approaches, the linear ansatzes between fluxes and thermodynamic forces [20, 24] or fluxes of momentum and energy derived from a triaxial kinetic approach [10, 35], is more appropriate for the description of the dynamics of planetary rings, which is done in Section 4.

In Section 5 we present new models for the collision dynamics [7, 15, 36].

2 Kinetic Theory of Granular Gases

With $6N$ equations of motion of N granular particles, including the collisional dynamics and initial conditions, all information is given to describe the system completely. However, the huge number of particles makes it impossible to obtain a solution. Under these conditions methods of statistical mechanics of non-equilibrium systems provide a useful tool to describe the collective dynamics of an assembly of granular particles.

Apart from the mass, size, and occupied volume fraction, the main difference between the molecules of gases and granular grains is the inelastic character of particle collisions. While molecules collide elastically, granular particles loose kinetic energy in every collision. The steady removal of energy from the granular gas causes a variety of non-equilibrium processes, e.g. the tendency of granular assemblies to form clusters [9, 27]. A kinetic description of granular gases must account for these differences to molecular gases.

2.1 The Boltzmann Equation and the Collision Integral

Keeping in mind the special properties of granular assemblies and considering only binary collisions and the molecular chaos assumption, the dynamics of the ensemble is described by the evolution of the one particle distribution function $\tilde{f}(R, \boldsymbol{r}, \boldsymbol{v}, t)$ which determines the number of particles

$$dN = \tilde{f}(R, \boldsymbol{r}, \boldsymbol{v}, t)\, dR\, d^3\boldsymbol{v}\, d^3\boldsymbol{r} \tag{1}$$

expected to be found in the size range between the particle radii R and $R + dR$ in a volume element $d^3\boldsymbol{r}$ at position $\boldsymbol{r}$ and in the volume element of the velocity space $d^3\boldsymbol{v}$ at velocity $\boldsymbol{v}$.

In the following we consider only a mono-disperse granular ensemble, i.e. all particles are of the same radius R_p according to $\tilde{f} = \delta(R - R_p)\, f(\boldsymbol{r}, \boldsymbol{v}, t)$. The evolution of the distribution function is described by

$$\partial_t f + \boldsymbol{v} \cdot \nabla f - \nabla \Phi \cdot \nabla_{\boldsymbol{v}} f = \mathbf{C}(f, f_1) \ . \tag{2}$$

with the external potential Φ. The right hand side of Eq. (2) is the collision integral which stands for the change of f due to particle collisions. We will just briefly report the form of the collision integral. For a rigorous derivation we refer to the monographs [3, 8, 29] or papers [2, 20, 24, 43].

In the case of spherical particles of radius R_p and considering only translational degrees of freedom one obtains as a result of the "Stoßzahlansatz" [3]

$$\mathbf{C}(f, f_1) = 4 \int d^3\boldsymbol{v}_1 d^2\boldsymbol{k}\, \sigma\, [\boldsymbol{g} \cdot \boldsymbol{k}]\, \Theta_H(\boldsymbol{g} \cdot \boldsymbol{k}) \left[Y_E\left(n(\boldsymbol{r} + R_p\boldsymbol{k})\right) \frac{f'(\boldsymbol{r})\, f_1'(\boldsymbol{r} + 2R_p\boldsymbol{k})}{e^2} - Y_E\left(n(\boldsymbol{r} - R_p\boldsymbol{k})\right) f(\boldsymbol{r}) f_1(\boldsymbol{r} - 2R_p\boldsymbol{k}) \right] . \tag{3}$$

This integral consists of a gain and a loss term describing particles which are scattered in and out of the $6D$ volume element $d^3\boldsymbol{v}\, d^3\boldsymbol{r}$ at $\boldsymbol{r}$ and $\boldsymbol{v}$, respectively. The subscript 1 denotes the collision partner of the particle under consideration and the primes mark quantities after a collision, e.g. for the distribution function one has $f_1' = f(\boldsymbol{v}_1')$. Here, n is the number density of the particles. The coefficient of restitution e measures the inelasticity of the collision, i.e. the reduction of the normal component of the relative velocity $\boldsymbol{g} = \boldsymbol{v}_1 - \boldsymbol{v}$ between the colliding particles, according to

$$\boldsymbol{g}' \cdot \boldsymbol{k}' = -\boldsymbol{g}' \cdot \boldsymbol{k} = e\, \boldsymbol{g} \cdot \boldsymbol{k} \quad \text{with} \quad e \in [0,1] \quad . \tag{4}$$

The unit vector $\boldsymbol{k}$ points from the centre of the collision partner 1 to the centre of the particle under consideration.

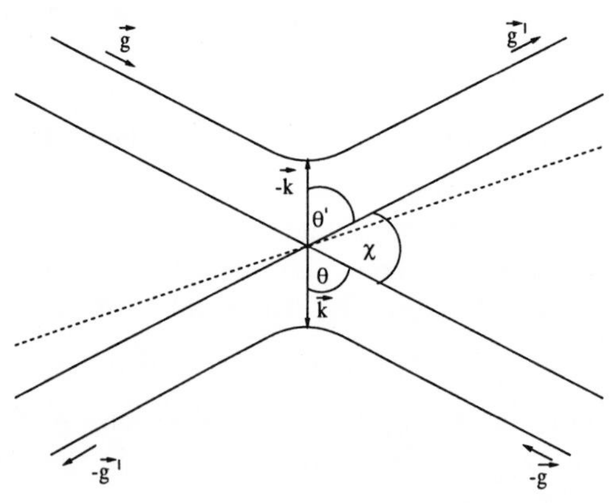

Fig. 1. The geometry of the collision including its inverse collision. The dashed line marks the reduction of the velocity component $\boldsymbol{g} \cdot \boldsymbol{k}$ due to inelastic collisions. This results in a reduction of the deflection angle χ and in a loss in reversibility of the collision process.

The unit vector $\boldsymbol{k}' = -\boldsymbol{k}$ belongs to the inverse collision process, obtained by exchanging the states after and before the collision and by setting $t \to -t$ (see Fig. 1). The factor e^{-2} in the gain term arises from a phase-space contraction in the case of inelastic collisions, and additionally, from the term $\boldsymbol{g} \cdot \boldsymbol{k}$ which differs by the factor e after and before the collision (see Eq. (1)). The value $\sigma = R_p^2$ is the cross-section of two colliding hard spheres having equal size.

The Enskog factor $Y_E \approx (2 - V_s n) / \left(2(1 - V_s n)^3\right)$ is proportional to the increase of the collision frequency caused by the reduction of the volume available for the kinematic particle motion due to the finite size of the colliding particles (V_s – volume of the granular spheres). The Boltzmann collision integral is then obtained by the neglect of the finite size of the particles ($R_p \to 0$) resulting in the assumption that the collision takes place at a point $\boldsymbol{r}$ in the limit $Y_E \to 1$ [8, 29].

With these assumptions the problem of kinetic theory is to find the solution $f(\boldsymbol{r},\boldsymbol{v},t)$ of Eq. (2). A further simplification of the problem is obtained by formulating the moment equations of the kinetic equation (2), provided a truncation of the resulting hierarchy of equations is justified. This is briefly sketched in the next paragraph.

2.2 The Hydrodynamic Approximation

The (infinite) hierarchy of the moment equations is obtained by applying the operators

$$\int d^3\boldsymbol{v}\ \ldots\ ;\ \frac{1}{n}\int d^3\,\boldsymbol{v}\,\boldsymbol{v}\ \ldots\ ,\quad \int d^3\boldsymbol{v}\,(\boldsymbol{v}-\boldsymbol{u})(\boldsymbol{v}-\boldsymbol{u})\ \ldots$$

to Eq. (2). The particle number density $n(\boldsymbol{r},t)$, mean velocity $\boldsymbol{u}(\boldsymbol{r},t)$ and the pressure tensor $\hat{\mathbf{P}}$ are given by

$$n(\boldsymbol{r},t)=\int d^3\boldsymbol{v}\quad f(\boldsymbol{r},\boldsymbol{v},t)\quad , \tag{5}$$

$$n\,\boldsymbol{u}(\boldsymbol{r},t)=\int d^3\boldsymbol{v}\quad \boldsymbol{v}(\boldsymbol{r},t)\ f(\boldsymbol{r},\boldsymbol{v},t)\quad , \tag{6}$$

$$\hat{\mathbf{P}}=p_{ij}\,\boldsymbol{e}_i\boldsymbol{e}_j=\int d^3\boldsymbol{v}\quad (\boldsymbol{v}-\boldsymbol{u})(\boldsymbol{v}-\boldsymbol{u})\ f(\boldsymbol{r},\boldsymbol{v},t)\quad . \tag{7}$$

The resulting first three equations of the hierarchy of the infinite number of relations are the conservation of particle number (we assume $m=1$ giving $\rho=n$), momentum and of the pressure tensor

$$\frac{\partial\rho}{\partial t}+\nabla\cdot\{\rho\,\boldsymbol{u}\}=\left\langle\frac{\partial\rho}{\partial t}\right\rangle_{coll}\quad , \tag{8}$$

$$\frac{\partial\,(\rho\,\boldsymbol{u})}{\partial t}+\nabla\cdot\left\{\hat{\mathbf{P}}+\rho\,\boldsymbol{u}\boldsymbol{u}\right\}-\rho\,\boldsymbol{F}=\left\langle\frac{\partial\,(\rho\boldsymbol{u})}{\partial t}\right\rangle_{coll}\quad , \tag{9}$$

$$\begin{aligned}\frac{\partial}{\partial_t}\{p_{ij}+\rho\,u_iu_j\}+\frac{\partial}{\partial x_k}\{p_{ijk}+u_j\,p_{ik}+u_k\,p_{ij}+u_i\,p_{jk}+u_iu_ju_k\,\rho\}\\ -\rho\,\{u_jF_i+u_iF_j\}=\left\langle\frac{\partial\,(p_{ij}+u_iu_j\,\rho)}{\partial_t}\right\rangle_{coll}\quad .\end{aligned} \tag{10}$$

These three equations are the beginning of the infinite hierarchy of higher moments. However, as long as the function f is close to the equilibrium distribution f_0, which is a Maxwellian, the infinite hierarchy can be truncated after Eq. (10).

A further simplification is achieved if one applies the model of hard indestructible spheres for the granular particles. In this case we have $\left\langle\frac{\partial\rho}{\partial t}\right\rangle_{coll}=0$ and $\left\langle\frac{\partial(\rho\boldsymbol{u})}{\partial t}\right\rangle_{coll}=0$, because the number of particles and their momenta do not change in course of a collision.

Furthermore, if the system is not far from equilibrium, the fluxes of momentum $\hat{\mathbf{P}}$ and the heat flux $\boldsymbol{Q}$ can be assumed to be linear functions of the related thermodynamic forces, as for instance, gradients of the mean velocity $\nabla \boldsymbol{u}$ and of the temperature ∇T, with $T = (1/3)\langle(\boldsymbol{v} - \boldsymbol{u})^2\rangle$

$$\hat{\mathbf{P}}(\hat{\mathbf{D}}) = -p + 2\eta(T)\hat{\mathbf{D}} + \zeta \nabla \cdot \boldsymbol{u}\, \hat{\mathbf{I}}\,, \tag{11}$$

$$\boldsymbol{Q} = -\kappa(T)\nabla T\,, \tag{12}$$

with the shear tensor

$$\hat{\mathbf{D}} = \frac{1}{2}\left(\nabla \boldsymbol{u} + \boldsymbol{u}\nabla - \frac{2}{3}\nabla \cdot \boldsymbol{u}\, \hat{\mathbf{I}}\right). \tag{13}$$

The coefficients η, ζ, and κ are the dynamic shear viscosity, the dynamic volume viscosity, and the heat conductivity, respectively, and $\hat{\mathbf{I}}$ denotes the identity tensor. With these linear dependencies and taking the trace of Eq. (10) one gets the hydrodynamic equations:

$$\frac{d\rho}{dt} = -\rho \nabla \cdot \boldsymbol{u} \tag{14}$$

$$\rho\left\{\frac{d\boldsymbol{u}}{dt} + 2\boldsymbol{\Omega}_0 \times \boldsymbol{u} - 3\Omega_0^2 y \boldsymbol{e}_y + \Omega_0^2 z\, \boldsymbol{e}_z\right\} = -\rho\, \nabla \Phi - \nabla \cdot \hat{\mathbf{P}} \tag{15}$$

$$\frac{3}{2}\rho\frac{dT}{dt} = -\nabla \cdot \boldsymbol{Q} - \hat{\mathbf{P}} : \nabla \boldsymbol{u} - \Gamma \quad . \tag{16}$$

The material derivative is given by $\frac{d}{dt} = \frac{\partial}{\partial t} + \boldsymbol{u} \cdot \nabla$ and Γ is the cooling due to inelastic collisions.

We want to describe a planetary ring as an ensemble of granular particles rotating according to Kepler's laws around a planet between their inelastic collisions. Thus, in Eq. (15) inertia forces are included that arise in a co-rotating Cartesian coordinate system orbiting the planet of mass M with an angular velocity $\Omega_0 = \sqrt{GM/r_0^3}$ at distance r_0 [38].

Using this, one gets a strain rate tensor $\hat{\mathbf{D}} = \frac{3}{2}\Omega_0\, \boldsymbol{e}_x \boldsymbol{e}_x$ and the mean circular motion around the planet $\boldsymbol{u} = \frac{3}{2} y \Omega_0 \boldsymbol{e}_x$, where $-y$ points to the planet and $-x$ is the direction of the orbital motion.

Then, the deviations from a circular motion account for the granular temperature T. The value Φ denotes force fields apart from the central force field, e.g. the self gravity of the ring or satellites.

In the case of planetary rings Goldreich and Tremaine [10] and Stewart et al. [40] formulated Eqs. (8) - (10) by using a triaxial Gaussian distribution function

$$f(\boldsymbol{r}, \boldsymbol{v}, t) = \frac{\rho}{\sqrt{(2\pi)^3 \det \hat{\mathbf{T}}}} \exp\left\{-\frac{1}{2}\boldsymbol{c} \cdot \hat{\mathbf{T}}^{-1} \cdot \boldsymbol{c}\right\}\,, \tag{17}$$

where $\boldsymbol{c} = \boldsymbol{v} - \boldsymbol{u}$ and $\hat{\mathbf{T}} = \hat{\mathbf{P}}/\rho$ are the random walk speed and the velocity dispersion tensor, respectively. For this triaxial Gaussian the set of equations (8) - (10) is automatically closed, because all uneven higher moments vanish. Using the collision integral derived by Trulsen [43], Goldreich and Tremaine calculated the eigenvalues and eigenvectors of $\hat{\mathbf{T}}$. They found that the larger the collision frequency ω_c is and the more dissipative the collisions are the more the main axes of $\hat{\mathbf{T}}$ deviate from the radial and azimuthal directions by a certain angle ψ. With the change of the vertical height z the angle ψ varies (for details see [34]). It is largest in the mid plane $z = 0$ and vanishes for $z \to \infty$. This is plausible as the distance from equilibrium is strongest in the ring plane, because of the maximum of the density and the collision frequency ω_c there and weaker for $|z| > 0$. Despite of this, in previous papers about ring kinetics [2, 10, 35] the variation of the angle $\psi(z)$ had not been taken into account.

Identifying the momentum flux $P_{r\phi} = \frac{1}{2}\rho\nu\Omega = \frac{1}{2}\eta\Omega$ (ν – kinematic viscosity, η – dynamic viscosity) Goldreich and Tremaine found for the local kinematic viscosity

$$\nu_l = K_1 \frac{tr\hat{\mathbf{T}}}{\Omega} \frac{\tau}{1+\tau^2} \ , \tag{18}$$

where K_1 is a constant of the order of unity. In this relation the optical depth τ is given by the vertical integration of the density

$$\tau = \int dR_p \, dz \; \tilde{n}(R_p, r, \phi, z) \; \pi \; R_p^2 \tag{19}$$

In equation (19) the density $\tilde{n}$ combines the volume density ρ with the size distribution function of the particles. For Saturn's rings the latter has been estimated to follow a power law $\propto R_p^{-\beta}$ [25, 44, 45]. The optical depth τ replaces the density ρ and characterises together with the temperature T (or more precisely with the velocity dispersion $\hat{\mathbf{T}}$) the ring completely, which is demonstrated in the subsequent section. The expression (18) corresponds to the kinematic part of ν where the particle motions alone are responsible for the momentum transport and the effects of the finite size of the particles are omitted. If the sizes of the particles are taken into account [2] an additional contribution to the transport processes appears, the so called nonlocal part of the viscosity given by [30, 31, 47]

$$\nu_{nl} = K_2 \frac{tr\hat{\mathbf{T}}}{\Omega} \tau^{\epsilon} \ , \tag{20}$$

with $\epsilon > 1$ and $K_2 \approx 1$. Although Goldreich and Tremaine came to the conclusion that stability of a planetary ring is only achieved, if the coefficient of restitution is a decreasing function of the impact velocity [10] their analytical analyses – and also subsequent work [2] – of the collisional changes of the component of the stress-tensor have been carried out under the assumption $e = const.$

Under force free conditions $\Omega_0 = 0$, Lun and colleagues [24] as well as Jenkins and Richman [20] have investigated Eqs. (15)-(16). They expanded the distribution function $f(\boldsymbol{c})$ with respect to the thermal velocity $\boldsymbol{c}$, and assumed a restitution coefficient $e = const.$ In contrast to the triaxial approach, which can not account for processes like heat conduction (which is a contraction of a third moment), this treatment covers all irreversible processes and Grad's moment method [20] results in the following expressions for the viscosity η, the heat conductivity κ, and the collisional cooling Γ:

$$
\begin{aligned}
\tilde{\eta} &= C_1 \, \frac{\sqrt{T}}{(1+e)(3-e)Y_E} \\
\tilde{\omega} &= C_2 \, (1+e) \, \rho^2 Y_E \sqrt{T} \\
\eta &= \tilde{\eta}\Big[1 + \frac{2}{5}\rho Y_E \, (3e-1)(1+e)\Big]\Big[1 + \frac{4}{5}\rho Y_E \, (1+e)\Big] + \frac{2\tilde{\omega}}{5} \\
\tilde{\kappa} &= C_3 \, \frac{\sqrt{T}}{(1+e)(49-33e)Y_E} \\
\kappa &= \tilde{\kappa}\Big[1 + \frac{3}{5}\rho Y_E (1+e)^2 (2e-1)\Big]\Big[1 + \frac{6}{5}\rho Y_E (1+e)\Big] \\
\Gamma &= C_4 \, (1-e^2) \, \rho^2 \, \sqrt{T^3} \quad .
\end{aligned}
\tag{21}
$$

Additionally to $m = 1$ we have normalised the bulk density $\rho_p = 1$, so that the density ρ equals the filling factor (the solid matter fraction). Then, the constants in Eqs. (21) read: $C_1 = 5\pi^{1/6}/\left(2^{8/3}3^{2/3}\right)$, $C_2 = 2^{7/3}/\left(3^{2/3}\pi^{5/6}\right)$, $C_3 = 5^2 3^{1/3}\pi^{1/6}/2^{5/3}$, and $C_4 = 2^{5/3}3^{2/3}/\pi^{1/6}$. Note that according to our normalisation, the particle radius must be written as $R_p = \sqrt[3]{3/(4\pi)}$.

Comparing the equations (18), (20) with the relations (21) for the viscosity, a difference in the temperature dependence is observed. One would expect the same transport coefficients for the same material, independent of the physical environment.

In the case of planetary rings this has been studied in detail [30, 31, 34, 47]. The simulations have well confirmed the above ansatz (17) as well as the relation (20) for the viscosity ν_{nl}, where in the case of a monodisperse granular ensemble it is obtained $\epsilon \approx \frac{5}{4}$ [30].

In order to check the expressions (21) we have carried out numerical experiments, where a shear cell is driven via two boundaries and the resulting momentum flux is measured. This is briefly presented in the next section.

3 Measurement of the Viscosity: The Force Free Case

To measure the viscosity of a force free granular gas, a shear cell – a cube of side length L – is initially homogeneously filled with granular grains of radius

R_p. Periodic boundary conditions are applied in the x and y directions. In the z direction, the boundary conditions are periodic, too, but with an offset $\Delta v/2$ in the velocity keeping the random walk speed. As we will show below, for the stationary state of the velocity field a linear shear $\boldsymbol{u}(z) = u(z)\boldsymbol{e}_x \propto z\,\boldsymbol{e}_x$ will develop.

With these assumptions Eqs. (15)-(16) read:

$$\partial_x(\rho\, u) = 0 \tag{22}$$

$$\partial_t u = -\frac{1}{\rho}\partial_z(\rho T) + \partial_z(\eta\, \partial_z u) \tag{23}$$

$$\frac{3}{2}\rho\, \partial_t T \;=\; \partial_z\left(\kappa\, \partial_z T\right) + \eta\left(\partial_z u\right)^2 - \Gamma \tag{24}$$

One stationary solution of Eqs. (23)-(24) is

$$u_0 \;=\; \frac{\Delta u}{L}\, z - \frac{1}{2}\Delta u\ ;\ \ \nabla\rho_0 = \nabla T_0 = 0 \quad . \tag{25}$$

Stability analyses [33] and numerical studies [17] have shown that this solution (25) is always unstable unless the particles collide elastically. However, here we are interested in a check of the Eqs. (21) which have been derived [20] for nearly elastically colliding spheres ($e \approx 1$). Because with increasing restitution e the unstable length-scales get larger according to $\lambda_c \propto (1-e^2)^{1/2}$ [9, 27, 38], we assume that for our purposes the solution (25) holds throughout the the simulations.

In order to analyse the data of the numerical experiments we have used the method presented by Wisdom and Tremaine [47], which was also applied in subsequent investigations [30, 31]. The local viscosity is given by

$$\eta_l = \rho\nu_l = \frac{L\, P_{xz}^{(l)}}{\Delta v} \Rightarrow \nu_l \propto \langle \dot{x}\dot{z}\rangle\,, \tag{26}$$

where the averaging is carried out over the box and over time. The nonlocal parts are caused by the transport of momentum over the finite extension of the granules given by [47]

$$\nu_{nl} \propto \sum\left(z_> - z_<\right)\, \Delta\dot{x}_> \quad . \tag{27}$$

With these simulations we attempt to answer the questions: Is the linear ansatz (13) valid? Can the dependence $\eta \propto \sqrt{T}$ be reproduced by the numerical experiments? How large will the deviations from the transport expressions (21) become if one decreases the restitution $e \to 0$ and if one increases the density ρ of the granular ensemble?

The first two questions can be answered by plotting $P_{xz}\,(u')^{-1}\, T^{-\frac{1}{2}}$ versus the shear rate u', where the primes denote derivatives with respect to z. If the dynamic viscosity is a linear function of u' and $\sqrt{T}$, the resulting plot should show a constant value. This is shown in Figure 2 where this dimensionless quantity is plotted against u' over *two* orders of magnitude. In this range of the variation of the thermodynamic force u' the Newtonian ansatz, as well as the temperature dependence $\sqrt{T}$, is valid within an accuracy of $< 5\%$. This even holds for smaller values of the restitution $e = 0.7$ and also for a wider range of u'. Figure 3 shows the relative changes of the dynamical

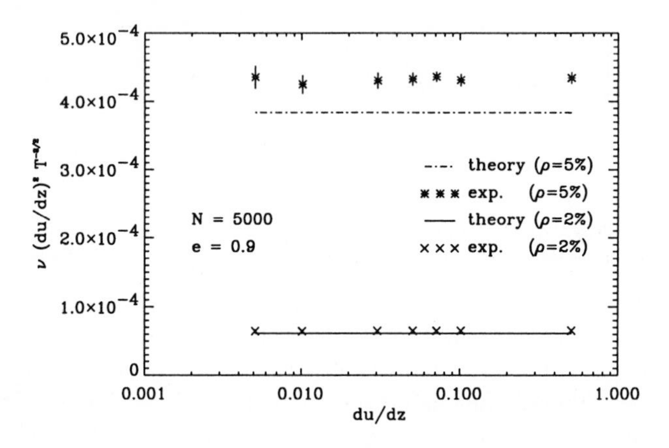

Fig. 2. The shear stress normalised by $(\partial u/\partial z)\sqrt{T}$ vs. the shear rate $\partial u/\partial z$ for $e = 0.9$ and densities $\rho = 0.02$ and 0.05.

viscosity $|\Delta\eta|/\eta$ (left part) as well as the deviation of the velocity profile $u(z)$ from the stationary linear one $\propto z$ (right part) as functions of the collisional dissipation $(1 - e)$ and the density ρ. As expected, in both cases the deviations get stronger the more dissipative the system gets. On the other hand, the theoretical expression for the viscosity as a function of the density $\eta(\rho)$ is well reproduced by the numerical experiments (see Fig. 3, right part).

Corresponding simulations have yielded analogous results for the heat conductivity, which, however, are not reported here because this transport is not of that much importance for planetary rings as the momentum transfer.

The main conclusion of this section is that the linear ansatz for the transport coefficients, if the anisotropy of $\hat{\mathbf{T}}$ and $\hat{\mathbf{P}}$ is neglected, hold for a wide range of the velocity shear u'. Furthermore, the temperature dependence $\sqrt{T}$ is valid as long as the restitution is not too low $(e < 0.6)$.

On the other hand, also the expressions (18) and (20) in a central force field, e.g. in planetary rings, have been well reproduced by the numerical experiments of Wisdom and Tremaine [47] and Salo [30].

These results imply that the viscosity of granular matter cannot simply be considered as a material-specific value. It is sensitively dependent on the physical environment which the granular ensemble is exposed to.

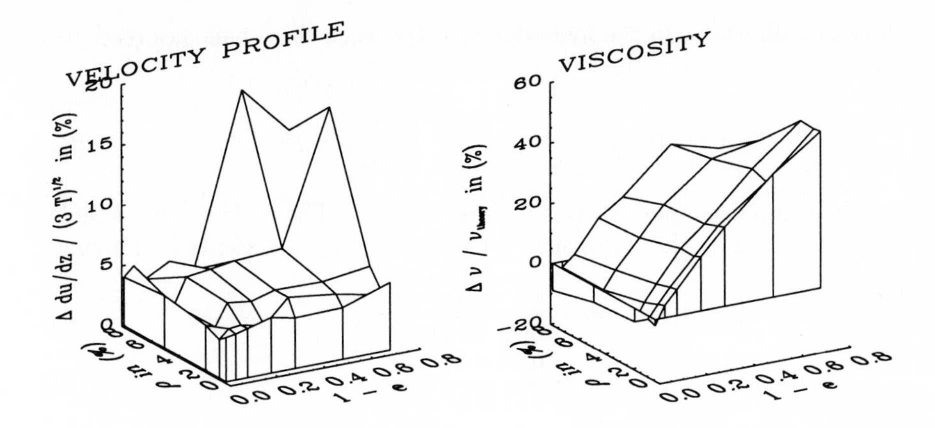

Fig. 3. *Left:* The deviation of the velocity profile from the linear one (see Eq. (25)) in dependence of the dissipation $\propto (1-e)$ and on the density ρ. *Right:* The relative difference between the measured and the theoretical (according to Jenkins and Richman [20]) viscosity.

4 Vertical Structure and Stability of Planetary Rings

In this section we study the vertical stratification of a planetary ring in the framework of a hydrodynamic description (Eqs. (15) - (16)) of the granular particle ensemble that constitutes the ring. Then, these results are compared with those of the triaxial approach (Eqs. (8) - (10)), see Schmidt et al. [34] or Simon and Jenkins [35]).

In order to judge whether the triaxial approach (8) - (10) or the hydrodynamical method (15) - (16) match the results better, we have carried out numerical particle simulations. We use an event driven code to simulate $N = 500$ to 4000 particles in a box in the gravitational field of a central mass M with periodic boundary conditions in radial and azimuthal direction. We compare the vertical stratification of density and temperature in the simulation box with the theoretical profiles of the hydrodynamic and the triaxial kinetic approach. A more detailed description of the numerical experiments is presented in Salo [30] or in Schmidt et al. [34].

In order to formulate the problem it is assumed that the ring is homogeneous in horizontal directions, i.e. in the radial and azimuthal ones (cylindrical coordinates) giving $\partial_r, \partial_\phi \to 0$, and only variations perpendicular to the mid plane of the ring (z direction) are considered.

Then, the stationary state in a planetary ring is determined by the balance between energy input due to viscous heating and collisional cooling combined with the balance between the vertical pressure gradient and the gravitational force in z direction. In the hydrodynamic approach these balances read

$$
\begin{aligned}
(\rho T)' + z\Omega^2 \rho = 0 \\
a(e)\left[T'\sqrt{T}\right]' + b(e)\Omega^2\sqrt{T} - \frac{c(e)}{4R_p^2}\rho^2\sqrt{T^3} = 0\ ,
\end{aligned}
\tag{28}
$$

where primes denote derivatives with respect to z. The coefficients a, b and c are determined by the expressions for the transport coefficients (21) giving

$$
\begin{aligned}
a &= \frac{C_3}{(1+e)(49-33e)} \\
b &= \frac{9}{4}\frac{C_1}{(1+e)(3-e)} \\
c &= C_4(1-e^2) \quad ,
\end{aligned}
\tag{29}
$$

with the constants $C_1 - C_4$ defined in the context with Eqs. (21).

In the case of the alternative theoretical treatment in the framework of the triaxial approach, the momentum and energy balances read

$$
\begin{aligned}
(\rho T)' + \delta(e)\, z\, \Omega_0^2\, \rho = 0 \\
\alpha(e)\left[T'\sqrt{T}\right]' + \frac{\beta(e)}{2R_p}\,\Omega_0\, T\, \rho - \frac{\gamma(e)}{4R_p^2}\,\rho^2\sqrt{T^3} = 0 \quad ,
\end{aligned}
\tag{30}
$$

with

$$
\begin{aligned}
\alpha &= \frac{[1-\frac{2}{7}(1-e)][\frac{1}{5}-\frac{1}{7}(1-e)]}{(1+e)\left\{49-33e+\frac{1-e}{2744}(5e-54)(237e-461)\right\}} \\
\beta &= \frac{1}{9-5e}\sqrt{\frac{1-e}{125\pi}(225e^2+4415e-2876)} \\
\gamma &= \frac{48}{25\pi}(1+\frac{(e-1)(195e-979)}{6272})(1-e^2) \\
\delta &= \frac{7}{2+5e} \qquad .
\end{aligned}
\tag{31}
$$

We observe a different temperature and density dependence of the heating term ($\sim T\rho$) in comparison to Eq. (28) ($\sim \sqrt{T}$). In the hydrodynamic description the heating is a consequence of the granular viscosity, which is a pure material property of the ring particles.

In planetary rings, however, the restriction of the mean free path l by the epicyclic motion, or equivalently, the adjustment of the vertical scale height of such systems leads to an expression as given in Eqs. (18) - (20).

An order of magnitude estimate for the kinematic viscosity is given by $\nu \approx \omega_c l^2$, with the collision frequency ω_c. For very dense systems, where $\omega_c \gg \Omega_0$, the well known relations $\omega_c \propto \rho\sqrt{T}$ and $l \propto \rho^{-1}$ hold, giving $\nu \propto \sqrt{T}/\rho$, and thus, for the dynamic viscosity $\eta = \rho\nu$ the density dependence cancels and we obtain the viscosity in Eq. (21).

If the collision frequency ω_c is comparable to the orbital frequency Ω_0, both kinetic scales, Ω_0 and l, change their dependencies on temperature and density. For simplicity, we assume an isothermal ring $T(z) = const.$ In this case the triaxial and the hydrodynamic approach yield $T\rho'/\rho \approx - \Omega_0^2 z$, which can be immediately integrated giving $\rho(z) \approx \rho_0 \exp\left[-(\Omega_0 z)^2/(2T)\right]$, with scale height of the ring $H \approx \sqrt{T}/\Omega_0$ [37]. Turning back to the collision frequency we find $\bar{\omega_c} \propto \bar{\rho}\sqrt{T} \approx \Sigma\sqrt{T}/H \approx \Sigma\,\Omega_0$, where $\Sigma \propto \tau$ is the vertically integrated density[1]. Furthermore, in planetary rings the mean free path cannot be only dependent on the density ρ, because for $\rho \to 0$ the value approaches its upper limit $l \approx \bar{v}/\Omega_0$, which is about the length of an orbital epicycle. The velocity $\bar{v}$ is the deviation from the Keplerian circular velocity, which is caused by eccentric and inclined particle orbits. This deviation velocity $\bar{v}$ can also be considered to be related to a temperature according to $\bar{v} \propto \sqrt{T}$. Using $\omega_c \approx \Sigma\,\Omega_0$ and $l^2 \approx T/\Omega_0^2$, the kinematic viscosity becomes $\nu \propto T/\Omega_0$. Then the density dependence of the heating term is retained and we find $\eta = \rho\nu \sim \rho T/\Omega_0$. In detail this modified viscosity η^* reads

$$\eta^* = \frac{C_1}{(1+e)(3-e)}\frac{T}{\Omega_0}\rho \ , \tag{32}$$

which shows the same dependence on temperature and density as the local and nonlocal part of the viscosity (18)-(20) derived in the framework of the triaxial approach [2, 10].

In Fig. 4 the numerical solutions of Eqs. (28) and (30) are shown, displaying different vertical stratification due to the varying heating contributions (for more details see Schmidt et al. [34]).

Figure 4 shows that the denser the system and the more elastic the collisions become the better the solutions of the hydrodynamic approach (28) with the corresponding transport coefficients (21) fit the simulations. This is plausible because, first, Jenkins and Richman [20] have derived the transports under the condition $e \approx 1$, and second, the denser the system the less important the curvature of the particle trajectories becomes. This means, that the influence of epicyclic motion can be neglected under these circumstances, and thus, the solutions of the triaxial method and hydrodynamic approach are very similar.

In contrast to this, the less dense and the more dissipative the system gets, the stronger the deviation between both methods becomes. While both theoretical ansatzes reproduce the simulated temperature profiles rather unsatisfactory, the density profiles are well reproduced by the theory, provided that the modified viscosity (32) is used in the hydrodynamic equations.

Summarizing, it can be concluded that both theoretical mean field approaches have been found to be comparable appropriate to describe the stratification of planetary rings, as long as the system is not too dissipative. The

[1] In this consideration all state-values have to be considered as mean values with respect to the vertical scale.

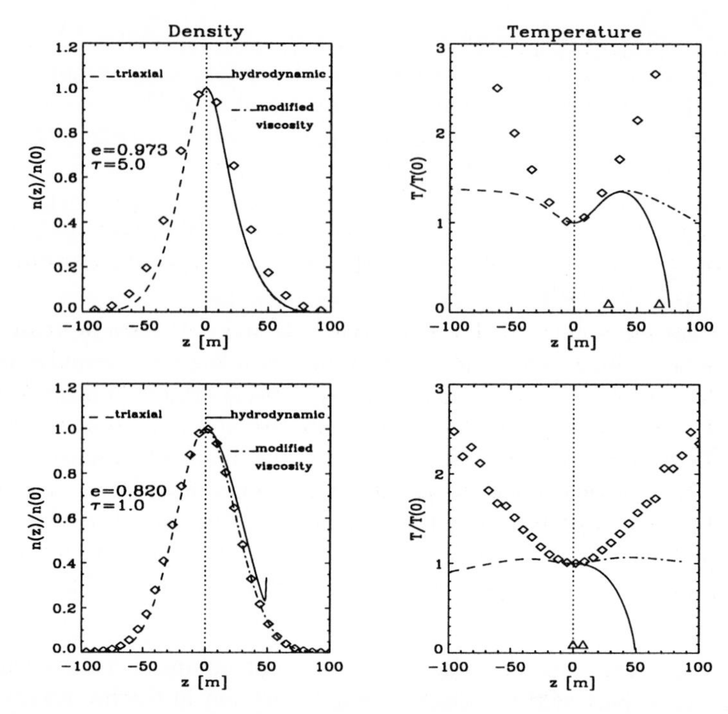

Fig. 4. The numerical solution of Eqs. (28) (solid lines) and (30) (dashed), applying the viscosity expressions (21), (20), and (32) (dashed-dotted). The values in the mid plane have been chosen close to the situation in Saturn's A ring ($T(0) \sim 10^{-4} m^2/s^2$, $\rho(0) \sim 0.01$, $\Omega_0 = 1.5 \cdot 10^{-4}$ and $R_p = 1m$). The diamonds show measured points from the simulations. The *upper plots* correspond to a higher density and nearly elastic collisions. The *lower plots* belong to less dense and more dissipative systems. This leads to stronger deviations of the theoretical results from the particle simulations. The reason for the correspondence between τ and e is discussed below in connection with the quasi-equilibrium in planetary rings.

performance of the hydrodynamic approach can be improved by introducing a modified viscosity for planetary rings [34] which matches well the temperature and density dependence of η in the frame of the triaxial ansatz. A drawback is that the triaxial approach cannot account for the heat conductivity (third contracted moment of f). In previous theories [35] this has been introduced by considering higher order corrections to the ansatz of an anisotropic Gaussian [48].

At this point we focus the attention to the problem of the stationarity and stability of planetary rings. Stationarity is achieved by the balance between viscous heating and collisional cooling, which is roughly expressed by the equality $9\,\eta\,\Omega_0^2/4 \approx K\,\omega_c\,(1-e^2)\,T$ (K is a constant of the order of unity). With the values $\omega_c \propto \tau\Omega_0$ and $\eta \approx \rho T/\Omega_0$ estimated above, this balance becomes in vertically integrated form [2, 10, 30, 40, 47]

$$(1 + \tau^2)(1 - e^2) \approx const. \tag{33}$$

$$\tau^{\epsilon - 1}(1 - e^2) \approx const. \tag{34}$$

which is the so called $e - \tau$ relation characterising this stationarity or quasi-equilibrium. The two equations correspond to the respective viscosities given by Eqs. (18) and (20).

Interesting is the fact, that both Eqs. (33) and (34) show no temperature dependence. This means, that for given optical depth one has only one specified restitution that leads to a stationary state of the system. Choosing a "wrong" restitution coefficient, the ring is heated up to $T \to \infty$ (e too high), or the ring cools down (e too low) until finally non-local transport effects due to the finite particle size establish an equilibrium temperature. In the latter case, the particles repel each other under the Keplerian shear due to their finite volume. In Fig. 5 the evolution of the temperature of a planetary ring for a given τ and various values of e, obtained from simulations, is shown.

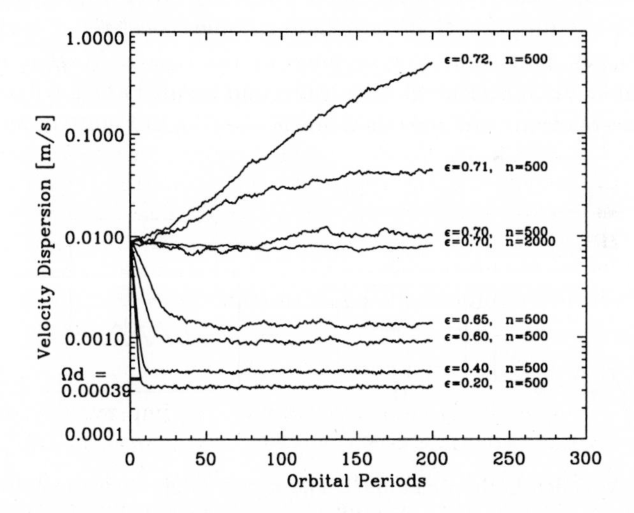

Fig. 5. The evolution of the velocity dispersion ($\bar{v} \propto \sqrt{T}$) plotted versus time obtained by numerical experiments (from [34]). The optical depth has been fixed to the value $\tau = 0.4$. In this plot $\epsilon = e$ is the restitution coefficient.

If the restitution e were constant, the thickness of a planetary ring should be in the range of one to two particle diameters. The resulting, forced value T is caused by the finite size effects, i.e. the particles in denser (collapsed regions) must repel each other, and thus, the limit $T \to 0$ is avoided.

In contrast to that forced temperature T, a variable restitution $e(v_{imp})$ ensures a natural adjustment of a quasi-equilibrium temperature T, a situation which is expected to be present in all planetary rings [5, 12, 41].

Because of the importance of the material property $e(v_{imp})$ for the stability of planetary rings, in the final section we briefly sketch our contributions to the collisional dynamics of visco-elastic bodies.

5 Model for Collisions – Normal Restitution

The dynamics of nonelastic collisions have been found to be crucial for the evolution of structures in granular systems, as discussed in the previous chapter concerning the stability of planetary rings. In this Section we give a brief review of the calculation of the coefficient of normal restitution.

In the most recent model not only the forces acting in the material are taken into account, but also effects on the surfaces due to layers of dust, frost, liquid or other soft material are considered. This new approach generalises earlier dynamical models of collisions [7, 36].

Experimental results which demonstrated that the normal restitution coefficient decreases with increasing impact velocity [5, 11, 13, 26] could be explained by the assumption that the colliding bodies consisted of a viscoelastic material [16, 22]. This can be represented by the Kelvin–Voigt model where an elastic and a viscous element are arranged in parallel as shown in the rheologic sketches of elastic and viscous elements λ and η in Figure 6, respectively.

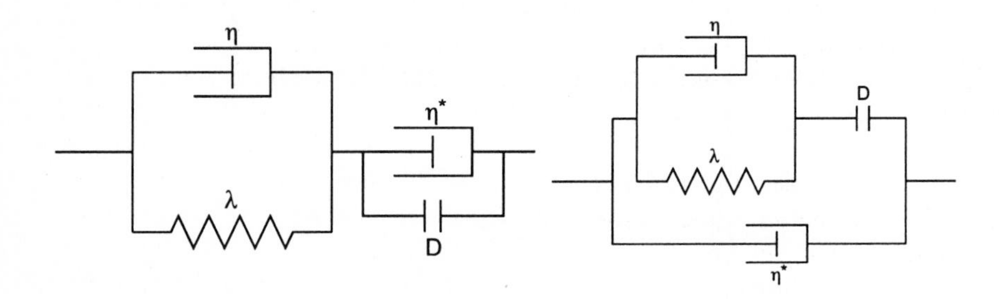

Fig. 6. *Left:* Modified Letherisch model. The stoppers represent the finite thickness D of the soft outer layer. *Right:* Simplified model with a very soft outer layer

A dynamical equation can be derived [7, 16, 36] for the mutual compression ξ of spherical particles – the deformation along the line connecting the centres of both spheres – similar to Hertz's theory of elastic compliance [14]:

$$\ddot{\xi} + \frac{2}{3}\frac{\sqrt{R_{\text{eff}}}}{m_{\text{eff}}}\left(\frac{Y}{1-\mu^2}\,\xi^{3/2} + \frac{\eta_1^2}{\eta_1 + 2\eta_2}\,\frac{1-2\mu}{\mu^2}\,\sqrt{\xi}\,\dot{\xi}\right) = 0 \quad . \tag{35}$$

Y denotes the Young modulus of the material and μ the Poisson ratio. η_1 is the "coefficient of internal friction" (shear viscosity) and η_2 the bulk viscosity of the material. R_{eff} and m_{eff} are the effective radius and the effective mass of the particles, respectively.

This model is based on a quasi-static approach, i.e. the velocity of impact and subsequent deformation is much lower than the speed of sound in the bulk of the material which the granular particles consist of. Then the parallel arrangement of the elastic and viscous element correspond to the sum of the elastic - and dissipative momentum fluxes, the viscous - and elastic stress tensors σ_{ij}^{vis} and σ_{ij}^{el}, respectively. The equation governing the visco-elastic state of the colliding spheres is then given by $\partial_{x_j}\left(\sigma_{ij}^{vis}+\sigma_{ij}^{el}\right) = 0$. Further analysis, together with the linear dependencies of the stresses on the deformation and their rates (Hooke's law), provides the equation (35) [7, 36].

Although the above model has been successful in explaining the decrease of the normal restitution coefficient e for higher velocities, other experiments [6, 12, 28, 46] indicate that the normal restitution coefficient is not always a monotonous decreasing function of the impact velocity, but that it takes lower values in slow collisions, too. In particular, in the contact of ice particles [6, 12], sticking behaviour has been observed leading to $e = 0$ below a certain impact velocity. Therefore it is necessary for a realistic model of collisions to incorporate surface effects. Most important among those are the presence of soft surface layers (dust, frost, liquids) and surface energy.

A soft surface layer can be regarded as a material with a very low elasticity, but a rather high viscosity. This appears to be reasonable in particular for powdery layers of dust or frost. Particles with such covers might be represented similar to the Letherisch or Zener model of viscoelasticity by a Kelvin–Voigt element and a viscous element arranged in series. However, one must take into account the finite thickness D of the surface layer. This will be represented by a pair of "stoppers" or "buffers" with this distance (see Fig. 6). Also, in the cases of greatest interest the surface does not consist of a purely viscous material which will simply flow under load. It will rather be compactified, thereby slowing down the impact, a phenomenon known in soil mechanics. Furthermore, the analysis of the Letherisch model is rather complicated due to coupling between the individual elements. Therefore a simplification is attempted in order to allow for the derivation of an equation of motion similar to Eq. (35). The liquid or dust–like surface layers are usually much more easily deformed than the bulk of the material. Thus, one may assume that the latter remains unaffected by the collision until the former is completely compressed. Only then, the bulk can be deformed viscoelastically. Thus, one is lead to a simplified model where one can imagine two independent stages of the collision: First, the viscous outer layer is compressed while the bulk material is not affected, then viscoelastic deformation of the bulk occurs which does not feel any more the influence of the soft surface layer. The corresponding rheological diagrams are shown in Fig. 6.

In order to describe the compression of the outer layer, the purely elastic term in eq. (35) is neglected, and appropriate values for the viscosity are used. The dependence of the compression ξ on time in the simplified model can now be formulated using results of the viscoelastic theory:

$$\ddot{\xi} + B_2 \sqrt{\xi}\,\dot{\xi} = 0 \text{ if } \xi \le D_1 + D_2 \tag{36}$$

$$\ddot{\xi} + A\,\xi^{3/2} + B_1 \sqrt{\xi}\,\dot{\xi} = 0 \text{ if } \xi > D_1 + D_2 \tag{37}$$

with the constants depending on the material properties:

$$A = \frac{2\,Y\sqrt{R_{\text{eff}}}}{3\,m_{\text{eff}}\,(1-\mu^2)}\,,\; B_1 = \frac{2}{3}\,\frac{\eta_{1b}^2}{\eta_{1b} + 2\eta_{2b}}\,\frac{1-2\mu_b}{\mu_b^2}\,,\; B_2 = \frac{2}{3}\,\frac{\eta_{1s}^2}{\eta_{1s} + 2\eta_{2s}}\,\frac{1-2\mu_s}{\mu_s^2}\,.$$

D_1 and D_2 are the thicknesses of the layers which cover the surfaces of the colliding bodies. Indices b and s refer to the bulk and the surface layer material, respectively.

The equations of motion (36) and (37) have been solved with a self–adjusting Runge–Kutta integrator. The particles were assumed to have an effective radius $R_{eff} = 1$ cm and to consist of ice with an average density of $\rho_p = 0.9$ g/cm^3. The material constants used in the simulation were [23]:

$$Y = 10 \text{ GPa}\,, \quad \mu_b = 0.3\,, \quad \eta_{1b} = 10^6 \text{ Pa s}\,, \quad \eta_{2b} = 10^{12} \text{ Pa s}\,.$$

Also for the surface layer, $\mu_s = 0.3$ was assumed. Because experimental data on viscosities of a powdery layer are very scarce, the quotient $\eta_{1s}^2/(\eta_{1s} + 2\eta_{2s})$ was assumed to yield either a lower value of 10^2 Pa s or a higher one of 10^3 Pa s, accounting for easy deformation, but sufficient energy dissipation. Results are shown in Figure 7.

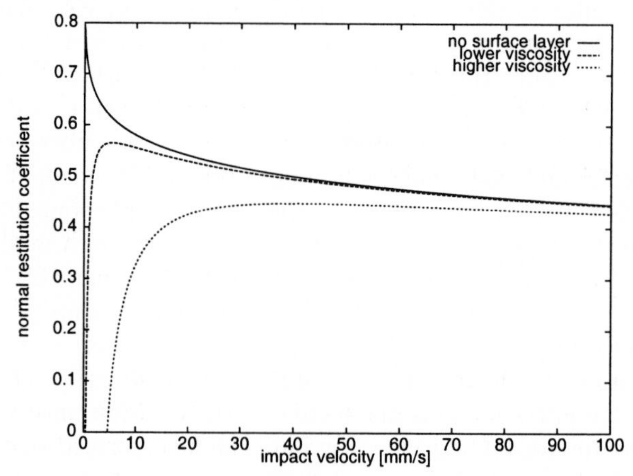

Fig. 7. Dependency $e(g_N)$ for different viscosities of the surface layer.

The chief effects of a viscous surface with respect to our former models [7, 16, 36] are:

- A critical velocity $g_{Nc} = [\boldsymbol{g} \cdot \boldsymbol{k}]_c$ exists which depends on the viscosity of the surface layer. Collisions with impact velocities below g_{Nc} are stopped.
- For impact velocities above g_{Nc}, the restitution coefficient is reduced compared to particles without cover.

This is in accord with experimental observations [6, 12, 42].

In order to provide a convenient formulation for use in simulations of many–particle systems, the numerically obtained velocity dependence (g_N - measured in $mm\ s^{-1}$) of the normal restitution coefficient can be fitted by a function

$$e(g_N) \approx \max\left[0\,,\, a g_N^b \left(1 - \exp\left(-c g_N^d\right)\right)\right] \tag{38}$$

where the constants take the values $a = 0.7$, $b = 0.08$ and $c = 4.5$, $d = 0.45$ for $\eta_{1s}^2/\left(\eta_{1s} + 2\eta_{2s}\right) = 10^2$ Pa s or $c = 1.8$, $d = 0.7$ for $\eta_{1s}^2/\left(\eta_{1s} + 2\eta_{2s}\right) = 10^3$ Pa s.

Another phenomenon which might affect the collisional behaviour is surface energy which causes adhesive forces between the particles in contact. Its influence can be expressed by [21]

$$F_1 = F_0 + 3\pi R_{\text{eff}} E_s + \sqrt{6\pi R_{\text{eff}} E_s F_0 + (3\pi R\ E_s)^2} \tag{39}$$

with $F_1 = -m_{\text{red}}\,\ddot{\xi}$ denoting the actual force on the particles in contact. F_0 is the external force unaffected by surface interactions and E_s the surface energy.

Assuming a typical value for the surface energy of $E_s = 0.1\ \text{J/m}^2$, the above equation of motion has also been solved for colliding viscoelastic spheres with an effective radius of 1 cm and the above material constants for ice. F_0 has been expressed by equation (35). The numerical solution suggests that at very low impact velocities the particles will indeed "stick" together. However, although the effects of surface energy are similar to those of soft surface layers, they are considerably less pronounced for the chosen values of the material constants. Still, surface energy may provide tensile strength in aggregates of very small particles.

For the stability of planetary rings, the consequences of the above presented results are: Ring systems are only stable in the region where the coefficient of restitution decreases with the normal impact velocity according to $de/dg_N < 0$. That means, the random walk speed, or correspondingly the temperature T of the ring particles in a quasi equilibrium must be high enough in order to achieve mean restitutions $\langle e \rangle$ in that falling range. Otherwise sticking would lead to accretion processes which would dissolve the ring systems with time.

6 Discussion

In this review we report on the transport of momentum in granular gases in different physical situations: in a force-free but sheared system and in a system under Keplerian shear, driven by the gravity of a central body.

The viscosities η in either cases differ, in particular with respect to the temperature dependence. This is mainly caused by the action of the central gravity which causes the particles to loop in epicycles. In plasma physics similar situations can have been found, e.g. if charged particles (ions, electrons) gyrate around the field-lines in a magnetic field. Then a similar dependence $\eta(T)$ has been found [39].

This means, the differences between the force-free case and the case in a central gravitational field is not an intrinsic property of granular matter, but merely due to the action of the gradient of a volume force, which bounds the motion of the particles. Thus, the mean free path can not diverge for infinite dilute systems. One may even state, that the difference in the expressions for η is caused by the fact, that besides pure material properties the granular viscosity in a planetary ring experiences additional contributions from volume forces.

Both expressions for the viscosity η – that one derived by Jenkins and Richman [20] and also that calculated by Goldreich and Tremaine [2, 10] – have been confirmed by numerical experiments (for rings see [30, 31, 47], for the force free case see section 3). Both formulations of the problem are appropriate for the description of the corresponding situation.

When volume forces produce a considerable curvature in the motion of the particle on a scale comparable to the mean free path, the limits of a hydrodynamic description are reached. This is the case in dilute plasmas in presence of a magnetic field as well as in dilute granular gases revolving a planet.

On the other hand, the combined constitutive relations Eqs. (21)-(27) provide useful tools for the description. However, it has to be noted once again that in these cases the transport processes reflect not only material properties but also effects due to the action of gradients of volume forces.

Another crucial point reported in this review is the stability of planetary rings in the dependence of the material properties of the granular particles during the collisions. It turned out that instability or quasi-equilibrium of such circumplanetary granular assemblies depends sensitively on the collisional dynamics - i.e. how two granular particles collide in dependence on their impact speed. It has been shown that the dissipation of kinetic energy must increase with increasing impact velocity (temperature). This is indeed the case in a certain range of collision velocities. Therefore, models have been developed for the collisional dynamics basing on the continuum mechanics of visco-elastic bodies [7, 16, 36]. On the other hand, these models did not account for the sticking of particles, which has been found in laboratory experiments of "soft" granules at quite low velocities [6, 42]. Therefore, we present new models for the dynamics of frost or regolith covered granular spheres, which account for these effects including the decrease of the restitution with the velocity.

References

1. S. Araki. The dynamics of particle disks. II. Effects of spin degrees of freedom. Icarus **76**, 182 (1988).
2. S. Araki and S. Tremaine. The dynamics of dense particle disks. Icarus **65**, 83 (1986).
3. L. Boltzmann *Vorlesungen über Gastheorie.* Verlag J.A. Barth, Leipzig (1896).
4. J. P. Bouchaud, P. Claudin, M. E. Cates, and J. P. Wittmer. Models of stress propagation in granular media. In H.J. Herrmann, J.-P. Hovi, and S. Luding, editors, *Physics of dry granular media*, p. 97. Kluwer Academic Publishers (1998).
5. F. G. Bridges, A.P. Hatzes, and D. N. C. Lin. Structure, stability and evolution of Saturn's rings. Nature **309**, 333 (1984).
6. F. G. Bridges, K. D. Supulver, D. N. C. Lin, R. Knight, and M. Zafra. Energy loss and sticking mechanisms in particle aggregation in planetesimal formation. Icarus **123**, 422 (1996).
7. N. Brilliantov, T. Pöschel, F. Spahn, and J.-M. Hertzsch. Model for collisions in granular gases. Phys. Rev. E **53**, 5382 (1996).
8. S. Chapman and Cowling T.G. *The mathematical theory of non-uniform gases.* Cambridge University Press, Cambridge (1970).
9. I. Goldhirsch and G. Zanetti. Clustering instability in dissipative gases. Phys. Rev. Lett. **70**, 1619 (1993).
10. P. Goldreich and S. Tremaine. Velocity dispersion in Saturn's rings. Icarus **34**, 227 (1978).
11. W. Goldsmith. *Impact. The theory and physical behaviour of colliding solids.* Edward Arnold, London (1960).
12. A. Hatzes, F. G. Bridges, and D. N. C. Lin. Collisional properties of ice spheres at low impact velocities. Mon. Not. R. Astr. Soc. **231**, 1091 (1988).
13. S. Haughton. On the dynamical coefficient of elasticity of steel, iron, brass, oak, and teak. Proc. Roy. Irish Acad. **8**, 86 (1864).
14. H. Hertz. Über die Berührung fester elastischer Körper. J. f. Reine u. Angew. Math. **92**, 156 (1881).
15. J.-M. Hertzsch. *Über nichtelastische Kollisionen und ihren Einfluß auf die Strukturbildung in planetaren Ringen.* PhD thesis, Universität Potsdam (1996).
16. J. M. Hertzsch, F. Spahn, and N. V. Brilliantov. On low-velocity collisions of viscoelastic particles. J. Phys. II (France) **5**, 1725 (1995).
17. M.A. Hopkins and M.Y. Louge. Inelastic microstructure in rapid granular flows. Phys. Fluids A **3**, 47 (1991).
18. K. Hutter and R. Rajagopal. On the flow of granular materials. Cont. Mech. Thermodyn. **6**, 81 (1994).
19. H. M. Jaeger and S. R. Nagel, and R. P. Behringer. Granular solids, liquids, and gases. Rev. Mod. Phys. **68**, 1259 (1996).
20. J. T. Jenkins and M. T. Richman. Grad's 13-moment system for a dense gas of inelastic spheres. Arch. Ration. Mech. Anal. **87**, 355 (1985).

21. K. L. Johnson, K. Kendall, and A. D. Roberts. Surface energy and the contact of elastic solids. Proc. R. Soc. Lond. A **324**, 301 (1971).
22. G. Kuwabara and K. Kono. Restitution coefficient in a collision between two spheres. Jap. J. Appl. Mech **26**, 1230 (1987).
23. Landolt-Börnstein. *Zahlenwerte und Funktionen aus Naturwissenschaft und Technik.* Springer, Berlin, Heidelberg, New York (1952).
24. C. K. K. Lun, S. B. Savage, D. J. Jeffrey, and N. Chepurniy. Kinetic theories for granular flow: inelastic particles in Cuette flow and slightly inelastic particles in a general flow field. J. Fluid Mech. **140**, 223 (1984).
25. E. A. Marouf, G. L. Tyler, and V. R. Eshleman. Theory of radio occultation by Saturn's rings. Icarus **49**, 161 (1982).
26. J. Ôkubo. Some experiments on impact. Sci. rep. Tohoku Imp. Univ. **11**, 455 (1922).
27. O. Petzschmann, U. Schwarz, F. Spahn, C. Grebogi, and J. Kurths. Lenght scales of clustering in granular gases. Phys. Rev. Lett. **82**, 4819 (1999).
28. C. V. Raman. The photographic study of impact at minimal velocities. Phys. Rev. **12**, 442 (1918).
29. P. Résibois and M. De Leener. *Classical kinetic theory of fluids.* Wiley & Sons, New York (1977).
30. H. Salo. Numerical simulations of dense collisional systems. Icarus **90**, 254 (1991).
31. H. Salo. Numerical simulations of dense collisional systems: II. Extended distribution of particle size. Icarus **96**, 85 (1992).
32. S. B. Savage. Modeling and granular material boundary value problems. In H.J. Herrmann, J.-P. Hovi, and S. Luding, editors, *Physics of dry granular media*, pp. 25–96. Kluwer Academic Publishers (1998).
33. S. B. Savage. Instability of an unbounded uniform granular shear flow. J. Fluid Mech. **241**, 109 (1992).
34. J. Schmidt, H. Salo, O. Petzschmann, and F. Spahn. Vertical distribution of temperature and density in a planetary ring. Astronomy and Astrophysics **345**, 646 (1999).
35. V. Simon and J.T. Jenkins. On the vertical structure of dilute planetary rings. Icarus **110**, 109 (1994).
36. F. Spahn, J.-M. Hertzsch, and N.V. Brilliantov. The role of particle collisions for the dynamics in planetary rings. Chaos, Solitons and Fractals **5**, 1945 (1995).
37. F. Spahn, O. Petzschmann, K.-U. Thiessenhusen, and J. Schmidt. Inelastic collisions in planetary rings: Thickness and satellite-induced structures. In *Physics of dry granular media.* NATO-ASI Series E350 (1998).
38. F. Spahn, U. Schwarz, and J. Kurths. Clustering of granular assemblies with temperature dependent restitution and under differential rotation. Phys. Rev. Lett. **78**, 1596 (1997).
39. I. Spitzer. *Physics of fully ionized gases.* Interscience, New York (1962).
40. G. R. Stewart, D. N. C. Lin, and P. Bodenheimer. Collision-induced transport processes in planetary rings. In Planetary Rings, p. 447, Univ. of Arizona Press, Tucson (1984).
41. K. D. Supulver, F. G. Bridges, and D. N. C. Lin. The coefficient of testitution of ice particles in glancing collisions: Experimental results for unfrosted surfaces. Icarus **113**, 188 (1995).

42. K. D. Supulver, F. G. Bridges, S. Tiscareno, J. Lievore, and D. N. C. Lin. The sticking properties of water frost produced under various ambient conditions. Icarus **129**, 538 (1997).
43. J. Trulsen. Towards a theory of jet streams. Astrophysics and Space Science **12**, 329 (1971).
44. G. L. Tyler, V. R. Eshleman, D. P. Hinson, E. A. Marouf, R. A. Simpson, D. N. Sweetnam, J. D. Anderson, J. K. Campbell, G. S. Levy, and G. F. Lindal. Voyager 2 radio science observations of the Uranian system Atmosphere, rings, and satellites. Science **233**, 79 (1986).
45. G. L. Tyler, E. A. Marouf, R. A. Simpson, H. A. Zebker, and V. R. Eshleman. The microwave opacity of Saturn's rings at wavelengths of 3.6 and 13 cm from Voyager 1 radio occultation. Icarus **54**, 160 (1983).
46. J.H. Vincent. Experiments on impact. Proc. Camb. Phil. Soc. **8**, 332 (1900).
47. J. Wisdom and S. Tremaine. Local simulations of planetary rings. Astronomical Journal **95**, 925 (1988).
48. C. Zhang. *Kinetic theory for rapid granular flows.* PhD thesis, Cornell University (1993).

V

Towards Dense Granular Systems

The Equation of State for Almost Elastic, Smooth, Polydisperse Granular Gases for Arbitrary Density

Stefan Luding and Oliver Strauß

Institute for Computer Applications 1,
Pfaffenwaldring 27, 70569 Stuttgart, Germany. e-mail: lui@ica1.uni-stuttgart.de

Abstract. Simulation results of dense granular gases with particles of different size are compared with theoretical predictions concerning the pair-correlation functions, the collison rate, the energy dissipation, and the mixture pressure. The effective particle-particle correlation function, which enters the equation of state in the same way as the correlation function of monodisperse granular gases, depends only on the total volume fraction and on the dimensionless width $\mathcal{A}$ of the size-distribution function. The *global equation of state* is proposed, which unifies both the dilute and the dense regime.

The knowledge about a global equation of state is applied to steady-state situations of granular gases in the gravitational field, where averages over many snapshots are possible. The numerical results on the density profile agree perfectly with the predictions based on the global equation of state, for monodisperse situations. In the bi- or polydisperse cases, segregation occurs with the heavy particles at the bottom.

1 Introduction

The hard-sphere (HS) gas is a traditional, simple, tractable model for various phenomena and systems like e.g. disorder-order transitions, the glass transition, or simple gases and liquids [1–3]. A theory that describes the behavior of rigid particles is the kinetic theory [1, 4], where particles are assumed to be rigid and collisions take place in zero time (they are instantaneous), exactly like in the hard-sphere model. When dissipation is added to the HS model, one has the most simple version of a granular gas, i.e. the inelastic hard sphere (IHS) model. Granular media represent the more general class of dissipative, non-equilibrium, multi-particle systems [5]. Attempts to describe granular media by means of kinetic theory are usually restricted to certain limits like constant or small density [6] or weak dissipation [7, 8]. Also in the case of granular media, one has to apply higher order corrections to successfully describe the system under more general conditions [9, 10]. Already classical numerical studies showed that the equation of state can be expressed as some power series of the density in the low density regime [2, 3, 11], whereas, in the high density case, the free volume theory leads to useful results [12]. In the general situation, the granular system consists of particles with different sizes, a situation which is rarely addressed theoretically [13–16]. However, the treatment of bi- and polydisperse mixtures is easily performed by means of numerical simulations [17–19].

In this study, theories and simulations for situations with particles of equal and different sizes are compared. In section 2 the model system is introduced and in 3 we review theoretical results and compare them with numerical results concerning correlations, collision rates, energy dissipation and pressure. Based on the numerical data, a global equation of state is proposed. This global equation of state is valid for arbitrary densities and mixtures with particles of different size, and it is used to explain the density profile in a dense system in the gravitational field in section 4. The results are summarized and discussed in section 5.

2 Model System

For the numerical modeling of the system, periodic, two-dimensional (2D) systems of volume $V = L_x L_y$ are used, with horizontal and vertical size L_x and L_y, respectively. N particles are located at positions $\boldsymbol{r}_i$ with velocities $\boldsymbol{v}_i$ and masses m_i. From any simulation, one can extract the kinetic energy $E = \frac{1}{2}\sum_{i=1}^{N} m_i \boldsymbol{v}_i^2$, dependent on time via the particle velocity $\boldsymbol{v}_i$. In 2D, the "granular temperature" is defined as $T = E/N$.

2.1 Polydispersity

The particles in the system have the radii a_i randomly drawn from size distribution functions $w(a)$ as summarized in table 1 where the step-function $\theta[x] = 1$ for $x \geq 1$ and $\theta[x] = 0$ for $x < 1$ is implied.

Table 1. Size distribution functions used in this study.

(i)	monodisperse	$w(a) = \delta(a - a_0)$
(ii)	bidisperse	$w(a) = n_1\delta(a - a_1) + n_2\delta(a - a_2)$
(iii)	polydisperse	$w(a) = \frac{1}{2w_0 a_0}\theta[a - (1 - w_0)a_0]\,\theta[(1 + w_0)a_0 - a]$

The parameter a_0 is the mean particle radius $\langle a\rangle$ in cases (i) and (iii). In the bidisperse situation (ii), one has $a_0 = \langle a\rangle = n_1 a_1 + n_2 a_2 = (n_1 + (1 - n_1)/R)a_1$, with the fraction $n_1 = N_1/(N_1 + N_2)$ of particles with size a_1 in a system with $N = N_1 + N_2$ particles in total and N_2 particles with radius a_2. Thus, besides n_1, only the size ratio $R = a_1/a_2$ is needed to classify a bidisperse size distribution. The total volume fraction $\nu = \nu_1 + \nu_2$ is the last relevant system parameter, since the partial volume fractions $\nu_{1,2} = N_{1,2}\pi a_{1,2}^2/V = n_{1,2}\nu a_{1,2}^2/\langle a^2\rangle$ can be expressed in terms of n_1 and R: Using the dimensionless moments

$$A_k = n_1 + (1 - n_1)R^{-k} = \frac{\langle a^k \rangle}{a_1^k} , \tag{1}$$

one has $\nu_1 = n_1\nu/A_2$ and $\nu_2 = (1 - n_1)\nu/(R^2A_2)$. Since needed later on, the expectation values for the moments of a and their combination, the dimensionless width-correction $\mathcal{A} = \langle a \rangle^2/\langle a^2 \rangle$, are summarized in table 2 in terms of a_1, n_1, and R for the bidisperse situations and in terms of a_0 and w_0 in the polydisperse cases. Different values of ν are realized by shrinking or growing either the system or the particles.

Table 2. Moments $\langle a \rangle$, $\langle a^2 \rangle$ and $\mathcal{A} = \langle a \rangle^2/\langle a^2 \rangle$ of the size distribution functions.

		$\langle a \rangle$	$\langle a^2 \rangle$	$\langle a \rangle^2/\langle a^2 \rangle$
(i)	monodisperse	a_0	a_0^2	1
(ii)	bidisperse	A_1a_1	$A_2a_1^2$	A_1^2/A_2
(iii)	polydisperse	a_0	$\left(1 + w_0^2/3\right) a_0^2$	$3 \;/\; \left(3 + w_0^2\right)$

2.2 Particle Interactions

The particles are assumed to be perfectly rigid and to follow an undisturbed motion until a collision occurs as described below. Due to the rigidity, collisions occur instantaneously, so that an event driven simulation method [20, 21] can be used. Note that no multi-particle contacts can occur in this model. For a review on possible, more physical extensions of this model see Ref. [21].

A change in velocity – and thus a change in energy – can occur only at a collision. The standard interaction model for instantaneous collisions of particles with radii a_i, mass $m_i = (4/3)\pi\rho a_i^3$, and material density ρ is used in the following. (Using the mass of a sphere is an arbitrary choice, however, using disks would not influence most of the results discussed below.) This model was introduced and also discussed for the more general case of rough particle surfaces in Refs. [7, 22–25]. The post-collisional velocities $\boldsymbol{v}'$ of two collision partners in their center of mass reference frame are given, in terms of the pre-collisional velocities $\boldsymbol{v}$, by

$$\boldsymbol{v}'_{1,2} = \boldsymbol{v}_{1,2} \mp \frac{(1+r)\, m_{12}}{m_{1,2}} \boldsymbol{v}_n , \tag{2}$$

with $\boldsymbol{v}_n \equiv [(\boldsymbol{v}_1 - \boldsymbol{v}_2) \cdot \hat{\boldsymbol{n}}]\, \hat{\boldsymbol{n}}$, the normal component of $\boldsymbol{v}_1 - \boldsymbol{v}_2$ parallel to $\hat{\boldsymbol{n}}$, the unit vector pointing along the line connecting the centers of the colliding particles, and the reduced mass $m_{12} = m_1m_2/(m_1 + m_2)$. If two particles collide, their velocities are changed according to Eq. (2) and any $r(v_n)$, dependent on $v_n = |\boldsymbol{v}_n|$, can be used. For a pair of particles, the change of the translational energy at a collision is $\Delta E = -m_{12}(1 - r^2)v_n^2/2$.

3 Simulation and Theory

In the following, we compare simulations with different polydispersity, i.e. different size distribution functions $w(a)$, as summarized in table 3.

Table 3. Simulation parameters for the simulations discussed below. Note that sets C and E have different $w(a)$ but almost identical values of $\mathcal{A}$.

		$w(a)$ parameters	particles	$\mathcal{A}$
A	monodisperse	$w_0 = 0$,	$N = 1628$	1
B	monodisperse	$n_1 = 1$, $R = 1$	$N = 576$	1
C	bidisperse	$n_1 = 0.517$, $R = 3/4$	$N = 576$	0.9798
D1	bidisperse	$n_1 = 0.781$, $R = 1/2$	$N = 576$	0.8968
D2	bidisperse	$n_1 = 0.799$, $R = 1/2$	$N = 6561$	0.8998
E	polydisperse	$w_0 = 0.25$	$N = 1425$	0.9796
F1	polydisperse	$w_0 = 0.5$	$N = 1425$	0.9231
F2	polydisperse	$w_0 = 0.5$	$N = 1521$	0.9231

3.1 Particle Correlations

In **monodisperse systems**, the particle-particle pair correlation function at contact,

$$g_{2a}(\nu) = \frac{1 - 7\nu/16}{(1-\nu)^2} , \tag{3}$$

depends on the volume fraction only [1, 3, 7, 8, 21]. The particle-particle correlation function is obtained from the simulations by averaging over M snapshots with N particles, normalized to the value $g(r \gg 2a) = 1$ for long distances in large systems, so that

$$g(r) = \frac{1}{M} \sum_{m=1}^{M} \frac{2V}{N(N-1)} \frac{1}{V_r} \sum_{i=1}^{N} \sum_{j=1}^{i-1} \theta[r_{ij} - r]\theta[r + \Delta r - r_{ij}] , \tag{4}$$

with $r_{ij} = |\mathbf{r}_i - \mathbf{r}_j|$, and where the two θ functions select all particle pairs (i, j) with distance between r and $r + \Delta r$. The weight $N(N-1)/2$ accounts for all pairs summed over, and the term $V_r = \pi(2r + \Delta r)\Delta r$ is the volume (area) of a ring with inner radius r and width Δr. In Fig. 1, simulation results from set B (see table 3) with different ν are presented. Typical values used for the averages are e.g. $M = 50$, and $\Delta r = a/10$. Besides fluctuations, the values at contact nicely agree with the theoretical predictions from Eq. (3), as indicated by the arrows – as long as the system is disordered (left panel in Fig. 1). In a more ordered system (right panel in Fig. 1), $g_{2a}(\nu)$ is not a good estimate. Instead, one obtains a long range order with peaks at $r/2a = 1, \sqrt{3}, 2, \ldots$, indicating the triangular lattice structure of the assembly.

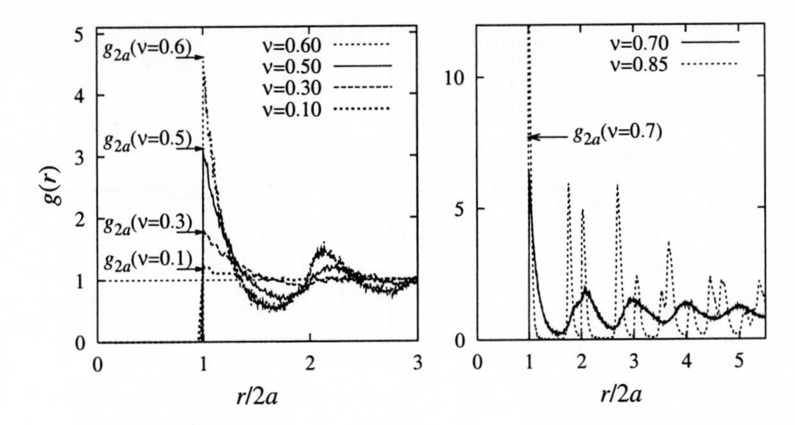

Fig. 1. Particle-particle correlation function $g(r)$ plotted against the normalized center-center distance $r/2a$. (Left) Disordered systems – the arrows indicate the values at contact from Eq. (3). (Right) Ordered systems with different axis scaling.

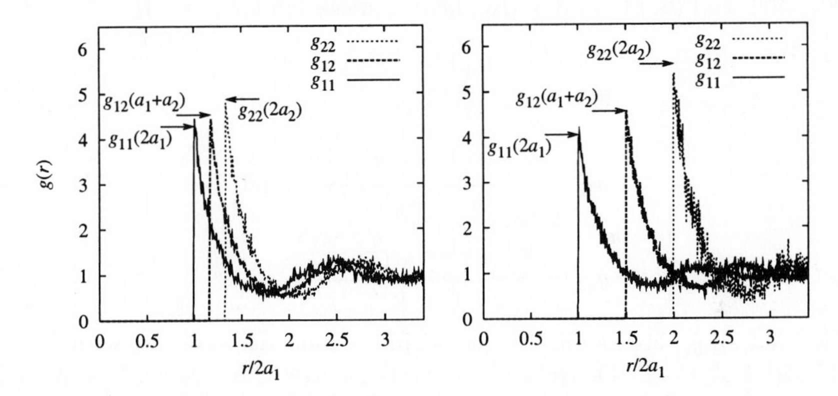

Fig. 2. Particle-particle correlation function $g(r)$ plotted against the center-center distance normalized by the radius a_1 of the smaller particles, at a volume fraction of $\nu = 0.60$. The arrows indicate the values g_{11}, g_{12}, and g_{22} at contact. (Left) Bidisperse simulation from set C with $R = 3/4$, and (Right) bidisperse simulation from set D1 with $R = 1/2$.

For **bidisperse situations**, the pair correlation functions for equal species g_{11} and g_{22} are obtained by replacing N in the first particle-sum in Eq. (4) by N_1 and N_2, respectively. For the correlation function g_{12}, it is necessary to perform the first particle-sum in Eq. (4) from $i = 1$ to N_1, the second sum from $j = 1$ to N_2, and to replace the weight $N(N-1)/2$ by N_1N_2 (in order to account for all pairs of different kind). In Fig. 2 simulation results from sets C and D1 for $\nu = 0.6$ are compared to the analytical expressions Eqs.

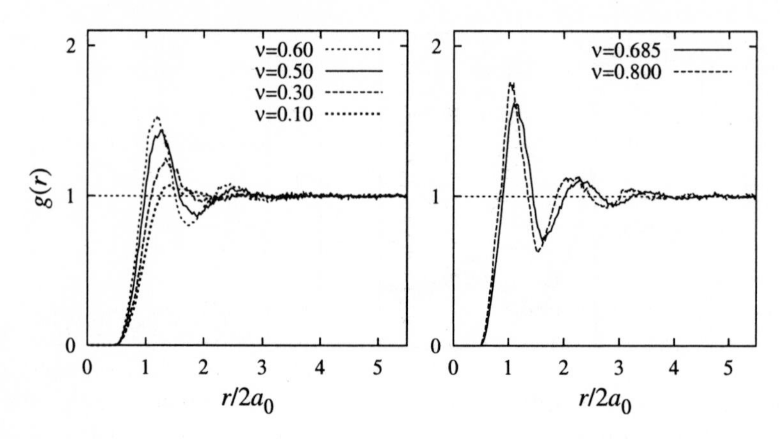

Fig. 3. Particle-particle correlation function $g(r)$ plotted against the normalized center-center distance $r/2a_0$. (Left) Low density systems and (Right) high density systems with the same axis scaling.

(90), (91), and (92) from Ref. [13], here expressed in terms of $A_{1,2}$, R, and ν:

$$g_{11} = \frac{1-\nu\left(1-\frac{9}{16}\frac{A_1}{A_2}\right)}{(1-\nu)^2}, \tag{5}$$

$$g_{22} = \frac{1-\nu\left(1-\frac{9}{16R}\frac{A_1}{A_2}\right)}{(1-\nu)^2}, \quad \text{and} \tag{6}$$

$$g_{12} = \frac{1-\nu\left(1-\frac{9}{8(1+R)}\frac{A_1}{A_2}\right)}{(1-\nu)^2}. \tag{7}$$

Note that all g_{ij} are identical to $g_{2a}(\nu)$ in the monodisperse case with $R=1$ and $A_1 = A_2 = 1$. The parameters used for averaging were $M = 50$ and $\Delta r = a_1/118$ (Left) and $\Delta r = a_1/64$ (Right). A finer binning leads to stronger fluctuations, a rougher binning does not resolve the values at contact, however, within the statistical error, the agreement between theoretical predictions and numerical results is reasonable.
Finally, in Fig. 3, particle correlation functions from **polydisperse simulations** (set F1) are presented. Due to the broad and continuous size distribution function, $g(r)$ is a smooth function with much less variety in magnitude than in the mono- and polydisperse situations discussed above. It more resembles the distribution function of a gas or liquid with a smooth interaction potential [3].

3.2 Collision Rates and Energy Dissipation

In order to estimate the rate of change of energy in the system $\dot{T} = t_E^{-1}\Delta T$, the collision frequency t_E^{-1} is needed. Rather than going into details concerning the calculation of t_E^{-1}, we will simply use the Enskog collision rate [2, 3, 21] for identical particles,

$$t_E^{-1} = \frac{4aN}{V}\sqrt{\pi}g_{2a}(\nu)\sqrt{T/m} \ , \tag{8}$$

and, equivalently, the inter-species collision rates

$$t_{ij}^{-1} = \frac{v_{ij}^{\mathrm{rel}}}{\lambda_{ij}} = \frac{2a_{ij}N_j}{V}\sqrt{\pi}g_{ij}\sqrt{T/(2m_{ij})} \ , \tag{9}$$

where all rates give the number of collisions of a particle per unit time, with $a_{ij} = a_i + a_j$. The temperature is here assumed to be independent of the particle species for the sake of simplicity. This is approximately true in the systems examined below, provided they stay rather homogeneous, but it is not true in general, since the cooling rates depend on the species.

In Eq. (9), the term $\sqrt{T/(m_{ij})}$ is proportional to the mean relative velocity v_{ij}^{rel} of a pair $(i,\ j)$, so that the remainder $t_{ij}^{-1}/v_{ij}^{\mathrm{rel}}$ can be seen as a measure for the inverse interspecies mean free path λ_{ij}. The mean collision rate in the system is

$$\mathcal{T}_{\mathrm{mix}}^{-1} = \sum_{i,j} n_i t_{ij}^{-1} = \frac{4a_1N}{V}\sqrt{\pi}\sqrt{T/m_1}\sum_{i,j} n_i n_j g_{ij} c_{ij} \ , \tag{10}$$

with $c_{11} = 1$, $c_{12} = (1+R)/(2R)\sqrt{(1+R^3)/2}$, and $c_{22} = \sqrt{R}$. Note that c_{12} and c_{22} depend on mass and density of the different species. The mean collision rate was tested for the monodisperse and bidisperse situations and showed the same quality of agreement as the pressure, which will be discussed in the next subsection. Therefore, we do not present data of the collision rates here, but perform a detailed numerical study of the mixture pressure below. However, we should remark that the interspecies collision rates are of the same order of magnitude, even if the species fluctuation velocities $v_i = \sqrt{T/m_i}$ strongly differ due to the differences in mass, also when the temperature $T = T_i = T_j = E/N$ is not species dependent. This means that the mean free distance between collisions compensates the speed; the distance traveled between collisions is proportional to the species velocity. In Fig. 4 several snapshots from simulation set D1 are presented. The grey-scale indicates the collision rates, dark particles collide more frequently than light particles, and the collision rate increases with the density, but as discussed above, the collision rates of the two species are comparable.

Knowing both the collision rates and the energy loss per collision

$$\Delta T_{ij} = -m_{ij}\frac{1-r^2}{2}(v_{ij}^{\mathrm{rel}})^2 = -\frac{1-r^2}{2}T \ , \tag{11}$$

it is straightforward to compute the decay of energy as a function of time

$$\frac{dT}{dt} = \sum_{i,j} n_i t_{ij}^{-1}\Delta E_{ij} = -\frac{1-r^2}{2}\mathcal{T}_{\mathrm{mix}}^{-1}(t)T(t) \ , \tag{12}$$

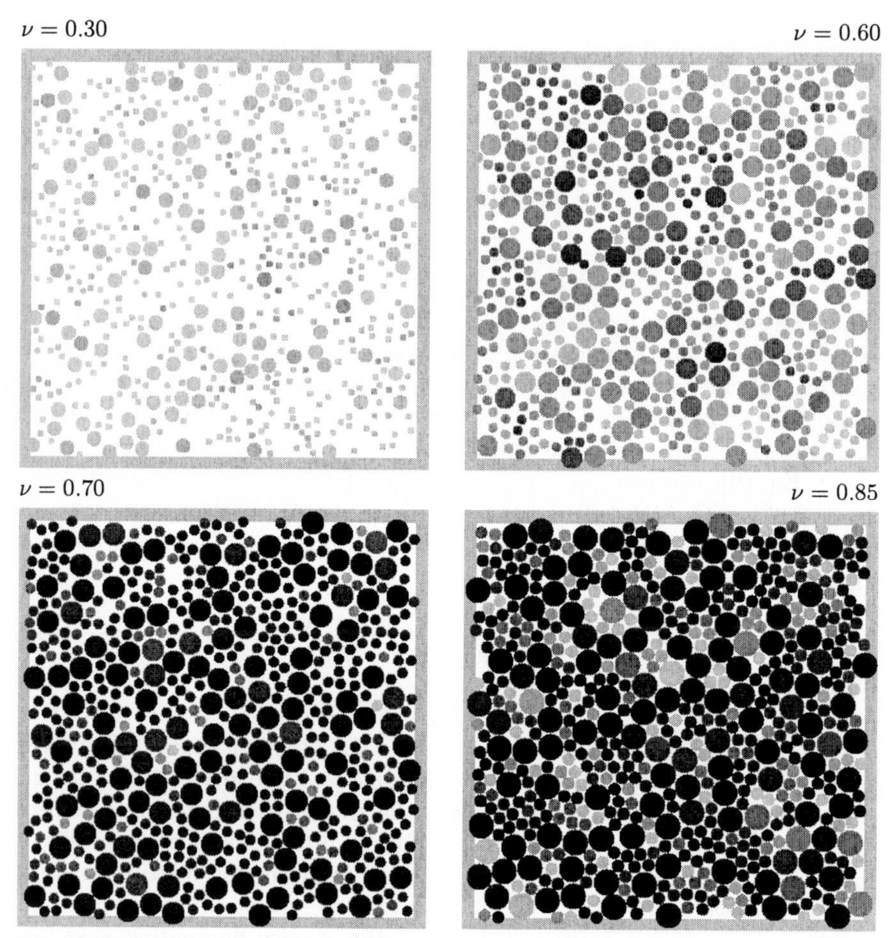

Fig. 4. Snapshots from the bisdisperse simulations (set D1) for different volume fractions ν. The grey-scale denotes the collision rate of the corresponding particle, the darkest particles had a collision rate of $1500\,\mathrm{s}^{-1}$, the lightest particles had a rate of less than $275\,\mathrm{s}^{-1}$. (These numbers have to be multiplied with 2/5 for $\nu = 0.30$ and with 4 for $\nu = 0.85$, in order to allow for a comparison of all pictures.)

where both $T(t)$ and $\mathcal{T}_{\mathrm{mix}}^{-1}(t) \propto \sqrt{T(t)}$ depend on time. The differential equation is easily solved and one gets the scaled temperature

$$\frac{T}{T_0} = \left(1 + \frac{1-r^2}{4}\mathcal{T}_{\mathrm{mix}}^{-1}(0)\,t\right)^{-2}, \tag{13}$$

identical to the solution of the homogeneous cooling state of monodisperse disks [25], where $t_E^{-1}(0)$ is replaced by $\mathcal{T}_{\mathrm{mix}}^{-1}(0)$ from Eq. (10).

In Fig. 5, simulations from set D2 are presented for different values of $r < 1$. The agreement between simulations and Eq. (13) is perfect for short times. With decreasing r, i.e. increasing dissipation, the deviations from the theory occur earlier due to the break-down of the homogeneity and the related simplifying assumptions of molecular chaos and Gaussian velocity distributions [9, 26].

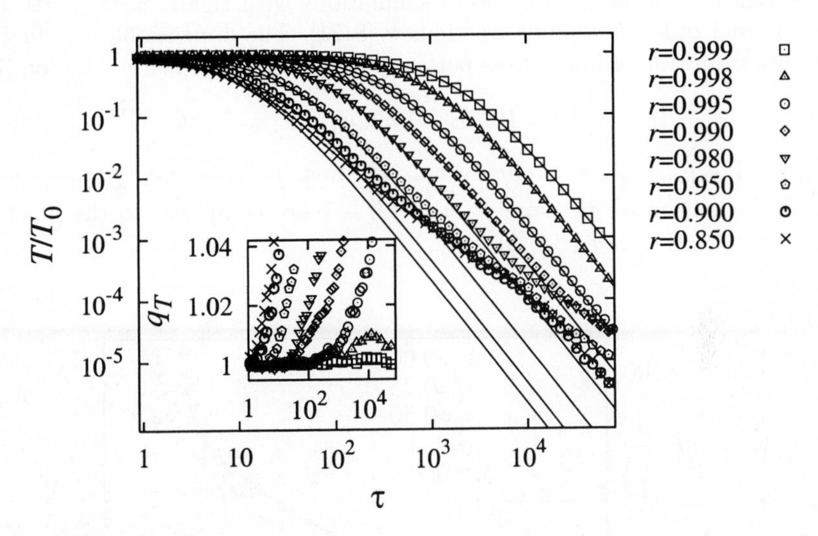

Fig. 5. Dimensionless temperature T/T_0 plotted against rescaled time $\tau = \mathcal{T}_{\rm mix}^{-1}(0)t$ for different r, with double logarithmic axis. The symbols are simulation results from set D2, the solid lines correspond to Eq. (13). In the inset the quality factor $q_T = T_{\rm sim}/T_{\rm theory}$ is plotted against the time τ.

3.3 Stress and the Equation of State

The stress tensor, defined for a test-volume V, has two contributions, one from the collisions and the other from the translational motion of the particles. Using a and b as indices for the cartesian coordinates one has the components of the stress tensor (where the sign is convention)

$$\sigma^{ab} = \frac{1}{V}\left[\sum_i m_i v_i^a v_i^b - \frac{1}{\Delta t}\sum_n \sum_{j=1,2} \Delta p_j^a \ell_j^b\right] , \tag{14}$$

with ℓ_j^b, the components of the vector from the center of mass of the two colliding particles j to their contact points at collision n, where the momentum Δp_j^a is exchanged. The sum in the left term runs over all particles i, the first sum in the right term runs over all collisions n occuring in the time-interval Δt, and the second sum in the right term concerns the collision partners of collision n – in any case the corresponding particles must be within the averaging volume V [21, 26–28]. Note that the results may depend on the choice of Δt and V [29], however, a discussion of different averaging procedures and parameters is far from the scope of this study.

The mean pressure $p = (\sigma_1 + \sigma_2)/2$, with the eigenvalues σ_1 and σ_2 of the stress tensor, can be obtained from simulations with rigid, elastic particles ($r = 1$) and different volume fractions ν [8, 21]. The dimensionless reduced pressure from simulations agrees perfectly with the theoretical prediction [7]

$$P_0 = PV/E - 1 = 2\nu g_{2a}(\nu) , \tag{15}$$

with the total energy $E = (1/2)\Sigma_{i=1}^{N} m_i v_i^2$. In Fig. 6 the simulation results for the dimensionless pressure $P = pV/E - 1$ are compared to the kinetic theory result $P_0 = 2\nu g_{2a}(\nu)$ [21].

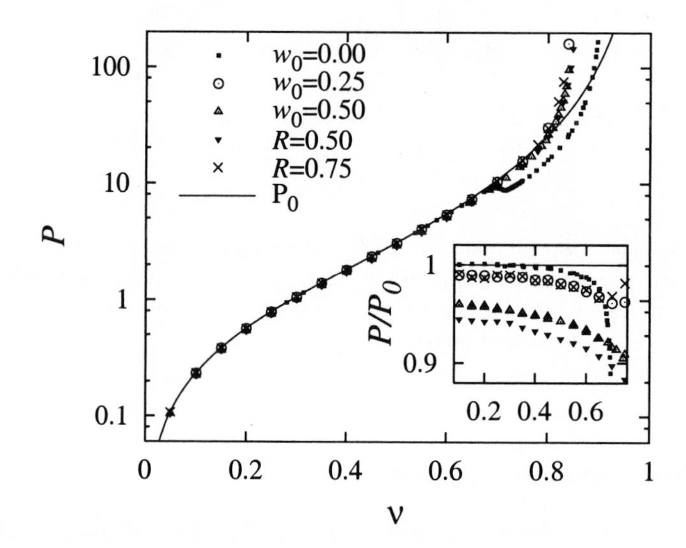

Fig. 6. Dimensionless pressure $P = pV/E - 1$ plotted against the volume fraction ν for different particle size-distribution functions. The bidisperse situation is identified by the size ratio R. The solid line corresponds to P_0. In the inset the quality factor P/P_0 is plotted for the same data. The wiggle of the $w_0 = 0$ data is discussed in more detail below.

When plotting P against the volume fraction ν with a logarithmic vertical axis, the results for the different simulations can not be distinguished for $\nu < 0.7$. In the monodisperse system, we obtain crystallization around $\nu = 0.7$, and the data clearly deviate from P_0, i.e. the pressure is strongly reduced due to crystallization and, thus, enhanced free volume. The monodisperse data diverge at the maximum packing fraction $\nu_{\max}^{\text{mono}} = \pi/(2\sqrt{3})$ in 2D.

Note that one has to choose the system size such that a triangular lattice fits perfectly in the system, i.e. $L_y/L_x = \sqrt{3}h/2w$ with integer h and w – otherwise the maximum volume fraction is smaller. All other simulations are close to P_0 up to $\nu \approx 0.8$, where they begin to diverge; the bidisperse data with $R = 1/2$, for example, diverge at $\nu_{\rm max}^{\rm bi} \approx 0.858$. The maximum packing fraction is smaller for the polydisperse size distributions used here and the crystallization, i.e. the pressure drop, does not occur for polydisperse packings with $w_0 \gtrapprox 0.15$ (the data which lead to this approximate result are not shown). Since the logarithmic axis hides small deviations, we plot also the quality factor P/P_0 of the data in the inset. In this representation, values of unity mean perfect agreement, while smaller values correspond to an overestimation of the data by a factor of P_0/P when P_0 would be used instead of the simulation results P. The deviations increase with increasing width of $w(a)$ and with increasing volume fraction. Note that there exists a deviation already for small ν.

A more elaborate calculation in the style of Jenkins and Mancini, see Eq. (60) in [13], leads to the partial translational pressures $p_i^t = n_i E/V$ for species i and to the collisional pressures $p_{ij}^c = \pi N_i N_j g_{ij} a_{ij}^2 (1 + r_{ij}) T/(4V^2)$ with the particle correlation functions from Eqs. (5)-(7) evaluated at contact, and $a_{ij} = a_i + a_j$. In the simulations from Fig. 6, the inter-species restitution coefficients are equal and elasticity is assumed, $r = r_{11} = r_{12} = r_{22} = 1$. Note that the species temperatures are equal, so that the corresponding correction term can be dropped. Thus, the global pressure in the mixture is

$$\begin{aligned} p^{\rm m} &= p_1^t + p_2^t + p_{11}^c + 2p_{12}^c + p_{22}^c \\ &= \frac{E}{V}\left[1 + (1+r)\frac{\nu}{A_2 a_{11}^2}(g_{11}a_{11}^2 n_1^2 + 2g_{12}a_{12}^2 n_1 n_2 + g_{22}a_{22}^2 n_2^2)\right] \\ &= \frac{E}{V}\left[1 + (1+r)\nu g_{\mathcal{A}}(\nu)\right] . \end{aligned} \tag{16}$$

Assuming a monodisperse system as a test case, i.e. inserting $R = A_1 = A_2 = 1$, into Eq. (16), leads to the monodisperse solution $p^{\rm m}V/E - 1 = P_0$, as expected. The effective correlation function $g_{\mathcal{A}}(\nu)$ can be expressed in terms of the width-correction $\mathcal{A}$ of the size distribution so that

$$g_{\mathcal{A}}(\nu) = \frac{(1+\mathcal{A}) - \nu(1 - \mathcal{A}/8)}{2(1-\nu)^2} , \tag{17}$$

with $\mathcal{A} = \langle a \rangle^2 / \langle a^2 \rangle$. Note that $\mathcal{A}$ is well defined for any size distribution function, so that Eq. (17) can also be applied to polydisperse situations. In the limit of small volume fraction $\nu \to 0$, one can estimate the normalized pressure by

$$P_1 = (1+r)\nu g_{2a}(\nu)\frac{1+\mathcal{A}}{2} , \tag{18}$$

as proposed by Zhang et al. [30], when disregarding the dependence of $g(r)$ on the types of the collision partners. The values of P/P_1 in the limit $\nu \to 0$

agree very well with the simulations. Using the effective particle correlations, one can define

$$P_2(\nu) = \frac{p^{\mathrm{m}}V}{E} - 1 = (1+r)\nu g_{\mathcal{A}}(\nu) \ , \tag{19}$$

and compare the resulting expected reduced pressure with the simulation results from Fig. 6. An almost perfect agreement between P and $P_2(\nu)$ is obtained for $\nu < 0.4$ and even for larger $\nu \approx 0.65$, the difference is always less than about two percent, and, the quality factors for *all* simulations collapse. Note that the quality is perfect (within less than 0.5 percent for all $\nu < 0.65$) if $P_2(\nu)$ is multiplied by the empirical function $1 - \nu^4/10$, as fitted to the quality factor P/P_2. Thus, based on our simulation results, we propose the corrected, nondimensional mixture pressure

$$P_4(\nu) = \frac{p^{\mathrm{m}}V}{E} - 1 = (1+r)\nu g_{\mathcal{A}}(\nu)\left[1 - a_g\nu^4\right] \ , \tag{20}$$

with the empirical constant $a_g \approx 0.1$, for the pressure for all $\nu < 0.65$. For larger ν the excluded volume effect becomes more and more important, leading to a divergence of P/P_4. Furthermore, in the high density regime, the behavior is strongly dependent on the width of the size distribution function, see Fig. 7.

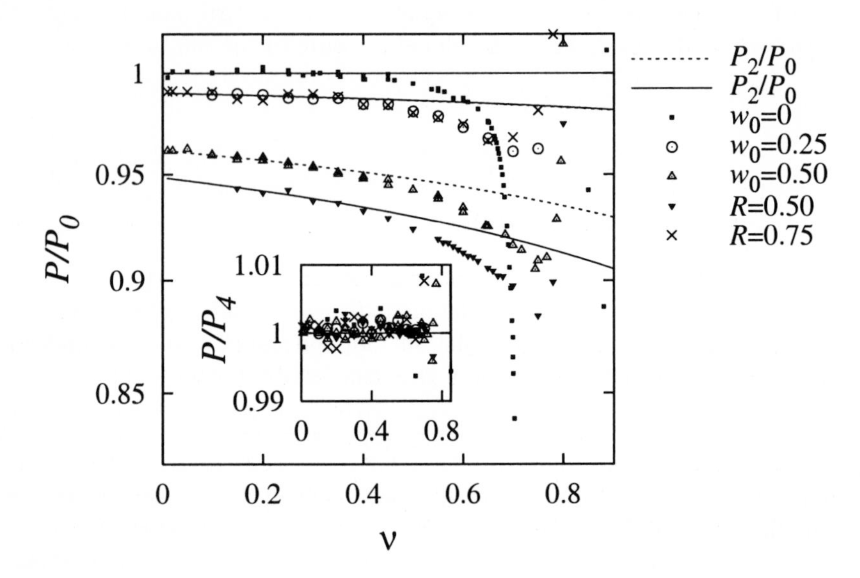

Fig. 7. Quality factor P/P_0 from the inset of Fig. 6. The lines give $P_2(\nu)/P_0$ from Eq. (19). In the inset, the simulation data for P are rescaled by $P_4(\nu)$ from Eq. (20).

3.4 Accounting for the Dense, Ordered Phase

The equation of state in the dense, ordered phase has been calculated by means of a free volume theory [12, 31, 32], that leads in 2D to the reduced pressure $P_{\text{fv}} = 1/(\sqrt{\nu_{\max}/\nu} - 1)$ with the maximum volume fraction $\nu_{\max}$. Based on our simulation results we propse the corrected high density pressure

$$P_{\text{dense}} = \frac{1}{\sqrt{\nu_{\max}/\nu} - 1} \left[1 + a_d(\nu_{\max} - \nu)^{a_p}\right] , \tag{21}$$

where the term in brackets [...] is a fit function with $a_d = 0.340$ and $a_p = 1.09$. The special case $a_d = 0$ leads to the theoretical result P_{fv}. In the left panel of Fig. 8, data from set B (see table 3) are presented, together with the "low" and "high" density predictions P_4 and P_{dense}, respectively (dashed and dotted lines).

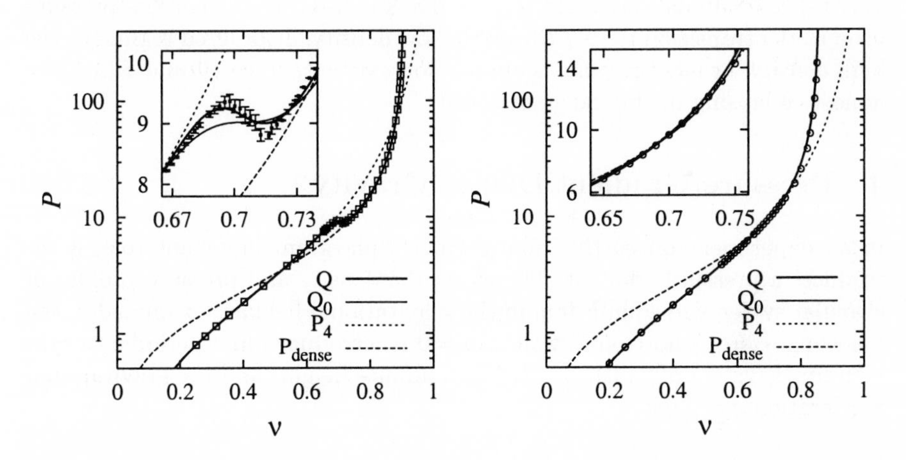

Fig. 8. Dimensionless pressure P from set B simulations (symbols) plotted against the volume fraction ν. (Left) The dashed lines are P_4 from Eq. (20) and P_{dense} from Eq. (21). The thick solid line is Q, the corrected global equation of state from Eq. (22) with the fit-parameters $a_g = 0.1$, $a_d = 0.340$, $a_p = 1.09$, $\nu_c = 0.701$, $\nu_{\max} = 0.9069$, and $m_0 = 0.00928$. The thin solid line is Q_0 without corrections, i.e. $a_g = 0$, $a_d = 0$, and $m_0 = 0.0015$ and $\nu_c = 0.7$, so that $Q_0 = P_2 + m(\nu)[P_{\text{fv}} - P_2]$. (Right) Analogous data from set D1, where only $\nu_{\max} = 0.858$ is different from the left panel.

To our knowledge, no theory exists, which combines the disordered and the ordered regime. Therefore, we propose a global equation of state

$$Q = P_4 + m(\nu)[P_{\text{dense}} - P_4] , \tag{22}$$

with an empirical merging function

$$m(\nu) = \frac{1}{1 + \exp\left(-(\nu - \nu_c)/m_0\right)} \tag{23}$$

which selects P_4 for $\nu \ll \nu_c$ and P_{dense} for $\nu \gg \nu_c$ with the width of the transition m_0. In Fig. 8, the fit parameters $\nu_c \approx 0.70$ and $m_0 \approx 0.009$ lead to qualitative and quantitative agreement between Q (thick line) and the simulation results (symbols). However, also a simpler version Q_0 (thin line) without numerical corrections leads to reasonable agreement when $m_0 = 0.015$ is used, except for the transition region. The pressure drop when ν is increased above ν_c is qualitatively reproduced but no negative slope occurs. Due to the latter fact, the expression Q_0 allows for an easy numerical integration of P. We selected the parameters for Q_0 as a compromise between the quality of the fit on the one hand and the treatability of the function on the other hand.

Remarkably, as one can see from Fig. 8 (Right), the dimensionless pressure Q from Eq. (22) describes, at least qualitatively, the behavior of the polydisperse simulations when $\nu_{\text{max}} = 0.858$ is used. Note that the pressure drop at the transition $\nu_c \approx 0.7$ from the low density, disordered regime to the high density, ordered regime, is almost non-existent, since P_4 and P_{dense} are almost collapsing in this range of density.

4 Pressure Gradient Due to Gravity

In an experiment on earth, usually gravity plays an important role, it introduces a pressure gradient. Therefore, the density and pressure profiles of granular systems in equilibrium in the gravitational field are examined in the following. Here, a horizontal wall at $z = 0$ is introduced in a periodic two dimensional system of width $L = l_x/(2a)$, infinite height, and the gravitational acceleration $g = 1\,\text{ms}^{-2}$.

4.1 Density Profile in Dilute Systems

In the special case of low density, one can use the equation of state of an ideal gas and express the pressure as a function of the energy density:

$$p = \frac{E}{V} = nT\ , \tag{24}$$

with the number density $n = N/V = n(z) = \nu(z)/(\pi a^2)$ and the "granular temperature" $T = E/N$ in two dimensions.

The gradient of pressure dp/dz compensates for the weight $nmgLdz$ of the particles in a layer of height dz, so that

$$\frac{dp}{dz} = \frac{dp}{d\nu}\frac{d\nu}{dz} = -nmg \tag{25}$$

Separation of variables and the assumption of a constant temperature leads to the density profile for an ideal gas

$$\nu(z) = \nu_0 \exp\left(-\frac{mg(z-z_0)}{T}\right) \quad \text{or} \quad z(\nu) = z_0 + z_T \ln\frac{\nu_0}{\nu} , \tag{26}$$

with $\nu < \nu_0$ and $z_T = T/(mg)$. In a system with a constant particle number N, one has

$$N \stackrel{!}{=} \frac{L}{\pi a^2}\int_{z_0}^{\infty} \nu(z)dz = \frac{L}{\pi a^2}\int_0^{\nu_0} (z(\nu) - z_0)\, d\nu . \tag{27}$$

Eq. (27) allows to determine analytically the volume fraction ν_d at the bottom z_0, in the dilute limit, by integration of $z(\nu)$

$$\nu_d = \frac{N\pi a^2 mg}{TL} = \frac{N\pi a^2}{z_T L} , \tag{28}$$

defined here for later use.

This case can be extended to dilute and weakly dissipative systems, since the temperature is almost constant except for the bottom boundary layer [8, 33]. In the following we rather extend it to arbitrary density, but keep $r = 1$.

4.2 Density Profile for a Monodisperse Hard Sphere Gas

In the dense case, Eq. (24) is modified to

$$p = nT\left[1 + 2\nu g_{2a}(\nu)\right] , \tag{29}$$

using Eq. (15) with $r = 1$, and inserting Eq. (29) into Eq. (25) leads to

$$\pi a^2 \frac{dp}{dz} = \frac{d}{d\nu}\left[\nu T(1 + 2\nu g_{2a}(\nu))\right]\frac{d\nu}{dz} = -\nu mg . \tag{30}$$

Assuming again that T is constant, one gets

$$\frac{d}{d\nu}[\ldots] = T\left\{1 + \frac{\partial}{\partial\nu}\left(2\nu^2 g_{2a}(\nu)\right)\right\} = T\,\frac{8 + 8\nu + 3\nu^2 - \nu^3}{8(1-\nu)^3} , \tag{31}$$

which, inserted in Eq. (30), allows integration from ν_0 to ν and from z_0 to z:

$$\int_{\nu_0}^{\nu}\left\{\frac{8}{\nu'} + \frac{7}{1-\nu'} + \frac{7}{(1-\nu')^2} + \frac{18}{(1-\nu')^3}\right\} d\nu' = -\frac{8mg}{T}\int_{z_0}^{z} dz' , \tag{32}$$

and leads to an implicit definition of $\nu(z)$:

$$\left[\ln\nu' - \frac{7}{8}\ln(1-\nu') + 2g_{2a}(\nu')\right]_{\nu_0}^{\nu(z)} = -\frac{z - z_0}{z_T} . \tag{33}$$

We express z as a function of the volume fraction

$$\frac{z(\nu)-z_0}{z_T} = \ln\frac{\nu_0}{\nu} - \frac{7}{8}\ln\frac{1-\nu_0}{1-\nu} + 2g_{2a}(\nu_0) - 2g_{2a}(\nu) , \tag{34}$$

with the unknown volume fraction ν_0 at z_0, which, however, is determined using Eq. (27):

$$\frac{N_0 \pi a^2}{z_T L} = \nu_d \stackrel{!}{=} \int_0^{\nu_0} \frac{z(\nu)-z_0}{z_T} d\nu = \nu_0 \frac{8+\nu_0^2}{8(1-\nu_0)^2} , \tag{35}$$

where N_0 is the number of particles above a given height z_0. Only if $z_0 = 0$, one has $N_0 = N$. This leads to a third order polynomial for ν_0,

$$\nu_0^3 - 8\nu_d \nu_0^2 + (16\nu_d + 8)\nu_0 - 8\nu_d = 0 , \tag{36}$$

which can be solved analytically [34], and always has at least one real solution. Note that the function $g_{2a}(\nu)$ is wrong at high densities $\nu > \nu_c$, so that also the pressure is not correct for high densities. This fact is discussed also by D. Hong [35], who performed the three dimensional calculations analogous to our 2D calculus in this section.

4.3 Comparison with Simulations

In this subsection, the theoretical density profile in Eq. (34), with the parameter ν_0 determined via Eq. (36), is compared to numerical simulations with the parameters as specified in table 4. In Fig. 9, the rescaled height z/z_T is plotted against the volume fraction ν, according to Eq. (34). Note that even when the simulation parameters are rather arbitrary, the data follow a master-curve from $\nu = 0$ to $\nu = \nu_0$ (or equivalently from $z = \infty$ to $z = 0$) only shifted vertically such that $z(\nu_0) = 0$. The agreement between simulation and theory is almost perfect, except for simulation IV where densities above $\nu \approx 0.65$ are observed, i.e. above the limit of validity of the equation of state. Therefore, the numerical values of z/z_T are systematically smaller than the theoretical line obtained for ν_0 from table 4.

Table 4. Simulation parameters for density profile measurements. In these simulations, the particle radius $a = 5 \times 10^{-4}$ m and the particle mass $m = 1.047 \times 10^{-6}$ kg were not changed.

	N	$L/(2a)$	T (kg m^2 s^{-2})	$z_T/(2a)$	ν_d	ν_0
I	1562	100	3.07×10^{-8}	29.4	0.418	0.240
II	3000	100	2.22×10^{-8}	21.2	1.110	0.396
III	1000	100	2.61×10^{-9}	2.49	3.151	0.567
IV	1000	10	6.13×10^{-9}	5.85	13.41	0.755

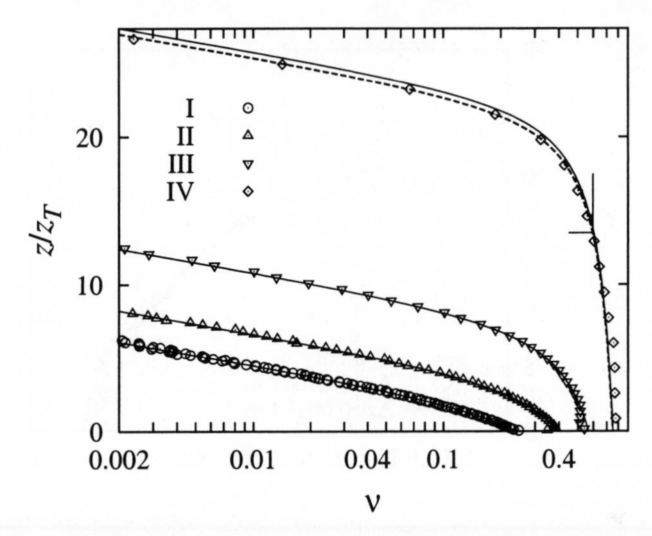

Fig. 9. Rescaled height z/z_T plotted against the volume fraction ν from four different simulations, see table 4, with logarithmic horizontal axis. Symbols are simulation data and the lines correspond to Eq. (34) with the respective value of ν_0.

The only way to get the correct theoretical density profile is for points with $\nu < 0.65$. The value of $z_0 = 13.5 z_T$, where $\nu(z_0/z_T) = 0.65$, is taken from the simulation data and the normalization accounts only for the N_0 particles above z_0. The resulting density profile (dashed line) nicely agrees with the simulations, and its limit of validity is inidcated by the angle at $z/z_T = 13.5$ and $\nu = 0.65$.

The reason for the increased density at low z/z_T in the case of simulation IV is the pressure drop due to crystallization in the equation of state, see Fig. 7. A higher density is necessary to sustain a given pressure when $\nu > 0.65$. The data for the lowest $z/z_T \approx 0$ are slightly off due to the wall induced ordering at $z/z_T = 0$.

If, instead of $2\nu g_{2a}(\nu)$ in Eq. (30), we use the more general form Q_0, we have to integrate the differential equation $dp/dz = \nu mg/(\pi a^2)$ numerically with $p = nT(1 + Q_0)$ and the condition that Eq. (27) is fulfilled. Simulation IV and the numerical solution are compared in Fig. 10. The qualitative behavior of the density profile is well reproduced by the numerical solution with $\nu(z_0 = 0) = 0.8016$. Note that the averaging result is dependent on the binning – we evidence strong coarse-graining effects in the dense, ordered region with densities $\nu > 0.70$.

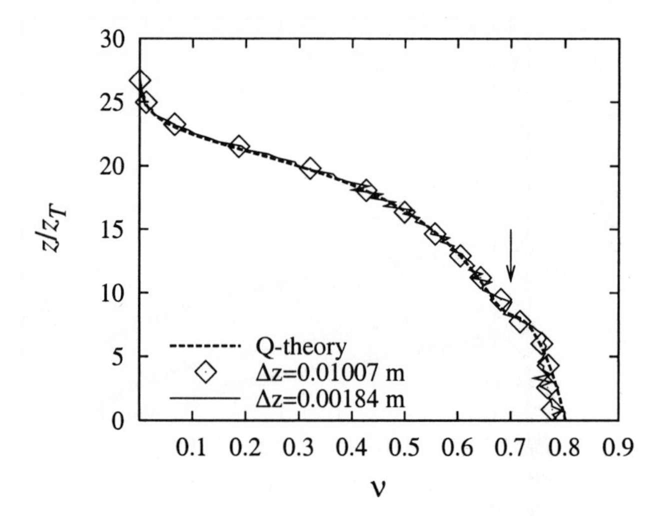

Fig. 10. Rescaled height z/z_T plotted against the volume fraction ν from simulation IV, see table 4. The symbols are simulation data of the particle-center with a rough binning $\Delta z = 0.01007\,\text{m}$. The solid line is the density from the same data but with a much finer binning $\Delta z = 0.00184\,\text{m}$. Both binnings start at $z_0 + \frac{1}{5}a$ with $z_0 = 0$. The dashed line corresponds to the numerical solution of Eq. (30) with $p = nT(1 + Q_0)$. The transition density $\nu_c \approx 0.7$ is indicated by an arrow.

4.4 Bidisperse Systems with Gravitation

In Fig. 11 the species volume fractions ν_1 and ν_2 are plotted against the vertical coordinate $z/(2a_1)$. The data are obtained after long equilibration, in a system with $N = 2000$ particles, width $L_x = l_x/(2a_1) = 100$, and the size distribution of set D1 in the previous section, with $a_1 = 0.0005\,\text{m}$. The particles are no longer mixed, as in the homogeneous, periodic systems of the previous section. Segregation takes place, the larger and heavier particles (squares) settle close to the bottom, whereas the gas of small and lighter particles (circles) extends to larger heights. Knowing both volume fractions, one could compute $\mathcal{A}$ as a function of the height, and insert it into Eq. (17) in order to compute the density profile.

5 Summary and Outlook

In summary, we reviewed existing theories for dilute and dense almost elastic, smooth, 2D granular gases. For mono-, bi-, and polydisperse systems, we compared theoretical predictions with numerical simulations of various systems. The collision frequency, the energy dissipation and the equation of state, i.e. the scaled pressure, are nicely predicted by the theoretical expressions up to intermediate densities. Especially, for arbitrary particle size

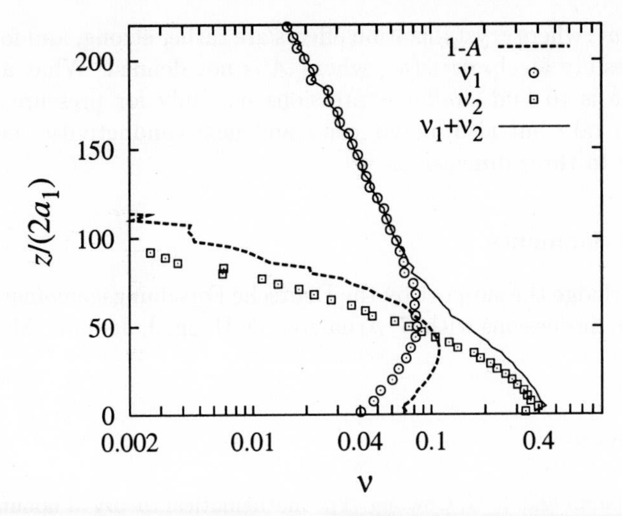

Fig. 11. Dimensionless height $z/(2a_1)$ plotted against the volume fraction (in logarithmic scale). The thick, dashed line gives $1-\mathcal{A}$ (which is zero when only the small species is present at large heights), while the solid line indicates the total volume fraction of the mixture $\nu^{\rm m} = \nu_1 + \nu_2$.

distribution functions, the equation of state can be written in a nice form which only contains the width-correction $\mathcal{A}$ of the size-distribution function. A small, empirical correction can be added to the theories to raise the quality even further. Finally, a merging function that connects the low and high density theories is proposed to give a *global equation of state* for *all* densities and size distribution functions.

The equation of state is used to compute analytically and numerically the density profile of an elastic, monodisperse granular gas in the gravitational field. When a mixture is simulated, segregation is observed, a case to which the theory cannot be applied. For densities below $\nu \approx 0.65$, the analytical solution works well, for higher densities close to the maximum density, one has to use a numerical solver, since the global equation of state cannot be integrated analytically. The strange shape of the density profile, as obtained from simulations, is nicely reproduced.

The simulations and the theories presented here were applied to homogeneous systems. The range of applicability may be reduced by the fact that already weak dissipation can lead to strong inhomogeneities in density, temperature, and pressure. In a freely cooling system, for example, clustering leads to all densities between $\nu \approx 0$ and $\nu \approx \nu_{\rm max}$. The proposed *global equation of state* is a necessary tool to account for such strong inhomogeneities with very high densities, above which the low-density theory fails. For another approach to handle the high density regions, see Ref. [35].

The proposed global equation of state is based on a limited amount of data. It has to be checked, whether it still makes sense in the extreme cases of narrow $w(a)$, where crystallization effects are rather strong, and for extremely broad, possibly algebraic $w(a)$, where $\mathcal{A}$ is not defined. What also remains to be done is to find similar expressions not only for pressure and energy dissipation rate but also for viscosity and heat-conductivity and to extend the theory to three dimensions

Acknowledgements

We acknowledge the support by the Deutsche Forschungsgemeinschaft (DFG) and helpful discussions with B. Arnarson, D. Hong, J. Jenkins, M. Louge, and A. Santos.

References

1. S. Chapman and T. G. Cowling. The mathematical theory of nonuniform gases. Cambridge University Press, London (1960).
2. J. M. Ziman. Models of Disorder. Cambridge University Press, Cambridge (1979).
3. J. P. Hansen and I. R. McDonald. Theory of simple liquids. Academic Press Limited, London (1986).
4. L. D. Landau and E. M. Lifschitz. Physikalische Kinetik. Akademie Verlag Berlin, Berlin (1986).
5. H. J. Herrmann, J.-P. Hovi, and S. Luding, editors. Physics of dry granular media - NATO ASI Series E 350. Kluwer Academic Publishers, Dordrecht 1998.
6. P. K. Haff. Grain flow as a fluid-mechanical phenomenon. J. Fluid Mech. **134**, 401 (1983).
7. J. T. Jenkins and M. W. Richman. Kinetic theory for plane flows of a dense gas of identical, rough, inelastic, circular disks. Phys. of Fluids **28**, 3485 (1985).
8. P. Sunthar and V. Kumaran. Temperature scaling in a dense vibrofluidized granular material. Phys. Rev. E **60**, 1951 (1999).
9. T. P. C. van Noije and M. H. Ernst. Velocity distributions in homogeneously cooling and heated granular fluids. Granular Matter **1**, 57 (1998).
10. N. Sela and I. Goldhirsch. Hydrodynamic equations for rapid flows of smooth inelastic spheres, to burnett order. J. Fluid Mech. **361**, 41 (1998).
11. B. J. Alder and T. E. Wainwright. Studies in molecular dynamics. I. General method. J. Chem. Phys. **31**, 459 (1959).
12. R. J. Buehler, Jr. R. H. Wentorf, J. O. Hirschfelder, and C. F. Curtiss. The free volume for rigid sphere molecules. J. of Chem. Phys. **19**, 61 (1951).
13. J. T. Jenkins and F. Mancini. Balance laws and constitutive relations for plane flows of a dense, binary mixture of smooth, nearly elastic, circular discs. J. Appl. Mech. **54**, 27 (1987).
14. B. Arnarson and J. T. Willits. Thermal diffusion in binary mixtures of smooth, nearly elastic spheres with and without gravity. Phys. Fluids **10**, 1324 (1998).
15. B. Arnarson. Simplified kinetic theory of a binary mixture of nearly elastic, smooth disks, preprint (1999).
16. J. T. Willits and B. Arnarson. Kinetic theory of a binary mixture of nearly elastic disks, preprint (1999).

17. E. Dickinson. Molecular dynamics simulation of hard-disc mixtures. The equation of state. Molecular Physics **33**, 1463 (1977).
18. S. Luding, O. Strauß, and S. McNamara. Segregation of polydisperse granular media in the presence of a temperature gradient. In T. Rosato, editor, IUTAM Symposium on Segregation in Granular Flows, Kluwer Academic Publishers (2000).
19. S. McNamara and S. Luding. A simple method to mix granular materials. In T. Rosato, editor, IUTAM Symposium on Segregation in Granular Flows, Kluwer Academic Publishers (2000).
20. B. D. Lubachevsky. How to simulate billards and similar systems. J. of Comp. Phys. **94**, 255 (1991).
21. S. Luding and S. McNamara. How to handle the inelastic collapse of a dissipative hard-sphere gas with the TC model. Granular Matter **1**, 113 (1998). cond-mat/9810009.
22. O. R. Walton and R. L Braun. Stress calculations for assemblies of inelastic spheres in uniform shear. Acta Mechanica **63**, 73 (1986).
23. C. K. K. Lun. Kinetic theory for granular flow of dense, slightly inelastic, slightly rough spheres. J. Fluid Mech. **233**, 539 (1991).
24. A. Goldshtein and M. Shapiro. Mechanics of collisional motion of granular materials. Part 1. General hydrodynamic equations. J. Fluid Mech. **282**, 75 (1995).
25. S. Luding, M. Huthmann, S. McNamara, and A. Zippelius. Homogeneous cooling of rough dissipative particles: Theory and simulations. Phys. Rev. E **58**, 3416 (1998).
26. S. Luding. Clustering instabilities, arching, and anomalous interaction probabilities as examples for cooperative phenomena in dry granular media. T.A.S.K. Quarterly, Scientific Bulletin of Academic Computer Centre of the Technical University of Gdansk **2**, 417 (July 1998).
27. J. D. Goddard. Microstructural origins of continuum stress fields - a brief history and some unresolved issues. In D. DeKee and P. N. Kaloni, editors, Recent Developments in Structered Continua. Pitman Research Notes in Mathematics No. 143, p. 179, New York, Longman, J. Wiley (1986).
28. F. Emeriault and C. S. Chang. Interparticle forces and displacements in granular materials. Computers and Geotechnics **20**, 223 (1997).
29. I. Goldhirsch. Kinetics and dynamics of rapid granular flows. In H. J. Herrmann, J.-P. Hovi, and S. Luding, editors, Physics of dry granular media - NATO ASI Series, p. 371, Dordrecht, Kluwer Academic Publishers (1998).
30. J. Zhang, R. Blaak, E. Trizac, J. A. Cuesta, and D. Frenkel. Optimal packing of polydisperse hard-sphere fluids. preprint (1999).
31. J. G. Kirkwood, E. K. Maun, and B. J. Alder. Radial distribution functions and the equation of state of a fluid composed of rigid spherical molecules. J. Chem. Phys. **18**, 1040 (1950).
32. W. W. Wood. Note on the free volume equation of state for hard spheres. J. Chem. Phys. **20**, 1334 (1952). Letters to the Editor.
33. J. Eggers. Sand as Maxwell's demon. Phys. Rev. Lett. **83**, 5322 (1999).
34. I. N. Bronstein and K. A. Semendjajew. Taschenbuch der Mathematik. Teubner, Leipzig (1979).
35. D. C. Hong. Fermi Statistics and Condensation, (in this volume, p. 429).

Experimental Observations of Non-equilibrium Distributions and Transitions in a 2D Granular Gas

Jeffrey S. Urbach and Jeffrey S. Olafsen

Department of Physics, Georgetown University, Washington, DC 20057.
e-mail: urbach@physics.georgetown.edu

Abstract. A large number (~10,000) of uniform stainless steel balls comprising less than one layer coverage on a vertically shaken plate provides a rich system for the study of excited granular media. Viewed from above, the horizontal motion in the layer shows interesting collective behavior as a result of inelastic particle-particle collisions. Clusters appear as localized fluctuations from purely random density distributions, as demonstrated by increased particle correlations. The clusters grow as the medium is "cooled" by reducing the rate of energy input. Further reduction of the energy input leads to the nucleation of a collapse: a close-packed crystal of particles at rest. High speed photography allows for measurement of particle velocities between collisions. The velocity distributions deviate strongly from a Maxwell distribution at low accelerations, and show approximately exponential tails, possibly due to an observed cross-correlation between density and velocity fluctuations. When the layer is confined with a lid, the velocity distributions at higher accelerations are non-Maxwellian and independent of the granular temperature.

1 Introduction

The development of a kinetic theory of granular gases, *i.e.* collections of large numbers of inelastically colliding particles, has proven to be a very challenging undertaking. While the equilibrium properties of elastically colliding gases are relatively well understood, the introduction of physically relevant levels of dissipation changes the dynamics dramatically. Considerable progress has been made in understanding the behavior of freely cooling granular gases, where the energy lost to collisions is not replaced. In order to model a variety of industrial processes where energy is added to the grains to enhance mixing and transport, a kinetic theory of forced granular gases is required. By analogy with the theory of equilibrium fluids, the goal is to solve a Boltzmann-like equation for the relevant 'microscopic' statistical distribution functions describing the behavior of individual grains. The rate at which energy is put into the fluctuating velocities is balanced by the rate at which it is lost in collisions. 'Macroscopic' transport coefficients, relating average fluxes (or currents) to gradients in local (coarse grained) variables, can then be calculated from the appropriate correlation functions.

Some progress has been made in building a continuum description of granular fluids from a microscopic kinetic theory, but has typically involved a number of approximations that have yet to be tested. A hydrodynamic description is only possible if the spatial and temporal fluctuations in the flow are sufficiently localized to permit coarse graining. This may be the case if the granular system is large enough and the energy in the flow is high enough, but making this statement quantitative is very much an ongoing effort. Precise experimental measurements of the steady state statistical properties in a variety of granular fluids are essential to developing and testing generally applicable descriptions. We have been investigating the fluidized state and the fluidization transition in a simple representation of a granular material: a single layer of identical spherical particles. Using identical particles simplifies analysis of the system and allows for comparison with experimental and theoretical results derived for atomic and molecular dynamics. The use of a single granular layer, in combination with high speed imaging technology, allows for a thorough description of the granular system, including particle velocity distributions, correlation functions, and transport properties. We have found a complex phase diagram that bears many similarities to equilibrium two-dimensional systems, but we have also directly measured particle velocity distributions that are non-Maxwellian, and density correlation functions that show non-equilibrium effects [1]. In addition, we have measured the cross-correlation between density and temperature fluctuations, and the effects of constraining the granular layer by placing a lid above it [2].

This paper is organized as follows: Section 2 describes the experimental setup and analysis techniques, and Section 3 describes the experimentally observed phase diagram. Our results on the statistical characterization of the granular gas are reported in 4, followed by a comparison with related work and a discussion of future directions in Section 5.

2 Experimental Setup and Methods

The experimental apparatus consists of a 20 cm diameter smooth, rigid aluminum plate that is mounted horizontally on an electromagnetic shaker that oscillates the plate vertically. The plate is carefully leveled, and the amplitude of acceleration is uniform across the plate to better that 0.5 %. The acceleration of the plate is monitored with a fast-response accelerometer mounted on the bottom surface of the plate. Two types of particles were used for the experiments described below: smaller spheres of 302 stainless steel with an average diameter of 1.191 $\pm$ 0.0024 mm, and larger spheres of 316 stainless with an average diameter of 1.588 $\pm$ 0.0032 mm. The coefficient of restitution, both particle-particle and particle-plate, is about 0.9. The particles are surrounded by an aluminum rim that occupies the outer 1.9 cm of the plate. The particles are illuminated by low angle diffuse light. This illumination produces a small bright spot at the top of each particle when viewed through a video camera mounted directly above the plate.

Two digital video cameras are used for data acquisition, a high-resolution camera for studying spatial correlations (Pulnix-1040, 1024 x 1024 pixels, Pulnix America, Inc., Sunnyvale, CA), and a high-speed camera for measuring velocity distributions (Dalsa CAD1, 128 x 128 pixels, 838 frames/second, Dalsa Inc., Waterloo, ON; Canada). A collection of images and movies is available for viewing at www.physics.georgetown.edu/~granular. The acquired digital images are analyzed to determine particle locations by calculating intensity weighted centers of bright spots identified in the images.

The frame rate of the high speed camera is much faster than the interparticle collision rate, so it is possible to measure particle velocities by measuring the displacement of the particle from one image to the next. The trajectories determined from the images are nearly straight lines between collisions. Although the frame rate is fast compared the collision time, the fact that it is finite introduces unavoidable systematic errors into the experimentally determined velocity distributions. Some particles will undergo a collision between frames, and the resulting displacement will not represent the true velocity of the particle either before or after the collision. Since fast particles are more likely to undergo a collision, the high energy tails of the distribution will be decreased. The probability that a particle of velocity v will not undergo a collision during a time interval Δt is proportional to $\exp(-v\Delta t/l_o)$, where l_o is the mean free path. The effect of this on the velocity distribution function reported here is quite small. In addition to taking velocities out of the tails of the distribution, collisions will incorrectly add the velocities of those particles elsewhere in the distribution. This effect can be minimized by filtering out the portion of particle trajectories where collisions occur. This is done by analyzing particle position in 3 images at a time. When the change in apparent velocity from the first two images differs from the change in the second two by more than a specified cutoff value, the points are neglected. Because a relatively small fraction of the particles undergo a collision between any two frames, the results are not sensitive to the precise value chosen. At the highest accelerations considered in this work, varying the cutoff value over more than an order of magnitude changes the measured granular temperature (*rms* velocity) by less than 10%, and the flatness of the distribution (Eq. 2) by less than 0.2. Similar issues are considered in a somewhat different manner in reference [3].

3 Phase Diagram

When the plate oscillation amplitude is not too large, the spheres rarely hop over one another; thus the system essentially two-dimensional. Nonetheless, there is sufficient energy in the horizontal velocity component to generate fascinating dynamic phenomena. At moderately large sinusoidal vibration amplitudes, a fully fluidized state is observed. The spheres are constantly in motion and there is no large-scale spatial ordering. Figure 1(a) shows an instantaneous image of part of a cell containing 8000 particles in this regime. (The number of particles in a single hexagonal close-packed layer is $N_{max} = 17275$ for the smaller particles, giving a reduced density $\rho = N/N_{max} = 0.463$.) The peak plate acceleration relative to the acceleration due to gravity, $\Gamma = A\omega^2/g$, is just over one. This phase is characterized by an apparently random distribution of particle positions and velocities. A sense of the dynamics can be gained from an average of 15 frames taken over a period of 1 s (Fig. 1(b)), which shows the lack of any stable structure. As the amplitude of the acceleration is slowly decreased, the average kinetic energy of the particles decreases, and localized transient clusters of low velocity particles appear. An instantaneous image in this regime (Fig. 1(c)) does not look very different from the one taken at higher acceleration, but in the time-averaged image (Fig. 1(d)) bright peaks are clearly evident, corresponding to low-velocity particles that have remained relatively close to each other over the time interval. In this regime, the clusters typically survive for 1-20 seconds. In the low density regions outside of the clusters, there are particles with anomalously high velocities, and these appear to be responsible for the breakup of the clusters. There are no attractive interactions between these particles; the cluster formation is a uniquely non-equilibrium phenomenon, resulting from the dissipation during inter-particle collisions, and the particle-plate dynamics.

When the amplitude of the vibration is decreased somewhat below that of Fig. 1 (c,d), the typical cluster size increases to 12-15 particles. Within a few minutes at this acceleration, one of these large clusters will become a nucleation point for a 'solid' phase, similar to what is referred to as 'inelastic collapse', where the relative velocity between two or more particles is completely dissipated by an infinite series of inelastic collisions [4]. The particles in the collapse are in contact with all of their neighbors, and form a perfect hexagonal lattice (Fig. 1(d)). The collapse is surrounded by a gas of the remaining particles. The sharp interface between the coexisting phases can be seen in the time-averaged image (lower panel of Fig. 1(e)) . The two-phase co-existence persists essentially unchanged for as long as the driving is maintained. At higher densities, instead of a transition directly from the clustering behavior to collapse, there is an intermediate phase with apparent long range order. Figure 1(g) shows a monolayer in this ordered state, where the spheres are arranged in a hexagonal lattice but are not at rest or in contact with one another. The disorder in the image is a consequence of the fluctuations induced by inter-particle collisions. When the particle positions in the ordered phase are averaged over a short time (Fig. 1(h)), the resulting image displays a nearly perfect lattice, with one unoccupied site. Measurements of the correlation functions for positional and orientational order parameters in this phase suggest that the transition to this ordered

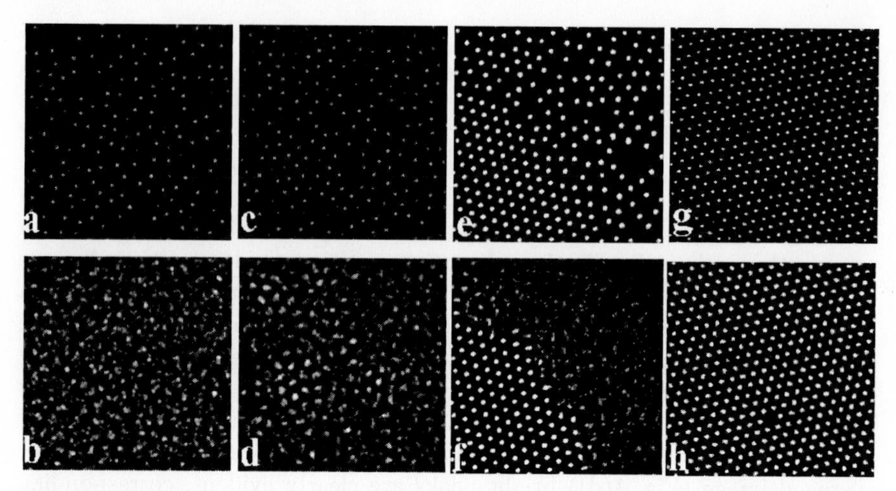

Fig. 1. Instantaneous (top row) and time-averaged (bottom row) photographs detailing the different phases of the granular monolayer. (a), (b), uniform particle distributions typical of the gas phase ($\rho = 0.463, \Gamma = 1.01, \nu = 70$ Hz). (c) The clustered phase ($\rho = 0.463, \Gamma = 0.8, \nu = 70$ Hz). The higher intensity points in a time-averaged image, (d), denote slower, densely packed particles. (e) A portion of a collapse for $\rho = 0.463$, $\Gamma = 0.76$, and $\nu = 70$ Hz. (f) The time-averaged image shows that the particles in the collapse are stationary while the surrounding gas particles continue to move. (g),(h) In a more dense system, $\rho = 0.839$, there is an ordered phase ($\Gamma = 1.0, \nu = 90$ Hz) where all of the particles remain in motion. The area shown in the images is approximately 4 cm^2.

phase is quantitatively similar to the liquid-solid transitions observed in a variety of equilibrium systems.

Figure 2 shows phase diagrams of the system with two different densities. The data for the nucleation points were taken by decreasing the plate acceleration in steps of about $0.003g$, and waiting 5 minutes at each step to see if the collapse nucleates. The precise location of the nucleation line depends on the waiting time, but only very weakly. Once the collapse forms, increasing the acceleration causes it to 'evaporate' as particles return to the gas phase. The open circles in Figure 2(a) indicate the acceleration required to fully evaporate the collapse. The evaporation line is omitted from the lower panel for clarity.

It is not immediately clear what causes the frequency dependence in the phase diagram. For an ideal spherical particle on an oscillating plate with a velocity-independent coefficient of restitution, the dynamical behavior depends only on Γ, and the frequency sets the timescale for the motion, and the length scale through g/ω^2, which is proportional to the distance a ball falls during one oscillation period. Thus as the frequency is reduced for fixed Γ, balls will bounce higher. Because the balls in this system interact with their neighbors, it is possible that the frequency dependence enters through

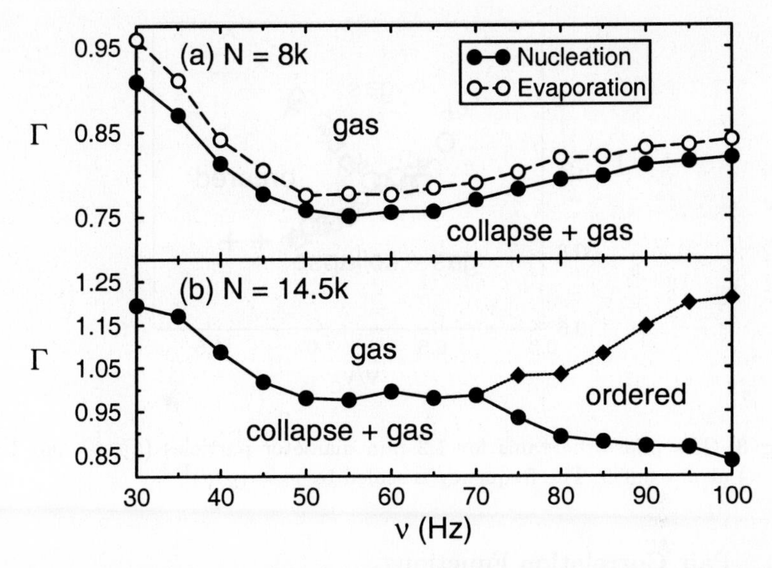

Fig. 2. The phase diagrams for (a) N = 8000 particles ($\rho = 0.463$) and (b) N = 14,500 ($\rho = 0.839$) particles. The filled circles denote the acceleration where the collapse nucleates. The open circles in (a) indicate the point where the collapse disappears upon increasing the acceleration. The diamonds in (b) show the transition to the ordered state as the acceleration is reduced.

the ratio of this length scale to the ball diameter. In fact, the rapid increase in the acceleration where collapse forms for frequencies below about 50 Hz (see Fig. 2) occurs when the particles begin to bounce high enough to hop over one another, resulting in a gradual transition from primarily 2D to 3D dynamics. This suggests that the frequency dependence comes from a characteristic frequency $\nu_c = (g/d)^{1/2}$, where d is the sphere diameter. Figure 3 shows the phase diagram measured for two different sets of particles determined at the same reduced density ρ, one with a diameter of 1.2 mm, and one with a diameter of 1.6 mm. The upturn at low frequencies and the appearance of the ordered phase occur at lower frequencies for the larger particles, but the scaled phase diagrams lie right on top of one another.

4 Statistical Characterization of the Granular Gas

In order to incorporate the non-equilibrium fluctuations observed in Fig. 1(c,d) into a kinetic theory of the granular fluids, quantitative measures are required. In the monolayer system, particle positions can be directly determined from images acquired with a digital video camera. These can be used to evaluate statistical measures, such as the pair correlation function and the velocity distribution function, that are essential components of kinetic theory.

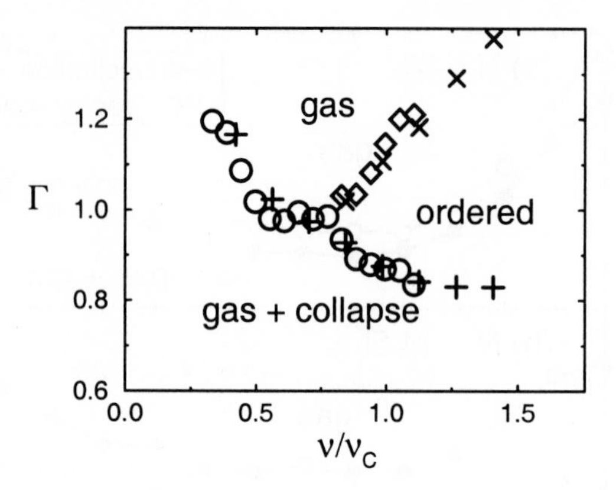

Fig. 3. The phase diagrams for 1.2 mm diameter particles (○, ⋄) and 1.6 mm (+,×) at $\rho = 0.839$. The frequency is scaled by $\nu_c = (g/d)^{1/2}$.

4.1 Pair Correlation Functions

Correlations in particle positions are most directly measured by the pair correlation function, $G(r)$:

$$G(r) = \frac{\langle \rho(0)\rho(r) \rangle}{\langle \rho \rangle^2}, \quad (1)$$

where ρ is the particle density. The correlations of a two-dimensional gas of elastic hard disks in equilibrium are due only to geometric factors of excluded volume and are independent of temperature [5].

The solid line in Fig. 4 shows $G(r)$ from a Monte Carlo calculation of a two-dimensional gas of elastic hard disks in equilibrium for a density of 0.463 [6]. The experimentally measured correlation function in the gas-like phase, shown by the open circles, is almost identical to the equilibrium result. There are no free parameters in the determination of the equilibrium result shown in Fig. 4. The agreement between the experimental correlation function and the equilibrium result suggests that the structure in the correlation function of the gas-like phase is dominated by excluded volume effects. As the granular medium is cooled, the correlations grow significantly. This is evident from the data for a lower vibration amplitude ($\Gamma = 0.774$, just above the acceleration where the collapse forms), shown as filled diamonds in Fig. 4. The increased correlations indicate that there are non-uniform density distributions in the medium: high density regions of relatively closely packed particles, which must coexist with relatively low density regions. In the kinetic theory of equilibrium fluids, the pair correlation function plays a central role, because thermodynamic quantities can be written as integrals with $G(r)$. The equation of state for an equilibrium hard disk system, as well as the Enskog modification to the Boltzmann equation (the correction

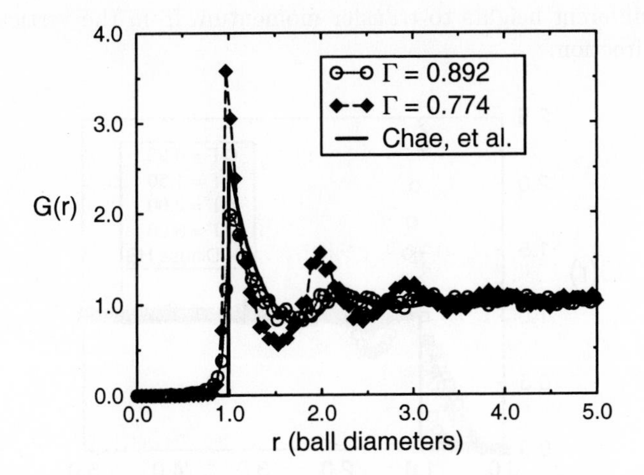

Fig. 4. The measured pair correlation function of a two-dimensional granular gas ($\rho = 0.463, \nu = 70$ Hz). The results are compared to the result from an equilibrium hard disk Monte Carlo calculation with no adjustable parameters. The legend gives the value of Γ.)

to account for excluded volume effects), depends on $G(r)$ at contact (r =1 diameter) [7]. Extensions of kinetic theory to inelastic gases [8] assume that $G(r)$ at contact is given by the Carnahan-Starling relationship [9], which works well for elastic gases. Thus the dramatic increase in correlations will likely have significant quantitative implications for transport coefficients in the granular gas.

At higher plate accelerations, the pair correlation function loses all of its structure. Fig. 5 shows $G(r)$ at $\Gamma = 0.93$, $\Gamma = 1.50$, and $\Gamma = 3.0$. Increasing the steady state kinetic energy of the granular gas by increasing the amplitude of the acceleration at constant frequency causes the gas to change from primarily two-dimensional, where the particles never hop over one another, to essentially three-dimensional. This transition can be observed in the pair correlation function, $G(r)$, by the increase in its value for $r < 1$. (The correlation function includes only horizontal particle separations.) This transition can affect the dynamics in several ways: the effective density is decreased, so that excluded volume effects are less important; the inter-particle collisions can occur at angles closer to vertical, affecting transfer of energy and momentum from the vertical direction to the horizontal; and the change in the dimensionality itself can have important consequences.

In order to separate these effects from the direct consequences of increasing the kinetic energy of the gas, a Plexiglas lid was added to the system at a height of 2.54 mm, or 1.6 ball diameters for the larger particles. For this plate-to-lid separation, the larger particles cannot pass over top of one another, although enough room remains for collisions between particles at sufficiently different heights to transfer momentum from the vertical to the horizontal direction.

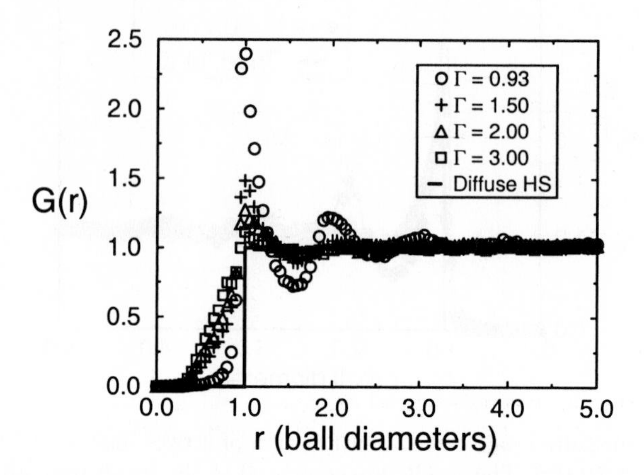

Fig. 5. (a) Pair correlation functions for larger accelerations in the unconstrained system. (○) $\Gamma = 0.93$, (+) $\Gamma = 1.5$, (△) $\Gamma = 2.0$, (□) $\Gamma = 3.0$.

Figure 6 shows the pair correlation functions measured with the lid on, and demonstrates that the particle-particle correlations persist and become independent of Γ when the system is constrained in the vertical direction. The correlations decrease slightly from $\Gamma = 0.93$ to $\Gamma = 1.50$, and then remains essentially constant up to $\Gamma = 3$. The small value of $G(r)$ for distances less than one ball diameter indicates that the system remains 2D as Γ is increased. The structure observed in the correlation function is essentially the same as that of an equilibrium elastic hard sphere gas at the same density, indicating that the correlations that exist are due to excluded volume effects. From $\Gamma = 1.50$ to $\Gamma = 3.0$, the granular temperature changes by more than a factor of 2 (see Fig. 10), yet there is no detectable change in the pair correlation function. Thus, unlike the low acceleration behavior shown in Fig 4, the particle correlations in the constrained system at higher accelerations behave like those of a system of elastic hard disks.

4.2 Velocity Distribution Functions

A crucial ingredient of a statistical approach to describing the dynamics in a granular system is the velocity distribution, which may show non-equilibrium effects as can the correlation function. As described in section 2, the horizontal components of the particle velocities between collisions can be determined with the use of a high speed camera. Extensive measurement of the velocity distributions in the plane of the granular gas demonstrate non-Gaussian behavior.

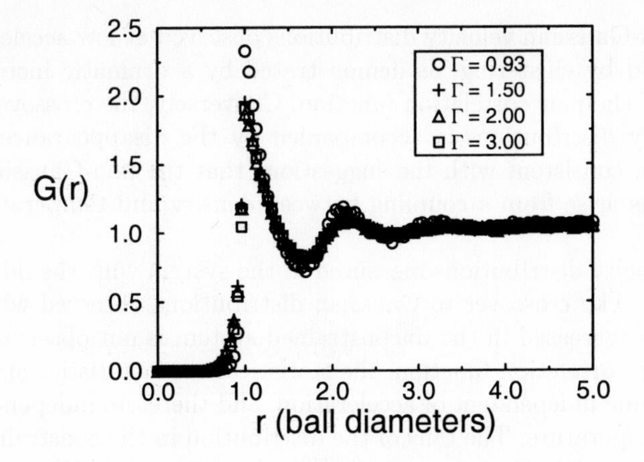

Fig. 6. Pair correlation functions for the velocity distributions where a lid constrains the system to remain two dimensional. The particle correlations remain as the shaking amplitude is increased. (○) $\Gamma = 0.93$, (+) $\Gamma = 1.5$, (△) $\Gamma = 2.0$, (□) $\Gamma = 3.0$.

The measured horizontal velocity distributions at $\Gamma = 0.93$, $\Gamma = 1.50$, and $\Gamma = 3.0$ are shown in Fig. 7. The distributions at low Γ are strongly non-Gaussian, showing approximately exponential tails [1]. As the acceleration is increased, the distribution crosses over smoothly to a Gaussian. This behavior is superficially similar to that of freely cooling granular media, where an initial Gaussian velocity distribution becomes non-Gaussian as the system cools, but these results are obtained in a steady state.

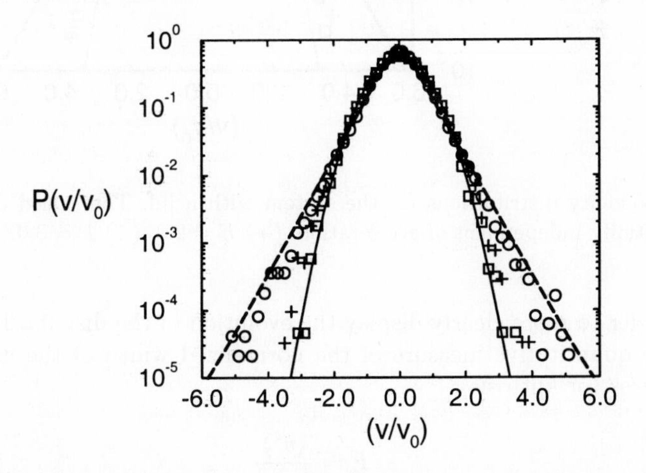

Fig. 7. Velocity distributions for the system without a lid. As the acceleration is increased, the distributions go from having nearly exponential tails to Gaussian tails. (○) $\Gamma = 0.93$, (+) $\Gamma = 1.5$, (□) $\Gamma = 3.0$.

The non-Gaussian velocity distributions observed at low accelerations are accompanied by clustering, as demonstrated by a dramatic increase in the structure of the pair correlation function. Conversely, the crossover to Gaussian velocity distributions is accompanied by the disappearance of spatial correlations, consistent with the suggestions that the non-Gaussian velocity distributions arise from a coupling between density and temperature fluctuations [10].

The velocity distributions measured in the system with the lid are shown in Figure 8. The crossover to Gaussian distributions observed when the acceleration is increased in the unconstrained system is not observed. Instead, like the pair correlation function, the statistical characteristics of the granular gas become independent of acceleration, and therefore independent of the granular temperature. The tails of the distribution in the constrained system are consistent with $P(v) \propto \exp(-|v|^{3/2})$, as observed in [3] and predicted by [11]. The relationship between our results and that work is discussed in section 5.2.

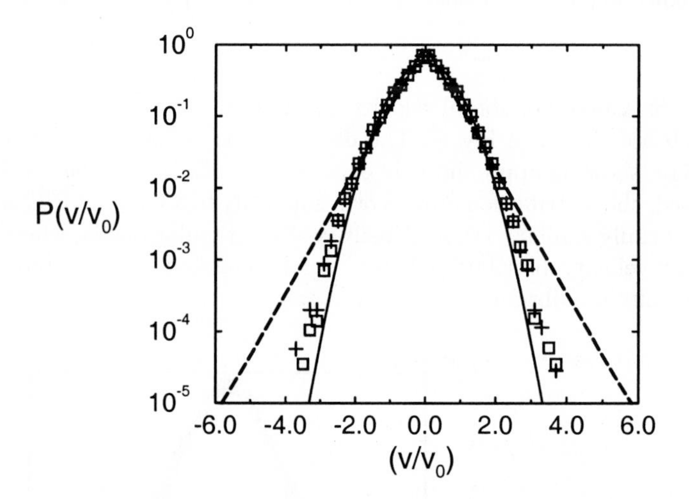

Fig. 8. Velocity distributions for the system with a lid. The scaled distributions are essentially independent of acceleration. (+) $\Gamma = 2.0$, (□) $\Gamma = 3.0$.

In order to more clearly display the evolution of the distributions, we use a simple quantitative measure of the normalized width of the distribution, the flatness (or kurtosis):

$$F = \frac{\langle v^4 \rangle}{\langle v^2 \rangle^2}. \tag{2}$$

For a Gaussian distribution, the flatness is 3 and for the broader exponential distribution, the flatness is 6. In the absence of a lid, the flatness demonstrates a smooth transition from non-Gaussian to Gaussian behavior as the acceleration is increased (Fig. 9), whether the smaller (circles) or larger (stars) particles are used. With the lid on, the velocity distributions remain more non-Gaussian than in the free system for identical accelerations (or identical granular temperatures) and density (diamonds).

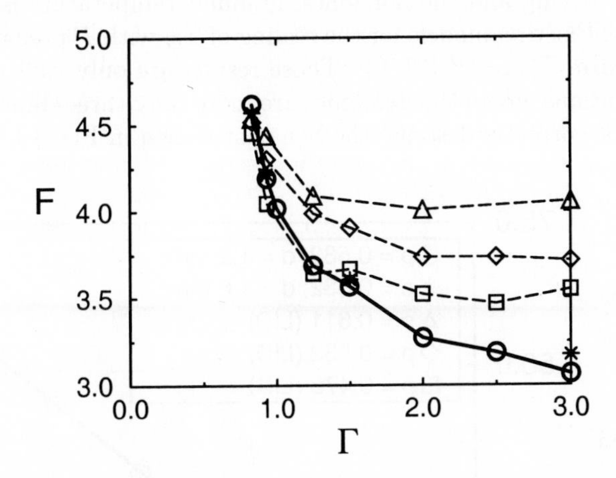

Fig. 9. Plot of the flatness as a function of granular temperature with and without a lid in the system. At low Γ the system is nearly 2D and the lid has no effect. Data for $\rho = 0.532$: (○) d = 1.2 mm and (∗) d = 1.6 mm without lid; (◇) d = 1.6 mm with lid. Data for $\rho = 0.478$ (△) and $\rho = 0.611$ (□) with a lid and d = 1.6 mm.

The crossover from Gaussian to non-Gaussian behavior observed without the lid is therefore not simply an effect of increasing the vertical kinetic energy of the particles, but rather related to the transfer of energy from the vertical to horizontal motion in the system via collisions, the change in the density, or the change in dimensionality of the gas.

To determine the relative contribution of density changes to the non-Gaussian velocity distributions in the gas, the number of particles on the plate was increased by 15% and decreased by 10% from the value of $\rho = 0.532$ and the lid was kept on. For all accelerations, the flatness decreased with increased density. This surprising result may be related to the fact that strongly non-Gaussian distributions observed at low Γ are accompanied by strong clustering [1]. If the average density is increased, the larger excluded volume means that less phase space remains for fluctuations to persist. The fact that increasing the density with the lid on makes the velocity distribution more Gaussian suggests that the crossover to Gaussian observed without the lid is not due to the decrease in density of the gas.

In addition to providing velocity distribution functions, measurements of the particle velocities can be used to calculate the granular temperature,

$T_G = (1/2)\langle v^2 \rangle$. Figure 10 shows the granular temperature versus plate acceleration for several different experimental configurations. At low acceleration ($\Gamma \leq 1$), very few particles strike the lid and the lid has no significant effect. At larger Γ it is clear that the horizontal granular temperature is reduced when a significant number of particles strike the lid. It is interesting to note that T_G approaches zero linearly at finite Γ, indicating that the relationship between the driving and the horizontal granular temperature is of the form $T_G \propto \Gamma - \Gamma_c$. Existing models for the scaling of T_G with Γ predict a relationship of the form $T_G \propto \Gamma^\theta$ [12–14]. Those results are only valid at relatively large accelerations where the dynamics are more fully three-dimensional, and clearly do not correctly describe the behavior shown in Fig. 10.

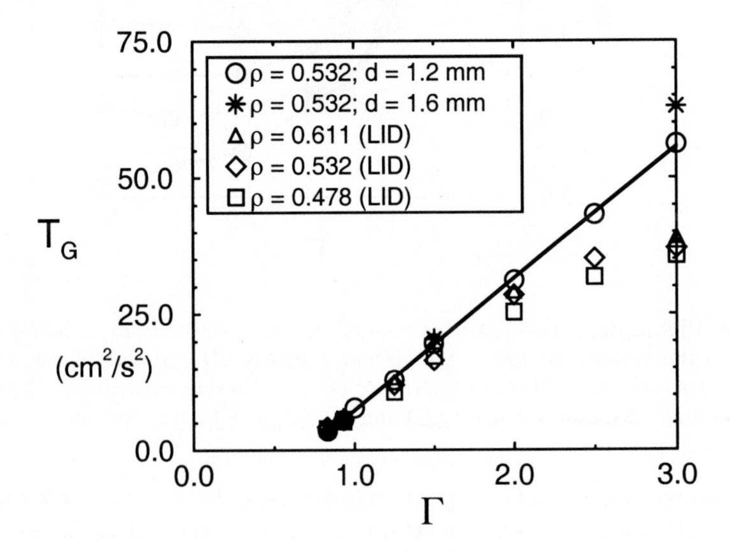

Fig. 10. Plot of the horizontal granular temperature, T_G, as a function of dimensionless acceleration, Γ, with and without a lid. Without the lid (○,∗) the system becomes three dimensional at high accelerations.

4.3 Density-Velocity Cross-Correlations

Some insight into the origins of the non-Gaussian velocity distributions can be obtained by investigating the relationship between the local *fluctuations* in density and kinetic energy. Puglisi *et al.* [10] have proposed a model which relates strong clustering to non-Gaussian velocity distributions in a driven granular medium. In their framework, at each local density the velocity distributions are Gaussian, and the non-Gaussian behavior arises from the relative weighting of the temperature by local density in the following manner:

$$P(v) = \sum_N n(N) e^{-(v^2/v_0^2(N))}, \qquad (3)$$

where N is the number of particles in a box, $v_0^2(N)$ is the second moment of the distribution of velocities for boxes with N particles, and $n(N)$ is the number of boxes that contain N particles. In this model, the local temperature is a decreasing function of the local density, and the velocity distributions conditioned on the local density are Gaussian.

We have investigated the velocity distributions conditioned on the local density by examining the distribution of velocities for data at a constant number of particles in the frame of the camera. In the strongly clustering regime, we do observe a direct correlation between local density and temperature. Figure 11 is a plot of the local temperature as a function of particle number in the camera frame normalized by the global granular temperature.

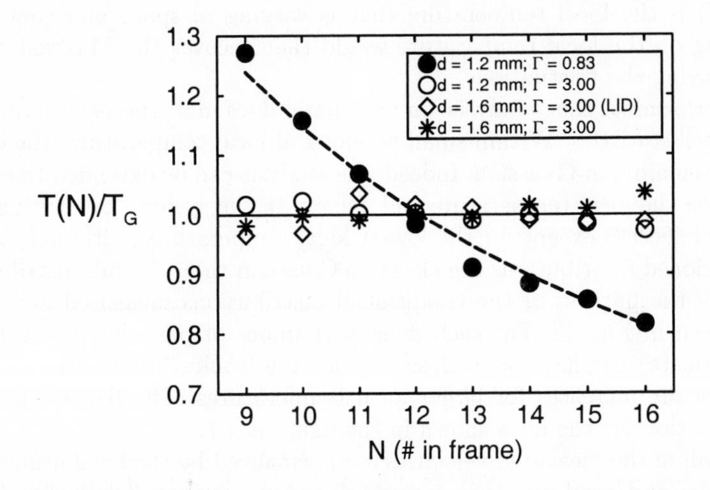

Fig. 11. Plot of $T(N)/T_G$ where N is the number of particles in the camera frame (see text). At low Γ, the local temperature and density are strongly correlated. At high Γ, the local granular temperature is independent of density, even when the velocity distributions remain non-Gaussian in the constrained system. $\rho = 0.532$. The area of the camera frame was approximately 24 ball areas.

The result at low accelerations is similar to the model of Puglisi *et al.* [10]: When the particle-particle correlations are strongest (and larger than those of an equilibrium hard sphere gas [1]), there is a density dependence to the granular temperature (filled circles). At $\Gamma = 3$, where all of the particles are essentially uncorrelated in a 3D volume in the absence of a lid, there is no density dependence (open circles, stars). However, even in the confined system at $\Gamma = 3$, where the distribution is still not Gaussian, no appreciable density dependence is observed (diamonds), suggesting that the non-Gaussian velocity distributions and density-dependent temperature are not as simply related as they are in the model of Puglisi *et al.* [10]. In fact, while there is a clear density dependence on the local temperature at low Γ, the measured velocity distribution conditioned on the local density is not Gaussian. At each density, the conditioned velocity distribution function is almost identical to that of the whole: when the entire distribution is non-Gaussian, the distribution at a single density is non-Gaussian, and when the whole is Gaussian, each conditional velocity distribution is Gaussian.

A more general form of Eq. 3 represents the total velocity distribution as a product of local Gaussian velocity distributions with a distribution of local temperatures:

$$P(v) = \int f(T(\boldsymbol{r},t))e^{-\left(\frac{v^2}{2T(\boldsymbol{r}(t))}\right)} d\boldsymbol{r}dt\,, \tag{4}$$

$T(\boldsymbol{r},t)$ is the local temperature that is varying in space and time. Conditioning on the local temperature would then recover the Maxwell statistics underlying the fluctuations.

Performing this analysis on our data does not succeed in producing Maxwell statistics. Within small windows of local temperature, the distributions remain non-Gaussian. Indeed, the analysis can be extended to condition on both the local temperature and density in the system, but with similarly limited success except for the lowest local temperatures, although all of the conditioned distributions are closer to Gaussian than the full distribution. A plot of the flatness of the conditioned distributions measured at $\Gamma = 0.83$ is shown in Fig. 12. For each density (number of particles in the frame of the camera) the flatness is close to 3 for the 'coolest' fluctuations, but systematically increases for larger local temperatures. The flatness of the full distribution for the data shown in the figure is 4.7.

If all of the measured velocities are normalized by the local granular temperature and combined into a single distribution, that distribution is Gaussian, as observed in a simulation of vibrated granular media [15]. This is a rather surprising result, and the origin of this behavior is not understood.

5 Discussion

5.1 Clustering and Collapse

An understanding of the interesting dynamics displayed in both the gas phase and the two-phase coexistence regions will require a better understanding of the flow of energy from the plate to the granular layer. Energy flow into thicker layers from a vibrating surface have been extensively studied, but the dynamics of the monolayer system are quite different [12–14]. Energy is

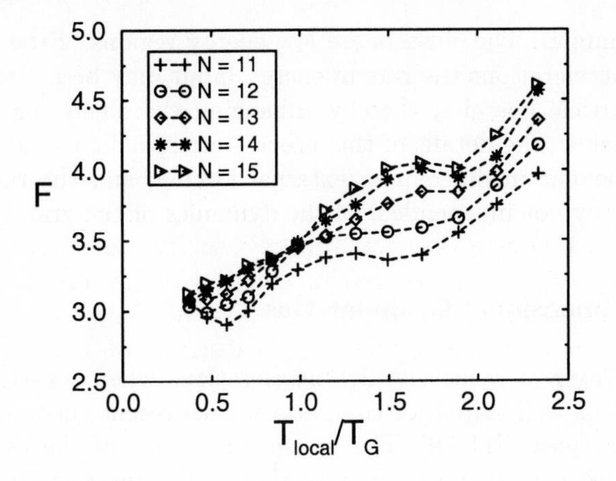

Fig. 12. Flatness of the velocity distribution conditioned on local granular temperature and local density. The velocity data is separated according to the number of particles in the frame of the camera (N), and the local granular temperature (T_{local}) determined from all of the particles in the frame. T_G is the granular temperature averaged over all of the frames. Data is for $\rho = 0.532$, $\Gamma = 0.83$, $\nu = 80$ Hz.

dissipated much more slowly than in thicker layers, and therefore the chaotic dynamics of particles on an oscillating plate must be considered [16, 17]. The net energy transferred to the vertical motion by the plate must balance the energy dissipated by inter-particle collisions. It is this balance that determines the steady-state horizontal granular temperature. In order to generate an equation of state, an expression for the energy input by the plate is required. The motion of a single ball on a plate displays the characteristics of low-dimensional chaos, but when coupled with a large number of similar systems through inter-particle collisions, the result is apparently a very regular rate of energy flow, producing a system that looks in many ways very much like an equilibrium system of a large number of interacting particles [17].

The interaction of the spheres with the plate, in particular the apparent lack of any periodic or chaotic attractors with average energies less than the period-one orbit, may partially explain the two-phase coexistence and hysteresis observed in this system [16]. If the kinetic energy of the spheres drops too low, they will fall into the 'ground state', where they remain at rest on the plate, and the energy input drops to zero. Collisions from neighboring spheres may keep a sphere from falling into the ground state, but also dissipate energy. Quantitative predictions of the conditions under which density fluctuations can nucleate a collapse, as well as the stability of the collapse-gas interface, may be derivable from considerations of the sphere-plate dynamics, coupled with the kinetic theory of dense, inelastic gases.

The appearance of strong clustering and associated nearly exponential velocity distributions at low accelerations may also be strongly influenced by ball-plate dynamics. The clusters are low energy regions of the system, and at low plate accelerations the rate of energy input may be a strong function of average particle energies, thereby enhancing the clustering tendency of inelastic particles. The details of this process are specific to our system, but all excited granular media require external forcing, and the rate of energy input is typically not independent of the dynamics of the grains themselves.

5.2 Two Dimensional Granular Gas

The granular layer at relatively high accelerations when constrained with a lid appears to be well suited for comparison with recent theoretical work on forced granular gases [11, 19]. The unusual behavior of the system at low accelerations appears to be intimately tied to the interaction of the balls with the plate, and the unconstrained system at high accelerations has a somewhat ill-defined density. In contrast, once the average kinetic energy of the particles in the constrained system is large enough that the particles can explore the entire volume of the cell, the effective density does not vary with acceleration. Our results show that the correlations and velocity distributions that are independent of Γ, and so do not appear to be very sensitive to the details of the energy input. (Although the way energy is transferred from the vertical direction to the horizontal, through collisions between particles at heights that are not too different, may have important implications.)

As described in section 4.2, the tails of the velocity distribution in this regime are reasonably well described by $P(v) \propto \exp(-|v|^{3/2})$, in agreement with recent experimental and theoretical results. However, the details of how energy is transferred to the horizontal motion in the shaking experiments are sufficiently complicated that the agreement with the theoretical calculation of van Noije, et al.,[11] is perhaps surprising. Furthermore, the experimental system of Losert, et al.,[3] was not constrained to be two-dimensional, but they do not obtain the crossover to Gaussian measured in our system without the lid.

In the calculations of van Noije, et al., the inelastic particles are forced by uncorrelated white noise. In the experiment, the energy is input into the horizontal velocities from the vertical velocity via collisions. Given the complex dynamics of the system, the collisional forcing may be reasonably well described by uncorrelated accelerations, but the forcing occurs only during inter-particle collisions. For accelerations that occur much less frequently than the inter-particle collision times, the velocity distribution should approach the freely cooling result ($P(v) \propto \exp(-|v|)$) [18], whereas white noise forcing is well modeled by accelerations that occur on timescales much faster than the inter-particle collision time [10]. However, a recent numerical simulation of white noise forcing [19] found empirically that the behavior of the system was independent of the ratio between the rate of accelerations and the rate of collisions as long as the ratio was greater than one. This result suggests that the van Noije, et al., calculations are in fact valid for the granular monolayer, and the agreement between theory and experiment is not simply fortuitous.

Both our experiments and the results of Losert, et al., are consistent with velocity distributions that behave like $P(v) \propto \exp(-|v|^{3/2})$ in the tails, but we obtain that result only when the system is tightly constrained so that particles cannot pass over one another, whereas the lid of the cell used by Losert et al., is 5 ball diameters above the plate. Furthermore, Losert et al., include measurements up to $\Gamma = 8$, whereas the our results for an unconstrained layer show a velocity distribution that becomes almost completely Gaussian when Γ is increased to 3. There are several differences between the systems (stainless steel spheres on an aluminum plate vs. glass beads on Delrin in Losert, et al.), that might effect the results, but the most significant difference may be that suggested by Figure 3, that the dynamics of the layer for a particular coverage are governed by Γ and the ratio of the driving frequency to $\nu_c = (g/d)^{1/2}$, where d is the sphere diameter. Since even our largest particles are 2.5 times smaller than the glass beads used by Losert, et al., our measurements at $\nu = 70$Hz are in a different regime than their measurements at $\nu = 100$Hz. In particular, our system at $\Gamma = 3$ may be more fully three dimensional than that of Losert, et al., and $\Gamma = 8$. Consistent with this picture, Losert, et al. do find that the $\exp(-|v|^{3/2})$ scaling fails for lower frequencies. However, their measured velocity distribution does not appear closer to Gaussian, as might be expected from our results.

In summary, the vibrated granular monolayer has proved to be a rich testbed for the non-equilibrium dynamics of excited granular media. The ability to precisely measure statistical distribution functions under well controlled experimental conditions should allow for careful tests of the predictions of kinetic theory. A complete understanding of the dynamics will require a thorough study of the phase space, as well as a more careful consideration of the dynamics of energy transfers in the system.

Acknowledgments

This work was supported by an award from the Research Corporation, a grant from the Petroleum Research Fund and grant DMR-9875529 from the NSF. One of us (JSU) was supported by a fellowship from the Sloan Foundation.

References

1. J. S. Olafsen and J. S. Urbach, Phys. Rev. Lett. **81**, 4369 (1998).
2. J. S. Olafsen and J. S. Urbach, Phys. Rev. E **60** R2468 (1999).
3. W. Losert, D. G. W. Cooper, J. Delour, A. Kudrolli, and J. P. Gollub, Chaos **9**, 682 (1999).
4. S. McNamara and W. R. Young, Phys. Rev. E **53**, 5089 (1996).
5. L. D. Landau and E. M. Lifshitz, *Statistical Physics* (Pergamon Press, Oxford, 1980).

6. D. G. Chae, F. H. Ree, and T. Ree, J. Chem. Phys. **50**, 1581 (1969); interpolated for $\rho = 0.463$.
7. J. P. Hansen and I. R. McDonald, *Theory of Simple Liquids*, 2nd ed. (Academic Press, New York, 1986).
8. J. T. Jenkins and M. W. Richman, Phys. Fluids **28**, 3485 (1985).
9. N. F. Charnahan and K. E. Starling, J. Chem. Phys. **51**, 635 (1969).
10. A. Puglisi, V. Loreto, U. M. B. Marconi, A. Petri, and A. Vulpiani, Phys. Rev. Lett. **81**, 3848 (1998); A. Puglisi, V. Loreto, U. M. B. Marconi, and A. Vulpiani, Phys. Rev. E. **59**, 5582 (1998).
11. T. P. C. van Noije and M. H. Ernst, Granular Matter **1**, 57 (1998).
12. S. McNamara and S. Luding, Phys. Rev. E **58**, 813 (1998).
13. J. M Huntley, Phys. Rev. E **58**, 5168 (1998).
14. E. Falcon, S. Fauve, C. Laroche, European Physical Journal B, **9**, 183 (1999).
15. C. Bizon, PhD. Thesis, University of Texas at Austin, 1998.
16. W. Losert, D. G. W. Cooper, and J. P. Gollub, Phys. Rev. E **59**, 5855 (1999).
17. J. S. Urbach and J. S. Olafsen, Proceedings of the 5th Experimental Chaos Conference, to appear.
18. S. E. Esipov and Th. Pöschel, J. Stat. Phys. **86**, 1385 (1997).
19. C. Bizon, M. D. Shattuck, J. B. Swift, and H. L. Swinney, Phys. Rev. E **60**, 4340 (1999).

Effect of Excluded Volume and Anisotropy on Granular Statistics: "Fermi Statistics" and Condensation

Daniel C. Hong

Physics, Lewis Laboratory, Lehigh University, Bethlehem, PA 18015, USA.
e-mail: dh09@lehigh.edu

Abstract. We explore the consequences of the excluded volume interaction of hard spheres at high densities and present a theory for excited granular materials. We first demonstrate that, in the presence of gravity, the granular density crosses over from Boltzmann to Fermi statistics, when temperature is decreased in the weak excitation limit. Comparisons of numerical simulations with our predictions concerning the scaling behavior of temperature with agitation frequency, gravity and particle-diameter show satisfying agreement. Next, within the framework of the Enskog theory of hard spheres, we interpret this crossover as a "condensation" of hard spheres from the dilute gas-state to a high density solid-like state. In the high density, low temperature limit Enskog theory fails because it predicts densities larger than the closed packed density below a certain temperature. We show how to extend the range of applicability of the Enskog theory to arbitrarily low temperatures by constructing a physical solution: all particles that are situated in regions with densities larger than a certain maximum density are assumed to be "condensed".

1 Introduction

This paper is a review of a recently proposed theory of granular dynamics [1, 2], which is based on the simple recognition that granular materials are basically a collection of hard spheres that interact with each other via a hard core potential [3]. For this reason, many of the properties of excited granular materials may be understood from the atomistic view of molecular gases, in particular from the point of view of kinetic theory [4]. There are, however, several distinctions between molecular gases and granular materials: First, the mean free path of the grains can be rather small – less than the particle diameter – even if the system is strongly forced. Second, granular material is made of macroscopic particles with finite diameter, and thus the material cannot be compressed indefinitely. When the mean free path vanishes, the density is the maximum, closed packed density. Third, gravity plays an important role in the collective response of granular materials to external stimuli because of the ordering of the particles according to their potential energy in the gravitational field.

For example, one of the notable characteristics of excited granular materials in a confined system under gravity is the appearance of a thin boundary layer near the surface that separates a fluidized region from a solid region. This effect is also seen in shearing experiments [7], avalanches [8], and grains subjected to weak excitations [9, 10]. In this limit, those grains in a solid region are effectively frozen, and thus do not participate in dynamical, diffusive processes. Hence, the conventional Boltzmann statistics, which is applicable in the limit of strong excitation and rapid flow where all the particles are dynamically active, certainly needs modification. Our first aim is to show that, in the weak excitation limit, the statistics where such a boundary layer appears, is analogous to the Fermi statistics. We will in fact demonstrate that the density profile of the grains is qualitatively well given by a Fermi function, from which we define the global temperature T and develop a thermodynamic theory of configurational statistics for excited granular materials. We present an explicit formula to relate the temperature T to the external control parameters such as the frequency ω and amplitude A of the excitation vibration, the diameter D of the grains, and the gravitational acceleration g. Next, we examine the microscopic basis of the crossover from Boltzmann to Fermi statistics based on Enskog theory for hard spheres [11], and demonstrate how the crossover proceeds as grains condense from the bottom toward the surface.

2 Configurational Statistics and Maximum Entropy Principle: Justification of a Thermodynamic Approach

It is quite well known that variational methods are not useful in determining the properties of nonequilibrium systems. In systems that show e.g. periodic patterns, it is evident that one cannot perform variations about a steady state situation. Since the system being studied here is a dense, dissipative, nonequilibrium granular system, we find it necessary to make some comments on this point. If the mean free path of the grains is much less than a particle diameter, each particle may be considered to be effectively confined in a cage as also assumed for the free volume theory of dense liquids [12]. In such a case, the basic granular state is not a gas, but a solid or a crystal [1]. Thus an effective thermodynamic theory based on the free energy argument may be more appropriate than the kinetic theory. Our argument is that the dense state can be assumed as a "steady state" for which we compute the "configurational statistics" by means of the usual variational method as the most probably state.

To be more specific, consider the excitation of disordered granular materials confined in a box with vibrations of the bottom plate. The vibrations will inject energy into the system which cause the "ground state" to become unstable. A new, excited state will emerge with an expanded volume. The time average of such configurations which have undergone structural distortions, may be deviating weakly from the ground state so that the use of an effective thermodynamic theory based on the variational principle could be justified.

Such a thermodynamic approach may be further justified by the following two recent experiments in both the weakly and the strongly excited regimes:

- *Weakly or moderately excited regime*: Clément and Rajchenbach(CR) [9] have performed an experiment with the vibrational strength, i.e. the dimensionless vibration acceleration $\Gamma = A\omega^2/g$, for a two dimensional vibrating bed, using inclined side walls to suppress convection. CR have found that the ensemble-averaged density profile as a function of height from the bottom layer obeys a universal function that is *independent* of the phase of oscillations of the vibrating plate. Namely, it is independent of the kinetics imposed on the system. One conceptually important point here is that the reference point of the density profile is not the bottom plate, but the bottom layer, which of course is fluidized.
- *Strongly excited regime*: Warr and Hansen (WH) [6] have performed an experiment on highly agitated, vertically vibrating beds of $\Gamma \approx 30-50$ using steel balls with a large coefficient of restitution. They have found that the collective behavior of this vibrated granular medium *in a stationary nonequilibrium state* exhibits strong similarities to those of an atomistic fluid in *thermal equilibrium* at the corresponding particle packing fraction, in particular, concerning the two-point correlation function [6].

The results of both experiments indicate that for both moderately and strongly excited systems, a one-to-one correspondence seems to exist between the *configurational* statistics of the *nonequilibrium* stationary state and the *equilibrium* thermal state. In fact, this is not so surprising considering the fact that upon vibration, the granular materials expand and consequently the volume of the system increases. In turn, this increase corresponds to a rise in the potential energy after the configurational average is appropriately taken. Then the problem reduces to the packing problem, and the temperature-like variable, T, may be associated to the vibrating bed. The existence of distinctive configurational statistics in the density profile of CR (and also in WH in a special case) appears to be fairly convincing evidence that kinetic aspects of the excited granular materials may be separated out from the statistical configurations. These observations are the basis of the thermodynamic theory proposed in [1].

3 Fermi Statistics of Weakly Excited Granular Materials

We first view the system of granular particles as a mixture of holes and particles as in the lattice gas or the diffusing void model [12], which is the simplest version of the free volume theory [13]. We now assign virtual lattice points by dividing the vibrating bed of width L and the height μD, with D typically the diameter of a grain and μ the number of layers, into cells of size $D \times D$. Each row, i, is then associated with the potential energy $\epsilon_i = mgz_i$ with $z_i = (i - 1/2)D$ and m the mass of the grain. Note that the degeneracy, Ω, of the each row is simply the number of available cells, i.e: $\Omega = L/D$. For a weakly excited system with $\Gamma \simeq 1$, the most probable configuration should be determined by the state that maximizes the entropy in the micro-canonical ensemble approach.

Taking into account the excluded volume effect which plays a similar role for dense granular systems as the Pauli principle in Fermi statistics, we derive the entropy S, defined as the total number of ways of distributing N particles into the system. Standard counting argument [14] yields,

$$S = \ln W = \ln \left(\prod_i \frac{\Omega!}{N_i!(\Omega - N_i)!} \right) , \tag{1}$$

where N_i is the number of particles occupying the $i-$th row. Since gravity orders the grains with respect to their potential energy, a grain can be seen as a "spinless Fermion", where the height z plays the role of the momentum variable (if one makes the connection to the electron gas). Maximizing S with the constraints, $\sum_i N_i = N$ and $\sum_i N_i \epsilon_i = \langle U(T(\Gamma)) \rangle$, the mean steady state system energy, we find that the density profile, $\phi(z)$, which is the average number of occupied cells at a given energy level, is given by the Fermi distribution [1]:

$$\phi(z) = N_i/\Omega = 1/[1 + \exp(\beta(z - \mu))], \tag{2}$$

where $\beta = mgD/T$, the height is $z = z_i/D$ and analogous to the Fermi energy, μ, measured in units of D, is the number of layers in the low temperature limit, i.e. in the system at rest. Note that both μ and T enter as Langrange multipliers introduced by the two constraints, i.e, the conserved number of particles and the mean energy. The global temperature T defined here is similar to the compactivity introduced by Edwards and his collaborators in their thermodynamic theory of grains [16], but is different from the kinetic temperature defined through the kinetic energy [17]. We point out that the Fermi statistics is essentially the macroscopic manifestation of the classical excluded volume effect and the anisotropy which causes the ordering of potential energy by gravity. In the spirit of the proposed analogy, the top surface of the granules at rest plays the role of the Fermi surface, and the grains in the thin boundary layer near the surface play the role of the excited electrons of a Fermi gas in metals.

For strongly excited systems, the exclusion principle does not apply. The Fermi analogy is valid when the zero temperature Fermi energy satisfies $\mu \gg n_l$, where n_l is the number of fluidized layers. Now, the energy E_i, injected into the system is of the order of $mA^2\omega^2/2$ and the potential energy E_p, needed to fluidize the particles in the top n_l layers is of the order of mgn_lD. Equating these two energies we find a necessary condition for the fluidization of the top n_l layers, namely: $n_l \sim \Gamma A/(2D)$. Hence, we expect the Fermi statistics to be valid for $\Gamma \sim 1$, if $\mu \gg A/D$.

4 Relation Between T and Γ

We now relate the temperature T to the external control parameters such as Γ. First, a thermal expansion. We determine the energy per column $\bar{u}(T) = \int_0^\infty \phi(z) mzgdz$, from which we can determine the shift in the center of mass per particle; $\bar{h}(T) = \bar{u}(T)/mg$, which is given by:

$$\bar{h}(T) = h(0)\left[1 + \frac{\pi^2}{3}\left(\frac{T}{mgD\mu}\right)^2\right] + \cdots \tag{3}$$

with $h(0) = D\mu^2/2$. Second, a kinetic expansion. We make an observation that for a weakly excited granular system, most excitations occur near the Fermi surface, and thus the volume expansion may be effectively well represented by the maximum height, $H_0(\Gamma)$ of a single particle bouncing on the vibrating plate assuming that the Fermi surface is in contact with the vibrating plate. The kinetic expansion, $H_0(\Gamma)$, is then determined by the maximum of $\Delta(t)$ in the following equation that describes the trajectory of a single ball on a vibrating plane with intensity $\Gamma = \omega^2/g$ with A the amplitude, ω the frequency of the vibrating plate, and g the gravitational constant:

$$\Delta(t) = \Gamma(\sin(t_0) - \sin(t)) + \Gamma\cos(t_0)(t - t_0) - \frac{1}{2}(t - t_0)^2 \tag{4}$$

in units of $g = \omega = 1$,where $t_0 = \sin^{-1}(1/\Gamma)$. Note that since Δ is the relative distance between the ball and the plate, it cannot be negative. More precisely, the particle is launched from the plate at $t = t_o$ by inertia, and then makes a free flight motion until it strikes the plate. It then stays on the plate until $t = t_o + 2\pi$, when it is launched again and repeats the same motion. Hence, Δ is a periodic function with period 2π. Since the Fermi distribution near $T = 0$ can be approximated by a piecewise linear function and $H_0(\Gamma)$ is thought to be the edge of the function, we expect $H_0(\Gamma) \approx \Delta h/2 = (\bar{h}(T) - h(0))/2$. By equating the thermal expansion,(3), to the kinetic expansion, $H_0(\Gamma)g/\omega^2$, in physical units, we now complete our thermodynamic formulation by presenting the explicit relation between T and Γ [1]:

$$T = \frac{mg}{\pi}\left(3D\frac{gH_0(\Gamma)}{\omega^2}\right)^{1/2}. \tag{5}$$

In MD simulation, one may measure the maximum height, $\bar{h}_o(\Gamma)$, of a single ball on a vibrating plate and replace $gH_o(\Gamma)/\omega^2$ with $2\bar{h}_o(\Gamma)/\alpha$ with α an adjustable parameter [15].

Now we compare our theoretical prediction with an experimental result of Clément and Rajchenbach [9]. Figure 1 shows the fitting of the experimental density profile for $\Gamma = 4$ of CR by the scaled Fermi distribution, $\phi(z) = \rho(z)/\rho_c$, with ρ_c the closed packed density. For hexagonal packing, $\rho_c \approx 0.92$. The fitting value of T/mg is 2.0 mm, while Eq.(5) yields $T/mg \approx 2.6$ mm. The agreement between the two is fairly good in spite of such a simple calculation. This expression also agrees with the simulation result [10, 15].

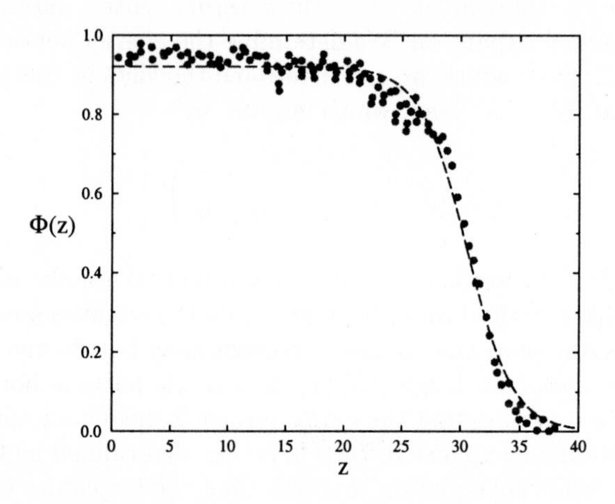

Fig. 1. Density $\phi(z)$ as a function of the height z. The symbols are the data by Clément and Rajchenbach and the dotted line is the Fermi distribution function, Eq.(2).

Note that the detailed expression of $H_0(\Gamma)$ depends on the manner by which the grains are excited and we expect that our main scaling prediction of Eq.(5), namely $T \propto g^{3/2}D^{1/2}/\omega$, will hold even for systems driven not by sinusoidal waves. Further, T has a gap at $T = 0$ because the time between the launching and landing of the ball is always finite for $\Gamma > 1$. Next, it is well known that the *specific heat* per particle, $C_v = d\bar{u}/dT$, can be written as the fluctuations in the energy, namely $\langle(\Delta\bar{u})^2\rangle = \langle(\bar{u}(z) - \bar{u})^2\rangle = T^2C_v$ [18]. Hence, our theory makes a nontrivial prediction for the fluctuations in the center of mass:

$$\langle(\Delta z)^2\rangle = \langle(z(T) - \langle z\rangle)^2\rangle = \frac{\langle(\Delta h)^2\rangle}{\mu_o^2} = \frac{\pi^2}{3}\left(\frac{T}{mgD}\right)^3\frac{D^2}{\mu_o^2} \tag{6}$$

while the center of mass is given by:

$$\Delta z(T) = z(T) - z(0) = \frac{D\mu_o\pi^2}{6}\left(\frac{T}{mgD\mu_o}\right)^2 \tag{7}$$

Note that the total expansion, $\Delta h(T) \equiv \mu_o \Delta z$ and its fluctuations $\langle(\Delta h)^2\rangle/D^2 = \langle\mu_o(\Delta z)^2/D^2\rangle$ are only a function of the dimensionless Fermi temperature

$T_f = T/mgD$ as expected. Furthermore, note that (6) is an indirect confirmation that the specific heat is linear in T as it is for the non-interacting Fermi gas.

5 Test of Fermi Statistics by Molecular Dynamics Simulations

In this section, we examine the configurational statistics of granular materials in a vibrating bed, in particular the density profile, and its fluctuations by MD simulations and compare the results with the predictions made in the previous chapter. The MD code was provided by the authors of [19] and the details of the code can be found in the literature. Simulations were carried out in two dimensional boxes with vertical walls for N particles each with a diameter $D = 0.2\,\text{cm}$ and a mass $m = 4\pi(D/2)^3/3$ with the degeneracy Ω using $(N, \Omega) = (100, 4), (200, 4)$ and (200,8) with a sine wave vibration. The dimensionless Fermi energy $\mu = N/\Omega$ is the system height at rest. For all cases, the inequality $\mu \gg A/D$ was satisfied. In Fig. 2 the temperature, obtained by the best fit of the density profile to the Fermi function (dots), is plotted against Γ and the values predicted by Eq.(5). Note the fairly good agreement between theory and simulations.

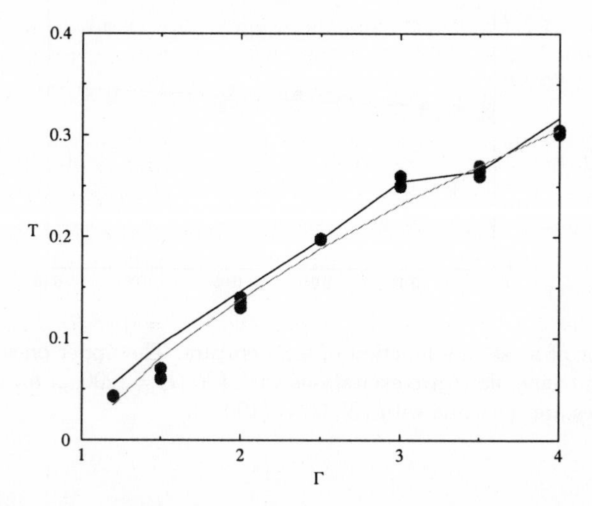

Fig. 2. Comparison between the measured temperatures by MD(dots) and the predicted ones. The lower line was obtained by Eq.(5), while the upper one was obtained by replacing gH_o/ω^2 by $2\bar{h}_o(\Gamma)/\alpha$ with $\alpha \approx 12.8$. $\bar{h}_o(\Gamma)$ is the maximum jump height of a single ball on a vibrating plated obtained by MD.

We also studied the temperature scaling against the frequency, gravity and diameter to further check the validity of Eq. (5). The scaling laws as determined by the simulations are:

$$T \approx \omega^{-m_1}$$

$$T \approx g^{m_2}$$

$$T \approx D^{m_3}$$

with $m_1 \approx 1.16$, $m_2 \approx 0.48$, and $m_3 \approx 0.53$. These values are close to the predicted ones by Eq. (5), i.e $m_1 = 1$, $m_2 = 0.5$, $m_3 = 0.5$. For detailed comparisons, see the original paper [15].

Finally, we have also checked the scaling of the center of mass, and the fluctuations against T^2 and T^3. Since the density profiles are well fitted by the Fermi function, we anticipate that the center of mass and its fluctuations obey the scaling as shown in Figs. 3 and 4. Note that the increase in the center of mass is *second* order in temperature T, which is contrary to the mean field prediction of the linear increase of volume in the compactivity X [16].

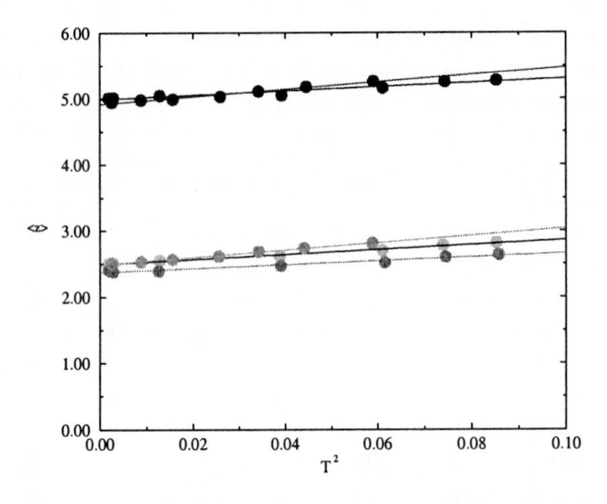

Fig. 3. Center of mass as s function of temperature. The upper ones are data for sine wave and trianglular wave excitations with $(N, \Omega) = (200, 4)$ for different Γ's. The lower ones are the same with $(N, \Omega) = (100, 4)$.

6 Condensation of Hard Spheres under Gravity

Our next goal is to examine the microscopic basis of Fermi statistics based on the kinetic theory, in particular the Enskog equation for hard spheres of mass m and diameter D, to explore whether or not the kinetic theory can describe the cross over from Boltzmann to Fermi statistics and if so, under what conditions it occurs.

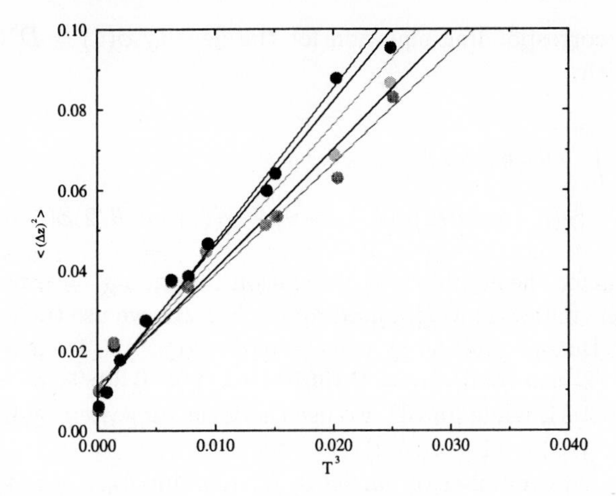

Fig. 4. Fluctuations in the center of mass as a function of temperature. Symbols used are the same as in Fig.3.

Our particularly interesting discovery [2] is that the prediction of the Enskog equation is only valid when $\beta\mu \leq \mu_o$, where μ is the dimensionless initial layer thickness of the granules (or the Fermi energy), $\beta = mgD/T$ with T the temperature, and the critical value, μ_o, is determined to be $\mu_o = 21.756$ in two dimensions (2D) and $\mu_o = 15.299$ in three dimensions (3D). (For maximum packing, $\mu_o = 151.36$ in 3D and $\mu_o = 144.6155$ in 2D.) When this inequality is violated, the Enskog equation does not conserve the particles, and we present a scenario of resolving this puzzle based on physical intuition, namely that the missing particles condense from the bottom toward the surface [2]. In this way, the hard sphere Enskog gas appears to contain the essence of Fermi statistics. We briefly describe how the density profile, ϕ as a function of dimensionless variable ζ, can be obtained from the Enskog equation. For details, see Ref. [2]. Note that ϕ introduced here is not the volume fraction ν but is scaled so that $\phi = 1$ at closed packing.

In a free volume theory, particles are confined in a cage. Hence, if we use a simple cubic lattice as a basic lattice, the close packed volume fraction $\rho_c = N/V = N/(D^2N) = 1/D^2$. If we define the dimensionless density $\phi(z) = G(z)/\rho_c = D^2G(z)$ or $\phi(\zeta,\beta) = D^2G(z)$ with $\zeta = z/D$, we then obtain the following exact dimensionless equation of motion for $\phi(\zeta,\beta)$:

$$\frac{d\phi(\zeta)}{d\zeta} + \beta\phi(\zeta) = \phi(\zeta)I_\zeta(\zeta) \tag{8}$$

where $\beta = mgD/T$ and

$$I_\zeta(\zeta) = \frac{1}{2}\int_0^{2\pi} d\theta \cos\theta[\chi(\zeta - \frac{1}{2}\cos\theta)\phi(\zeta - \cos\theta) - \chi(\zeta + \frac{1}{2}\cos\theta)\phi(\zeta + \cos\theta)] \tag{9}$$

For 3D, the corresponding equation for the density $\phi(\zeta) = D^3 G(z)$ is given by Eq.(8) with:

$$I_\zeta(\zeta) = \pi \int_0^\pi d\theta \sin\theta \cos\theta$$
$$[\chi(\zeta - \cos\theta/2)\phi(\zeta - \cos\theta) - \chi(\zeta + \cos\theta/2)\phi(\zeta + \cos\theta)] \quad (10)$$

Several forms for the equilibrium correlation function χ have been proposed, but we use the following widely used forms: For 2D, we use the form proposed by Ree and Hoover [20]: $\chi(\phi) = (1 - \alpha_1\phi + \alpha_2\phi^2)/((1 - \alpha\phi)^2$ with $\alpha = 0.489351 \bullet \pi/2 \approx 0.76867, \alpha_1 = 0.196703 \bullet \pi/2 \approx 0.30898, \alpha_2 = 0.006519 \bullet \pi^2/4 \approx 0.0168084$, while for 3D, we use the form suggested by Carnahan and Starling [21]: $\chi(\phi) = (1 - \pi\phi/12)/(1 - \pi\phi/6)^3$

Since the total number of particles, N, remains fixed, the following normalization condition should be satisfied for both 2D and 3D.

$$\int_o^\infty d\zeta\phi(\zeta;\beta) = \mu \quad (11)$$

where $\mu \equiv N/\Omega_x$ (or $\mu \equiv N/\Omega_x\Omega_y$ in 3D) is the Fermi energy and Ω_x, Ω_y are the degeneracies along the x and y axes. We now perform the gradient expansion of (9) and (10) and retain only the terms to first order in $d\chi/d\zeta$. The 3D solutions for the first order differential equation can be obtained easily, and are given by [2]:

$$-\beta(\zeta - \bar{\mu}) = ln\phi - 1/(1 - \alpha\phi)^2 + 2/(1 - \alpha\phi)^3 \quad (12)$$
$$\beta\bar{\mu} = ln(\phi_o) - 1/(1 - \alpha\phi_o)^2 + 2/(1 - \alpha\phi_o)^3 \quad (13)$$
$$\beta\mu = \phi_o - \frac{2\phi_o}{1 - \alpha\phi_o} + \frac{2\phi_o}{(1 - \alpha\phi_o)^3} \quad (14)$$

where $\alpha = \pi/6$. For given values of β and μ, $\phi_o \equiv \phi(\zeta = 0)$ will be determined by Eq.(14) with the condition that ϕ_o cannot in anyway greater than the closed packed density. As we will see, we need care to proceed. First, since the right hand side of (14) is an monotonically increasing function of ϕ_o and ϕ_o cannot be greater than the closed packed density, which is 1 in our units, $\beta\mu$ must have an upper bound μ_o; namely, $\mu_o = 15.299$ in 3D, ($\mu_o = 21.756$ in 2D), which is the value obtained by setting $\phi_o = 1$ in the right hand side of (14). Note that this upper limit depends on the underlying basic lattice structure. (For a hard sphere gas in a continuum space, at the closed packed density, $\eta = \frac{4\pi}{3}(\frac{D}{2})^3 = \frac{\pi}{6}\phi \approx 0.74$ in 3D and $\eta = \frac{\pi}{4}\phi = \pi/2\sqrt{3} \approx 0.907$ in 2D hexagonal packing, in which case, the upper limit $\mu_o = 151.36$ in 3D and $\mu_o = 144.6155$ in 2D.) Considering the fact that both the temperature T and the Fermi energy μ are *arbitrary* control parameters, the existence of such bounds is a puzzle: if $\beta\mu$ is less than μ_o, then the density profile given by Eq.(12) is well defined, but if $\beta\mu$ is greater than μ_o, then ϕ_o must be one at the bottom, and the particle conservation appears to break down, namely

$$\int_o^1 d\phi\zeta(\phi) = \int_0^\infty d\zeta\phi(\zeta) \equiv \mu_o/\beta < \mu \tag{15}$$

The central question is: where does the rest of the particles go? In order to gain some insight into this question, consider first the case of point particles under gravity, in which case the density profile is given by: $\rho(\zeta) = \rho(0)exp(-mg\zeta/T)$. If we put more particles into the system, we simply need to increase $\rho(0)$ because the point particles can be compressed indefinitely, and the profile simply shifts to the right. We now replace these point particles with hard spheres, which cannot be compressed indefinitely and thus the maximum density at any point is the closed packed density. Suppose we have a system of hard spheres at a certain temperature T, where the density is closed packed at the bottom layer and smoothly decreases toward the surface. At this point, if we add more hard spheres, say the amount of one layer, to the system, how does the density profile modify? Since the hard spheres cannot be compressed, our intuition tells us that after the system reaches the equilibrium, the density of the first(bottom) and second layer becomes closed packed, forming a rectangle, which is then followed by the original Enskog profile. If we keep adding more particles, we obtain the density profile that is the combination of the rectangle (we term this the Fermi rectangle) beginning at the bottom and the smooth original Enskog profile, which adjoins the Fermi rectangle at its upper edge. The total number of hard spheres in the Fermi rectangle should be the same as that of the added particles. One may obtain the same picture in a reverse way. Suppose we start from a high temperature where all the particles are active. We then slowly decrease the temperature to suppress the thermal motion. There will be a temperature where the density at the bottom is the closed packed density. Now, let us lower the temperature further. What happens? The next layer will become closed packed and is thus effectively frozen. As we keep lowering the temperature, the freezing of the particles will occur from the bottom and the frozen region will then spread out, until at T=0 all the particles are frozen. Note that the frozen particles in the closed packed region behave like a solid with $\phi_o = 1$. Such an observation helps us to resolve the puzzle associated with the disappearance of particles. The missing particles should form the condensate in the Fermi rectangle spanning the bottom to the lower part of the fluidized layer. We term the surface that separates the frozen or a closed packed region with $\phi = 1$ from the fluidized region with $\phi < 1$ the Fermi surface. The location of the Fermi surface, ζ_F, is determined by the number of the missing particles, namely, $\zeta_F = \mu - \mu_o/\beta$, and is thus a function of temperature. For nonzero ζ_F, we must put the missing particles below the Fermi surface and shift the bottom layer from $\zeta = 0$ to ζ_F. From Eq.(15), we find that in 2D the number of missing or condensed particles, $N_o(T)$, at $T < T_c$ is $N_o(T) = \Omega_x(\mu - \mu_o/\beta) = N(1 - T/T_c)$, where N($\equiv \Omega_x\mu$) is the total number of particles, and T_c is the condensation temperature defined as the point where the particle conservation, Eq.(15), breaks down, namely

$$T_c = mgD\mu/\mu_o \tag{16}$$

Hence, the fraction of condensed particles is given by:

$$N_o(T)/N = 1 - T/T_c \tag{17}$$

How is this picture modified in the presence of dissipation? With dissipation, solving the Enskog equation becomes a nontrivial task for three reasons: First, we do not yet know the precise functional form of the velocity distribution function for inelastic particles. Experimentally observed profiles [23] definitely indicate that they are not Gaussian. Second, even if we somehow have empirical formulas for the velocity distribution functions, carrying out the integral with non-Gaussian profiles for the Enskog integro differential operator becomes a non-trivial task. Third, in the presence of dissipation, the temperature profile is nonuniform [24], and thus one has to solve the equation for the density profile, i.e., Eq.(8), along with the energy equation [25]. Analytic solution for this case is difficult to obtain. However, if the dissipation is small, then one might assume that the velocity distribution function is still Gaussian and the temperature remains uniform except near the heat reservoir. Under such assumption, we only need to solve the Eqs.(8)-(10) with the corrected pressure term [4], which is given by:

$$P = \rho T[1 + \gamma\rho D^d\chi]$$

where $\gamma = \frac{\pi}{4}(1+\epsilon)$ when $d = 2$ and $\gamma = \frac{\pi}{3}(1+\epsilon)$ when $d = 3$. Since the solution of the force balance equation, $dP/dz = -\rho g$ yields the same result, in the presence of dissipation, we use the force balance equation instead of solving the Enskog equation as was done in the elastic case [2]. eDefining $\phi = \rho D^d$ as above and using the above forms of P in the force balance equation yields the differential equation

$$\beta\phi(\zeta) + d\phi(\zeta)/d\zeta = -\gamma[\phi d\chi(\phi)/d\phi + 2\chi(\phi)d\phi(\zeta)/d\zeta] \tag{18}$$

Upon integration we find in the two dimensional case:

$$\beta(\zeta - \bar{\mu}) = -\log\phi + c_1\phi + c_2\log(1-\alpha\phi) + c_3/(1-\alpha\phi) + c_4/(1-\alpha\phi)^2 \tag{19}$$

where

$$c_1 = -2\gamma\alpha_2/\alpha^2,$$

$$c_2 = \gamma(\alpha_1/\alpha^2 - 2\alpha_2/\alpha^3)$$

$$c_3 = -c_2$$

$$c_4 = \gamma(-1/\alpha + \alpha_1/\alpha^2 - \alpha_2/\alpha^3)$$

Then $\beta\bar{\mu}$ is the negative of the right hand side of (19) with ϕ replaced with $\phi_0 \equiv \phi(\zeta = 0)$. Integrating $\beta\zeta$ from zero to ϕ_0 yields

$$\beta\mu = c_5 + c_6\phi_0 - c_2\phi_0^2 + c_7/(1-\alpha\phi_0) + c_8/(1-\alpha\phi_0)^2 \tag{20}$$

where

$$c_5 = \gamma(-2\alpha_1\alpha + \alpha^2 + 3\alpha_2)/\alpha^4$$

$$c_6 = -c_2 + 1$$
$$c_7 = -\gamma(2\alpha^2 - 3\alpha_1\alpha + 4\alpha_2)/\alpha^4$$
$$c_8 = -c_4/\alpha$$

This result may be shown to be equivalent to the elastic case when $\epsilon = 1$. Substituting the numeric values for the constants α, α_1, α_2 and evaluating at $\phi_0 = 1$ yields

$$\mu_0 = 1 + 10.3779(\epsilon + 1) \tag{21}$$

In three dimensions we find

$$\beta(\zeta - \bar{\mu}) = -\log\phi - \frac{1}{4}\frac{\gamma}{\alpha(1-\alpha\phi)^2} - \frac{1}{2}\frac{\gamma}{\alpha(1-\alpha\phi)^3} \tag{22}$$

where $\beta\bar{\mu}$ is again given by the negative of the right hand side of (22) with ϕ_0 replacing ϕ. Then we integrate $\beta\zeta$ between zero and ϕ_0 to get

$$\beta\mu = \phi_0 + \frac{1}{2}\frac{\gamma\phi_0^2}{(1-\alpha\phi_0)^2} + \frac{1}{2}\frac{\gamma\phi_0^2}{(1-\alpha\phi_0)^3} \tag{23}$$

where $\alpha = \pi/6$. When $\epsilon = 1$, this again reduces to the profile for the elastic hard spheres (12). We find upon evaluating $\beta\mu$ at $\phi_0 = 1$ that

$$\mu_0 = 1 + 7.14964(\epsilon + 1) \tag{24}$$

We note that in general, μ_0 takes the form $1 + C(\epsilon + 1)$, and we recall that $T_c = mgD\mu/\mu_0$. Hence we may write

$$T_c(\epsilon) = T_c(\epsilon = 1) + \frac{mgD\mu C(1-\epsilon)}{(1+2C)(1+C(\epsilon+1))} \tag{25}$$

Letting $\delta = (1-\epsilon) \ll 1$, we may recast the above result as

$$T_c(\epsilon) = T_c(\epsilon = 1) + \frac{mgD\mu C}{(1+2C)^2}\delta + \frac{mgD\mu C^2}{(1+2C)^3}\delta^2 + \ldots \tag{26}$$

In summary, we conclude that the condensation phenomenon persists in the presence weak of dissipation. Note that in the above analysis, we set $\phi_o = 1$ as the upper bound for a cage model. For maximum packing, ϕ_o can exceed unity, as discussed before, i.(in 3D, $\phi_o \approx 0.74\,\pi/6$ and $\phi_o = 2/\sqrt{3}$.) The clustering of particles near the bottom wall in two dimensional experiments [27] may be a strong confirmation of this scenario.

7 Conclusion

In this review, we have explored simple consequences of excluded volume interaction for dense granular materials, and shown that the granular statistics cross over from Boltzmann to Fermi statistics as the strength of the excitation is reduced. We have also advanced a theory that such a crossover may be understood as the condensation of hard spheres below a condensation temperature. Since the Enskog equation breaks down at high densities, comparable to the closed packed density, we have constructed physical solutions beyond this regime by forming the Fermi rectangle near the bottom. Those particles in the Fermi rectangle are assumed to be condensed [29]. This way, we have extended the applicability of the Enskog theory to arbitrarily low temperatures.

As demonstrated in Eq.(14), the Enskog solution based on a virial expansion from the dilute to the dense case conserves particles only when we allow the density profile greater than the maximum closed packed density at the bottom, which is clearly unphysical. This is due to the fact that the equation of state, i.e. the pressure as function of density, is not known for all densities. If one would use the correct equation for the pressure (which is not analytically known at this point, except for a numerically determined one [30]) that must *diverge* at the close packed density, it may be possible that the strange condensation phenomenon associated with the Enskog equation may be absorbed by an advanced theory [30]. We point out, however, that even with this empirical pressure profile the formation of a "Fermi rectangle" is clearly visible. The density profile above the Fermi rectangle was shown to be nicely fit by the Enskog profile. Thus, our picture of condensation, which is an alternative way to extend Enskog theory results up to the maximum possible density, seems to capture the essence of physics – and we have shown analogies to the Fermi statistics.

For a polydisperse system, the upper limit for η is usually larger than that of the monodisperse system. Thus, we expect the freezing temperature to be much lower for a polydisperse system. Note, however, that as long as the upper limit for η is smaller than unity, Enskog theory for a gas predicts the condensation phenomenon, in the sense that unphysical densities are predicted. For the extreme limit of an Appolonian packing [28], for which the close packed density can be unity, $\eta = \alpha\phi_o$ in (14) can be arbitrarily close to unity and thus the right hand side of (14) can be arbitrarily large. In this case, the condensation phenomenon discussed in this paper disappears.

The original experiment done in Ref. [9] used a rectangular box with tilted side walls, but one may use a different shape such as a parabola to mimic the density profile of the electron gas in 3D, which may help one to visualize the excitation spectrum near the Fermi surface. Finally, we point out that while it is trivial to drive hard sphere grains by a thermal reservoir in MD simulations, it becomes a nontrivial task to do this experimentally. The normal way of exciting the granular system experimentally is through vibration, yet it remains an unresolved issue to determine the precise relation between the vibration strength and the effective temperature of the reservoir. Eq.(5), which is based on a single ball picture, may be a first step in this direction. The discovery of the Fermi statistics and the associated condensation phenomena may provide an avenue to apply the methods developed in equilibrium statistical mechanics to the study of granular dynamics, notably from the point of view of elementary excitations such as the Fermi liquid theory in condensed matter physics.

Acknowledgments

I wish to thank H. Hayakawa for collaboration and J. A. McLennan for helpful discussions on the Enskog equation. I also wish to thank Paul Quinn for carrying out MD simulations and Joseph Both for checking some of the algebra in section 6, as well as S. Luding for discussions on the Appolonian packing and the puzzle associated with the condensation phenomenon.

References

1. H. Hayakawa and D. C. Hong, Phys. Rev. Lett., **78**, 2764 (1997).
2. D. C. Hong, Physica A **271**, 192 (1999).
3. H. Jaeger, S. R. Nagel and R. P. Behringer, Physics Today, **49**, 32 (1996), Rev. Mod. Phys. **68**, 1259 (1996); H. Hayakawa, H. Nishimori, S. Sasa and Y-h. Taguchi, Jpn. J. Appl. Phys. Part 1, **34**, 397 (1995) and references therein.
4. J. Jenkins and S. Savage, J. Fluid. Mech. **130**, 197 (1983); S. Chapman and T. G. Cowling, *The Mathematical Theory of Nonuniform Gases* (Cambridge, London, 1970); J. A. McLennan, *Introduction to Non-Equilibrium Statistical Mechanics*, Prentice Hall (1989).
5. This is similar to the case of electron gas subjected to thermal excitation, in which case only electrons near the Fermi surface are excited. Note that while the exclusion principle operates in momentum space for an electron gas, for grains, the corresponding exclusion principle operates in real space.
6. S. Warr and J. P. Hansen, Europhys. Lett. Vol. 36, no.8 (1996). Urbach also found that the two point correlation function of excited granular system is the same as that of the equilibrium hard sphere gas at the equivalent packing density (See J. S. Urbach, *Experimental observations of non-equilibrium distributions and transitions in a 2D granular gas* (in this volume, page 410.)
7. D. M. Hanes and D. Inman, J. Fluid. Mech. **150**, 357 (1985); S. Savage and D. Jeffrey, *ibid*, **110**, 255 (1981).
8. For example, see: J. P. Bouchaoud, M. E. Cates, R. Prakash, and S. F.Edwards, J. Phys. France **4**, 1383 (1994).
9. E. Clément & J. Rajchenbach, Europhys. Lett. **16**, 133 (1991).
10. J. A. C. Gallas, H. J. Herrmann & S. Sokolowski, Physica A **189**, 437 (1993).
11. D. Enskog, K. Sven. Vetenskapsaked, Handl. **63**, 4 (1922).
12. H. Caram & D. C. Hong, Phys. Rev. Lett., 67, 828 (1991); Mod. Phys. Lett. B **6**, 761 (1992); For earlier development, see J. Litwinyszyn, Bull. Acad. Polon. Sci., Ser. Sci. Tech. **11**, 61 (1963); W. W. Mullins, J. Appl. Phys. **43**, 665 (1972).

13. T. L. Hill, *Statistical Mechanics*, Chap.8, New York, Dover (1985).
14. For example, see Eq.8.41 in Huang, Statistical Mechanics, Second Edition, Wiley (1987).
15. P. Quinn and D. C. Hong, cond-matt/9901113 (To appear in Physica A, 1999).
16. S. F. Edwards and R. B. S. Oakeshott, Physica A **157**, 1080 (1989); A. Mehta and S. F. Edwards, Physica A (Amsterdam) **168**, 714 (1990).
17. For other thermodynamic theories of grains, see: B. Bernu, F, Delyon,and R. Mazighi, Phys. Rev. E **50**, 4551 (1994); J. J. Brey, F. Moreno, and J. W. Dufty, Phys. Rev. E **54**, 445 (1996).
18. Nowak et al. used such a fluctuation formula to experimentally measure the compactivity of the excited grains. See: E. R. Nowak, J. B. Knight, E. Ben-Naim, H. M. Jaeger and S. R. Nagel, Density Fluctuations in Vibrated Granular Materials, Phys. Rev. E 57, 1971-1982 (1998).
19. Jysoo Lee, cond-mat/9606013; S. Luding, E. Clément, A. Blumen, J. Rajchenbach, and J. Duran, Phys. Rev. E. **50** R1762 (1994).
20. F. R. Ree and W. G. Hoover, J. Chem. Phys. **40**, 939 (1964).
21. N. F. Carnahan and K. E. Starling, J. Chem. Phys. **51**, 635 (1969).
22. A. Kudrolli, M. Wolpert and J. P. Gollub, Phys. Rev. Lett. **78**, 1383 (1997).
23. J. S. Olafsen and J. S. Urbach, Phys. Rev. Lett. **81**, 4369 (1998); J. Delour, A. Kudrolli, and J. Gollub, cond-matt/9806366; J. Delour, A. Kudrolli, and J. Gollub, cond-matt/9901203.
24. E. Grossman, T. Zhou, and E. Ben-Naim, cond-matt/9607165.
25. P. Haff, J. Fluid. Mech. vol. 134, 401 (1983).
26. B. J. Alder , W. G. Hoover, and D. A. Young, J. Chem. Phys. **49**, No.8, 3688 (1968).
27. S. Luding, *Models and simulations of granular materials*, Ph.D thesis, Albert-Ludwigs Universät Freiburg, Germany (1994). See Fig.19.
28. For Appolonian packing, see B. Mandelbrot, "The Fractal Geometry of nature,"(W. H. Freeman and Company, New York, 1982).
29. Such a condensation is an intrinsic phenomenon associated with a system where the excluded volume is present. The formation of a solid below a critical temperature is due to a massive occupation of low energy states at the low temperature. For details see cond-mat/0005196.
30. The 2D simulation results by Luding and Strauß (S. Luding and O. Strauß, *The equation of state for almost elastic, smooth, polydisperse granular gases for arbitrary density*, (in this volume, page 389)) in this book are not inconsistent with our scenario. See Fig.9 (curve IV) and Fig.10 of their paper for the formation of the "Fermi rectangle" near the bottom layer. Note that in the paper by Luding and Strauß, a density is obtained where a disorder-order transition occurs which is noteably smaller than the maximum, close packed density. For other MD simulations for dense granular gases, see also, P. Sunthar and V. Kumaran, Phys. Rev. E. **60**, 1951 (1999)

How Much Can We Simplify a System of Grains?

Matthias Müller and Hans J. Herrmann

University of Stuttgart, Institute for Computer Applications 1, Pfaffenwaldring 27, D-70569 Stuttgart, Germany. e-mail: hans@ica1.uni-stuttgart.de

Abstract. Realistic collisions between sand grains are very difficult to simulate correctly because many physical and mechanical nonlinearities occur. In order to simulate larger systems, many proposals have been made to treat collisions in a simplified form. Two of these models and some of the attached pitfalls are presented as well as success stories.

1 Introduction

The simulation of granular systems has become more and more important in the last years because they allow to explore regimes in density and velocity for which theories like the kinetic gas theory or the plasticity theory in soil mechanics cannot be applied. Other examples are the study of mixing and segregation of different species and in general the treatment of grains with complicated or varying shapes and sizes. Although the first computer simulations of granular systems were performed over 30 years ago by Walton, it is only in the last 10 years that a large community of physicists has become interested in simulating granular systems. Many different ways have been approached and it is the purpose of this chapter to present the main difficulties, the various attempts to cope with them, and to present a certain number of tests which allow to determine if an algorithm gives realistic results.

The two requirements for a program to simulate many particle systems like granular materials are, besides the necessity that the algorithm is correct, the speed and the number of particles that can be simulated. Both criteria go hand in hand, because large systems can only be simulated if the CPU-time needed to simulate a single particle is very small. In fact, the time consuming part of a program for granular media is the treatment of collisions. On the one hand, one has to determine which two particles are going to collide at the next time step and on the other hand one must calculate the outcome (for instance velocities and angular velocities after collision).

The most realistic techniques are molecular dynamics (MD) simulations in which one tries to solve Newton's equations of motions for each particle simultaneously. Many tricks are known and they have been well reviewed, on the one hand in the book by Allen and Tildesley [1] which treats the general case and on the other hand by several authors [8, 9, 18], specifically for granular media. Two variants of MD are typically applied to granular media.

On the one hand with one can use a constant time step and soft particles and on the other hand one can consider pointlike collision of perfectly rigid particles (event driven, ED [see the paper by Luding and Strauß in this book]). In the first case the limitations are given by the size of the time step (typically $10^{-4} - 10^{-6}$ sec) which depends on the stiffness of the particles. The second method is very efficient for dilute systems but becomes very slow when the system is dense and can even come to a complete standstill (inelastic collapse or finite time singularity [12]).

MD methods have been pushed to simulate up to $1.4{\cdot}10^9$ particles (for very short times, however). Typical simulations on a workstation deal with 10^4 particles and can simulate such systems for several minutes real time in a day or two. These sizes are too small for many purposes, in particular in the three dimensional case. Examples are for instance simulations of three dimensional heaps, silos and segregation under vibration. Also long time effects like axial segregation in a rotating drum or compaction under vibration as well as effects containing a wide distribution of sizes cannot be dealt with in MD-techniques.

However, both MD methods tend to produce much more information than is actually needed to compare with experiments. One could actually think of other techniques for which the efficiency is larger, that means that for comparatively less effort one avoids to calculate unnecessary information in techniques which will be presented later in this article. In addition to the well established MD and the methods presented here, other techniques like Boltzmann Lattice Gases [17] have been applied to granular media.

2 Examples for Simplified Models

2.1 Lattice Gas Automata

A Lattice Gas Automaton (LGA) proceeds in discrete time steps. The time can therefore be labeled with integers $t = 0, 1, 2, ...$ The particles are located at the sites of a two dimensional triangular lattice of size L. Gravity acts downward in the vertical direction and forms an angle of 30^0 with the closest lattice axis. At each site there are seven bit variables which refer to the velocities $v_i(i = 0, 1, 2, ...6)$. Here $v_i(i = 1, ...6)$ are the nearest neighboring (NN) lattice vectors and $v_0 = 0$ refers to the rest state (zero velocity). Each state can be either occupied by a single particle or empty. Therefore, the number of particles per site has a maximal value of seven and a minimal value of zero.

The time evolution of the LGA consists of a collision step and a propagation step. In the collision step particles can change their velocities due to collisions and in the propagation step particles move in the direction of their velocities to the NN sites where they will collide again. The system is updated in parallel. Only the collisions specified in Figure 1 can deviate the trajectories of particles from straight lines with probabilities depending on

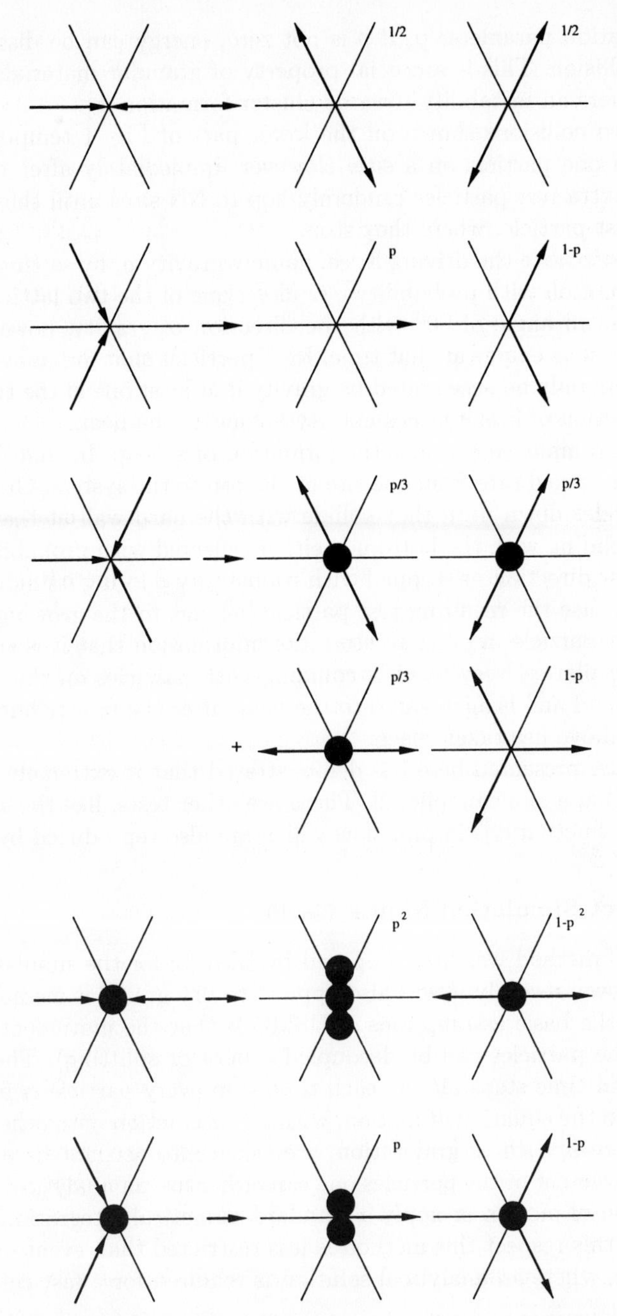

Fig. 1. Collision rules of the LGA. Arrows represent moving particles and full dots rest particles. The number next to each configuration is the probability for that transition.

the dissipation parameter p. If p is not zero, energy can be dissipated during the collision. This is a crucial property of granular materials and yields among others an instability towards cluster formation [6].

The two collisions shown on the lower part of Fig. 1 temporarily allow more than one particle on a site. However, immediately after the collision step, the extra rest particles randomly hop to NN sites until they find a site with no rest particle, where they stop.

We incorporate the driving force, namely gravity g, by setting a rest particle into motion with probability $g/2$ along one of the two lattice directions which form an angle of 30^0 with the direction of gravity, however, only if the site below is empty at that time. Rest particles that are above other rest particles can only be accelerated by gravity if at least one of the two NN sites in the direction of gravity does not yet belong to the heap.

As an example we consider the formation of a heap. In that case we add particles at a fixed rate from one site at the top to the system. Gravity moves these particles down until they collide with the hard wall at the bottom. A particle colliding with the bottom is either reflected with probability $1-p$ in the specular direction or stopped with probability p losing its momentum. In the second case the resulting rest particle belongs to the growing heap, and we tag that particle in order to store the information that it is sustained by the bottom plate. Every particle colliding with particles on the heap looses its momentum and is aggregated to the heap after the redistribution process described above has taken place.

The LGA presented here has demonstrated that it extremely well reproduces the shape of a sandpile [2]. There are other tests, like the spectrum of the density fluctuations in pipe flow which are also reproduced by LGA [16].

2.2 Direct Simulation Monte Carlo

The DSMC method was first proposed by Bird [3] for the simulation of rarefied gas flows, recently it was also applied to dry granular media [4, 13].

One of the basic assumptions of DSMC is that the movement and interaction of the particles can be decoupled (operator splitting). The system is integrated in time steps Δt. At each time step every particle is first moved, according to the equation of motion, without interaction with other particles. External forces, such as gravitation, are taken into account here. To calculate the movement of the particles one can either use an analytical solution of the equation of motion or apply a standard numerical integration scheme to solve it. In this respect this method is less restricted than event driven (ED) simulations, where an analytical solution is required for a fast calculation of a collision.

Next, we take the particle-particle interactions, *i.e.* the collisions into account. In contrast to ED simulations, the exact times and places of these collisions are not calculated, but a stochastic algorithm is applied as described in the following:

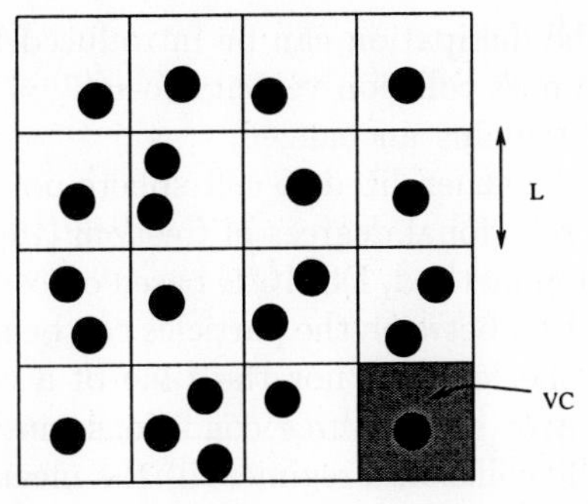

Fig. 2. The system is divided into cells of length L and Volume V_c.

The particles are sorted into spatial cells of linear size L and volume $V_c = L^{\mathcal{D}}$, where $\mathcal{D}$ is the dimensionality of the system. Collisions occur only between the particles in the same cell, which ensures that only particles which are close to each other may collide. In every cell with more than one particle, we choose randomly

$$M_c = \frac{N_c(N_c - 1)\sigma v_{max}\Delta t}{2V_c} \tag{1}$$

pairs of particles. Here, N_c is the number of particles in the cell, σ the scattering cross section (for spherical particles, $\sigma_{2D} = 4R$, $\sigma_{3D} = 4\pi R^2$) and v_{max} is an upper limit for the relative velocity between the particles. To get v_{max} we sample the velocity distribution from time to time and set v_{max} to twice the maximum particle velocity found. In order to determine the correct number of collisions, we apply an acceptance-rejection method: For a pair of particles i and j the collision is performed if

$$\frac{|\boldsymbol{v}_i - \boldsymbol{v}_j|}{v_{max}} < Z, \tag{2}$$

where Z is an independent uniformly distributed random number in the interval $[0;1]$. This method leads to a collision probability proportional to the relative velocity of the particles.

Since the collision takes place regardless of the particle positions in the cell, we have to choose an impact parameter b in order to calculate the post collision velocities. Molecular chaos is assumed here; b is drawn from a uniform distribution in the interval $[-2R, 2R]$ in 2D or in a circle with radius $2R$ in 3D. The post collision velocities $\boldsymbol{v}_i'$ and $\boldsymbol{v}_j'$ are now calculated as if the two particles collided with that impact parameter, i.e. like in event driven simulations so that

$$\boldsymbol{v}_i' = \boldsymbol{v}_{cm} + \frac{m_j}{m_i + m_j}|\boldsymbol{v}_{rel}|\boldsymbol{e} \quad \text{and} \quad \boldsymbol{v}_j' = \boldsymbol{v}_{cm} - \frac{m_i}{m_i + m_j}|\boldsymbol{v}_{rel}|\boldsymbol{e}, \tag{3}$$

where $\boldsymbol{e}$ is the unit vector pointing in the direction of the relative velocities after the collision. For hard spheres $\boldsymbol{e}$ is uniformly distributed over the unit sphere [14]. Finally, the dissipation can be introduced by changing the normal component of the post collision velocity to $\boldsymbol{v}'^{(n)} = -\tau \boldsymbol{v}^{(n)}$, whereas the tangential component remains unchanged.

The algorithm can be generalized to soft sphere potentials. It is also possible to introduce the rotational degrees of freedom [15].

Like every simulation method, DSMC is based on certain approximations. One is that the interaction between the particles can be modeled by collisions. Furthermore, neither the location nor the time of a collision is calculated exactly. To keep the error small, three conditions must be fulfilled: (i) the system should be in the collisional regime, (ii) the mean free path should be larger than the cell size, (iii) the mean time between two collisions should be larger than the time step.

The major disadvantage of DSMC is, that there is no limitation in the particle density. This is a consequence of the fact that during the advection step the particles move independently of each other. During the interaction step only stochastic collisions take place. One example where the straightforward application of DSMC will result in unrealistic high particle densities is a heap of dissipative particles. In a heap the collisions need to be highly correlated in order to support the particles at the top. A further problem is that with vanishing kinetic energy and therefore velocity the momentum transfer Δp in a collision also disappears. Because the force $F = \frac{\Delta p}{\Delta t}$ needs to balance gravity, the number of collisions per time diverges. To overcome this restriction we modified the DSMC algorithm in the following way: because it combines ideas from several algorithms we refer to the modified algorithm as HSMC (Hybrid Simulation Monte Carlo). The idea of HSMC is to change the advection step of the DSMC algorithm to introduce an excluded volume. A real excluded volume would however result in an algorithm similar to MD or ED, because every particle and its neighbors have to be inspected. We therefore check the constrains only on the coarse grained level of the cells that are used by the original DSMC. Whenever a particle tries to enter a full cell, the move is rejected since it would result in a unrealistic high density from the physical point of view.

3 Applications to Different Flows

In the following we will present several test cases which can be used to check if the typical behavior of granular materials is reproduced. They can be used to decide if a certain technique is suited to simulate granular materials and within which limits. The test cases presented here of course represent only a limited subset of the wide range of phenomena visible in granular matter.

3.1 Homogeneous Cooling

In the homogeneous cooling state [5, 7, 12] we expect from kinetic gas theory and computer simulations that the kinetic energy $E(t)$ of the system decays with time following the functional form

$$\frac{E}{E_0} = \left(\frac{1}{1+t/t_0}\right)^2 . \tag{4}$$

The theoretically expected time scale

$$t_0 = \frac{\sqrt{\pi} d s_*(\nu)}{\sqrt{2}(1-r^2)\nu\bar{v}} \tag{5}$$

is a function of the initial energy $\bar{v} = \sqrt{2E_0/Nm}$, the particle diameter d, the restitution coefficient r, the volume fraction ν, with $s_*(\nu) = (1-\nu)^2/(1-7\nu/16)$. Eq. (4) holds as long as the system stays homogeneous. In section 3.3 differences from this behavior are described.

Fig. 3 shows the expected agreement between kinetic theory and simulation method. DSMC is based on this theory and should therefore reproduce its results.

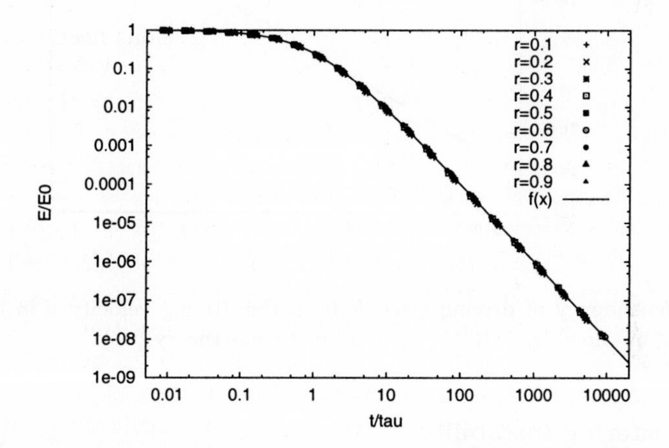

Fig. 3. Homogeneous cooling of a freely evolving granular gas for different restitution coefficients r. f(x) refers to Eq. (4).

3.2 Bagnold Shear Flow

One of the standard experiments to gain insight into the behavior of materials is the shear experiment. In its classic form it is used to define the viscosity μ. The force F needed to drive two plates of Area A separated a distance l with relative velocity v follows the law

$$F = \mu A \frac{v}{l} . \tag{6}$$

For non dissipative particles in a system without walls this system does not have a steady state: due to the viscous heating the temperature of the system is increasing. For dissipative particles the temperature of the system will scale with the velocity of the driving plate and so does μ. Therefore the driving force F will follow Bagnold's law and scale with v^2:

$$F \sim v^2. \tag{7}$$

In its general form $\frac{v}{l}$ is replaced by the velocity gradient. The special case of a constant gradient is called uniform shear. This special case is homogeneous in a Lagrangian frame. Recent studies have shown that choosing the correct reference frame is crucial in shear flows. Simulation methods that are not invariant under Galileian transformations will not yield the correct results.

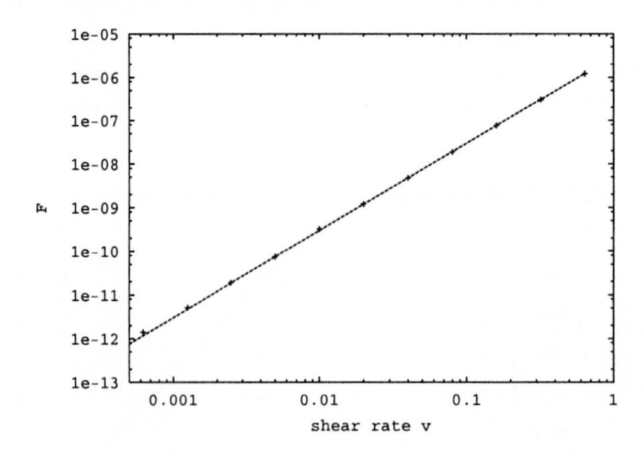

Fig. 4. Dependency of driving force F from the driving velocity v in a Bagnold shear flow, obtained by DSMC (+) and kinetic gas theory (—).

3.3 Clustering Instability

In the homogeneous cooling state (see section 3.1) we expect that the energy $E(t)$ of the system decays with time and follows the functional form of Eq. (4). This holds as long as the system stays homogeneous and HSMC was able to reproduce this result for all densities. In this section we will focus on a system which will undergo the so called cluster instability.

In Fig. 5 we present the normalized kinetic energy $K(t)/K(0)$ as a function of the normalized time t/t_0. At the beginning of the simulation we observe a perfect agreement between the theory for homogeneous cooling and the simulations. At $t/t_0 \approx 2$ all three simulation methods show substantial deviations from the homogeneous cooling behavior, and only at $t/t_0 \approx 10$ we evidence a difference between ED and DSMC. After that time, the kinetic energy obtained from the DSMC simulation is systematically smaller than $K(t)$ from the ED simulation. We relate this to the fact that the molecular chaos assumption of a constant probability distribution of the impact parameter b

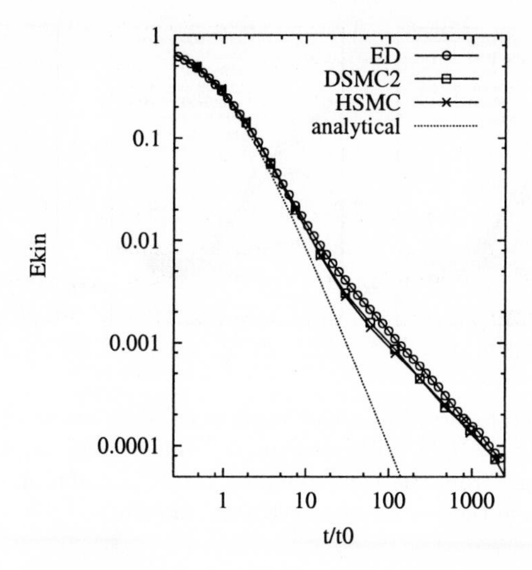

Fig. 5. Normalized kinetic energy vs. normalized time from an ED and a DSMC simulation in 2D with $N = 99856$, $\nu = 0.25$, and $r = 0.8$. The dotted line represents Eq. (4).

is no longer valid[11]. Since dissipation acts only at the normal component of the relative velocity, DSMC dissipates more energy than ED as soon as the number of central collisions is overestimated. One difference between ED and DSMC simulations is the handling of excluded volume by the two methods. While the ED method models hard spheres with a well defined excluded volume, the DSMC method models point particles and excluded volume is introduced by the approximations described in subsection 2.2. As expected we obtain differences in the particle-particle correlation function $g(r)$: At large times ED simulations lead to a $g(r)$ with a rich structure for short distances, indicating a rather close triangular packing of the monodisperse spheres. In contrast, the DSMC simulations show no short range correlations between particle positions throughout the whole simulation. In other words: the DSMC method cannot describe depletion effects.

We would like to know if this difference has consequences at larger length scales. The formation and growth of large clusters [5, 6, 12] is quantified by $g(r)$ at large r, or equivalently, the structure factor $S(k)$ at small k. We calculate $S(k)$ by a direct FFT of the two-dimensional density.

We plot the structure factors obtained by ED,DSMC and HSMC in Fig. 6. Different symbols correspond to different times. We observe an increase of $S(k)$ for short wavenumbers $k < 25$, until the structure factor ceases to change for $t \geq 20$.

The structure factor agrees reasonably well for all three simulation methods. This proves that the DSMC simulation is capable to reproduce the more realistic, but computationally more expensive, ED results that account for

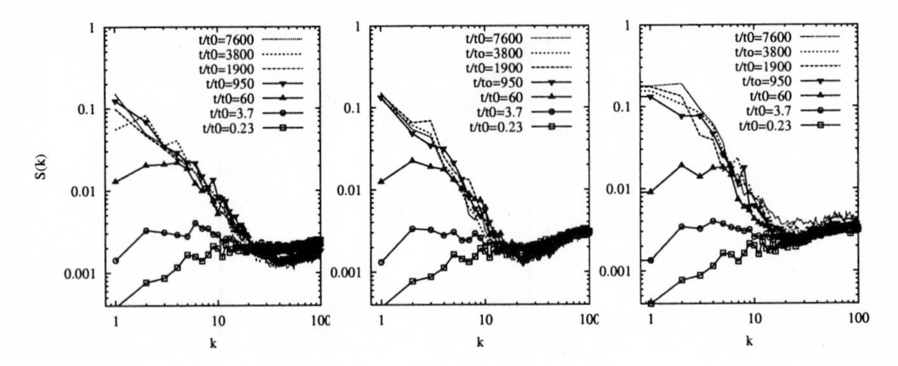

Fig. 6. (left) Structure factor obtained from the ED simulations of Fig. 5 as function of the wavenumber $k = L/\lambda$, with wavelength λ and system size L. (middle) Structure factor obtained from the corresponding DSMC simulation. (right) Structure factor obtained from the corresponding HSMC simulation.

the excluded volume by construction. Even without short-range correlations, the information about large wavelengths is well reproduced by DSMC simulations.

3.4 Heaps

HSMC does not include static friction and the question is to what extent the formation of heaps can be reproduced. Using hard sphere molecular dynamics it has been shown that heap formation is possible without static friction [10] due to the steric effects caused by the excluded volume. Because the later is implemented on the coarse grained grid level it should also be possible to reproduce heaps with HSMC. DSMC on the other side is not capable to reproduce the behavior of granular matter in this case. This is a consequence of the limitations described in section 2.2.

In order to build a heap, particles were put into the system at a constant rate (one per time unit). The freshly created particle is put at a fixed height above the highest particle in the system. The disadvantage is of course that the simulation time grows linearly with the number of particles N. But this algorithm resembles more closely the experimental way of pouring particles out of a funnel. To reach an equilibrium state we wait 500 time units after the last particle has been added. Fig. 7 shows a typical result. On a first glance the major feature of a heap like the finite slope are reproduced well. In the following we will have a closer look at the shape of the heap.

Although the surface of a heap is close to a straight line, it is known that the tail of the heap deviates from that line [2] following

$$x = \frac{h_m - h}{m} + \frac{l}{r}\ln\left(\frac{h_m}{h}\right). \tag{8}$$

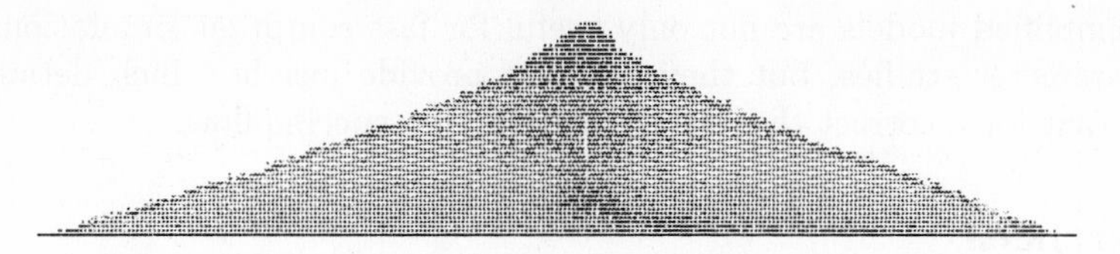

Fig. 7. Heap of 30500 particles simulated with HSMC.

Where x is the horizontal distance from the top, h the height, h_m the maximum height and m the slope of the heap Fig. 8 shows a comparison of the simulation with Eq. (8). The LGA of sect. 2.1 also reproduces extremely well Eq. (8) even with the right prefactor [2].

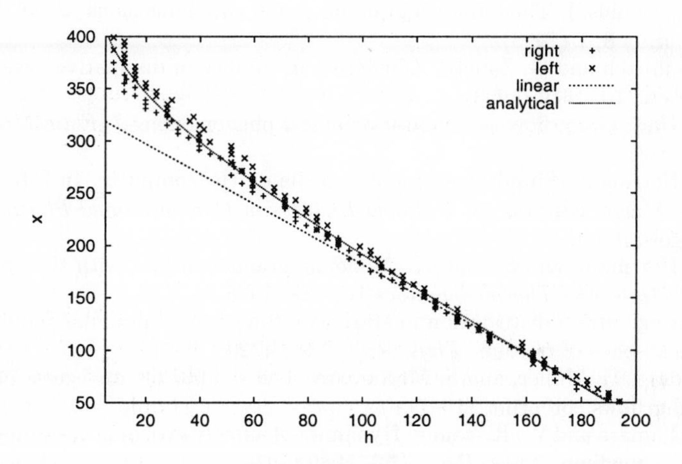

Fig. 8. Comparison of the shape of a heap on a plane with an analytical prediction according to Eq. (8). Symbols denote the results of HSMC simulations with $N = 30500$ and $r = 0.9$

4 Conclusion

We have discussed various alternative simulation methods faster than traditional molecular dynamics and have seen how they perform in four tests (homogeneous cooling, shear flow, clustering and the shape of the heap). In these tests the DSMC or better its variant HSMC gives very good results. The good agreement with theory and the results of more precise methods demonstrate that a precise calculation of the particle interaction is often not needed in order to obtain the expected behavior on a larger scale. Therefore this simplified models are not only useful for fast computer simulations, i.e. for parameter studies, but they can also provide insight which details are important for a correct theoretical analysis of a specific flow.

References

1. M. P. Allen and D. J. Tildesley. *Computer Simulation of Liquids.* Clarendon Press, Oxford (1987).
2. J. J. Alonso and H. J. Herrmann. Shape of the tail of a two-dimensional sandpile. *Phys. Rev. Lett.* **76**, 4911 (1996).
3. G. A. Bird. *Molecular Dynamics and the Direct Simulation of Gas Flow.* Oxford Science Publications, Oxford (1994).
4. J. J. Brey, M. J. Ruiz-Montero, and D. Cubero. Homogeneous cooling state of a low-density granular flow. *Phys. Rev. E* **54**, 3664 (1996).
5. I. Goldhirsch, M.-L. Tan, and G. Zanetti. A molecule dynamical study of granular fluids. I. The unforced granular gas in two dimensions. *J. of Scientific Computing* **8**, 1 (1993).
6. I. Goldhirsch and G. Zanetti. Clustering instability in dissipative gases. *Phys. Rev. Lett.* **70**, 1619 (1993).
7. P. K. Haff. Grain flow as a fluid-mechanical phenomenon. *J. Fluid Mech.* **134**, 401 (1983).
8. H. J. Herrmann. Simulating granular media on the computer. In P.L. Garrido and J. Marro, editors, *3rd Granada Lectures in Computational Physics*, p. 67, Springer (1995).
9. H. J. Herrmann and S. Luding. Modeling granular media with the computer. *Cont. Mech. and Thermodynamics* **10**, 189 (1998).
10. S. Luding. Stress distribution in static two dimensional granular model media in the absence of friction. *Phys. Rev. E* **55**, 4720 (1997).
11. S. Luding, M. Müller, and S. McNamara. The validity of "molecular chaos" in granular flows, preprint, (1997).
12. S. McNamara and W. R. Young. Dynamics of a freely evolving, two-dimensional granular medium. *Phys. Rev. E* **53**, 5089 (1996).
13. M. Müller, S. Luding, and H. J. Herrmann. Simulations of vibrated granular media in 2D and 3D. In D. E. Wolf and P. Grassberger, editors, *Friction, Arching and Contact Dynamics.* World Scientific, Singapore (1997).
14. Kenichi Nanbu. Direct simulation derived from the Boltzmann equation. *J. Phys. Soc. Japan* **49**, 2042 (1980).
15. Kenichi Nanbu. Direct simulation derived from the boltzmann equation. II. Rough sphere gases. *J. Phys. Soc. Japan* **49**, 2050 (1980).
16. G. Peng and H. J. Herrmann. Density waves of granular flow in a pipe using lattice-gas automata. *Phys. Rev. E* **49**, 1796 (1994).
17. Harald Puhl. *Eine Studie zur Simulation granularer Medien.* PhD thesis, Universität Stuttgart (1998).
18. G. H. Ristow. Granular dynamics: A review about recent molecular dynamics simulations of granular materials. In D. Stauffer, editor, *Annual Reviews of Computational Physics I.* World Scientific, Singapore (1994).
19. Stefan Luding and Oliver Strauß, *The Equation of State for Almost Elastic, Smooth, Polydisperse Granular Gases for Arbitrary Density*, (in this volume, page 389).

Author Index